Revolutions and Continuity in Greek Mathematics

Science, Technology, and Medicine in Ancient Cultures

Volume 8

Revolutions and Continuity in Greek Mathematics

Edited by
Michalis Sialaros

DE GRUYTER

ISBN 978-3-11-056365-8
e-ISBN (PDF) 978-3-11-056595-9
e-ISBN (EPUB) 978-3-11-056527-0
ISSN 2194-976X

Library of Congress Control Number: 2018934241

Bibliographic information published by the Deutsche Nationalbibliothek
The Deutsche Nationalbibliothek lists this publication in the Deutsche Nationalbibliografie; detailed bibliographic data are available on the Internet at http://dnb.dnb.de.

Typesetting: Compuscript Ltd., Ireland
Printing and binding: CPI books GmbH, Leck

MIX
Papier aus verantwortungsvollen Quellen
FSC www.fsc.org FSC® C083411

www.degruyter.com

Preface

Back in 2013, I had a lengthy discussion on the current status and possible future of the discipline of ancient Greek mathematics with someone in the know. “The field,” concluded the famous scholar, “is, I am afraid, practically dead.” This statement—backed by my mother’s unceasing question: “Why don’t you get a proper job?”—does not seem to be so very far from the truth. The so-called ‘community’ of scholars who study ancient Greek mathematical texts is conspicuously small in number and heterogeneous. Its few members, coming from notably different backgrounds, are scattered throughout the world in all types of departments. As a consequence, they rarely meet. And when they do, these meetings usually take place in the margins of conferences on broader thematic areas.

In May 2014, I was employed as a British Academy Postdoctoral Fellow at Birkbeck, University of London, when a unique opportunity emerged: by the generous support of the British Academy, the Institute of Classical Studies, and the British Society for the History of Science, I was able to invite several scholars to come to London and present their thoughts on the notions of ‘revolution’ and ‘continuity’ in history and historiography of Greek mathematics. And, as surprising as it may sound, they did come. The conference lasted for two days, and the list of participants is rather long: Sabetai Unguru (who gave a keynote speech), Serafina Cuomo, Jean Christianidis, Henry Mendell, Courtney Roby, Andrew Gregory, Vassilis Karasmanis, Gabriele Galluzzo, Claas Lattmann, Stylianos Negrepontis, Naoya Iwata, Ellen Harlizius-Klück, Michael Weinman, A E L Davis, and Elizabeth Kosmetatou. I would like to thank all of them for traveling from all corners of the world to participate in this meeting. My special thanks go to Charles Burnett and Sarah Hart for chairing sessions and leading the discussions, as well as to Alison Watson, Danae Baffa, and Ioanna Zouli, who were present to make sure things ran smoothly. Above all, I am deeply indebted to Danae Karydaki, the heart and soul of this enterprise.

Although all of the papers offered genuinely exciting and stimulating insights, by the end of the conference, it was evident to me that a number of contributions could formulate a specific and coherent overall thematic unit. Bearing this in mind and with an eye towards publication, I asked some other scholars to write down their own thoughts on the matter. Fabio Acerbi, Jeffrey Oaks, and Geoff Lehman kindly agreed to participate, and I would like to thank them for greatly enriching the scope of this collection. Finally, I would like to wholeheartedly thank Markus Asper who agreed to publish a volume on a ‘dead’ discipline. He is, undoubtedly, a brave man.

Michalis Sialaros
Berlin, September 2017

https://doi.org/10.1515/9783110565959-202

Contents

Notes on Contributors

Sabetai Unguru is Professor Emeritus in the Cohn Institute for the History and Philosophy of Science and Ideas at the University of Tel-Aviv. His research areas include the History of Ancient Greek Mathematics and its Historiography and the History of Medieval Optics. He has published books and articles in these areas, as well as numerous essay reviews in the field.

Jean Christianidis is Professor of History of Mathematics at the National and Kapodistrian University of Athens (NKUA) and Chercheur Associé at the Centre Alexandre Koyré in Paris, specializing in the History of Greek Mathematics. He is the Head of the Department of History and Philosophy of Science (NKUA), a member of the International Commission on the History of Mathematics, and editor-in-chief of *Neusis*, a Greek journal in the History and Philosophy of Science and Technology.

Courtney Ann Roby is Associate Professor in the Department of Classics at Cornell University. Her research focuses on the literary aspects of scientific and technical texts from the ancient Mediterranean world. Her first book (*Technical Ekphrasis in Greek and Roman Science and Literature: The Written Machine between Alexandria and Rome*, Cambridge University Press, 2016) traces the literary techniques used in the textual representation of technological artifacts from Hellenistic Greece to late-ancient Rome. Her current work includes a book project on Hero of Alexandria and articles on concepts of distributed cognition and scientific models in antiquity, as well as the early modern reception of ancient science and technology.

Michalis Sialaros is Assistant Professor of History of Science in Antiquity in the Department of History and Philosophy of Science at the National and Kapodistrian University of Athens. He has published on Ancient Greek Mathematics and he has taught courses on the History of Ancient Science and Philosophy. He is currently preparing a new English commentary on Euclid's *Elements*.

Claas Lattmann (Dr. phil. Kiel 2008) is Wissenschaftlicher Mitarbeiter at the Institut für Klassische Altertumskunde at Christian-Albrechts-Universität zu Kiel (Germany). His main research interests are Greek literature, ancient science and philosophy, and model and metaphor theory.

Fabio Acerbi (CNRS, UMR8560 Centre Alexandre Koyré, Paris) specializes in editing ancient Greek and Byzantine mathematical and astronomical texts, and in studying their transmission.

Henry Mendell has taught Philosophy at California State University, Los Angeles for several decades. His principal research interests lie in Ancient Philosophies of Mathematics and Science, Aristotle's, in particular. Ancillary to these interests, he has also worked on Greek astronomy and mathematics, especially on Aristotle's contemporary, Eudoxus.

Geoff Lehman teaches Art History at Bard College Berlin. His research interests include Renaissance Painting, the Theory and History of Perspective, and the Phenomenology of Art.

Michael Weinman is Professor of Philosophy at Bard College Berlin. His research interests include Ancient Greek Philosophy, Political Philosophy, and the intersection of the two. Lehman and Weinman are co-authors of a monograph that analyses the Parthenon as a study in the liberal arts, set to appear in the SUNY Series in Ancient Greek Philosophy in late 2017.

Jeff Oaks received his PhD in differential geometry in 1991 and has taught Mathematics at the University of Indianapolis since 1992. In 1999, after writing a book on the History of Railroad Tie Preservation, he turned to History of Arabic Mathematics. Since that time he has published a number of articles and one book on Algebra and Arithmetic, extending the range to include Diophantus and Medieval Italian Algebra.

Gabriele Galluzzo is Lecturer in Ancient Philosophy in the Department of Classics and Ancient History at the University of Exeter. He has previously held positions at the LMU in Munich and the Scuola Normale Superiore of Pisa. He has published extensively on Aristotle's Metaphysics and its reception in the Middle Ages. He also has broad interests in contemporary analytic metaphysics.

Vassilis Karasmanis is Professor of Philosophy at the National Technical University of Athens. He specializes in Ancient Philosophy and the History and Philosophy of Technology. He is editor of the books *Socrates: 2400 Years from his Death* and (with L. Judson) *Remembering Socrates*, as well as author of several papers on Ancient Greek Philosophy and Mathematics.

Stelios Negrepontis is Professor Emeritus in the Department of Mathematics at the National and Kapodistrian University of Athens. Previous positions include Indiana U., Bloomington and McGill U., Montreal, visiting positions at Texas University, Austin, University of Ohio, Columbus, and the University of Cyprus. He has served as Vice-Rector of Athens University and as Chairman of the Department of Mathematics, University of Athens. His research interests include Ultrafilters, Banach Spaces and Topology, History of Greek Mathematics, and Plato's Philosophy. He has published papers in Annals of Math, Bulletin AMS, Transactions AMS, and books in Springer Grundlehren series, Cambridge University Press. He has supervised twenty-five PhD's.

Ὥσπερ ξένοι χαίρουσιν ἰδεῖν πατρίδα,
καὶ οἱ θαλαττεύοντες εὑρεῖν λιμένα,
οὕτως καὶ οἱ γράφοντες βιβλίου τέλος.

Michalis Sialaros

Introduction: Revolutions in Greek Mathematics

Revolution! The mere presence of this word in the title of an edited volume on ancient Greek mathematics may cause justified confusion. After all, any dictionary would inform the reader that the term 'revolution', which more often than not sparks images of violence, is usually employed in a political context. So, how could the (deceptively) peaceful discipline of ancient Greek mathematics be associated with revolutions? Or, to be more precise, when was the term introduced into the field and why? How do contemporary historians respond to it, and to what degree do they consider their work 'revolutionary'? These are the questions that form the nucleus of this volume. Furthermore, this publication also has a threefold, more general, aim: first, to make the majority of the papers presented at the international conference 'Revolutions and Continuity in Greek Mathematics' accessible to a wider audience; second, to give to scholars of ancient Greek mathematics the opportunity to publish their latest research in a volume exclusively devoted to this subject; third, to allow the non-specialist to perceive and appreciate the complexity associated with contemporary research on ancient Greek mathematics in terms of historical questions, methodological approaches, and historiographical debates. Through this prism, Section I of the introduction outlines those key historiographical debates that led to the entry of the term 'revolution' into the standard vocabulary of contemporary historians of Greek mathematics, while Section II provides a summary of the contents of this volume.

I Historiographical Background[1]

1975 was a year marked by a series of events that changed our world: the Vietnam War came to an end, Microsoft was founded, Margaret Thatcher became the first woman leader of the Tories, and Agatha Christie (so unforgivingly) killed off Hercule Poirot. Especially for historians of Greek mathematics—unaffected as they might appear to be by global developments—1975 was a turning point in their discipline: two papers that were independently published that year (Michael Crowe's "Ten 'Laws' Concerning Patterns of Change in the History of Mathematics" and Sabetai Unguru's

1 For comprehensive surveys on recent research on ancient Greek mathematics, see Sidoli 2014, Saito 1998, Knorr 1990, and Berggren 1984. For the latest collective histories, the reader should consult both Acerbi 2010 and Cuomo 2001. The histories of Heath 1921, van der Waerden 1950, and Neugebauer 1952 are still useful, although significantly outdated.

https://doi.org/10.1515/9783110565959-001

"On the Need to Rewrite the History of Greek Mathematics") triggered a series of heated debates which critically shaped the contemporary historical agenda of the field.

I.1 From 'Scientific Revolutions' to 'Revolutions' in Historiography of Mathematics

The publication of Thomas Kuhn's *The Structure of Scientific Revolutions* in 1962 is rightly considered by many as a landmark in the historiography and philosophy of science.[2] Even though the book attracted much criticism (and, in fact, Kuhn was eventually forced to reconstruct some of his original views),[3] Kuhn's view of science, as progressing via occasional revolutions, dominated discussions in the historiography and philosophy of science in the decade following its publication.[4]

In spite of the popularity of *The Structure of Scientific Revolutions*, the question of the applicability of Kuhn's ideas to mathematics did not immediately attract much attention. Reflecting on this state of affairs, Brush (1970, 115–116) argued: "Anyone who examines the recent contents of the major periodicals devoted to the history of science in general might well infer that the history of science does not include the history of mathematics," and Crowe (1975, 161): "no one has announced a revolution in the historiography of mathematics". In Grattan-Guinness's eyes (1976, 325), this lack of interest was fault characteristic of historians who "seem to be terrified by substantial amounts of symbolism". Yet, the fear of 'substantial symbolism'—which every sane human being is justified in having—is not the only reason why Kuhn's ideas seem, at first sight, inapplicable to mathematics. According to Kuhn himself (1970b, 92), a scientific community, having fallen into crisis, rejects its established paradigm and chooses a new one, after a successful 'noncumulative developmental episode'; namely, a 'scientific revolution'.[5] As a result of this radical change, some parts of the first paradigm might become incoherent, unexplained, unintelligible, or simply wrong, when seen through the prism of the second paradigm. But how could a new mathematical paradigm turn proven mathematical propositions into non-explicable—or, even worse, false—statements? Would mathematical truths die? And, how should a mathematician understand Kuhn's remark (1970b, 148) that "the competition between paradigms in not the sort of battle that can be resolved by

2 Three further editions (1970, 1996, 2012) of *The Structure of Scientific Evolutions* have appeared.

3 See Kuhn 1970b, postscript, and Kuhn 1970a; cf. Laudan et al. 1986, 180–195.

4 See the bibliography in Lakatos & Musgrave 1970. For the idea that the concept of 'scientific revolutions' can be traced back to the eighteenth century, see Bernard Cohen 1976. For a recent discussion, see Kindi & Arabatzis 2012.

5 Due to criticism, Kuhn eventually replaced the term 'paradigm' with 'disciplinary matrix'; see Kuhn 1970b, 182.

proofs"? Or, stated in a different way, how does mathematics evolve, if not by resolving its ambiguities via proofs?

All these questions were reflected in Michael Crowe's 1975 seminal paper "Ten 'Laws' Concerning Patterns of Change in the History of Mathematics", the aim of which, according to Crowe himself, was to use 'the new historiography of science' to examine the historiography of mathematics.[6] Admittedly influenced by the works of Imre Lakatos and Raymond Wilder,[7] Crowe proposed ten 'laws' that supposedly explain how mathematics develops and (1975, 162) "touch on issues that will have to be considered if a new historiography of mathematics is to develop." The final 'law' claims (1975, 165): "Revolutions never occur in mathematics". To defend this 'law', Crowe argued that an essential characteristic of any revolution is that some established authority (a king, a constitution, a theory) must be overthrown and "irrecoverably discarded". Based on this criterion, Crowe concluded that, contrary to scientific revolutions (e.g. the Copernican) which might satisfy this condition, old mathematical theories are never 'irrecoverably discarded' (for example, non-Euclidean geometries do not seem to cancel Euclid), and thus "revolutions never occur in mathematics".

Not surprisingly, Crowe's 'laws' triggered a series of replies, some of which were supportive whereas others highly critical. Herbert Mehrtens, for example, in support of Crowe, argued (1976, 312): "The general pattern of T. Kuhn's theory...seems to be not applicable to mathematics".[8] Conversely, Joseph Dauben challenged Crowe's 'essential criterion' by noticing that 'the old regime' is not always 'irrecoverably discarded' in political revolutions (such as the French Revolution or the Glorious Revolution).[9] On this basis, he suggested that a 'radical innovation' or a 'definite break' are better descriptions of a 'revolution', and thus he concluded (1992b, 229): "To the question of whether or not revolutions occur in mathematics, my answer is an emphatic 'yes'."

In 1992, Donald Gillies, fascinated by the ongoing debate, published *Revolutions in Mathematics*, an edited volume which included, among other contributions, Crowe's, Mehrtens's, and Dauben's original papers, accompanied by afterwords.[10] Gillies himself tried to offer a solution to the debate by addressing what he thought to be the heart of the problem, namely that the discussion on revolutions in mathematics was

6 By "new historiography of science", Crowe (1975, 161) refers to the works of T. Kuhn, J. Agassi, N. R. Hanson, K. Popper, and S. Toulmin (relying on Buchdahl's 1965 authority).

7 See Lakatos 1963–1964; Wilder 1968.

8 Mehrtens rejected the overall use of the concept of 'revolution' and promoted a terminology based on the sociology of a scientific community; also see Bos & Mehrtens 1977. For Kuhn's description of a scientific community, see Kuhn 1970b, 177–178.

9 See Dauben 1984, 51.

10 See Crowe 1992, Mehrtens 1992, and Dauben 1992a. All three authors remain quite in agreement with their younger selves.

just "the application of a political metaphor".[11] With this connection in mind, Gillies classified political revolutions into two categories: (a) the 'Russian' type, in which previous paradigms become 'irrecoverably discarded'; and (b) the 'Franco-British' type, in which previous paradigms are relegated to a significantly lesser position. Through this prism, Gillies concluded that, although in science we may have revolutions of both types, it is not possible to have 'Russian' type revolutions in mathematics.[12]

It goes well beyond the scope of this introduction to examine (all versions of) Kuhn's ideas, or the details of the Crowe-Dauben debate, or to try to settle the question of whether revolutions may happen in mathematics.[13] Nonetheless, whatever one's position might be on these issues, the following remarks seem fairly established: (a) the growing interest in Kuhn's ideas in the first decades after the publication of the *Structure of Scientific Revolutions* urged historians of mathematics to examine the applicability of his scheme to mathematics.[14] (b) Within the framework of these discussions, examples from the history of Greek mathematics were often employed to support the arguments of both sides.[15] (c) Although an agreement was not reached on whether revolutions (defined in one way or another) may happen in mathematics, there emerged some consensus regarding the usefulness of employing the concept of 'revolution' on a meta-mathematical level. In this line of thought, Mehrtens (1976, 312) argued: "Kuhn's theory... seems to be not applicable to mathematics. But many of Kuhn's concepts remain valuable for the historiography of science, even if the basic pattern of the theory is rejected,"[16] and Dunmore (1992, 212) concluded: "Revolutions do occur in mathematics, but they are confined entirely to the metamathematical component of the mathematical world". Crowe himself, the most symbolic defender

---

11 In addition, Gilles (1992, 1–2) makes the interesting remark that three scientific revolutions happened at about the same time as three political revolutions: (a) the Copernican and the Glorious Revolutions; (b) the Chemical and the French Revolutions; and (c) the Einsteinian and the Russian Revolutions. Nonetheless, he provides no arguments in the direction of establishing a direct correlation between these events.

12 Gillies 1992, 6. It is worth noting that the participants in the Crowe-Dauben debate often seem to lose sight of Kuhn's description of a 'scientific' revolution. Mehrtens (1992, 43) is careful enough to notice that: "[political] Revolutions are no longer what they once were". And, thus, "one must press the metaphor hard to make it work in [mathematics and science]." On this basis, Mehrtens proposes the term 'epistemological rupture', as employed by Bachelard 1938 and Foucault 1969.

13 Gillies compromising solution was, in fact, short-lived. Only a year later, Corry (1993, 95) noticed that Gillies's only real achievement was to reduce the Crowe-Dauben debate into a discussion on the best definition of 'revolution'. In the same line of thought, Pourciau 2000 claimed that (a) no real controversy exists between the apparently opposing views, and that (b) Kuhnian revolutions are possible in mathematics—Luitzen Brouwer's intuitionism is, according to Pourciau, such an example. For a recent summary of the Crowe-Dauben debate, see Francois & van Bendegem 2010. Also, see Ausejo & Hormigón 1996.

14 On this, see Bernard Cohen 1985.

15 The examples are numerous; see, for instance, Bos & Mehrtens 1977, 22 and Dauben 1984.

16 Also, see Mehrtens 1992, 43.

of the anti-revolution view, wrote (1975, 166): "revolutions can occur in mathematical nomenclature, symbolism, metamathematics..., methodology..., and perhaps even in the historiography of mathematics", and later on, in his 1992 afterword (1992, 316): "*A revolution is underway in the historiography of mathematics*, a revolution that is enabling a discipline that dates back to Eudemus to attain new and unprecedented levels of insight and interest"[17] (my emphasis).

I.2 A Revolution in Historiography of Greek Mathematics?

Coined by Hieronymus Georg Zeuthen and Paul Tannery around the end of the nineteenth century,[18] the term 'geometrical/geometric algebra' denotes the idea that ancient Greek mathematicians applied underlying algebra in their geometrical proofs. From a historiographical perspective, what is particularly interesting is that the term was uncritically adopted and systematically employed by the majority of twentieth-century historians, most notably by those who wrote general histories of the field; e.g. Thomas Heath (1921), Otto Neugebauer (1952), and Bartel Leendert van der Waerden (1950).

In 1975, the publication of Sabetai Unguru's "On the Need to Rewrite the History of Greek Mathematics" set off a series of heated debates among leading historians. In this controversial paper, Unguru emphatically criticized the concept of 'geometrical algebra', by arguing that algebra and modern algebraic symbolism, being products of modern mathematics, could not have been known to the Greeks. On this basis, he accused previous scholars of systematic anachronism and urged the new generation of historians (1975, 22) "to rewrite that history on a new and historically sane basis".[19]

Unguru's paper provoked the polemical responses of van der Waerden, Hans Freudenthal, and André Weil, who tried to defend the authority of the established historiography. Van der Waerden attacked what he identified as the heart of the problem—namely, Unguru's definition of 'algebra'—and concluded (1976, 203): "We...feel that the Greeks started with algebraic problems and translated them into geometric language".[20] The replies of Freudenthal and Weil were far more emotional (and far less polite):

> "If history of mathematics is to be rewritten, I would like to have it done by somebody who shows higher esteem for ancient mathematicians."[21]

17 For the same view, see Francois & van Bendegem 2010, 113.

18 See Zeuthen 1886; 1902; and Tannery 1903; 1912.

19 For earlier attempts to reproach 'geometrical algebra', see Schneider 2016, 250–255.

20 On the ahistorical way in which the term 'algebra' was employed in the debate on 'geometrical algebra', see Sialaros & Christianides 2016.

21 Freudenthal 1977, 200.

> "When a discipline, intermediary in some sense between two already existing ones (say A and B) becomes newly established, this often makes room for the proliferation of parasites, equally ignorant of both A and B, who seek to thrive by intimating to practitioners of A that they do not understand B, and vice versa. We see this happening now, alas, in the history of mathematics. Let us try to stop the disease before it proves fatal."[22]

In the following years, Unguru further elaborated his thesis in a series of publications[23] and the controversy eventually went well beyond 'geometrical algebra', touching upon historiography of Greek mathematics as a whole.[24] A thorough examination of these discussions, however, is well beyond the scope of this introduction.[25] And neither is it easy to evaluate the outcome of this debate in black-or-white terms; nevertheless, a gradual shift of historiographical perspective towards Unguru's thesis is easily attested. The following testimonies are indicative of this gradual change:

(a) Mueller (1981, 54): "Attempts to argue against the algebraic interpretation...by Unguru have not met with much success."
(b) Knorr (1990, 211): "Scholars have since expressed a variety of commitments on both of the issues—the validity of the conception of geometric algebra, and the continuity of the Mesopotamian and Greek mathematical traditions—so they must be considered open at this time."
(c) Artmann (1991, 46): "What Unguru says is certainly correct."[26]
(d) Saito (1998, 137): "With the changes in research approach described above, the belief that algebraic interpretation could provide a sufficient description of the treatment of magnitudes in Greek geometry has become less convincing."
(e) Rowe (2012, 37): "Unguru's position could now be regarded as the accepted orthodoxy."
(f) Corry (2013, 638): "Unguru's view became essentially a mainstream interpretation accepted by most historians."

To conclude this brief overview; for more than four decades, historians of Greek mathematics have debated the legitimacy of Unguru's call to "rewrite the history of Greek mathematics", some supporting it, others criticizing it, but few (if, indeed, any)

22 Weil 1978, 93.

23 See Unguru 1979; Unguru & Rowe 1981; Unguru & Rowe 1982; Unguru 1994b; and Fried & Unguru 2001, 17–55.

24 The examples are numerous. See, for instance, the discussions on the notion of 'induction' in ancient Greek mathematics in Unguru 1991, Unguru 1994a, and Fowler 1994.

25 The details of the aforementioned discussions are very well documented in the literature of the field; see Sialaros & Christianides 2016; Corry 2013, 638–650; Saito 1998, 136–137; Grattan-Guinness 1996, 358–360; Høyrup 1996; Rowe 1996, 6–8; Artmann 1991, 45–47; Vitrac 1990–2001, I 366–385; Berggren 1984, 397–398. Cf. Bernard 2003 and Bashmakova & Vandoulakis 1994.

26 Quoting Unguru 1975, 91.

denying its influence.[27] In fact, historiographical distinctions between 'old' and 'new' historiography along such lines are nowadays commonplace. But is this change of historiographical perspective dramatic enough to justify a (Kuhnian) revolution in the field? Perhaps not. Be that as it may, the question can be translated into another, more meaningful, form: 'Is it possible to apply some of Kuhn's fundamental ideas to the historiography of Greek mathematics? And, if so, are we able to derive from this perspective, aspects of Greek mathematics which would be difficult to attain through others?' The idea that the notion of 'revolution' could be a useful historiographical tool in a meta-mathematical context has gained currency after the debate on Crowe's 'tenth law', and I believe that most authors collected in this volume, if not all, would agree with this.

II Contents

This edited volume contains twelve essays from experts in the field of ancient Greek mathematics from around the globe. In their respective area of specialization, each contributor offers an approach to questions of historical and historiographical 'revolutions' and 'continuity'. Besides Unguru's latest essay on the debate on 'geometrical algebra', the reader will find here a number of papers which offer new insights into the controversial relationship between Greek and pre-Hellenic mathematical activity. Some contributions place emphasis on the other edge of the historical spectrum, by exploring historical and historiographical lines of 'continuity' between ancient Greek, later Greek, and post-Hellenic mathematics. The terminology employed by Greek mathematicians, along with various non-textual and material elements, is another topic upon which some essays place emphasis. Finally, the last three papers in this volume focus on a traditionally rich source on ancient Greek mathematics; namely, Greek philosophy, and particularly the works of Plato and Aristotle. Given that almost all papers address more than one topic, utilize more than one methodology, and (most importantly) employ the term 'revolution' in a variety of ways, the reader should feel free to read the essays in any order they wish.

Sabetai Unguru, in his introductory paper ("Counter-Revolutions in Mathematics"), picks up the threads of the aforementioned discussions on the Crowe-Dauben and geometrical algebra. In a partly confessional manner, the paper entails a reply to Weil's 1978 criticism of Unguru's 1975 paper "On the Need to Rewrite the History of Greek Mathematics", based on the coinage of the historiographical term

27 Notice that the debate on 'geometrical algebra' is far from settled. For a recent defence of 'geometrical algebra' see Blåsjö 2014 and Blåsjö 2016. For a reply, see Fried 2014. Schneider 2016 provides an interesting contextualization of Unguru's 1975 paper.

'counter-revolution'. Unguru's starting point is a sharp distinction between what he calls the "two polar-opposites" in the way history of mathematics has traditionally been approached by mathematicians and historians. On the one hand, he sees those who "discard the historical development" of mathematics and, on the other hand, those who "jettison the difference betwixt the body of mathematical knowledge and its historico-philosophical components and valuations, adding to this a watered-down definition of revolution". Between these two options, Unguru proposes a third road, that of distinguishing between internal mathematical developments, in which accumulation is the prevailing phenomenon, and the images of mathematical knowledge, in which revolutionary breakthroughs do occasionally occur. Within this framework, Unguru draws a sharp line between axiomatic-and-deductive mathematics and other forms of mathematical justification in order to defend the traditional view that Greek mathematics comprises a revolutionary break compared to pre-Hellenic traditions. Similarly, he marks a clear distinction between Greek and modern mathematics, arguing that the latter is profoundly unlike the former.

The second paper of this volume, Jean Christianidis's "Diophantus and Premodern Algebra: New light on an Old Image", is also primarily concerned with historiography. This time, however, a historical period largely untouched by Unguru and the discussions on 'geometrical algebra' comes into perspective: the apparently idiosyncratic case of Diophantus. Based on the new historiographical notion of 'premodern algebra'—a term recently introduced by Jeffrey Oaks to portray the algebraic activities of Arabic mathematicians—, Christianidis's meticulous analysis shows that the same concept can describe Diophantus' method of solving arithmetical problems. Based on Greek and Arabic mathematical examples, Christianidis draws a sharp line between 'modern' and 'premodern' algebra, regarding goals, scope, and the conceptual tools employed by mathematicians. Within this context, the author emphatically criticizes those historians who use the term 'algebra' to describe both 'modern' and 'premodern' algebraic activities for being "blatantly ahistorical". What is particularly interesting here, from a historiographical perspective, is that Christianidis's conclusions seem to stretch far beyond Diophantus and Greek mathematics, constituting a new call for historians to rewrite the history of ancient and medieval mathematics.

Courtney Roby belongs to the next promising generation of scholars of ancient Greek mathematics. Thus, it comes as no surprise that her paper ("Geometer, in a Landscape: Embodied Mathematics in Hero's *Dioptra*") introduces a new perspective; linguistical analysis. Roby's case study is Hero's *Dioptra*, a treatise that presents the reader with a hybrid landscape, part mathematical and part material. Based on recent work on embodied cognition, the author places emphasis on Hero's phraseology and describes how he transforms conventions of 'pure' mathematics into a new technical "phenotype". This approach, Roby argues, allows understanding of Hero's problem-solving from the point of view of a person standing on the ground, not just the synoptic "bird's-eye view" of geometry. Building upon Kuhn's ideas on conceptual incommensurability between two paradigms, as well as on the concept that the transition from one paradigm to another involves work of interpretation, not just translation, Roby maintains that Hero's *Dioptra* carries out a "conceptual revolution". Thus, the author concludes, the division of Greek mathematics into 'pure' and 'applied' is, by and large, artificial.

As several papers in this volume demonstrate, the discipline of ancient Greek mathematics does not live in literary isolation. On the contrary, it often borrows conceptual tools and methodologies from other fields in order to offer new perspectives. In this line of thought, the main aim of my contribution ("How Much Does a Theorem Cost?") is to examine the available anecdotal information on Euclid as a case study to highlight the importance of employing new digital tools in the field. More particularly, I show how the digitalization of the ancient Greek literature—an effort which started in the 1970s and constitutes only a small part of what is usually described today as 'digital humanities'—allows us to show as false an idea developed by Taylor about a century ago (and later on used by Morrow in his commentary on Proclus' *Commentary on Book I of the Elements*). With this approach, I argue that the extant Euclid anecdotes are much more than brief, entertaining stories about a famous mathematician. They are the outcome of a multilevel interaction between several traditions (among which, Early and Late Pythagoreanism, Neopythagoreanism, Aristotelean, and Hellenistic editorial practices), emerging in different contexts and responding to various problems. As such, they compose a complex and multifaceted literary phenomenon.

The fifth paper in this volume ("Diagrammatizing Mathematics: Some Remarks on a Revolutionary Aspect of Ancient Greek Mathematics") comes by Claas Lattmann. Over the last few decades, there has been a vast increase in the attention paid to the so-called non-textual aspects of Greek mathematical activity, and especially diagrams. Lattmann revisits this topic, this time from the revolutionary viewpoint of contemporary semiotic model theory. Based on Charles Peirce's interpretation of diagrams as a specific type of iconic signs, Lattmann examines the epistemic validity of diagrams and their interaction with the textual part of mathematical proofs. He concludes that a diagram, consisting of 'diagrammatical atoms', stands as a holistic model for a general mathematical structure, and that the aim of the textual part of the proof is to show that each particular model has the specific requested mathematical properties. On this basis, Lattmann argues that Greek mathematical diagrams are conceptually unique when compared to either pre-Hellenic or modern diagrammatic reasoning, and are thus incommensurable to both, and so 'revolutionary' in a Kuhnian sense.

One of the several misconceptions about Greek mathematics is the idea that Late Antiquity marks its end. As a result of this mistaken belief, the later Byzantine period has not received much scholarly attention. Fabio Acerbi's essay ("Composition and Removal of Ratios in Geometric and Logistic Texts from the Hellenistic to the Byzantine Period") comprises the author's first attempt to communicate some of the results of his recent editorial work on several sources, among which some scholia to the *Almagest*, the *Isagoge* of George of Trebizond, the treatises of Theodorus Metochites and John Chortasmenos, and a short text by George Gemistus Plethon. By focusing on a specific conception of Greek mathematics (the operations of 'composition' and 'removal' of ratios), Acerbi explores a variety of sources in order to trace its pattern of evolution. Special emphasis is placed on the transition from proportion theory to logistic, in which these operations were modified by introducing the crucial notion of πηλικότης. Within this framework, Acerbi's paper (and his project in general) constitutes a call for historians to further examine largely neglected later sources, and, so reconsider the whole spectrum of ancient and medieval Greek mathematics.

The title of Henry Mendell's contribution ("Why Did the Greeks Develop Proportion Theory? A Conjecture") speaks for itself. Mendell begins with the observation that while the early Greek mathematicians developed a number theory based on whole numbers and proportion theory (like the one we find in Euclid's *Elements*), their neighbors, the Egyptians and Babylonians, did not. In order to explain this phenomenon, Mendell examines the way in which Greek numerical systems dealt with fractions and provides a distinction between ratios, on the one hand, and arithmetical calculations on the other.[28] Using this methodology, the author draws on a variety of sources, among which Plato's *Republic*, Euclid's *Sectio Canonis*, Ptolemy's *Harmonica*, the *Anthologia Palatina*, and Archimedes' 'Cattle Problem'. The kind of number theory we find in Euclid's *Elements*, Mendell concludes, represents the culmination of one of the several revolutions in mathematics that took place in the fifth and fourth centuries BC and was certainly fundamental to the development of a general theory of proportion and irrationals.

Geoff Lehman and Michael Weinman's paper ("Recursive Knowledge Procedures Informing the Design of the Parthenon: One Instance of Continuity between Greek and Near Eastern Mathematical Practices") deals with the much debated, yet extremely difficult question, of the interaction between Greek and pre-Hellenic mathematical cultures. On a methodological level, this paper promotes an exciting new perspective; that material evidence—in this case, the Parthenon itself—may be used in order to establish conclusions related to the foggy area of the origins of Greek mathematics. Most shockingly, the authors suggest that the sophisticated construction of the Parthenon entails a properly theoretical understanding of the notions of continuous proportions and anthyphairesis, tools that the Greeks might have borrowed from neighboring cultures. On this basis, Lehman and Weinman call into question a more general historiographical point which holds that one can find more discontinuity than continuity between Greek mathematics and its neighboring Eastern predecessors.

Jeffrey Oaks's paper ("Diophantus, al-Karajī, and Quadratic Equations") is unique in the framework of this volume in the sense that it places emphasis mostly on Arabic mathematics. The author's basic historiographical methodology is simple, yet shockingly revealing: in order to evaluate Greek mathematics, one must sometimes take into

---

28 On latest research on Greek numeracy, see Sidoli 2014, 39 and Berggren 1984, 398–400. For the latest comprehensive study, see Cuomo, forthcoming.

consideration what followed it. In other words, a historical evaluation of Greek mathematics can benefit from a multilevel contextualization. Oaks's essay studies the so-called 'three-term quadratic equations'. Set against the background of the new historiographical category of 'premodern algebra' (on this, also see Christianidis's paper he reconstitutes a connection between the works of Diophantus and al-Karajī. At the beginning of his *Arithmetica*, Diophantus promises to show his readers how to solve this kind of equation. Unfortunately, the extant part of the *Arithmetica* does not fulfil this undertaking. Al-Karajī, who made ample use of an Arabic translation of Diophantus' work, gives derivations of solutions to these equations by what he calls "the method of Diophantus", while at the same time giving a rule that corresponds to the one employed on such equations in the *Arithmetica*. Oaks hypothesizes, then, that Diophantus derived his rule by this "method" in a lost part of the work. Oaks notes that continuity in algebraic practice from Diophantus to al-Karajī was more likely maintained through oral tradition than through the chain of a few written texts, and a fortuitous kink in the transmission of the method of solution helps identify Diophantine elements in al-Karajī's book.

The paper of Gabriele Galluzzo ("Substantiae sunt Sicut Numeri: Aristotle on the Structure of Numbers"), discusses Aristotle's conception of numbers. In his *Metaphysics*, Aristotle claims that not only material objects but numbers too can be analyzed in terms of matter (the units that compose a number) and form (the structure that keeps the units together). On this basis, Galluzzo argues that Aristotle's philosophy of mathematics stems not only from his critical attitude towards Plato (as it is usually thought), but that it should be seen as a sophistical metaphysical philosophy of mathematics. If Galluzzo is right, then Aristotle's understanding of numbers offers a new paradigm in the realm of philosophy of mathematics, a paradigm for understanding the notion of number, since in the traditional picture, numbers are pluralities of physical objects considered in a certain way, for instance as countable or numerable. Although it does not entirely reject this picture, the hylomorphic model invites us to think of numbers in a different way, that is, as pluralities of countable physical objects (units) *arranged in a certain numerical structure*.

Vasilis Karasmanis's paper ("The Axiomatization of Mathematics and Plato's Conception of Knowledge in the *Meno* and the *Republic*") challenges the common idea that all Platonic dialogues present the same theory of knowledge. In the *Republic*, Plato describes mathematics as a science organized in a system where one is able to see what is fundamental, what is derived from the basics, and how each conclusion is derived from a previous proposition; in other words, it presents significant similarities with what one finds in Euclid's *Elements*. Now, in the *Meno*, Plato seems to offer a different model that is very close to modern coherent theories of justification: there exist no particular starting/ending points and every knot may serve as a first principle on occasion. In other words, the system is coherent, yet not hierarchically structured. After exposing the differences between Plato's image of knowledge in these two dialogues, Karasmanis puts forth the hypothesis that a mathematical revolution (the axiomatization of mathematics) took place at some point between the composition of the *Meno* and that of the *Republic*, a radical change which deeply influenced Plato's thinking. If Karasmanis is right, this would not only allow us to trace different stages in the evolution of Plato's thought; on a more general level, it would offer us valuable insights into one of the most interesting historical periods with regards to the interaction between Greek mathematics and philosophy.

The final paper of this volume by Stelios Negrepontis ("The Anthyphairetic Revolutions of the Platonic Ideas") places emphasis on the so-called 'lesson of Theodorus', a passage in Plato's *Theaetetus* that has led generations of scholars to serious puzzlement. Negrepontis draws attention to the technique of anthyphairesis and offers some new arguments for the view that Theodorus' proofs must have relied on it. In addition, he draws parallels between the method of 'division and collection' (which appears in Plato's *Sophist* and *Statesman*) and Theaetetus' "periodic anthyphairesis". Through this framework, Negrepontis criticizes modern scholars for neglecting to the fundamental role of periodic anthyphairesis in Plato's philosophy as a whole and calls for a radical new reading of Plato. Finally, Negrepontis argues that the so-called 'Palindromic Periodic Theorem' for the anthyphairesis of quadratic irrationals, usually attributed to Euler and Lagrange, is described in the *Statesman*, and can be proved by the tools provided in Euclid's book X of the *Elements*.

Afterword

This edited volume book has taken a long time to prepare—perhaps too long, if one takes into consideration the extensive institutional support I have received from the Research Training Group 'Philosophy, Science, and the Sciences' at the Humboldt University of Berlin in preparing it. Markus Asper, the chief editor of this project, has proved himself as a man of great patience, and for this I am grateful. A final word: By its very nature, the discipline of Greek mathematics may give the impression of an isolated discipline, untouched by modern evolutions in other fields, and thus doomed to extinction. In fact, the Preface of this volume might have also given this impression to the reader. Yet, now, upon completion of this project and while going through these lines, I have to confess that I am much more optimistic about the future. As we speak, a new generation of scholars is bringing innovative historical and historiographical perspectives into the field. At the same time, there seems to be a growing demand among historians to reconsider some of the fundamental hypotheses and methodological principles upon which the discipline was originally established. In other words, the field is undergoing rapid change—a 'scientific revolution' if you agree with Kuhn—the results of which remain to be seen and evaluated. The laborers are few, but the harvest is plentiful.[29]

29 I apologize to the Evangelist for the paraphrase.

Bibliography

Acerbi, F. 2010. *Il Silenzio delle Sirene: la Matematica Greca Antica*. Roma.

Artmann, B. 1991. "Euclid's *Elements* and its Prehistory." In: I. Mueller (ed.), *ΠΕΡΙ ΤΩΝ ΜΑΘΗΜΑΤΩΝ (Peri tōn Mathēmatōn), Apeiron XXIV (4)*. Edmonton, 1–47.

Ausejo, E. & M. Hormigôn (eds.) 1996. *Paradigms and Mathematics*. Madrid.

Bachelard, G. 1938. *La Formation de l' Esprit Scientifique*. Paris.

Bashmakova, I. G. & I. Vandoulakis. 1994. "On the Justification of the Method of Historical Interpretation." In: K. Gavroglu *et al.* (eds.), *Trends in the Historiography of Science*. Dordrecht, 249–264.

Berggren, L. J. 1984. "History of Greek Mathematics: A Survey of Recent Research." In: *Historia Mathematica* 11, 394–410.

Bernard, A. 2003. "Ancient Rhetoric and Greek Mathematics: A Response to a Modern Historiographical Dilemma." In: *Science in Context* 16, 391–412.

Bernard Cohen, I. 1976. "The Eighteenth-Century of the Concept of Scientific Revolution." In: *Journal of the History of Ideas* 37, 257–288.

Bernard Cohen, I. 1985. *Revolution in Science*. Cambridge, MA.

Blåsjö, V. 2014. "A Critique of the Modern Consensus in the Historiography of Mathematics." In: *Journal of Humanistic Mathematics* 4, 113–123.

Blåsjö, V. 2016. "In defence of geometrical algebra." In: *Archive for the History of Exact Sciences* 70, 325–359.

Bos, H. J. M. & H. Mehrtens 1977. "The Interactions of Mathematics and Society in History: Some Exploratory Remarks." In: *Historia Mathematica* 4, 7–30.

Brush, S. G. 1970. "History for Scientists." In: *Isis* 61, 115–118.

Buchdahl, G. 1965. "A Revolution in the Historiography of Science." In: *History of Science* 3, 55–69.

Corry, L. 1993. "Kuhnian Issues, Scientific Revolutions and the History of Mathematics." In: *Studies in History and Philosophy of Science* 24, 95–117.

Crowe, M. J. 1975. "Ten 'Laws' Concerning Conceptual Change in Mathematics." In: *Historia Mathematica* 2, 161–166.

Crowe, M. J. 1992. "Afterword (1992): A revolution in the Historiography of Mathematics?" In: D. Gillies (ed.), *Revolutions in Mathematics*. Oxford, 306–316.

Cuomo, S. 2001. *Ancient Mathematics*. London.

Cuomo, S. forthcoming. *Ancient Numeracy*. Cambridge, MA.

Dauben, J. 1984. "Conceptual Revolutions and the History of Mathematics: Two Studies in the Growth of Knowledge." In: E. Mendelsohn (ed.), *Transformation and Tradition in the Sciences*. Cambridge, 81–103 (reprinted In: D. Gillies (ed.), *Revolutions in Mathematics*. Oxford, 49–71).

Dauben, J. 1992a. "Appendix (1992): Revolutions Revisited." In: D. Gillies (ed.), *Revolutions in Mathematics*. Oxford, 72–82.

Dauben, J. 1992b. "Are There Revolutions in Mathematics?" In: J. Echeverria *et al.* (eds.), *The Space of Mathematics: Philosophical, Epistemological, and Historical Explorations*. Berlin, 205–229.

Dunmore, C. 1992. "Meta-level Revolutions in Mathematics" In: D. Gillies (ed.), *Revolutions in Mathematics*. Oxford, 209–225.

Foucault, M. 1969. *L'Archéologie du Savoir*. Paris.

Fowler, D. H. 1994. "Could the Greeks Have Used Mathematical Induction? Did They Use It?" In: *Physis* 31, 253–265.

Francois, K. & J.-P. Van Bendegem. 2010. "Revolutions in Mathematics. More than Thirty Years after Crowe's 'Ten Laws'. A New Interpretation." In: B. Löwe & T. Müller (eds.), *PhiMSAMP: Philosophy of Mathematics: Sociological Aspects and Mathematical Practice*. London, 107–120.

Freudenthal, H. 1977. "What Is Algebra and What has it Been in History?" In: *Archive for the History of Exact Sciences* 16, 189–200.

Fried, M. N. 2014. "The Discipline of History and the 'Modern Consensus in the Historiography of Mathematics'." In: *Journal of Humanistic Mathematics* 4, 124–136.

Fried, M. N. & S. Unguru. 2001. *Apollonius of Perga's Conica: Text, Context, Subtext.* Leiden.

Gillies, D. (ed.) 1992. *Revolutions in Mathematics*. Oxford.

Grattan-Guinness, I. 1976. "History of Mathematics and Histories of Other Sciences." In: *Historia Mathematica* 3, 325–327.

Grattan-Guinness, I. 1996. "Numbers, Magnitudes, Ratios, and Proportions in Euclid's Elements: How did he Handle them?" In: *Historia Mathematica* 23, 355–375.

Heath, T. L. 1921. *A History of Greek Mathematics*. Oxford.

Høyrup, J. 1996. "Changing Trends in the Historiography of Mesopotamian Mathematics: An Insider's View." In: *History of Science* 34, 1–32.

Kindi, V. & Th. Arabatzis (eds.) 2012. *The Structure of Scientific Revolutions Revisited*. London.

Knorr, W. R. 1990. "New readings in Greek mathematics: Sources, Problems, Publications." In: *Impact of Science on Society* 159, 207–218.

Kuhn, T. S. 1970a. "Reflections on My Critics." In: I. Lakatos & A. Musgrave (eds.), *Criticism and the Growth of Knoweldge*. Cambridge, 231–278.

Kuhn, T. 1970b. *The Structure of Scientific Revolutions, 2nd ed, enlarged*. Chicago.

Lakatos, I. 1963–1964. "Proofs and Refutations." In: *British Journal for the Philosophy of Science* 14, 1–25, 120–139, 221–243, 296–342.

Lakatos, I. & A. Musgrave (eds.) 1970. *Criticism and the Growth of Knoweldge*. Cambridge.

Laudan, L. *et al.* 1986. "Scientific Change: Philosophical Models and Historical Research." In: *Synthese* 69, 141–223.

Mehrtens, H. 1976. "T.S. Kuhn's Theories and Mathematics: A Discussion Paper on the "New Historiography" of Mathematics." In: *Historia Mathematica* 3, 297–320.

Mehrtens, H. 1992. "Appendix (1992): Revolutions Reconsidered." In: D. Gillies (ed.), *Revolutions in Mathematics*. Oxford, 42–48.

Mueller, I. 1981. *Philosophy of Mathematics and Deductive Structure in Euclid's Elements*. Cambridge, MA.

Neugebauer, O. 1952. *The Exact Sciences in Antiquity*. Princeton.

Pourciau, B. 2000. "Intuitionism as a (Failed) Kuhnian Revolution in Mathematics." In: *Studies in History and Philosophy of Science* 31, 297–329.

Rowe, D. 1996. "New Trends and Old Images in the History of Mathematics." In: R. Calinger (ed.), *Vita Mathematica: Historical Research and Integration with Teaching*. Washington, 3–16.

Rowe, D. E. 2012. "Otto Neugebauer and Richard Courant: On Exporting the Göttingen Approach to the History of Mathematics." In: *Mathematical Intelligencer* 34, 29–37.

Saito, K. 1998. "Mathematical Reconstructions Out, Textual Studies In: 30 Years in the Historiography of Greek Mathematics." In: *Revue d' histoire de mathematics* 4, 131–142.

Schneider, M. R. 2016. "Contextualizing Unguru's 1975 Attack on the Historiography of Ancient Greek Mathematics." In: V. R. Remmert *et al.* (eds.), *Historiography of Mathematics in the 19th and 20th Centuries*. Basel, 245–267.

Sialaros, M. & J. Christianidis 2016. "Situating the Debate on 'Geometrical Algebra' Within the Framework of Premodern Algebra." In: *Science in Context* 29, 129–150.

Sidoli, N. 2014. "Research on Ancient Greek Mathematical Sciences, 1998–2012." In: N. Sidoli & G. Van Brummelen (eds.), *From Alexandria, Through Baghdad: Surveys and Studies in the Ancient Greek and Medieval Islamic Mathematical Sciences in Honor of J. L. Berggren*. Berlin, 25–50.

Tannery, P. 1903. *Notions de mathématiques*. Paris.

Tannery, P. 1912. "De la solution géométrique des problèmes du second degré avant Euclide." In: J. L. Heiberg & H. G. Zeuthen (eds.), *Sciences Exactes dans L'Antiquité*. Paris, 254–280.

Unguru, S. 1975. "On the Need to Rewrite the History of Greek Mathematics." In: *Archive for the History of Exact Sciences* 15, 67–114.

Unguru, S. 1979. "History of Ancient Mathematics: Some Reflections on the State of the Art." In: *Isis* 70, 555–565.
Unguru, S. 1991. "Greek Mathematics and Mathematical Induction." In: *Physis* 28, 273–289.
Unguru, S. 1994a. "Fowling after induction." In: *Physis* 31, 267–272.
Unguru, S. 1994b. "Is Mathematics Ahistorical? An Attempt to an Answer Motivated by Greek Mathematics." In: K. Gavroglu *et al.* (eds.), *Trends in the Historiography of Science*. Dordrecht, 203–220.
Unguru, S. & D. E. Rowe 1981. "Does the quadratic equation have Greek roots? a study of 'Geometric Algebra', 'Application of Areas', and related problems (part I)." In: *Libertas Mathematica (ARA)* 1, 1–49.
Unguru, S. & D. E. Rowe 1982. "Does the quadratic equation have Greek roots? a study of 'Geometric Algebra', 'Application of Areas', and related problems (part II)." In: *Libertas Mathematica (ARA)* 2, 1–62.
Van der Waerden, B. L. 1950. *Ontwakende Wetenschap*. Groningen.
Van der Waerden, B. L. 1976. "Defence of a Shocking Point of View." In: *Archive for the History of Exact Sciences* 15, 199–210.
Vitrac, B. & M. Caveing (eds.) 1990–2001. *Euclid: Les Éléments*. 4 vols. Paris.
Weil, A. 1978. "Who betrayed Euclid." In: *Archive for the History of Exact Sciences* 19, 91–93.
Wilder, R. L. 1968. *Evolution of Mathematical Concepts*. New York.
Zeuthen, H. G. 1886. *Die Lehre von den Kegelschnitten im Altertum*. Kopenhagen.
Zeuthen, H. G. 1902. *Histoire des mathématiques dans l'antiquité et le moyen âge*. Paris.

Sabetai Unguru

Counter-Revolutions in Mathematics

Abstract: The paper deals with two polar-opposite approaches to the study of the history of mathematics, that of the mathematician, tackling the history of his discipline, and that of the historian. It argues for, something denied by the typical mathematician, the existence of revolutions in the development of mathematics, accompanied by upheavals in the history and, historically grounded, philosophy of mathematics. The main historical examples it analyzes are drawn from the history of Hellenistic mathematics, showing their fundamental disparity with their modern analogues. It rejects the conception of notation as a mere, neutral, contentually-indifferent, abbreviatory device in the development and elucidation of mathematical cultures.

Keywords: Euclid, Greek, Mathematics, History, Historiography, Revolutions, Number, Algebra, Diophantus, Babylonian, Neugebauer, Weil.

I

Since 'revolution' and 'counter-revolution' are correlative terms, speaking of the latter, as I do in the title of my paper, implies the existence of the former. Indeed, I shall try to argue that revolutions in the history and philosophy of mathematics bring in their wake counter-revolutions *in* mathematics.

In 1975, Michael Crowe published in *Historia Mathematica* an article titled "Ten 'Laws' Concerning Patterns of Change in the History of Mathematics". The tenth 'Law' reads: "Revolutions never occur in mathematics".[1] I think that, *sub specie* mathesis, Crowe is right, despite the numerous articles written as rebuttals to his argument—the most important of which were collected in *Revolutions in Mathematics*, a book edited by Donald Gillies.[2] This is what I shall try to convince you of in this paper.

What makes mathematics unique among the so-called exact sciences, let alone among scientific disciplines at large, is its exclusive reliance on proof to back up its claims. Without proof, there is no mathematics. It is in this sense that pre-Hellenic mathematics was not, strictly speaking, mathematics. And it is this very concept of proof that ultimately enables mathematicians to absorb past mathematics *in toto*, without any qualms or insuperable difficulties, in the growing, changing body of the discipline. Any true mathematical achievement of the past is also a true mathematical

1 Ibid., 165.
2 Gillies 1992.

https://doi.org/10.1515/9783110565959-002

achievement of the present and the future. Euclidean geometry is as true now, after the discovery/invention of non-Euclidean geometries, as it was in Greek antiquity. Diophantus' so-called 'algebra' has not been discarded by Bartel van der Waerden's *Moderne Algebra*.[3] Archimedes' quadrature of the parabola is as valid now, after the discovery/invention of calculus, as it was in the third century BC. Indeed, both Diophantus and Archimedes have been translated into modern analytical language and incorporated into the body of modern mathematics. Such examples can be multiplied *ad nauseam*.

Here is how some well-known, and lesser well-known mathematicians, express themselves on the issue: First, Godfrey Harold Hardy, in his rightly famous *A Mathematician's Apology*:[4]

> Archimedes will be remembered when Aeschylus is forgotten, because languages die and mathematical ideas do not. 'Immortality' may be a silly word, but probably a mathematician has the best chance of whatever it may mean.

Next, a lean chrestomathy from a contemporary, recently very active, internet list, run by Julio Gonzalez Cabillon, HISTORIA MATEMATICA (sic), in which many mathematicians, as well as historians and philosophers of mathematics participate on a daily basis:

> Date: Mon, 19 Jul 2004 09:42:50 +0300
> From: "Yaakov S. Kupitz" <kupitz@math.huji.ac.il>
> Subject: Re: [HM] Eudoxus-Archimedes' Lemma
>
> What Ken Pledger writes is incredible, but he seems to be right that in the 1900's still people of the calibre of Heath didn't realize that Euclid 5.4 and its direct consequence 10.1 (the principle of exhaustion) and its use in 12.2 (circular area is proportional to radius^2) and Archimedes' various quadratures (e.g., Kuklos Meitriris Prop. 1 and his quadrature of the parabola) are rigorous up to the most modern standards of epsilon, and that in modern times only Cauchy's definition restored ancient rigorosity.
> Yaakov S. Kupitz
> The Hebrew University of Jerusalem
>
> Date: Tue, 20 Jul 2004 08:34:53 -0500
> From: "Matt Insall" <montez@fidnet.com>
> Subject: Re: [HM] Eudoxus-Archimedes' Lemma
>
> Kupitz wrote:
> "...are rigorous up to the most modern standards of epsilon, and that in modern times only Cauchy's definition restored ancient rigorosity."
>
> Actually, Abraham Robinson restored the rigour to the methods of Leibniz and Newton, by providing "nonstandard analysis" as a logical foundation for infinitesimal calculus. It is

3 Van der Waerden 1930–1931.
4 Hardy 1967, 81.

simpler than Cauchy's approach, but for some reason Robinsonian Analysis has never been accepted as a replacement for Cauchy-style Analysis.

Dr. Matt Insall
Associate Professor of Mathematics
Department of Mathematics and Statistics
University of Missouri - Rolla
Rolla MO 65409-0020

And:

Roberval, one of the first to grasp Cavalieri's approach in the mid 1630's, also used the expression "tous les sinus", equivalent to "Omnes linea", in his (almost forgotten) integration of the sine curve, but he clearly meant a sum of rectangles, i.e., a Riemann sum approaching the area under the sine curve. This result of Roberval was used later by Mylon to rectify the cycloid using the language of indivisibles (AFTER Wren proved it in 1658 using exhaustion in a completely rigorous way) in the annex to his letter to Huygens from January 1659. One cannot understand Roberval's result as well as Mylon's proof of Wren's result without interpreting expressions like "Omnes linea" or "Omnes sinus" (used by Mylon and Roberval in these works) as sums of rectangles, i.e., as Riemann sums.
Yaakov Kupitz
Institute of Mathematics
The Hebrew University of Jerusalem

Date: Wed, 15 Sep 2004 11:26:58 +0300
From: "Yaakov S. Kupitz" <kupitz@math.huji.ac.il>
Subject: Re: [HM] The word "class"

Quoting Samuel S Kutler <reltuk@comcast.net>:

In ancient Greek the words for kind and form: idea (iota, delta, epsilon, alpha) and eidos (epsilon, iota, delta, omicron, sigma) are very important—especially in Plato. The idea of the good is said by Plato's Socrates to be beyond being! Idea is not like our word which is something we have, rather it is what gives intelligibility to things. Cantor thought that his word for set was like those Greek words!

Plato's Theory of Ideas is nothing but an ancient formulation of what modern mathematics calls equivalent classes, nicely described in Herman Weyl's "Philosophy Of Mathematics" (he doesn't mention Plato). For Plato the Idea of Bed is the equivalent class of all beds, so for Shuttle etc. A. N. Whitehead connects Plato's Theory of Ideas to modern mathematics in his "Mathematics and The Modern World". In modern algebra a quotient group G/H is a group whose elements are equivalent classes of conjugations with respect to the normal subgroup H. A point in a projective space P^d(F) is the equivalent class of all scalar multiples of a fixed vector. Thinking in terms of equivalent classes needs a considerable power of abstraction. Plato recognized the import of equivalent classes to our thinking, and he made it the vehicle of "noos" ("intelligence" in Gr.).

Enough!

There is *one* mathematics, from its pre-historical beginnings to the end of time, irrespective of its changing appearances over the centuries. This mathematics grows by accumulation and by a sharpening of its standards of rigor, while, its rational, ideal, Platonic Kernel, remaining unaffected by the historical changes mathematics undergoes, enjoys, as Hardy put it, immortality.

In short, proven mathematical claims remain proven forever, no matter what the changes are that mathematics is undergoing. And since it is always possible to present past mathematics in modern garb, ancient mathematical accomplishments can be easily made to look modern and, therewith seamlessly integrated into the growing body of mathematical knowledge. That this is a historical calamity is not the mathematician's worry. The very, and simple, fact that it can be done is sufficient: What can be done should be done! Indeed, doing it is its own justification! Never mind that this procedure is tantamount to the obliteration of the history of mathematics as a *historical* discipline. Why, after all, should this concern the mathematician?

What I have said up to this point is enough to cast serious doubt on the position of those who argue for the existence of revolutions *in* mathematics. How is it, then, possible to adopt such a position? By emptying the concept of revolution of its specific, particular, idiosyncratic, content; by using the concept metaphorically; by endowing breakthroughs and other significant changes in the development of mathematics with the *mathematically* undeserved title of revolutionary. Precisely as mathematicians, by overlooking the historical component of their field, can argue cogently and convincingly for the cumulative, non-revolutionary character of mathematics, historians, philosophers, sociologists of mathematics, and others, by overlooking the distinction between mathematics and meta-mathematics—understood as the discourse about mathematics—and by using a very soft definition of the concept of revolution, can argue (not cogently and convincingly, to my mind) for the revolutionary character of the growth of mathematics. If what counts as revolutionary for you is the way mathematics develops, then, clearly, there are revolutions in mathematics. As a matter of fact, mathematics is undergoing a (Trotskyite) continuous revolution!

Between these two extreme positions (i.e. the Scylla of the mathematician, who discards the historical development of his/her discipline, and the Charybdis of those who jettison the difference betwixt the body of mathematical knowledge and its historico-philosophical components and valuations, adding to this a watered down definition of revolution), there is a third position—which I adopt—distinguishing amongst the body of mathematical knowledge, internal mathematical developments taken *in abstracto*, the way mathematicians do, in which accumulation is the prevailing phenomenon, and the images of mathematical knowledge (historical, philosophical, sociological, metaphysical), in which revolutionary breakthroughs do occur occasionally, all this assessed in light of a bold definition of revolution.

As a historian of mathematics, I deem the transition from pre-Hellenic to Hellenic mathematics revolutionary. As a matter of fact, I think that even dyed in the wool mathematicians would agree with me that a change that transformed mathematics from practical computation to a postulatory-deductive discipline, granting it its true identity, is a revolutionary change. This is perhaps the only revolution *in* mathematics, mathematicians *qua* mathematicians would grant, though I am not at all sure, since even Egyptian, Babylonian, Chinese, etc. mathematics has been translated into modern algebraic symbolism and incorporated into the body of mathematics.

An example would perhaps be in order. Otto Neugebauer classified Babylonian mathematical texts into 'table' and 'problem' texts. All these texts contain are numbers, accompanied occasionally by a few words and diagrams. Still, Neugebauer calls repeatedly these texts 'algebraic' and provides an interpretation that is not only heavily symbolical, but is also couched in terms pregnant with the theory of equations, the beginning of which, however, can be traced, at the earliest, to Descartes. Here is a typical instance of a so-called 'algebraic' problem:

> I have subtracted the [side] of the square from the area, and 14,30 is it.

And here is the solution, as it appears in tablet BM 13901:[5]

> Take 1, the unity. Divide 1 into two equal parts: 0;30 times 0;30 is 0;15. Add this to 14,30 and it is 14,30;15 which has 29;30 as a square root. Add the 0;30 which you have multiplied by itself to 29;30, and 30 is the [side of the] square.

On the basis of such texts, including so-called 'normal forms', translated as canonical systems of simultaneous equations, all of which Neugebauer calls without qualms 'algebraic' despite the fact that these texts contain only series of arithmetical operations, he provides an analysis of the 'quadratic problems', allegedly solved by the Babylonians, in terms of "the elementary symmetric functions of the two roots"![6] Specifically, all problems fitting the first normal form, in which the sum and product of two numbers are given, $x + y = a$, $x \cdot y = b$, follow a common solution procedure in the texts:[7]

> One squares one-half the value of the sum, subtracts from the result the value of the product, and takes the square root of the remainder. The result is then added to one-half the sum to get one unknown and subtracted from it to get the other. $\left[x = \pm\sqrt{\left(\frac{x+y}{2}\right)^2 - x \cdot y} + \frac{x+y}{2} = x;\ y\right]$

This purely arithmetical procedure, however, fits, as Mahoney has shown:[8]

> two different interpretations equally well. As a solution of a simultaneous system, it computes $\frac{x-y}{2}$ from the data *via* the identity $\left(\frac{x+y}{2}\right)^2 - x \cdot y = \left(\frac{x-y}{2}\right)^2$ and then obtains x and y from the combinations $\frac{x+y}{2} \pm \frac{x-y}{2}$. But the solutions so obtained also result from computing the double

5 Neugebauer 1935–1937, III 6. The problem is also discussed in van der Waerden 1976, 200–201.

6 Mahoney 1970, 375.

7 Ibid., 375–376.

8 Ibid.

> root of the single quadratic equation $x^2 - ax + b = 0$ by the quadratic formula. Moreover the latter computation would coincide step for step with the former. That is computationally one cannot distinguish between [the] solution of the pair of simultaneous equations *via* the identity and [the] solution of a quadratic equation by the quadratic formula. *There is, however, a world of difference between the two derivations. The first yields two solutions, because there are two unknowns; the second yields two roots because of the fundamental theorem of algebra.* (My emphasis.)

What all this discussion shows is that even in the case of Babylonian pre-demonstrative mathematics—and this applies equally well to the case of ancient Egyptian mathematics (though I shall refrain from specific examples, for reasons of space)—that is, even before mathematics acquired its particular, unique identity, mathematicians, and mathematically oriented historians, like Neugebauer, do not hesitate in analyzing its textual products in terms of modern mathematics, as if to underscore the fact that there is just one universal, atemporal mathematics that grows by addition only, and to which revolutionary breaks are totally foreign.

And yet, there is a true chasm separating Babylonian and Egyptian mathematics from Greek mathematics, a chasm which is clearly the result of a revolution in mathematical thinking. From computational, practical, empirical, mathematics becomes, with the Greeks, theoretical, abstract, demonstrative, i.e., postulatory-deductive. From merely doing mathematics, the Greeks start thinking about doing mathematics, while doing it, and doing it in a manner which is uniquely Greek, not Oriental. Quoting again Mahoney, in his essay review of the unrevised second printing of Neugebauer's *Vorlesungen ueber Geschichte der antiken mathematischen Wissenschaften*, Band I, *Vorgriechische Mathematik* (1969): [9]

> If... the Babylonians *did* mathematics, and even if they did it remarkably well, there is absolutely no evidence that they *thought about* mathematics. ... But thinking about mathematics the way the Greeks did is a necessary prerequisite to doing what today is called 'metamathematics', that is, to talking about mathematics mathematically. ... There is nothing metamathematical about the cuneiform texts.

Hence, the unavoidable conclusion, it seems to me, is that the transition from pre-Greek to Greek mathematics is a revolutionary transition, if there ever was one! Mathematicians, however, treat it as an entirely normal, evolutionary development.

Greek mathematics presents the historian, for the first time, with a mathematical culture that we recognize as ours. It is not a culture of approximation, empirical calculations accompanied occasionally by a verification of the results, innocent of generalization, abstraction, and proof. On the contrary! Whereas Egyptian and Babylonian mathematics consists of specific exercises, attended sometimes by verification of the results, but never by any attempt to justify the method employed, Greek mathematics, from its beginnings in the sixth century BC, displays a quest for general propositions,

9 Ibid., 377.

demonstrated rigorously and, therefore, persuasive beyond reasonable doubt. As Serafina Cuomo says in a recently published book on *Ancient Mathematics*:[10]

> (W)here the Egyptians had been able to calculate the volume of a certain cylinder and verify that the result was correct, or at least suitable for their, usually practical, purposes, the Greeks found the general formula [I would say 'expression' or 'procedure'] for the volume of any cylinder and proved why that formula [i.e., expression] was right.

Greek mathematics, therefore, is ours. Its emergence marks the beginnings of Western mathematics. It is a constitutive, fundamental component of our culture, as are Greek astronomy and science, Greek literature and philosophy, Greek drama, mythology, and history. This does *not* mean, however, that there are no fundamental differences between Greek and modern, post-Renaissance mathematics, differences as the historian, but not the mathematician, necessarily registers. The existence of such differences should not surprise us. After all, as Leslie Hartley put it succinctly: "THE PAST is a foreign country; they do things differently there."[11] This is, of course, not the place for a survey of ancient Greek mathematics. It is enough to say that, despite Greek mathematics being the progenitor of modern mathematics, the two are not identical! (*Warning*: This is the *historian* speaking!) After the discovery of incommensurable magnitudes, an event, by the way, which some historians of mathematics, like Joseph Dauben, call a revolution in mathematics, (incommensurability being referred by mathematicians as the discovery of irrational numbers), Greek mathematics became largely Greek geometry, the geometry known to us, primarily, from Euclid's *Elements*, but which reached its zenith with the works of Archimedes and Apollonius. This discovery led to a separation of arithmetic from geometry; this separation, and, in general, the respect of disciplinary boundaries, being characteristic of Greek mathematics and science. To treat magnitudes in all their generality, the only instrument available to the Greek mathematician was Greek geometry, not algebra, which did not yet exist. Despite claims to the contrary, from the mathematicians' camp, there is no Greek algebra (the case of Diophantus's *Arithmetica*, third century AD, is special and I shall not deal with him now) and, consequently, no Greek analytic geometry. Also, the Greek concept of number is not the modern one. Finally, the Greeks, and Archimedes in particular, though he was the greatest mathematician of Antiquity and one of the greatest of all times, were not the inventors of the calculus.

The transition from Greek to modern mathematics, mainly in the sixteenth and seventeenth centuries in Western Europe—though Islamic mathematics also played a role in this transition—is, to my mind, a genuine revolutionary transition. The heroes of this revolution are Viète, Fermat, and Descartes, Newton, and Leibniz, to mention

---

10 Cuomo 2001, 5.
11 Hartley 1953, prologue.

the most important of them. These heroes, however, mostly saw their contributions as not only growing out of Greek mathematical achievements, which is true, but also as springing conceptually from the Greek conception of the nature of mathematics, which is historically inaccurate. Here is how Jacob Klein puts the matter:[12]

> (N)either Descartes nor Fermat believed he had advanced beyond Apollonius on any essential points. What we take to be the enormous achievement of Descartes and Fermat they themselves believed they had learned in essence from Apollonius or Pappus. Fermat finds fault with Apollonius only because he did not present matters 'generally enough'... He says very cautiously that his general procedure for constructing geometrical loci 'was *perhaps* not known to Apollonius... And Descartes is quite convinced that the Ancients—he expressly names Pappus along with Diophantus—deliberately erased the traces of their true knowledge out of a kind of perverted cunning...and divulged to us, not their own art, but only a few of their results.

What this shows is precisely my point that *in* mathematics revolutionary transitions are treated as normal, evolutionary growth. That we are indeed dealing with a revolution in mathematical thought in the sixteenth and seventeenth centuries is abundantly clear when one compares a Hellenistic mathematical text, say Apollonius's *Conica*, with one of its modern counterparts, say Descartes' *Geométrié*. Though we cannot do this here in detail, some circumscribed comparative remarks about the content and style of the *Conica* and *La Géométrie* are in order. First, despite the fact that Apollonius speaks of abscissas and ordinates in his treatise, there is *no* real system of coordinates in the *Conica*, a system which appears for the first time only with Fermat and Descartes. The so-called 'axes' in the *Conica* do not constitute a system of lines on their own, as Moritz Cantor had already remarked,[13]

> (B)ut, like other auxiliary geometrical lines, make their appearance only in connection with the conic section; they are brought into existence by the theorem to be proved in each instance.

In each case, Apollonius has in view the particular ellipse, or parabola, or hyperbola, which is obtained on the surface of a particular cone by the intersection of particular planes. The diagram gives a true image of this particular cone, these particular lines and planes, and this particular conic section. The genesis procedure of the conics is general, but there is no general object since the diagram represents no symbol!

"The characteristic of the μαθηματικα, mathematical objects in the Greek sense, is precisely that they can be grasped by the senses only in images, while they themselves, in their unalterable constitution, are accessible only to the discursive intellect; however, there are infinitely many of these objects."[14] This last quotation from Klein points to the fundamental difference in intentionality between Greek and modern

12 Klein 1985, 13–14.

13 Cantor 1922–1924, I 337.

14 Klein 1985, 16–17.

science, and specifically their different intentionalities of mathematical objects. The concepts of Greek science and mathematics intend always individual objects; indeed, they are, scholastically speaking, *intentiones primae*—that is, concepts which refer directly to individual objects. This fits the way Greek mathematics (and science) represents its objects. The geometrical diagram exhibits the object, it "images" it, as Klein felicitously puts it.

Against this, the modern presentation and mathematical analysis embody a fundamentally different relation to mathematical objects and the analysis of their relations. The presentation is symbolic and the objects and their relations are *intentiones secundae*, concepts which refer directly to other concepts, i.e., to *intentiones primae*, and only through the latter, that is, indirectly, to objects. The modern, Cartesian system of coordinates (a real system, unlike Apollonius's lines in the *Conica*) symbolizes the absolute generality of its application (its intersecting lines are no more specific, real lines, generated by the figure in question), its axes standing for the general concepts of abscissa and ordinate and not for the line-segments directly intended by these concepts in each individual instance. For Klein, whom I closely follow, abscissas and ordinates are, in the language of modern mathematics, concepts of the 'Variable n':[15]

> For this reason the abscissa and ordinate axes can be detached from the realm of objects. All the curves investigated with their help are from now on nothing but symbolic exhibitions of various possible relations, or of the different 'functional' relations, between two (or more) variables.

Summing up this part, then, modern mathematics violates all strictures of Greek mathematics, starting with the fundamental Aristotelian interdiction of *metabasis*, by mixing unmixables (that is, geometry and generalized arithmetic; i.e., the newly created symbolic algebra, thus inventing the new discipline of analytical geometry), it welcomes the infinite (banished by the Greek mathematician and philosopher, and creates thereby the totally new discipline of differential and integral calculus), it explodes the limiting confines of the Greek concept of number, accepting irrationals and transcendents as legal residents of the new mathematical world, and shows that it is possible to deal rigorously, i.e., mathematically, with uncertainty and approximation, through the newly created calculus of probabilities. This is by no means an exhaustive list of the achievements of the new mathematics, all of which are predicated on violations of one or more of the 'thou shalt nots' of Greek mathematics. The new mathematics of the sixteenth and seventeenth centuries is profoundly unlike Greek mathematics. I consider this a revolution in mathematics. Remember, however, that I am a historian and speak as one! My own career has been affected by this claim, in many unsavory ways. I do not want to make this paper confessional, but the responses I got from my critics to my contention that Greek mathematics

---

15 Ibid., 19.

is idiosyncratic and cannot be understood by applying to its analysis foreign, i.e., modern, mathematical criteria and standards, which lead necessarily to distortions and false quandaries, illustrate beautifully the refusal of the mathematician and of the mathematically minded historian to accept the existence of revolutionary breaks in the development of mathematics.

All of these responses, by prominent mathematical figures, as Van der Waerden, Freudenthal, and Andre Weil,[16] boil down to the statement that had I known more mathematics (indeed, I wish I did! Mathematics, mind you, not history!), I could not have made the absurd claims I did, which simply stem from my mathematical ignorance. This accusation was accompanied by statements, largely without justification, to the effect that it is clear that what the Greeks did cumbersomely, and in an awkward way, is the same mathematically with what we do nowadays with our more advanced mathematical means. Occasionally, Greek propositions were translated into modern algebraic symbolism, this serving as proof that both the Greeks and us inhabit precisely the same mathematical universe. Among other things, my critics spoke of group theory in connection with Euclid (Andre Weil), of the solution of quadratic equations in the *Elements* (van der Waerden), and used functional language in describing Euclid's *Data* (Freudenthal), all in order to show that there is just one universal, atemporal mathematics, in which, of course, Greeks, Babylonians, and modern West Europeans live all happily with one another. *Viva la mathematica*!...

Assuming I have convinced you of the existence of at least two historico-mathematical revolutions, accompanied by the corresponding counter-revolutions *in* mathematics, let me use the remaining time to present to you the world of the historian of mathematics, as I see it. I shall do this under the heading 'Betrayals, Indignities, and Steamroller Historiography: André Weil and Euclid.' There is a close connection between what follows and what preceded it. Let me also say, before I proceed, that various historians of mathematics, who think there are revolutions *in* mathematics, speak of transfinite set theory, non-standard analysis, non-Euclidean geometries, and Fregean logic as revolutions *in* mathematics. This may not be an exhaustive list.

II

Proposition IX.8 of the *Elements* reads:[17]

> If as many numbers as we please beginning from an unit be in continued proportion (hexes analogon), the third from the unit will be square, as will all those which successively leave out one; the fourth will be cube, as will all those which leave out two; and the seventh will be at once cube and square, as will also those which leave out five.

16 Cf. van der Waerden 1976; Freudenthal 1977; Weil 1978.
17 Euclid, *Elem.* IX.8.

A faithful paraphrase of the Euclidean proof follows: Let there be given as many numbers as one pleases, starting from the unit, like A, B, C, D, E, and F.

A.
B.
C.
D.
E.
F.

By the definition of continued proportion, it follows that as the unit is to A, so is A to B. But the unit measures A; so, as many times as the unit measures A, A measures B (by Def. VII.20). Now, since the unit measures A according to the units in A, A measures B also according to the units in A. Therefore, A by multiplying itself makes B, and B is a square. Furthermore, since B, C, D are in continued proportion and B is a square, it follows (by VIII.22) that D is also a square. For the same reason, F is a square, and so are all those which leave out one. Now, since as the unit is to A, so is B to C, the unit measures A the same number of times that B measures C. But, again, the unit measures A according to the units in the latter; thus B measures C according to the units in A, i.e., A by multiplying B makes C. Now, since A by multiplying itself makes B, and by multiplying B makes C, it follows that C is a cube. Furthermore, since C, D, E, and F are in continued proportion and C is a cube, it follows (by VIII.23) that F is also a cube. But F was proved to be also a square, so that the seventh from the unit is both a cube and a square. In a similar fashion, one can prove that all other numbers which leave out five are both cubes and squares, *q.e.d.*

So much for Euclid. This proposition can serve as a beautifully striking example of the inherent limitations and the built-in chronic inadequacies of the beloved method practiced by mathematicians posing as historians of automatically and ahistorically transcribing Euclid's language into the modern symbolism of algebra. A proposition for the proof of which Euclid has to toil subtly and painstakingly, and in the course of whose proof he had to rely on many previous propositions and definitions (e.g., VIII.22 and 23, def. VII.20) becomes a trivial commonplace, which is an immediate outgrowth, a trite after-effect, of our symbolic notation: 1, a, a^2, a^3, a^4, a^5, a^6, a^7, ... As a matter of fact, if we use modern symbolism, this ceases altogether to be a proposition and its truthfulness is an immediate and trivial application of the definition of a geometric progression in the particular case when the first member equals 1 and the ratio, q=a, is a positive integer (for Euclid)!

These conclusions, emphasizing the differences between Euclid's way of thinking and the modern algebraic approach, are not to Weil's liking.[18] What does he

18 Weil 1978.

propose instead? Overlooking abusive language and personal affronts, the very few non *ad hominem* passages contained in his unprecedented letter to the editor of the *Archive for History of Exact Sciences* are entirely repetitive of the arguments of van der Waerden and Freudenthal, who, in their rejoinder to my "On the Need to Rewrite the History of Greek Mathematics" in the same *Archive* (1975), made it abundantly clear that they failed to understand the nature and import of the arguments criticizing the mathematician's approach to history and the value of a historically sensitive analysis of ancient texts.[19] An adequate, and undeservedly civilized, answer to Weil's hate letter has appeared in "History of Ancient Mathematics: Some Reflections on the State of the Art," and in the monographic study "Does the Quadratic Equation Have Greek Roots? A Study of 'Geometric Algebra,' 'Application of Areas,' and Related Problems," contained in two consecutive volumes of *Libertas Mathematica*.[20] This answer, obviously, cannot be reproduced here. Due to its length, complexity, and the technically detailed *mathematical* arguments in the latter study, it cannot even be appropriately summarized. All that can be done now is to call attention to it and provide here only a drastically succinct statement of the positions involved.

Let us glean, then, from Weil's agape-suffused epistle the few substantive passages it contains, and try to understand what lies behind them. Weil thinks that the 'obvious' content of proposition IX.8, its substantive lesson "for one who [has] at least learnt school-algebra", but, better still, for "any mathematician" who knows how to make the distinction (a subtle one to the layman) between the obvious and the trivial", and who is "trained to know the difference between a definition, a notation and a theorem", is that it represents "the rule $a^{pq} = (a^p)^q$" for "the special case q=2". (He deals only with the case of squares, not cubes.) He goes on to say:[21]

> Perhaps the *modern* mathematician finds it easier, in this case, to perceive the truth of the matter, because nowadays the exponential notation x^α is used in many situations where x, α are not numbers. For instance, the exponent α may be taken from a non-commutative group; some care is then needed in the choice of definitions and notations if the rule $x^{\alpha\beta} = (x^\alpha)^\beta$ is still to hold true. However that may be, one who thinks that the rules governing the use of the exponential notation are trivial [who does?] must be lacking, not only in mathematical understanding, but also in historical sense. Let him read Euclid's book IX, then Archimedes' *Sandreckoner*, then pages 132 to 166 of J. Tropfke's excellent *Geschichte der Elementar-Mathematik*, volume II. There he will learn that the development of the exponential *notation* and the realization of its *properties* went hand in hand for almost twenty centuries before they reached perfection.

What is one to say to that? Simply and pointedly that, though he was no layman, but rather a mathematician who knew how to distinguish between the obvious and the

19 Cf. Unguru 1975; van der Waerden 1976; Freudenthal 1977.
20 Cf. Unguru 1979; Unguru & Rowe 1981; Unguru & Rowe 1982.
21 Weil 1978, 92.

trivial, a man fully aware of the differences between definitions, notations, and theorems, Euclid had not learnt school-algebra and so could not have understood the rule $a^{pq} = (a^p)^q$. Proposition IX.8 is not for Euclid about the exponential notation, but rather about the properties of continued proportions. Not having been a *modern* mathematician, "the truth of the matter" for him, the truth, the whole truth, and nothing but the truth, was wholly contained in the *protasis* (enunciation) of the proposition and not in wildly anachronistic connections between its *hidden* message and the features of non-commutative groups, laying in store for some 2,300 years, to be unearthed in a future that must have been rather dim in Euclid's eyes.

Nobody in his right mind thinks that "the rules governing the use of...exponential notation are trivial," but not everybody is inclined to trace them all the way back to Euclid, applying to the *Elements* misshapen and ahistorical interpretive standards. Though, inescapably, all history is retrospective history, the approach described and decried by Detlef Spalt as "Resultatismus,"[22] that takes its bearings and criteria from what it sees as the *modern* outcome of a lengthy, linear, and necessary evolution of concepts and operations, looking always back to the past in light of its modern offspring, is necessarily a highly distorting approach, since it adopts unashamedly the perspective of the present to judge and understand the past. Taken at face value, it is simply not true that "the past has meaning only in terms of the present".[23] The historian's stance is rather the opposite: the present has meaning only in terms of the past. The Platonic outlook embodied in Weil's statements, according to which (1) mathematical entities reside in the world of Platonic ideas and (2) mathematical equivalence is tantamount to historical equivalence, is inimical to history.

History focuses on the idiosyncratic rather than the nomothetic. It is impossible for a modern man to think like an ancient Greek. Historical understanding, however, involves the attempt at faithful reconstruction of the past. In intellectual history, this necessarily means the avoidance of conceptual pitfalls and interpretive anachronisms. Though it is impossible to think like Euclid, it is rather facile to think obtrusively unlike him. We cannot know what went through Euclid's mind when he wrote the *Elements*. But we can determine what Euclid could not have thought when he compiled his great work. He, most likely, did not employ concepts or operations for which there is no genuine evidence either in his time or in the works of his predecessors. This much is safe to conclude. Furthermore, he clearly could not have foreseen what mathematicians and historians of mathematics were going to do in the long run to his *Elements*; he could not have used mathematical devices and procedures which were invented many hundreds of years after his death. This much is obvious too. Given that we cannot think like Euclid, we should nevertheless strive to avoid thinking unlike him when elucidating or commenting on his writings. This is—and

22 Spalt 1981, 71.
23 Cf. Bridgman 1950, 64.

must remain—the historian's goal. One way of thinking *unlike* Euclid is to use the algebraic approach in interpreting his works.

It continues to be habitual among historians of mathematics to say that what really lies behind Euclid's geometrically couched statements are algebraic reasonings, appearing in geometrical garb because of the lack of an appropriate algebraic symbolism. This strikes me as both inaccurate and unilluminating. To see this, let us ask the following questions: "How illuminating would it be to propose that Euclid really thought in Sanskrit but, because of his ignorance of the Sanskrit alphabet, he had to use the Greek one and consequently expressed himself in Greek?"! Greek mathematics must be understood in its own right. This can be done by refusing to apply to its analysis foreign, anachronistic criteria. The only acceptable meta-language for a historically sympathetic investigation and comprehension of Greek mathematics seems to be ordinary language, not algebra. "To berate Heath and others for betraying Euclid," goes on André Weil,[24]

> when all they do is to use a certain amount of notation to clarify the contents of his writings does not merely indicate a lack of mathematical sense; it argues a deficiency in logic. As everyone knows, words, too, are symbols. The content of a theorem does not change greatly, whether it is expressed in words or in formulas; the choice, as we all know, is mostly a matter of taste and of style. 'Euclid's *numbers*,' we read in Z's article, 'are *given* line segments, not abstract symbols' ... What are A, B, Γ, Δ, E, Z in the proof of IX.8, if not symbols?

True, words are symbols, indeed so is the very word 'symbol'; but you cannot extract the square root from it! The *abstract* symbolism of modern mathematics is an *operational* symbolism. The content of a theorem does change, sometimes significantly, when its rhetorical content and appearance are abandoned in favor of formulaic expression. In "On the Need to Rewrite the History of Greek Mathematics" and the *Libertas Mathematica* monograph, many examples of this very fact were given. Thus, it was shown that the use of formulas to transcribe Euclid's proofs transforms a string of consecutive and *different* propositions of Book X (112, 113, 114) into essentially the same proposition! Furthermore, formulaic translation of geometric propositions leads, quite often, to richer results than those obtained in the original geometric proof, some of those results wholly unobtainable in the genuine geometric setup. And so the choice between words and formulas is a fateful choice and not merely "a matter of taste and of style"!

Finally, there is a fundamental difference between Euclid's letters, A, B, Γ, etc., and genuine algebraic symbols. Algebraic symbols are monosemic, while ordinary language constructs are polysemic. The letters of Euclidean proofs are not really symbols; they are rather proper names and thus both much richer and more circumscribed than the dry, exact, abstract algebraic symbols used to transcribe them, and

24 Weil 1978, 92.

the proofs in which they are embodied. The Greek language is richer and, therefore, more ambiguous, mysterious, equivocal, and much less precise than the symbolism of algebra used to "translate" it into.

The remainder of Weil's letter, purged of its poisonous juices, asserts the total distinction between numbers and the lines used to represent them, indeed the irrelevance of the diagrammatic linear representation of numbers in the *Elements*, the purely algebraic *cum* number-theoretical character of Bools VII, VIII, and IX, and the need to "distinguish between mere routine and creative reasoning" in our historical sources, without which "there can be no worthwhile history of mathematics."[25] Let us now deal succinctly with these matters.

'Number' (*arithmos*) for the Greeks meant *positive integer*. Numbers are represented by Euclid as line segments. After the discovery of incommensurability, it became obvious that it is not the case that all line segments could be associated with numbers. However, it does not follow that the converse of the false statement (that all lines are numbers) is also false, and, indeed, Euclid associates numbers with line segments throughout Books VII, VIII, IX, the 'arithmetical' books. Surely, these books are indeed arithmetical, not algebraic, in character, and they do contain many propositions belonging nowadays to the theory of numbers.

The Greek concept of *arithmos*, number, i.e., a number of things, was replaced in the sixteenth century by a new concept of number as an abstract symbol. Instrumental in this change was François Viète (1540–1603), who transformed the concept of *arithmos* into the modern concept. This transformation marks the beginning of modern mathematics. Greek *arithmos* and modern *number* do not mean the same thing. As Klein has it, the two concepts differ in "Begrifflichkeit," i.e., conceptualization and intentionality—by the latter, Klein understands "the mode in which our thought, and also our words, signify or intend their objects."[26] For the Greeks, *arithmos* always meant a *number of things* (Anzahl), although 'things' didn't have to be mentioned explicitly. For modern mathematics after Viète, on the other hand, number is a concept; it is the concept of quantity. As numbers came to be regarded as abstract and symbolic entities, a 'new' mathematics—and, by the same token, a 'new' science, in the long run—came into being, the mathematics in which the *symbolic form* of a statement is inseparable from its content; indeed, in an obvious sense, the form is the content. *Mutatis mutandis*, this separation is also, to a very great extent, impossible in modern physical science, where *mathematical form* and *physical content* are irreducibly intertwined and hopelessly enmeshed.

With Viète and his successors (Stevin, Descartes, Wallis, etc.), then, a radical conceptual change has occurred. It is, therefore, historically unwarranted to apply mechanically to Greek mathematics the manipulations and jugglings of modern

---

25 Ibid., 93.
26 Klein 1992, 118.

mathematical symbolism. Historians of mathematics, however, have been doing exactly this. Being themselves immersed in modern ways of thought, they have been reading Greek mathematical texts through modern glasses and, to nobody's surprise, were rather successful in identifying in those texts an inexistent Greek algebra. They could achieve such a fantastic result by betraying Greek mathematics, only by applying to it foreign categories of post-Renaissance mathematical thinking.

Modern algebra, the only properly befitting the unqualified name, is a creation of the sixteenth and seventeenth centuries. Its great protagonists are Viète, Descartes, and Fermat. It marks the passage from an old way of thinking in mathematics, the *geometrical* way, the *mos geometricus*, to a new way, the symbolic way, the *mos symbolicus* or *mos per symbola*. Its historical development is rightly connected with the reintroduction into the West of the great works of classical Greek mathematics which, however, contained the old way of thinking, to be discarded by modern mathematics. With Viète algebra becomes the very language of mathematics; in Diophantus's *Arithmetica*, on the other hand, we possess merely a refined, ingenious and novel, quasi-rhetorical tool for the solution of *arithmetical* problems. In the seventeenth century, algebra was called *ars analytica*, a pregnant name indeed. It shows the difference between the Greek approach and that of the seventeenth century. For the Greeks, mathematics was not an *art*, a manipulative technique, a *techne*, but a science, an *episteme*, a *scientia*. Furthermore, for the Greeks *analysis* was merely a means of discovery, a heuristic tool. Mathematics, *episteme*, was limited to *synthesis*. In the seventeenth century, on the other hand, one is faced with algebraic analysis without any synthesis. This new approach meant, inter alia, a certain loosening of the Greek strictures of rigor and a new mathematical style.

One should not apply modern symbolism to Greek mathematics with impunity, as if modern symbolism were nothing but a temporally universal, i.e., historically neutral, means for organizing and simplifying *any* conceptual content. The fact that it is *modern* symbolism that one applies is in itself the best evidence for the ahistoricity of such a procedure. Klein has shown successfully that symbolic formalism is at the *core* of the modern concept of number, and that to translate Greek mathematics into its terms obscures completely both the meaning of the Greek concept and the genuine Greek achievement in the theory of numbers.[27]

Do we find, then, any algebra in Euclid? *No*! Euclid's numbers are given line segments, no abstract symbols, and Euclid's presentation is not symbolic. It always deals with determinate numbers of units of measurement, which are *not* seen as representing specific illustrations, instances of a concept of general magnitude. As Klein put it, and it merits quoting in *extenso*:[28]

---

27 Cf. Klein 1992, dust jacket.

28 Ibid., 123–124.

> In *illustrating* each determinate number of units of Measurement by measures of distance it [i.e., the Euclidean presentation] does *not* do two things which constitute the heart of the symbolic procedure: It does *not* identify the object represented with the means of its representation, and it does *not* replace the real determinateness of an object with a *possibility* of making it determinate, such as would be expressed by a sign which, instead of *illustrating* a determinate object, would *signify* possible determinacy ... [W]hen in the arithmetical books an arithmetical, or more exactly a logistical proposition is demonstrated *generally* with the aid of lines, this does not in the least mean that there exists either a general number or the concept of a 'general', i.e., indeterminate, number corresponding to this general proof ... the *general* 'linear approach'...intends only *determinate* numbers ... Since in Euclid...the single illustrative lines are additionally identified by a letter, the possibility arises of representing the numbers intended by those letters. This does not, however, in the least amount to the introduction of symbolic designations. Letters for indicating magnitudes and numbers seem to have been used already by Archytas ... [As Tannery put it, however,] the *letter does not symbolize the value of a number, and does not lend itself to being operated on*. Aristotle, too, made use of such mathematical letters, e.g. in the *Physics* and in *On the Heavens*; and he even introduced Them into his 'logical' and ethical investigations. But such a letter is never a 'symbol' in the sense that that which is signified by the symbol is in itself a 'general' object.

It simply cannot be said better, and it disposes effectively of Weil's extraneous objections to the distinction between numbers and lines in the context of the arithmetical books, of the specious statements concerning the irrelevance of diagrammatic linear representation, and of the gratuitous and distorting assertions about the algebraic character of Books VII–IX.

All that still remains to be done in order to dispose *entirely* of Weil's unwelcome 'contribution' to the debate on the nature of Greek mathematics is simply to point out that the exhortation to distinguish between routine and creative reasoning in the work of ancient mathematicians is a further instance of the application of modern anachronistic insights in order to sift the ancient mathematical material and to separate, as it were, the wheat from the chaff. Another injunction that the historian of mathematics worth his/her salt better overlook. *Sic transit...*

Bibliography

Texts quoted

Euclid, *Elem.*: Heiberg, J. L. & H. Menge 1896–1916. *Euclidis Opera Omnia*. Leipzig.

Works quoted

Bridgman, P. W. 1950. "Impertinent Reflections on the History of Science." In: *Philosophy of Science* 17, 63–73.

Cantor, M. 1922–1924. *Vorlesungen uber die Geschichte der Mathematik* (orig. 1894–1908). 4th ed. 4 vols. Leipzig.

Crowe, M. J. 1975. “Ten ‘Laws’ concerning Patterns of change in the History of mathematics.” In: *Historia Mathematica* 2, 161–166.

Cuomo, S. 2001. *Ancient Mathematics*. London.

Freudenthal, H. 1977. “What is algebra and what has it been in history.” In: *Archive for the History of Exact Sciences* 16, 189–200.

Gillies, D. (ed.) 1992. *Revolutions in Mathematics*. Oxford.

Hardy, G. H. 1967. *A Mathematician's Apology* (orig. 1940). Reprinted, with a foreword by C. P. Snow. Cambridge.

Hartley, L. P. 1953. *The Go-Between*. London.

Klein, J. 1985. “The World of Physics and the Natural World.” In: Williamson, R. B. & E. Zuckerman (eds.), *Lectures and Essays*. Transl. D. R. Lachterman. Annapolis, 1–34.

Klein, J. 1992. *Greek Mathematical Though and the Origin of Algebra* (orig. Berlin 1934–1936). New York.

Mahoney, M. S. 1970. “Babylonian Algebra: Form vs. Content (Essay review of Neugebauer 1969).” In: *Studies in History and Philosophy of Science* 1, 369–380.

Neugebauer, O. (ed.) 1935–1937. *Mathematische Keilschrifttexte*. 3 vols. Berlin.

Neugebauer, O. 1969. *Vorlesungen über Geschichte der antiken mathematischen Wissenschaften, Band I: Vorgriechische Mathematik* (orig. 1934). Berlin.

Spalt, D. 1981. *Vom Mythos der Mathematischen Vernunft*. Darmstadt.

Unguru, S. 1975. “On the Need to Rewrite the History of Greek Mathematics.” In: *Archive for the History of Exact Sciences* 15, 67–114.

Unguru, S. 1979. “History of Ancient Mathematics: Some Reflections on the State of the Art.” In: *Isis* 70, 555–565.

Unguru, S. & D. E. Rowe 1981. “Does the Quadratic Equation have Greek Roots? A Study of ‘Geometric Algebra’, ‘Application of Areas’, and Related Problems (part I).” In: *Libertas Mathematica (ARA)* 1, 1–49.

Unguru, S. & D. E. Rowe 1982. “Does the Quadratic Equation have Greek Roots? A Study of ‘Geometric Algebra’, ‘Application of Areas’, and Related Problems (part II).” In: *Libertas Mathematica (ARA)* 2, 1–62.

van der Waerden, B. L. 1930–1931. *Moderne Algebra*. 2 vols. Berlin.

van der Waerden, B. L. 1976. “Defence of a shocking point of view.” In: *Archive for the History of Exact Sciences* 15, 199–210.

Weil, A. 1978. “Who betrayed Euclid.” In: *Archive for the History of Exact Sciences* 19, 91–93.

Jean Christianidis

Diophantus and Premodern Algebra: New Light on an Old Image

Abstract: As a theme of historical research, Diophantus' *Arithmetica* (ca 300 AD) raises two main issues that have been most intensively debated: the first concerns the proper understanding of Diophantus' practice, while the other relates to the identification of the mathematical tradition to which this practice belongs. Since the time of medieval Islam, through the Renaissance and the early modern period, the thesis that the work of Diophantus belongs to the history of algebra has enjoyed broad consensus among mathematicians, despite the fact that the term 'algebra' was introduced in the language of mathematics five centuries after Diophantus. Thus, as early as the Middle Ages, the work of Diophantus was recognized by mathematicians as a work on algebra, *avant la lettre*. The consensus was maintained during the nineteenth and the most part of the twentieth century—this time among historians of mathematics. It is essential to stress, however, that, when associating Diophantus' work with algebra, premodern mathematicians on the one hand and modern historians of mathematics on the other did not start from the same understanding of algebra. Those mathematicians understood algebra with its premodern meaning and, accordingly, characterized the *Arithmetica* as 'algebraic' *in the premodern meaning of the term*. In contrast, modern historians of mathematics approach the *Arithmetica* mostly through the viewpoint of *a loose understanding of modern algebra* and, precisely for this reason, their accounts are often exposed to anachronism. This explains why some contemporary historians of ancient mathematics are reluctant to accept the conclusions of the traditional historiography, while others deny without reservation any relation of Diophantus' practice to algebra. However, criticizing the methodology by which the traditional historiography has reached its conclusion does not necessarily mean that the conclusion itself was wrong. This paper discusses some crucial issues related to Diophantus' problem solving practice, thus, giving support to the traditional view of the algebraic character of his work, but put in a totally new framework of ideas.

Keywords: Diophantus, Algebra, Premodern Algebra, History of Algebra, Arithmetic, Greek Mathematics.

I Introduction: The Category of 'Premodern Algebra' in the History of Mathematics

"Modern algebra and modern formalism grew out of Vieta's direct occupation with Diophantus." This sentence of Jacob Klein epitomizes the key idea underlying the core of his

https://doi.org/10.1515/9783110565959-003

historical narrative about the creation of the formal mathematical language at the end of the sixteenth century, a process which he identifies with "the foundation of modern algebra" (Klein 1992, 4). The sentence appears in the article "The World of Physics and the 'Natural' World" (Klein 1981, 29), but its essence pervades the whole of Klein's very influential book *Greek Mathematical Thought and the Origin of Algebra* (Klein 1992). The word in the aforementioned phrases I want to draw attention to is the adjective 'modern', which accompanies, in both cases, the word 'algebra'. Indeed, Klein is very careful—throughout his work—not to omit, whenever he speaks of algebra from Viète's day onwards, to characterize *this particular algebra* as 'modern'. In this way, he alludes to a 'premodern' algebraic activity, which (of course) had preceded modern algebra. Occasionally, Klein refers to this premodern algebraic tradition by using expressions like "the origin of algebra," as in the title of the aforementioned book, "primitive stage of algebra"[1] (Klein 1992, 133), or even "traditional algebra" (Klein 1981, 29).

Whatever problems there might be with Klein's account of the development of algebra "from Diophantus over the Arabs and the Italian algebraists of the sixteenth century to Vieta" (Klein 1992, 139), there must be little doubt that the historically interesting idea that 'modern algebra' was not the only historical type of algebra was, by no means, foreign to his thought. For if it were, the very word 'modern', regularly occurring before the term 'algebra', in his writings, would have been superfluous. Quite the contrary! Klein is very consistent in adding the adjective 'modern' whenever he speaks about the algebraic achievements of Viète and his successors in the seventeenth century. In contrast, modern historians and commentators do not always demonstrate similar attentiveness in characterizing algebra in their historical investigations. Quite often, they speak about *algebra tout court*, thus portraying algebra as an uncontextualized category, in which any distinction between 'modern' and 'premodern' is either altogether ignored or is considered merely as a question of the extent to which a homogeneous, largely ahistorical 'algebraic essence' was manifested in the 'algebraic' compositions of the one or the other historical period.[2]

The above-mentioned attitude deserves some further comments. It would be a matter of fact to say that Klein's 'modern algebra' has become, in the hands of some contemporary historians of mathematics, *algebra tout court*, without further qualification. Yet, it must be stressed that when used carelessly, such an attitude may be misleading; especially, when premodern material is to be considered. Instances in which premodern algebraic sources have been interpreted from the conceptual framework of post-Vietan algebra—that is from the viewpoint of modern algebra—thus leading to misrepresentations (or even to biased interpretations), are not rare in past and current scholarship. Especially, regarding some recent studies, one may easily recognize the inconsistency by which some modern historians approach premodern

1 According to Klein, Diophantus' *Arithmetica* represents the primitive stage of algebra (ibid.).
2 See Corry 2004, 398 and Corry 2013, 639, from where I have borrowed the expression "algebraic essence."

algebraic sources: on the one hand, there have been over-interpretations of premodern algebraic sources, which are thus represented retroactively as meeting the conceptual presuppositions that underlie modern algebra; on the other hand, interpreted from the viewpoint of modern algebra, other premodern algebraic texts are not considered to be algebraic at all. The undertakings of al-Khwārizmī and Diophantus are two striking examples: two texts, both belonging to premodern algebra, are sometimes represented as being conceptually so far away from each other that the former becomes a work on 'algebra',[3] while the latter a "definitely" 'non-algebraic' work.[4]

'Premodern algebra' is a recently introduced historiographical term, aiming at describing a stage in the history of algebra that precedes the stage of 'modern algebra' and which is clearly distinguished from it regarding the goal, the scope, and the conceptual tools it employs. A growing number of studies has been produced over the last few years to elucidate this new historiographical category.[5] These studies enrich the scholarship of the history of algebra, by endowing it with a new dimension. Furthermore, they can be proven to be fruitful from a methodological standpoint, in the sense that they urge historians of mathematics to historicize the concept of algebra and conduct their explorations of ancient, medieval or early modern sources not through the historically problematic category of *algebra tout court*, but through the one of the two categories of 'modern' and 'premodern' algebra, respectively. Such an approach presents significant advantages: First, it avoids the drawbacks the concept of *algebra tout court* creates, when employed as a historiographical tool. So, ancient and medieval texts will not be discussed through an impertinent conceptual framework, like the one of modern algebra, which obscures rather than elucidates the mathematical practices they convey. Second, it allows the affinities between texts of premodern algebra to be disclosed. Finally, it allows the conceptual distance that separates modern algebraic texts from the corresponding ancient and medieval ones to be portrayed in a more lucid manner.

As it has already been stated, the features, the conceptual presuppositions, and the scope of premodern algebra have already been investigated in a number of recent studies. The main conclusions of these studies are summarized below:[6]

1. The core characteristic that distinguishes premodern from modern algebra is the way in which the key concepts of 'polynomial' and 'equation' are mutually conceived. In premodern algebra, the polynomial is an aggregation of the names of

---

3 Cf. the title of Rashed 2007: *Al-Khwārizmī: Le commencement de l'algèbre*.

4 I borrow again from Rashed (1994, 208): "If, in the course of his solution, Diophantus proceeded with the substitution, elimination and displacement of species, in short using algebraic techniques, the *Arithmetica* is not a treatise on algebra. According to our terminology, it is definitely a book on arithmetic, not in the ring of relative integers, but the half-field of positive rational numbers."

5 Cf. Oaks 2007, 2009, 2010a, 2010b, 2011a, 2011b, 2012, 2014, forthcoming; Oaks & Alkhateeb 2005, 2007; Christianidis & Oaks 2013; Christianidis & Skoura 2013; Christianidis 2007; Sialaros & Christianidis 2016.

6 Some of the conclusions will be demonstrated in concrete examples in the following parts of this study.

the powers of the unknown (a 'multi-nomial' in the strict sense of the word), with no operations involved. On the other hand, in modern algebra, the polynomial is conceived as a linear combination of the powers of the unknown. The polynomial equation in premodern algebra is a static relation between two aggregations, and thus it is free on any operation. In modern algebra, in contrast, operations are a critical component of the equation.

2. Taking into account the above characteristics, two other differences stand out:
   i. Premodern algebraists insisted on working out all operations called for by the problem before setting up an equation, where we today often begin by first writing the equation and then working out the operations in the context of the equation.
   ii. Being incompatible with the aggregate conceptualization, premodern algebraists did not allow for irrational 'coefficients' in their polynomials.[7]
3. Besides dissimilarities of conceptual order, premodern algebra differs from modern algebra in terms of their scope and purpose. Premodern algebra was conceived and practiced principally as a method of solving problems,[8] whereas modern algebra (in its full-fledged development) is no longer a mere method for solving problems.
4. More precisely, as far as the structure of solution is concerned, problem solving by algebra, in the premodern meaning of the term, entails the following tasks: (a) naming the unknown(s)—the finding of which the enunciation of the problem stipulates; (b) working through the operations stated in the enunciation by using the named terms; (c) setting up an equation, framed in the vocabulary of the names, as the outcome of the two previous processes; (d) simplifying and solving the equation; and (e) answering the problem, by means of the solution to the equation.
5. The technical vocabulary used in premodern algebra differs from one language to another. The Table below summarizes some of the most common terms used in different traditions:

---

7 This matter is investigated in the study "Irrational 'Coefficients' in Renaissance Algebra" of Jeffrey Oaks, 2017.

8 Instances of using algebra for theoretical purposes, like proving theorems or justifying statements, are scarce in the premodern period. Only two examples of such uses are known, both coming from Abū Kāmil's *Algebra*. The first is discussed in Oaks 2011a, the other is mentioned in Oaks 2014, 46. No other examples have been pointed out so far in Arabic, Latin, or Italian algebra. Besides, another application of a piece of the machinery of premodern algebra has been preserved by Theon of Alexandria, who provides a method of multiplication between sexagesimal numbers, modelled in the way Diophantus multiplies aggregations of species, that is, premodern polynomials; cf. Skoura & Christianidis 2014.

Table 1: Technical terms of premodern algebra

Diophantus	Arabic	Latin	Italian	Degree
monas	*dirham* / *ʿadad*	*dragma* / *numerus*	*dramma* / *numero*	
arithmos	*jidhr* / *shayʾ*	*radix* / *res*	*radice* / *cosa*	1
dynamis	*māl*	*census*	*censo*	2
kybos	*kaʿb*	*cubus*	*cubo*	3

In premodern algebra, the algebraic calculations are carried out through these terms, which function in problem solving as 'names' assigned to the unnamed sought-after numbers, the finding of which the problem calls for. The above terms correspond, respectively, to the unit and the terms x, x^2 and x^3 of modern algebra, although they are not identical with them.

6. Despite the fact that the problems solved by algebra may have been stated in terms of arithmetic, mensuration, commerce etc., the unknowns were always numerical measures of quantifiable objects. Accordingly, it is not surprising that premodern algebra was considered by its practitioners as part of arithmetic.[9]
7. By virtue of (3) and (6) premodern algebra should be best described as a numerical problem solving method.
8. In Late Antiquity and throughout the Middle Ages, algebra coexisted with other methods that were used in numerical problem solving, such as the simple arithmetical reasoning, rule of three, single false position, double false position, etc. However, compared to these methods, algebra is remarkably different, in the sense that it follows an entirely different pattern and, moreover, has a specific technical vocabulary with its own syntax (see points 4 and 5 above).

As already said, the term 'premodern algebra' has been recently introduced in the history of mathematics to denote a specific historiographical category, the fundamental characteristics of which are outlined in the above mentioned features. The notion itself was created to describe primarily the medieval Arabic, Latin, and Italian algebra. By adopting this notion (mainly from the works of Jeffrey Oaks), my aim in the present study is to show that the notion of 'premodern algebra' provides a suitable contextual framework for conceptualizing Diophantus' *Arithmetica* as well. In this sense, this study could be inscribed in the long tradition of studies that argue for the algebraic character of Diophantus' undertaking. At the same time, it intends to revisit this tradition by transferring the discussion into the context of premodern algebra, instead of the one shaped by a loose understanding of what algebra has become from Viète's time onwards.

9 On this, see especially Oaks (forthcoming).

II Diophantus' *Arithmetica*, a Work of Premodern Algebra

Since the time of medieval Islam, when the very term 'algebra' was introduced in mathematics through the expression 'al-jabr waʿl-muqābala',[10] identifying Diophantus' *Arithmetica* as an algebraic work was very common in the history of mathematics. This is easily seen, for example, in the work of Qusṭā ibn Lūqa, the ninth-century translator of Diophantus into Arabic, who refers to the *Arithmetica* by calling it "Treatise of Diophantus on *al-jabr waʿl-muqābala*" (i.e. on 'algebra'),[11] and also by considering how straightforward it was for Qusṭā ibn Lūqa to translate the technical terms of Diophantus treatise (shown in the first column of Table 1) into the vocabulary of the Arabic algebra of his day (shown in the second column of Table 1). It is further confirmed by the testimony of the tenth-century biobibliographer Ibn al-Nadīm, who in his *al-Fihrist* ascribes to his contemporaneous mathematician Abu'l-Wafā' "a Commentary on the Book of Diophantus on Algebra." The same statement is repeated three centuries later by the biobibliographer Ibn al-Qifṭī and the Syrian scholar Barhebraeus (Sesiano 1982, 10).[12] Another testimony comes from Regiomontanus who, in his 1464 Padua oration on the history and utility of the mathematical arts, praises Diophantus' book in which, he writes, "lie the very flower of all arithmetic, namely

---

10 The expression 'al-jabr waʿl-muqābala' was the name given to the art of algebra in medieval times; cf. Oaks & Alkhateeb 2007, 46.

11 In the end of the seventh book, where we read: "End of the seventh book of the treatise of Diophantus on *al-jabr waʿl-muqābala*, and it contains eighteen problems;" cf. Rashed 1984, IV, 120; Sesiano 1982, 171; also, cf. Oaks 2011b, 69–70. The word 'algebra' occurs in appellations of the treatise of Diophantus by several authors who wrote in Arabic. For a detailed list, see Sesiano 1982, 13.

12 On this occasion, it is important to notice that the same Ibn al-Nadīm also attributes a book on algebra to Hipparchus: "Among his books there were: The Art of Algebra (K. Ṣināʿat al-jabr), known as Definitions (al-Ḥudūd). This book was translated, and Abu'l-Wafā' corrected the book and also gave explanations and criticisms with geometrical proofs; Division of Numbers (K. Qismat al-aʿdād);" Oaks' translation, adjusted from Dodge 1970, 642. This entry appears right before the entry for Diophantus, whose book is also called "The Art of Algebra." In view of this, some historians have expressed doubts regarding the reliability of the testimony, suspecting confusion on the part of al-Nadīm (Steinschneider 1897, 348–350; Suter 1900, 213; Heath 1921, ii, 256). However, in my view, this thesis presents some serious problems: al-Nadīm mentions Hipparchus' book on algebra once again in the *Fihrist*, namely in the entry on Abu'l-Wafā', who wrote a commentary on it. In fact, among the works of Abu'l-Wafā', al-Nadīm reports "A commentary on the book of al-Khwārizmī on Algebra (K. Tafsīr kitāb al-Khwārizmī f?'l-jabr wa'l-muqābala); a Commentary on the Book of Diophantus on Algebra (K. Tafsīr kitāb Diyufanṭus f?'l-jabr); a Commentary on the Book of Hipparchus on Algebra (K. Tafsīr kitāb Ibarḫus f?'l-jabr); introduction to the Arithmetica, one section; Proofs of the Propositions Which Diophantus Employed in His Book and of the Things he Used in the Commentary;" Oaks' translation, adjusted from Dodge 1970, 668. I believe that the fact that Abu'l-Wafā' is reported as having written commentaries on the books of both Diophantus and Hipparchus strongly suggests that there was no confusion in al-Nadīm's report.

the art of assessing and accounting, which today is called algebra after its Arabic name" (Byrne 2006, 54–55).[13] Rafael Bombelli is another example of a premodern algebraist who recognizes algebra in Diophantus. As he writes in the preface of his *Algebra* (1572), "in recent years, a Greek work on this discipline [i.e. of algebra] has been discovered in the Library of our Lord in the Vatican, composed by a certain Diophantus of Alexandria" (Fauvel and Gray 1987, 263).

Traced back to the algebraists of the premodern period, the idea that Diophantus' work is part of the history of algebra has enjoyed broad consensus among historians of mathematics from the eighteenth century onwards. The names abound: Abbé C. Bossut, Marquis de Condorcet, J.-É. Montucla, H. T. Colebrooke, G. H. F. Nesselmann, H. Hankel, G. Wertheim, M. Cantor, P. Tannery, H. G. Zeuthen, T. L. Heath, D. E. Smith, S. Gandz, J. Klein, P. Ver Eecke, C. B. Boyer, B. L. van der Waerden, E. S. Stamatis, I. G. Bashmakova, to mention only a few leading figures. Despite differences in style and the arguments used, which partly reflect the attitudes that characterized the historiography of mathematics in the course of time, historians of mathematics, as a rule, do associate the *Arithmetica* with algebra. This, so to speak, *communis opinio* is typified in the title of Thomas Heath's well-known book, *Diophantus of Alexandria: A Study in the History of Greek Algebra* (Heath 1964).[14]

13 It has been claimed that Regiomontanus' failure to mention any Arabic text in the sources of algebra, while he identifies as such the work of Diophantus, has to be interpreted as reflecting a kind of humanist anti-Arabism together with an effort to establish an entirely classical genealogy of algebra (cf. Cifoletti 1996; Høyrup 1998). However, the fact that in other occasions, within the same oration, Regiomontanus does mention the Arabic contributions in the field of mathematical sciences does not corroborate such a claim. This is true in particular with regard to astronomy, for which, as Byrne points out, "He portrays Arabic and Latin sources in a positive light, implying that they can supplement and even supersede ... the body of knowledge left behind by classical authorities" (Byrne 2006, 54). According to Byrne, Regiomontanus' mentioning of Diophantus should rather be explained as reflecting a general pattern in the way he presents the history of the quadrivial sciences, namely as intended to establish the 'actual disciplines' on the basis of the *corresponding* texts that actually survive. As he writes (Byrne 2006, 55), "Diophantus's *Arithmetic*, which was not widely read ... is an interesting case, because it seems to be an example of Regiomontanus trying to establish the same kind of pattern for algebra, which until then had been considered an Arabic art."

14 In recent years, the thesis I described as the 'common view' has been challenged predominantly by Roshdi Rashed. Indeed, for more than three decades, Rashed has devoted great effort in an attempt to show that the method used by Diophantus for solving arithmetical problems cannot be regarded as algebra at all, and that the *Arithmetica* is, in fact, a work on arithmetic (cf. footnote 4). According to him, algebra was invented by a single mind (al-Khwārizmī), and it was conceived from the very beginning as a theoretical discipline, namely as a theory of equations—thus reducing the problem solving to a mere 'application of the theory'. Moreover, its creation was made possible because of the Islamic character of the society, in which al-Khwārizmī lived. I do not intend to argue here why Rashed's claims about the history of algebra, and more specifically about the relationship between Diophantus' mathematical activity and algebra, entail a highly biased view of history. Some of these claims, have been convincingly criticized in the past (cf. Jaouiche 1987; 1990), and more recently (cf. Oaks 2011b; 2014).

It is important, however, to stress that, in associating Diophantus' *Arithmetica* with algebra, premodern algebraists, on the one hand, and modern historians of mathematics, on the other, do not start from the same understanding of algebra. Premodern algebraists understand algebra with its premodern meaning—how could it be otherwise? Premodern algebra was *their* algebra. After the eighteenth century, historians of mathematics did not understand the premodern algebra, since the latter had become an alien concept to them; however, by approaching Diophantus' treatise through the viewpoint of *a loose understanding of the algebra of their time* and finding that it somehow fits their modern notions, they called it 'algebra' anyway. For that reason, precisely, their accounts are exposed to anachronism. Heath is, again, a typical example: the anachronism in his historical account is easily detected, even by a mere glance at the conspectus of the problems of the *Arithmetica* he has appended to his book (Heath 1964, 260–266). There, for example, one may observe the first problem of the *Arithmetica* to be represented by the simultaneous equations: $x + y = a$, $x - y = b$, and classified under the category "Determinate systems of equations of the first degree." Yet, the first problem of the *Arithmetica* reads as follows: "To divide a prescribed number into two having a given excess" (Tannery 1893–95, i, 16.9–10).[15] Does the enunciation of the problem speak really about equations? My answer, is 'certainly not': no word for 'equal' (or its cognates) appears in the enunciation. The enunciation does not state an equation at all; it only states a problem in arithmetic. Therefore, interpreting the problem of Diophantus through the algebraic equations $x + y = a$ and $x - y = b$ is misleading, since it converts a problem in arithmetic into an algebraic problem.[16] But algebraic problems of this kind can only exist in

15 Note on the system of referencing: The dot '.' separates the page number from the line numbers; thus, the above reference indicates that the quoted passage is found on page 16 of the volume of Tannery's edition of the works of Diophantus that contains the *Arithmetica*, that is, in the first volume of the *Opera omnia*, published in 1893, and it covers lines 9–10 of the printed text. Similarly, a reference like Tannery 1893–95, ii, 59.15–23 refers to the corresponding page and lines of the second volume of the *Opera omnia*, published in 1895; the former reference will be often given simply as 16.9–10.

16 Cf. for example Tannery 1893–95, ii, 287; Stamatis 1963, 21. Not only Diophantus' work has been exposed to misrepresentations of this kind. Similar anachronisms are innumerable, for example in Rashed's editions of the texts of Arabic algebra. More generally, translating problems in terms of algebra is very common in the history of premodern mathematics. The danger that is lurking in deducing historical conclusions by employing such an attitude is pointed out nicely by Warren Van Egmont (1996, 388): "Note that the problem is stated as concisely as possible, giving only the essential information, and the mathematical summary uses 'X' in place of the number to be found. This should not, however, be taken to imply that the problem is algebraic since any problem can be summarized in this way. Likewise, the mathematical form should not be presented in a way that is too closely tied to modern analysis. For example, it is not usually helpful to call a problem a 'determinate equation in one unknown' or an 'indeterminate equation in two unknowns' when it is obvious the authors themselves did not conceive the problem in this way. If it is to be any help in identifying the history of the problem, the form must reflect the way in which the composers themselves conceptualized it."

modern algebra. No evidence of such an algebraic problem has been surfaced in any premodern algebraic source so far.[17]

Heath's misrepresentation of Diophantus' practice is also obvious in the way he analyses the methods[18] of solution that Diophantus employs. In the lengthy chapter he has devoted to this matter bearing the title "Diophantus' methods of solution," although he starts by promising to "give an account in detail of the different methods which Diophantus employs for the solution of his problems" (Heath 1964, 54), he does not distinguish problems from equations in his analysis. Thus, most of his account falls under the title "treatment of equations" (Heath 1964, 58), where equations are classified into determinate and indeterminate, and furthermore, within each category, according to their degrees. The problem with such a working method is that it intends to a subject which is irrelevant to a great extent with what Diophantus takes as his principal theme. For, the principal subject taught by Diophantus in the *Arithmetica*—at least as far as we can judge from the preserved part of the work—is not to show how to solve equations (let alone indeterminate ones), but how to solve problems by converting them into equations. More precisely, Diophantus explains how to solve arithmetical problems, the vast majority of which ask to find more than one unknown numbers, by employing principally a single method, which is articulated around the idea of converting each problem into an equation, the latter being in modern terms always determinate. Since the time of medieval Islam, such a method of solution has been called algebra. In a solution by algebra, the treatment of the equation arising from the problem is just one part of the solution of the problem, not the entire solution procedure. Accordingly, the written text that the treatment of the equation covers in each problem in Diophantus is almost always only a small portion of the textual unit the solution of the problem in its entirety covers. Thus, by interpreting the Diophantine solutions as solutions of algebraic equations instead of solutions of problems in arithmetic, Heath (and with him the old historiography) disregards an entire (and indeed very important) stage of the solutions, which is precisely the conversion into equations. However, for conducting this particular stage, Diophantus had created a rich 'toolbox' of techniques.[19]

17 Of course, problems in a loose sense, asking, for example, to perform an operation with algebraic terms or to demonstrate the equality of two algebraic expressions, do occur in premodern algebra sources. Already, the advice of Diophantus to his addressee Dionysius to "acquire practice in the addition, subtraction, and multiplication of the various species," and to master "how to add present and lacking species, not of the same multitude, to other species, themselves either present or likewise present and lacking, and how from present and other lacking species to subtract other, either present or likewise present and lacking" (Tannery 1893–95, i, 14.3–10), may be seen as referring to such problems.

18 The plural number is Heath's.

19 This is shown with regard to the first three books of the *Arithmetica* in Bernard and Christianidis 2012. See below, section II.4. In part IV, we shall see that the same mistake is made by some modern historians, who suggest the interpretation of premodern algebraic texts by means of the tools of modern algebraic geometry.

This being the case, it does not come as a surprise that even the method of 'double-equation', occasionally employed by Diophantus in his solutions, is classified by Heath as falling into the category of indeterminate equations. Thus, the only one among the methods of solution that Heath discusses in his account that deserves to be described as a method of problem solving,[20] is annulled as such, and becomes a method for solving two simultaneous indeterminate equations! On the issue of the allegedly indeterminate equations that Diophantus solves, I shall come back later in this study.

On a relevant issue, Diophantus' notation has also caused confusion among past and modern historians. In the beginning of the first book of the *Arithmetica*, Diophantus introduces a list of abbreviations for the terms of what he calls "the arithmetical theory." The importance of these terms in Diophantus' problem solving was already identified by scholars like Theon of Alexandria, who names them "Diophantine numbers."[21] As Jacob Klein has shown, this notation has nothing to do with the corresponding modern algebraic symbolism. Yet, it is on account of the notation that some historians have argued in favor of the algebraic character of the *Arithmetica*—the particular notation used by Diophantus being considered as representing a stage in the history of algebra referred to as "syncopated," after Nesselmann (Nesselmann 1842)—while other historians deny the algebraic character of the *Arithmetica*, just because its notation is not a genuine algebraic symbolism![22] In my view, both approaches are equally misguided. Symbolism is not an essential characteristic of premodern algebra. For example, Arabic algebra was originally predominantly rhetorical, and only later, in the late twelfth century, a notation specific to algebra was developed in Maghreb (Abdeljaouad 2002; Oaks 2012). But this notation is not an essential but an accidental characteristic of Arabic algebra. In the same way, Diophantus' notation is an accidental, not essential characteristic of his method. The algebraic character of Diophantus' method lies not in his notation, but in the

---

20 In fact, the solution by 'double-equation' is a solution by algebra, for it entails naming the unknowns, operating on the names, and laying out the plan for equation. The difference with a typical algebraic solution is that in some problems the plan for equation is not necessary to be completed for one can instead take advantage of the fact that the difference of squares is the product of the sum and the difference of the roots in order to solve the problem.

21 See Christianidis & Skoura 2013.

22 Cf. the following excerpt from Rashed: "So Diophantus' work is in the final count interpreted in terms of variables, powers, parameters and general solutions. [...] This interpretation may no doubt enlighten the historian embarking on an examination of the internal cohesion and organization of the *Arithmetica*. When attributed to the author himself, however, it raises at least two problems in the field of historiography; it may substantiate the idea that the introduction of Diophantus was a source of algebra. Moreover, it precludes understanding a second trend whereby mathematicians took Diophantus' work for what it was in fact: a book on arithmetic" (Rashed 1994, 205–206).

features and modalities it demonstrates, as they are briefly described in the end of part I. Below, I give an example, Diophantus' problem II.33, illustrating the aforesaid features.[23]

> [*Enunciation*]
>
> To find three numbers such that the square of any one of them less the next following makes a square.
>
> [*Setting up the equation*]
>
> And since, whenever a number is the double of <another> number less one unit, the square on the lesser, less the greater, makes a square, I assign the first <number> to be 1 *arithmos*, 1 unit [$1x + 1$], the second, similarly, 2 *arithmoi*, 1 unit [$2x + 1$], and the third, 4 *arithmoi*, 1 unit [$4x + 1$], and it results that the square on the first, less the second, makes a square, and moreover the square on the second, less the third, makes a square.
>
> It remains also the square on the third less the first to make a square. But the square on the third less the first makes 16 *dynameis*, 7 *arithmoi* [$16x^2 + 7x$]; these are equal to a square.
>
> I form the square from <a side consisting of> 5 *arithmoi* [$5x$].
>
> Therefore, 25 *dynameis* are equal to 16 *dynameis*, 7 *arithmoi* [$25x^2 = 16x^2 + 7x$].
>
> [*Solving the equation*]
>
> And the *arithmos* becomes $\frac{7}{9}$.
>
> [*Answering the problem*]
>
> And the first <number> will be $\frac{16}{9}$, the second $\frac{23}{9}$, and the third $\frac{37}{9}$; and what is said in the proposition holds.

On the basis of the above text, let us now discuss the features of Diophantus' problem solving practice.

1 *The Enunciation of the Problem and the Equation*

The first point that must be stressed is that the enunciation of the problem must not be identified with the equation. The enunciation reads as follows: "To find three numbers (arithmoi) such that the square (tetragônos) of any one of them less the next following makes a square (tetragônos)." The equation, on the other hand,

23 For the Greek text, see Tannery 1893–95, i, 132.23–134.11. In my translation, I do not make use of Diophantus' notation, which, as I said, is not essential for my argument. Accordingly, all words which are abbreviated in the published text are here written in full, though, for the sake of simplicity, without respecting rigorously the rules of the Greek grammar, as regards the inflection.

is: "25 *dynameis* are equal to 16 *dynameis*, 7 *arithmoi*." There are crucial differences between the two, both textual and conceptual:

First, the word 'equal' that appears in the equation does not appear in the enunciation. Moreover, the enunciation is expressed in an algorithmic mode, whereas the equation is expressed as a static relation between two parts (aggregations of powers of the unknown). The enunciation states a problem in arithmetic; the equation does not.

Second, and most important, the enunciation and the equation are formulated in two different languages. In the enunciation, we find the word 'number', which is the translation of the Greek word 'arithmos' (in plural, 'arithmoi'), and the word 'square', which is a translation of the Greek word 'tetragônos' (in plural 'tetragônoi'). In the equation, the word 'square' (tetragônos) does not appear. Instead, we find the word '*dynamis*' (in plural '*dynameis*'). The word 'arithmos', on the other hand, appears in the equation too. But there is a significant difference in the way the word 'arithmos' is used in the two sentences: In the enunciation, the word is used as a common noun. Indeed, in the enunciation, there are three 'arithmoi', and so 'arithmos' cannot be the name of one single unknown the way it is in the equation (and the way our 'x' is)—hence, the translation 'number' I have used for the word 'arithmos' in the enunciation. In contrast, the same word in the equation is used to indicate a proper name, and, because of that, I left the word untranslated and in italics. The terms '*arithmos*' (when used in the equation) and '*dynamis*' are technical terms; the words 'arithmos' (when used in the enunciation) and 'tetragônos' are not technical terms. The former belong to what Diophantus calls in the introduction "arithmetical theory," and are the entities Theon refers to as "Diophantine numbers." The latter belong to everyday language.

The aforementioned distinction in the way terms are used—as common nouns (substantives) in the enunciations and as proper names in the solutions—entails another fundamental distinction: The substantive 'arithmos' differs from the proper name '*arithmos*' in the sense that only the latter intends the numerical value which makes a number be a particular number. So, in the example above, what the name "1 *arithmos*, 1 unit" [$1x + 1$], that Diophantus assigns to the "first <sought-after number>," names, is precisely its numerical value, its *hypostasis*. On the level of the enunciation, where the conditions are formulated to which the sought-after numbers of the problem are subjected, the *hypostaseis* of the sought-after numbers are not intended at all. In a Diophantine problem, a twofold determination is proposed for the numbers sought, with information given about their kind as well as the operations and relations to which they are subjected, and another type of determination is required, namely the concrete multitude of units that makes up each number. The former is a determination bearing on properties attributed to numbers, whereas the latter is a determination which specifies the numbers that possess these properties, which *hypostasize* the properties. This being the case, the whole point of a Diophantine problem can be stated as follows: how to get quantitatively determinate numbers, when given a combination of qualitative determinations to which they are subjected; in other words, how a combination of qualitative determinations of numbers results in quantitatively determinate numbers that 'do the problem'. In the above example, by assigning the name “1 *arithmos*, 1 unit” [$1x + 1$] to the “first <sought-after number>,” Diophantus, in fact, names its numerical value which is as yet unknown, and it will be known only at the end of the solution (in the present case, it is found $\frac{16}{9}$). The same holds true for the terms ‘tetragônos’ (square) and ‘*dynamis*’. The latter is used to denote the name assigned to the temporarily unknown numerical value of the former. Therefore, the terms ‘tetragônos’ and ‘*dynamis*’ should by no means be considered as simple synonyms, as the substantive ‘arithmos’ must not be considered as synonym with the proper name ‘*arithmos*’.[24] Thus, ‘tetragonos’ is a kind of number, like ‘perfect square’, to which numbers like 25, 81, $\frac{4}{49}$ belong, while ‘1 *dynamis*’ is the specific number that is the square of the set unknown ‘1 *arithmos*’. ‘1 *dynamis*’ is the *specific tetragonos* that solves the problem.

Within the context described above, it becomes clear that Diophantus uses two sets of conceptually different terms in his problem solving.[25] The two sets are listed together in the Table below:

Table 2: The terms of Diophantus’ problem solving

Numbers in enunciations (common nouns)	**Numbers in equations (proper names)**	**Corresponding algebraic symbol**
arithmos / pleura	*arithmos*	x
tetragônos	*dynamis*	x^2
kybos	*kybos*	x^3
tetragônos multiplied by itself	*dynamodynamis*	x^4
tetragônos multiplied by kybos (same side)	*dynamokybos*	x^5
kybos multiplied by itself	*kybokybos*	x^6
	arithmoston	x^{-1}
	dynamoston	x^{-2}
	kyboston	x^{-3}
	dynamodynamoston	x^{-4}
	dynamokyboston	x^{-5}
	kybokyboston	x^{-6}

2 *The Setting Up of the Equation*

The setting up of the equation covers the most part of the textual unit of problem II.33. In fact, this is not a peculiarity of this particular problem. The same pattern occurs almost in every worked out problem in Diophantus’ treatise. Indeed, Diophantus is quite detailed in explaining how the equation arises out of the problem, while, on the contrary, he is most often laconic in dealing with the task of manipulating the

24 For more on this issue, see Christianidis 2015.

25 This matter is amply discussed in Christianidis & Oaks 2013.

equation. In the present example, the task of setting up the equation covers ten lines in Tannery's edition (out of fifteen lines of the whole problem), while the solution of the equation takes only half a line. This *modus operandi* led me in a previous publication (Christianidis 2007, 299) to advance the hypothesis that, within the broad project to demonstrate how algebra should be practiced in problem solving, Diophantus' main goal in the *Arithmetica* is not primarily to explain how the equations are solved, but first and foremost to demonstrate how the problems are converted to equations. To state it in a different way, Diophantus' aim was to exhibit the way in which one should work in order to transfer an arithmetical problem into the 'arithmetical theory'.

The 'counterpart', so to speak, of the problem in the setting of the 'arithmetical theory' being the equation, there are two tasks to be conducted in order to arrive at the equation:

1. Assignment of names to the unnamed sought-after numbers that appear in the enunciation of the problem;
2. and, with the named terms, working through the arithmetical operations stated in the enunciation.

Let us see how these two tasks are accomplished in problem II.33.[26] As far as the task of naming the unknowns is concerned, Diophantus assigns four names. Namely,

The 1st sought-after number is set $1x + 1$,
the 2nd sought-after number is set $2x + 1$,
the 3rd sought-after number is set $4x + 1$, and
the side of the square produced once the square of the 1st sought-after number is diminished by the 3rd sought-after number, as the problem stipulates, is set $5x$.

With the named terms, Diophantus performs the following operations (the last one being implied in the text):

$$(4x + 1)^2 - (1x + 1) \rightarrow 16x^2 + 7x,$$
$$(5x) \cdot (5x) \rightarrow 25x^2.$$

At this point, it must be noticed that the representation of the operations by means of modern algebraic symbolism raises a subtle point of potential misrepresentation. For example, in problem 10 of book I, Diophantus adds 20 units to '1 *arithmos*' ($1x$) to get '1 *arithmos* and 20 units' (Tannery 1893–95, i, 28.14–15) and also subtracts '1 *arithmos*' ($1x$) from 100 units to get '100 units less 1 *arithmos*' (Tannery 1893–95, i, 28.15). Using modern symbolism, the two operations should be written $1x + 20 \rightarrow 1x + 20$ and $100 - 1x \rightarrow 100 - 1x$, respectively. The potential confusion with this writing lies in the fact that only in the left part in each expression the signs «+» and «–» denote operations, *which are to be performed*; the right part describes *the outcome* of each operation. Thus, when in expressions like the above, the sign appears in the left side,

26 For convenience, the terms of the 'arithmetical theory' have been replaced by the corresponding modern algebraic symbols in what follows.

it denotes an operation, while the same sign, in the right side of the same expression, merely links terms, very much like the words 'and' and 'less' do. A way to avoid the confusion would be to substitute «&» and «*l*» (initial letter of 'less') for «+» and «–», respectively, when no operation is intended, and retain the common signs of «+» and «–» only for operations. With this convention, the first operation that Diophantus performs in problem II.33 is written as follows: $(4x \,\&\, 1)^2 - (1x \,\&\, 1) \rightarrow 16x^2 \,\&\, 7x$. In the followings, and throughout the present paper, I will employ this notation.

Now, there is a crucial question regarding, more specifically, the task of assigning names to the unnamed numbers: How did Diophantus find the names mentioned above? More generally, how the process of assigning names to the unnamed sought-after numbers is to be conducted in each individual problem? The *Arithmetica* contains hundreds of problems, some of which are extremely difficult, requiring three, four, or even more numbers to be found. In view of the complexity of the problems, the issue of how to carry through the task of assigning names to the unnamed sought-after numbers plays a critical role in the solutions, and its mastering is of high importance for anyone who wants to acquire competence in solving problems by algebra.

As shown, the problem we are now discussing requires four namings. The four names, however, have not been found in the same way. Regarding how he found the first three names, Diophantus provides the following heuristic explanation in his text: "Since, whenever a number is the double of <another> number less one unit, the square on the lesser, less the greater, makes a square, I assign the first <number> to be 1 *arithmos*, 1 unit, the second, similarly, 2 *arithmoi*, 1 unit, and the third, 4 *arithmoi*, 1 unit, and it results that the square on the first less the second, makes a square and moreover the square on the second, less the third, makes a square." What is meant here is the following: If u and v are two numbers such that $v = 2u - 1$, then $u^2 - v$ makes a square. Accordingly, if $u = m + 1$, and $v = 2m + 1$, then $(m + 1)^2 - (2m + 1)$ is a square, while if $u = 2m + 1$ and $v = 4m + 1$, then $(2m + 1)^2 - (4m + 1)$ is also a square. But the generic relation "$u^2 - v$ makes a square" and the two relations which emerge from this have the same structure with the conditions of the problem. In this sense, the relations can be regarded as 'simulators' for the conditions of the problem.[27] The simulation is shown in the Table below:

Table 3: The correspondence between conditions and simulators

1st condition	(1st sought-after number)^2 – (2nd sought-after number) is a square
1st simulator	$(m + 1)^2 - (2m + 1)$ is a square
2nd condition	(2nd sought-after number)^2 – (3rd sought-after number) is a square
2nd simulator	$(2m + 1)^2 - (4m + 1)$ is a square

27 The term 'simulator' was first proposed in Bernard & Christianidis 2012.

On account of the structural similarity of the first two conditions of the problem with the corresponding simulators, one easily perceives how straightforwardly the first three names that Diophantus assigns are deduced. Obviously, what we have here is a method. It is a method for assigning names to unnamed sought-after numbers. This method, though, is not a method of solution. It is a method for naming the unknowns. As already said, naming the unknowns does not coincide with the solution to the problem. It merely constitutes a part of the solution procedure, namely its first stage. Therefore, a clear distinction should be made between the method of solution, on the one hand, and the method (or methods) by which Diophantus assigns names to the unknowns, thus conducting the first stage of the solution, on the other.

The fourth name in problem II.33 has not been found by simulation, but *via* another method. This method, which is repeatedly used by Diophantus,[28] is easily recognized by the verb 'plassô' (I form), by which it is introduced. Because of that reason, the method has been called the 'plassô method' (Bernard and Christianidis 2012)

We see, therefore, that in problem II.33, Diophantus uses two different methods of naming, by which he assigns four names to the unnamed terms of the problem. As explained in Bernard and Christianidis 2012, these methods—called there 'methods of invention'—are part of a real 'toolbox' that Diophantus had elaborated for conducting the task of assigning names. Section II.4 below outlines the methods of naming used in the first three books of the *Arithmetica*, according to the conclusions of the aforementioned study of 2012.

3 *Manipulation of the Equation*

In problem II.33, the manipulation of the equation is altogether skipped. After the formulation of the equation, Diophantus provides its root directly.[29] Normally, this stage includes two distinct processes: simplification of the equation and solution of the simplified equation. The first process involves two steps, which although are widely known by the names the medieval Arabic algebraists gave them, i.e. 'al-jabr' and 'al-muqābala', they are fully described in Diophantus' introduction (Tannery 1893–95, i, 14.11–24). Regarding the simplified equation, in most cases, its solution

---

28 In the first three books of the *Arithmetica* the method is used more than 40 times; see Bernard & Christianidis 2012, 39–40.

29 The equation to which Diophantus arrives is, in modern notation, $25x^2 = 16x^2 + 7x$. It is an equation between two aggregations, with no operations involved. All operations have been made before the statement of the equation. As it has been explained before, this is a clear-cut indication that the solution of Diophantus is premodern. Students today would be instructed to solve the problem differently, by setting up the equation $(4x + 1)^2 - (x + 1) = (5x)^2$ and then perform the operations *within* the equation. Diophantus, by contrast, works out the squares and the subtraction first.

is trivial, because Diophantus manages to arrive at a two-term equation.[30] This is due to the rich 'toolbox' Diophantus has developed, which allowed him to make a clever assignment of names, thus yielding two-term equations. However, he definitely knew how to solve three-term equations and, in fact, in some cases he did solve such equations.[31]

4 Method of Solution and Methods of Naming

From what has been said so far, it is evident that the method of solution should not be confused with the methods of naming the solution involves: the method of solution is algebra, in the premodern meaning of the term, while the methods of naming are methods Diophantus had elaborated in order to carry through one of the two tasks which are involved in the first stage of a solution by algebra, that is, the assignment of names—the other task being the working through the operations outlined in the enunciation on these names. As we have already mentioned, Diophantus had a rich 'toolbox' with such methods at his disposal. Part of this 'toolbox', I summarize in Table 4 below. The first column displays the problem. The second column contains the phrases by which Diophantus assigns the names in each particular problem; it is through these phrases that the different methods of naming are identified. Finally, the third column shows how the corresponding method is called in the aforementioned study. Capital letters like A, B, A^2 denote unnamed terms appearing in the enunciation of a problem mostly as sought-after numbers, while lower case letters like x, x^2 denote technical terms assigned as names to the former. For the conditions of a problem the sign '$\rightrightarrows$' is used, while the action of assigning the name $1x$ to the unnamed term A is denoted by $A := 1x$.[32]

Table 4: The methods of naming used in books I–III of the *Arithmetica*

Problem	Phrases by which Diophantus assigns names	Method of naming
Pb I.1	Let the lesser be assigned to be '1 *arithmos*' [$B := 1x$]	Simple method
$A + B \rightrightarrows 100$ $A - B \rightrightarrows 40$	Then the greater will be '1 *arithmos*, 40 units' [$A := 1x$ & 40]	Derivative method

30 This is another reason, in addition to the complexity of most of the problems, that makes the first part of the solution, the setting up of the equation, the most difficult one, and therefore the part to which Diophantus pays closer attention.

31 See more on this in the paper of Jeffrey Oaks in this volume.

32 On this notation see also part IV below.

Table 4: (Continued)

Pb I.16 $A + B \Rightarrow 20$ $B + C \Rightarrow 30$ $C + A \Rightarrow 40$	Let the three <together> be assigned to be '1 *arithmos*' $[A + B + C := 1x]$	All-together method
	And since the first and the second make 20 units, if therefore I subtract 20 units from 1 *arithmos*, I shall have that the third is '1 *arithmos* less 20 units' $[C := 1x\ l\ 20]$	Derivative
	In the same manner, the first will be '1 *arithmos* less 30 units' $[A := 1x\ l\ 30]$,	Derivative
	and the second '1 *arithmos* less 40 units' $[B := 1x\ l\ 40]$	Derivative

Pb II.8 $A^2 + B^2 \Rightarrow 16$	Let the first <square> be assigned to be '1 *dynamis*' $[A^2 := 1x^2]$	Simple
	Then the other will be '16 units less 1 *dynamis*' $[B^2 := 16\ l\ 1x^2]$	Derivative
	Then, it is required that 16 units less 1 *dynamis* are equal to a square. I form the square from <a side consisting of> any <number> of *arithmoi* less as many units as there are in the side of 16 units; let it be '2 *arithmoi* less 4 units' $[B := 2x\ l\ 4]$	Plassô method

Pb II.19 $(A^2 - B^2)$ $\Rightarrow 3(B^2 - C^2)$	Let the lesser be assigned to be '1 *dynamis*' $[C^2 := 1x^2]$, and the middle '1 *dynamis*, 2 *arithmoi*, 1 unit' $[B^2 := 1x^2\ \&\ 2x\ \&\ 1]$, that is from <a side consisting of> '1 *arithmos*, 1 unit' $B := 1x\ \&\ 1]$	Two-at-once method
	Then the greatest will be '1 *dynamis*, 8 *arithmoi*, 4 units' $[A^2 := 1x^2\ \&\ 8x\ \&\ 4]$	Derivative
	Therefore it is requited that $1x^2\ \&\ 8x\ \&\ 4$ are equal to a square; I form the square from ... '1 *arithmos* and 3 units' $[A := 1x\ \&\ 3]$	Plassô method

Pb I.27 $A + B \Rightarrow 20$ $A \cdot B \Rightarrow 96$	Let their excess be assigned to be '2 *arithmoi*' $[A - B := 2x]$	Sum-against-difference method
	Let the greater be assigned to be '1 *arithmos* and 10 units' $[A := 1x\ \&\ 10]$	Derivative
	Then the lesser will be '10 units less 1 *arithmos*' $[B := 10\ l\ 1x]$	Derivative

Pb II.20 $B + A^2 \Rightarrow S^2$ $A + B^2 \Rightarrow T^2$	Let the first be assigned to be '1 *arithmos*' $[A := 1x]$, and the other '1 unit, 2 *arithmoi*' $[B := 1\ \&\ 2x]$, so that the square on the first, if it receives the second, makes a square	Method of simulation Simulator: $m^2 + (1 + 2m)$ $\rightarrow (m + 1)^2$
	... it makes '4 *dynameis*, 5 *arithmoi*, 1 unit' $[4x^2\ \&\ 5x\ \&\ 1]$; these are equal to a square. I form the square from '2 *arithmoi* less 2 units' $[T := 2x\ l\ 2]$	Plassô method

Pb II.26 $AB + A \rightrightarrows S^2$ $AB + B \rightrightarrows T^2$ $S + T \rightrightarrows 6$	So, since if there are two numbers the greater of which is by one unit less than the quadruple of the lesser, their product if it receives the lesser makes a square, I assign the lesser to be ‘1 *arithmos*’ [$A \coloneqq 1x$], the greater ‘4 *arithmoi* less 1 unit’ [$B \coloneqq 4x \, l \, 1$] and it results, similarly, that their product, if it receives the lesser, makes a square	Method of simulation Simulator $m \cdot (4m - 1) + m \rightarrow (2m)^2$

III Problem Solving in Premodern and in Modern Algebra: Discriminations and Affinities

Names, operations with names, and formulation of equations (in the form of equalities between two aggregations of names of powers of the unknown) are three key notions of problem solving in premodern algebra. All three of them, are clearly described in the introduction to the *Arithmtetica*, where Diophantus not only discusses these notions, but, most significantly, explains how the corresponding tasks are interconnected so as to form a general problem-solving method. At this point, it must be noted that these notions are fundamental not only in premodern algebra, but in modern algebra as well. Below, I reproduce a passage from Isaac Newton's *Universal Arithmetick*, in which he deals precisely with this topic (Newton 1720, 67). Interestingly, the title Newton gave to the section this passage belongs to is “How a question may be brought to an æquation,” a clear indication that Newton comprehends the transition from problem to equation as part of the subject-matter of algebra.

> *How a question may be brought to an æquation*
> After the learner has been some time exercised in managing and transforming æquations, order requires that he should try his skill in bringing questions to an æquation. And any question being proposed, his skill is particularly required to denote all its conditions by so many æquations. To do which, he must first consider whether the propositions or sentences in which it is expressed, be all of them fit to be denoted in algebraick terms, just as we express our conceptions in Latin or Greek characters. And if so, (as will happen in questions conversant about numbers or abstract quantities) then let him give names to both known and unknown quantities, as far as occasion requires. And the conditions thus translated to algebraick terms will give as many æquations as are necessary to solve it.

Next, Newton provides three examples and 16 worked-out problems, all solved by algebra, before proceeding to another section to discuss how problems in geometry are solved by algebra. I quote below the third example with Newton's corresponding comments (Newton 1720, 68–70):

> Take another example. A certain merchant encreases his estate yearly by a third part, abating 100 l. which he spends yearly in his family; and after three years he finds his estate doubled. *Query*, What he is worth? To resolve this, you must know there are [or lie hid] several propositions, which are all thus found out and laid down.

In English	*Algebraically*
A merchant has an estate	x.
Out of which the first year he expends 100 l.	$x - 100$.
And augments the rest by one third	$x - 100 + \frac{x-100}{3}$, or $\frac{4x-400}{3}$.
And the second year expends 100 l.	$\frac{4x-400}{3} - 100$, or $\frac{4x-700}{3}$.
And augments the rest by a third.	$\frac{4x-700}{3} + \frac{4x-700}{9}$, or $\frac{16x-2800}{9}$.
And so the third year expends 100 l.	$\frac{16x-2800}{9} - 100$, or $\frac{16x-3700}{9}$.
And by the rest gains likewise one third part	$\frac{16x-3700}{9} + \frac{16x-3700}{27}$, or $\frac{64x-14800}{27}$.
And he becomes [at length] twice as rich as at first.	$\frac{64x-14800}{27} = 2x$.

Therefore the question is brought to this æquation, $\frac{64x-14800}{27} = 2x$, by the reduction whereof you are to find x; *viz.* Multiply it by 27, and you have $64x - 14800 = 54x$; subtract $54x$, and there remains $10x - 14800 = 0$, or $10x = 14800$, and divining by 10, you have $x = 1480$. Therefore, 1480 l. was his estate at first, as also his profit or gain since.

You see therefore, that to the solution of questions which only regard numbers, or the abstracted relations of quantities, there is scarce any thing else required than that the problem be translated out of the English, or any other tongue it is proposed in, into the algebraical language, that is, into characters fit to denote our conceptions of the relations of quantities. But it may sometimes happen, that the language [or the words] wherein the state of the question is expressed, may seem unfit to be turned into the algebraical language; but making use of a few changes, and attending to the sense rather than the sound of the words, the version will become easy. Thus, the forms of speech among [several] nations have their proper idioms; which, where they happen, the translation out of one into another is not to be made literally, but to be determined by the sense. But that I may illustrate these sorts of problems, and make familiar the method of reducing them to æquations; and since arts are more easily learned by examples than precepts, I have thought fit to adjoin the solutions of the following problems.

Here is another example, this time from Euler's *Elements of Algebra* (Euler 1972, 187–188):

Suppose twenty persons, men and women, go to a tavern; the men spend 24 shillings, and the women as much: but it is found that the men have spent 1 shilling each more than the women. Required the number of men and women separately?
Let the number of men be represented by x.
Then the women will be $20 - x$.
Now, the x men having spent 24 shillings, the share of each man is $\frac{24}{x}$. The $24 - x$ women having also spent 24 shillings, the share of each woman is $\frac{24}{20-x}$.

But we know that the share of each woman is one shilling less than that of each man; if, therefore, I subtract 1 from the share of a man, we must obtain that of a woman; and, consequently $\frac{24}{x} - 1 = \frac{24}{20-x}$. This is, therefore, the equation, from which we are to deduce the value of x.

The differences of Newton's and Euler's problem solving from that of Diophantus (in fact, of the premodern algebraists in general) are visible to the naked eye. At bottom, they are of conceptual order. Newton's and Euler's solutions are symbolical; symbols are used to denote not only unknowns but given quantities as well—as Newton explicitly says; operations may appear in an equation, as we see in Euler's solution, in contrast with premodern algebra, in which all operations the enunciation of the problem calls for must be worked out prior to arriving at the equation; polynomials are not mere aggregations as in premodern algebra, but have their modern meaning of linear combinations of powers, with additions, subtractions and scalar multiplications involved; accordingly, and despite the fact that the two examples do not illustrate it, irrational coefficients are allowed in modern polynomials.[33] Moreover, above all, modern algebra is no longer a mere problem solving method as its premodern counterpart is. A mere look at the table of contents of either Newton's or Euler's books, compared with the contents of any premodern algebra book, makes this manifest.

The above-mentioned differences notwithstanding, the excerpts from Newton's and Euler's books illustrate the common elements of all problems worked-out by algebra. First, and foremost, the solutions follow the same overall pattern, whether in modern algebra or in premodern. Thus, the structure of Diophantus' solution in problem II.33 is maintained in all solutions. Indeed, it could not be otherwise. The algebraic solution presupposes translation of the problem (which is formulated in everyday language) to a technical language, emergence of an equation framed entirely in the technical vocabulary, and manipulation of the equation instead of the problem itself. This is the "essence" of a solution by algebra, as Lagrange said.[34] Significantly, even the language used by Diophantus and Newton exhibits the affinity of the modus operandi. Newton speaks about "names" which are assigned, and yet the same word ('epônymiai', meaning 'names' or 'surnames') is used twice by Diophantus to characterize the technical terms of the 'arithmetical theory' (4.13; 6.23) which are to be assigned to sought-after numbers.

---

33 For an in depth analysis of these matters, I refer the reader to the works of Jeffrey Oaks mentioned in the bibliography.

34 The passage of Lagrange (1877, 219) is very illuminating and deserves to be quoted in full: "Diophante peut être regardé comme l'inventeur de l'Algèbre; en effet, par un mot de sa préface, ou plutôt de son épître d'envoi (car les anciens Géomètres envoyaient leurs Ouvrages à quelques-uns de leurs amis, comme on le voit aussi par les préfaces des Ouvrages d'Apollonius et d'Archimède); par un mot, dis-je, de sa préface, on voit qu'il a été le premier à s'occuper de cette partie de l'Arithmétique qui a été nommée depuis Algèbre. Son Ouvrage contient les premiers éléments de cette science: il y emploie, pour exprimer la quantité inconnue, une lettre grecque qui répondait à l'*st*; et que dans la traduction on a remplacée par N; pour les quantités connues, il n'emploie que des nombres, car pendant longtemps l'Algèbre n'a été destinée qu'à résoudre des questions numériques mais on voit qu'il traite également les quantités connues et les inconnues pour former l'équation d'après les conditions du problème. Voilà ce qui constitue proprement l'essence de l'Algèbre: c'est d'employer des quantités inconnues, de les calculer comme les connues, et d'en former une ou plusieurs équations d'après lesquelles on puisse déterminer la valeur de ces inconnues."

IV Interpreting Premodern Algebra *via* Modern Algebraic Geometry: a Historically Inappropriate Methodology

In the mid-'60s, Isabella Grigorievna Bashmakova set forth the idea that the expressions that Diophantus assigns in the course of the solutions to the sought-after numbers could be explained by means of the cord and tangent methods of modern algebraic geometry,[35] thus 'answering' positively the notorious problem about the generality of Diophantus' methods of solution, which had troubled historians of mathematics from the mid-nineteenth century. The problem is epitomized in the famous saying of Hermann Hankel, "It is on that account difficult for a modern mathematician, even after studying 100 Diophantine solutions to solve the 101st problem" (Hankel 1874, 165; cf. Heath 1964, 54–55). From the mid-'80s onwards, the idea of Bashmakova was taken up by Roshdi Rashed, and reached in his hands the highest level of abstractness and generality. In fact, Rashed applies the interpretation *via* modern algebraic geometry not only to Diophantus but to medieval Arabic algebraists as well, such as Abū Kāmil (Rashed 2012).[36]

I do not intend here to argue in detail for what is a matter of course: algebraic geometry does not provide the appropriate basis for a historical reconstruction of Diophantus' solutions, and those who do that are exposed to blatant anachronism. I shall confine myself to demonstrating how the interpretation *via* algebraic geometry distorts the *modus operandi* of Diophantus, and more generally of the premodern algebraists who were occupied with indeterminate problems. I shall do this by providing a sample problem that appears in two different sources. It is problem II.9 of Diophantus, which also appears in the late ninth-century *Book on algebra* (Kitāb fī al-jabr wa'l-muqābala) of Abū Kāmil (problem <12> in Rashed's edition of 2013). Despite the fact that Abū Kāmil was not *directly* influenced by Diophantus,[37] we shall see that the two solutions are effectuated by the same method: algebra. For the text of Diophantus (92.16–94.9), the English translation is mine, while for Abū Kāmil's text I am using the French translation of Rashed (Rashed 2012, 604), slightly adapted.

---

35 See, for example, Bachmakova 1966; Bašmakova 1974; Bashmakova 1981.

36 See Oaks' (2014) criticism of Rashed's methodology, which I approve fully. As regards Diophantus, Rashed's interpretational pursuit is quite interesting. Starting with the standard algebraic reading, in his first publications after the rediscovery of the Arabic books of the *Arithmetica* (Rashed 1974; 1975), he moved on to a double reading, by algebra and by algebraic geometry in his edition of these books (Rashed 1984), to end up with the interpretation *via* algebraic geometry alone in the recent (Rashed & Houzel 2013). As time goes by, it seems that Rashed manifests increasingly aversion for connecting the *Arithmetica* with the history of algebra.

37 See Sesiano 1977, 90. Nevertheless, it is quite probable that ultimately Abū Kāmil was *indirectly* influenced by the late antique Alexandrian milieu where Diophantus' activity was placed.

Table 5: Diophantus' and Abū Kāmil's solutions of the same problem

Diophantus' problem II.9	**Abū Kāmil's problem <12>, first solution**
[Enunciation of the problem]	
To divide a given number, which is composed of two squares, into two other squares. Let the <number> 13, which is composed of the squares 4 and 9, to be divided anew into two other squares.	Sache que tout nombre partagé en deux parties telles que chacune ait une racine peut être partagé en deux autres parties telles que chacune ait une racine, et encore en deux autres parties, et ainsi indéfiniment. Je vais t'expliquer cela pour que tu comprennes, si Dieu le veut. Si on te dit: divise cinq dirhams en deux parties dont chacune a une racine, tu sais alors que l'une des deux parties est quatre et l'autre est un.
[Setting up the equation]	
Let the sides, 2 units and 3 units, of the aforesaid squares be taken, and let the sides of the requested squares be assigned to be 1 *arithmos*, 2 units [$1x$ & 2], the one, and any <number> of *arithmoi* less as many units as there are in the side of the other, the other. Let it be 2 *arithmoi* less 3 units [$2x$ / 3]; and the squares become 1 *dynamis*, 4 *arithmoi*, 4 units, the one, 4 *dynameis*, 9 units less 12 *arithmoi*, the other. It remains the two added together to make 13 units. But the two added together make 5 *dynameis*, 13 units less 8 *arithmoi*; these are equal to 13 units.	Si tu veux le diviser en deux autres parties différentes de ces deux parties et telles que chacune ait une racine, prend alors la racine de l'une des deux parties, ajoute-lui une *chose* et multiplie la somme par elle-même; ce que tu obtiens est l'une des deux parties; enlève-la de cinq, ce qui reste est la seconde partie. C'est comme si nous prenions l'une des parties, c'est-à-dire un; nous prenons sa racine qui est un à laquelle nous ajoutons une *chose*; il vient un et une *chose* [1 & $1x$]; nous multiplions cette somme par elle-même, il résulte un *carré* et un dirham et deux *choses*, ce qui est l'une des deux parties que tu cherches. Nous la retranchons de cinq; il reste quatre dirhams manquant un *carré* manquant deux *choses*, qui est la seconde partie; il est nécessaire qu'elle ait une racine. Or nous savons que sa racine est plus petite que deux dirhams; nous la posons deux dirhams manquant une *chose* et un tiers $\left[2 \,/\, \left(1x \,\&\, \frac{1}{3}x\right)\right]$, nous la multiplions par elle-même, on aura quatre dirhams et un *carré* et sept neuvièmes de *carrés* manquant cinq *choses* et un tiers de *chose* égal à quatre dirhams manquant un *carré* manquant deux *choses*.
[Solving the equation]	
And the *arithmos* [x] becomes $\frac{8}{5}$.	Réduis par ceci, comme je t'ai expliqué; la *chose* [x] sera un dirham et un cinquième.

Table 5: (Continued)

[*Answering the problem and test-proof*]	
To the numerical values: I assigned the side of the first to be 1 *arithmos*, 2 units [1x & 2]; it will be $\frac{18}{5}$; and the side of the other, 2 *arithmoi* less 3 units [2x / 3]; it will be $\frac{1}{5}$. The squares themselves will be $\frac{324}{25}$, the one, $\frac{1}{25}$, the other. And the two added together make $\frac{325}{25}$, which yield the proposed 13 units.	Mais nous avons posé la racine de l'une des deux parties une *chose* et un dirham [1x & 1], alors la racine de l'une des deux parties sera deux dirhams et un cinquième; nous la multiplions par elle-même, on aura quatre dirhams et quatre cinquièmes et un cinquième d'un cinquième, ce qui est l'une des deux parties que tu cherches; l'autre est ce qui reste de cinq, c'est-à-dire quatre cinquièmes de cinquième.

Let us see now the two solutions in abbreviated tabular form. For it, a convenient notation will be employed. Thus, besides the signs already introduced in part II above, the capital letters A and B are used to denote the sought-after numbers,[38] the symbol '=' is used only to denote equations, the single arrow '→' is used for the outcome of an operation, while for the arithmetical operations involved in the solutions the standard symbols are used. By virtue of the above, the emergence of the equation in the two solutions appears as follows:

In Diophantus:

Problem (instantiated version): $2^2 + 3^2 = 13 \Rightarrow A^2 + B^2$		
Naming	**Working through the operations**	**Equation**
$A := 1x \,\&\, 2$		
$B := 2x \,/\, 3$		
	$(1x \,\&\, 2) \cdot (1x \,\&\, 2) \rightarrow 1x^2 \,\&\, 4x \,\&\, 4$	
	$(2x \,/\, 3) \cdot (2x \,/\, 3) \rightarrow 4x^2 \,\&\, 9 \,/\, 12x$	
	$(1x^2 \,\&\, 4x \,\&\, 4) + (4x^2 \,\&\, 9 \,/\, 12x)$ $\rightarrow 5x^2 \,\&\, 13 \,/\, 8x$	
		$5x^2 \,\&\, 13 \,/\, 8x = 13$

38 It must be stressed here that I do not use A and B as names in my transcription; they merely stand for the unnamed numbers the finding of which the enunciation of the problem calls for. Thus, A^2 and B^2 stand for the expressions 'two other squares' in Diophantus and 'deux autres parties telles que chacune ait une racine' in Abū Kāmil, all articulated with the unnamed terms 'tetragônoi' and '<number> having a root' respectively. These expressions are expressions of the common language and they do not have any technical (algebraic) meaning.

In Abū Kāmil:

Problem (instantiated version): $\mathbf{1^2 + 2^2 = 5 \rightrightarrows A^2 + B^2}$		
Naming	**Working through the operations**	**Equation**
$A := 1 \,\&\, 1x$		
	$(1 \,\&\, 1x) \cdot (1 \,\&\, 1x) \rightarrow 1x^2 \,\&\, 1 \,\&\, 2x$	
	$5 - (1x^2 \,\&\, 1 \,\&\, 2x) \rightarrow 4 \; l \; 1x^2 \; l \; 2x$	
		$4 \; l \; 1x^2 \; l \; 2x$ must be sq. (B^2)
$B := 2 \; l \left(1x \,\&\, \frac{1}{3}x\right)$	$\left[2 \; l \left(1x \,\&\, \frac{1}{3}x\right)\right] \cdot \left[2 \; l \left(1x \,\&\, \frac{1}{3}x\right)\right] \rightarrow$ $4 \,\&\, 1x^2 \,\&\, \frac{7}{9}x^2 \; l \left(5x \,\&\, \frac{1}{3}x\right)$	
		$4 \,\&\, 1x^2 \,\&\, \frac{7}{9}x^2 \; l \left(5x \,\&\, \frac{1}{3}x\right) =$ $4 \; l \; 1x^2 \; l \; 2x$

The tabular setting clearly shows that the two solutions follow the same overall pattern: names are assigned to the sought-after numbers; the operations are worked out on the named terms; and a determinate equation is created, framed entirely in the language of the names. There are, though, minor differences: for the naming, Diophantus and Abū Kāmil do not use the same method. Diophantus assigns the names to the sides of the two unnamed squares *at once*. Abū Kāmil assigns a name to the side of the first unnamed square, then he squares and subtracts from 5, after that he observes that the result must be the other square, and *then* he assigns the name to its side. Clearly, two different methods of naming are being employed, both found in Diophantus' treatise (see Table 4). Diophantus uses the method called 'two-at-once' while Abū Kāmil uses consecutively the 'simple' method and the 'plassô method'. Nevertheless, these differences bear on the method of naming; not on the method of solution. The method of solution is in both cases the same: i.e. algebra.

Let us see now how the two solutions are represented by using the cord method of modern algebraic geometry.[39]

39 For Diophantus' solution, see Bashmakova 1981, 396–397; Rashed 1984, III, cviii; Rashed & Houzel, 2013, 187–188. For Abū Kāmil's solution, see Rashed 2012, 151–152.

Table 6: The two solutions *via* the cord method of algebraic geometry

Diophantus	Abū Kāmil
Diophantus' problem is written	Abū Kāmil's problem is written
(*) $x^2 + y^2 = 13 = 2^2 + (-3)^2$	(*) $x^2 + y^2 = 5 = 1^2 + 2^2$
The equation (*) defines a circle and the point $(2, -3)$ is a rational point of that circle.	The equation (*) defines a circle and the point $(1, 2)$ is a rational point of that circle.
Diophantus sets	Abū Kāmil sets
$x = 2 + t$ and $y = -3 + ut$,	$x = 1 + t$ and $y = 2 + ut$,
that is, he draws a straight line passing through the point $(2, -3)$, with u rational ($u = 2$).	that is, he draws a straight line passing through the point $(1, 2)$, with u rational $\left(u = -\frac{4}{3}\right)$.
By substituting in (*) one gets the resulting equation	By substituting in (*) one gets the resulting equation
(**) $t[(u^2 + 1)t + 2(2 - 3u)] = 0$	(**) $t[(u^2 + 1)t + 2(2u + 1)] = 0$
the solution of which is $t = \frac{2(3u - 2)}{u^2 + 1}$.	the solution of which is $t = -\frac{2(2u + 1)}{u^2 + 1}$.
For $u = 2$ we get $t = \frac{8}{5}$.	For $u = -\frac{4}{3}$ we get $t = \frac{6}{5}$.
By substituting in the positions we get $x = \frac{18}{5}$ and $y = \frac{1}{5}$.	By substituting in the positions we get $x = \frac{11}{5}$ and $y = \frac{2}{5}$.

From the discussion above, one deduces that there are not major differences between the two interpretations (by algebra and by the cord method of algebraic geometry), as far as the structure of the solution is concerned. This becomes even more evident in Table 7 below, which displays the two interpretations of the solution to Diophantus' problem facing each other. The middle column of the Table contains a representative selection of key sentences of the Diophantine text, which provide a sketch of the solution, while the columns on the two sides display the respective interpretations by algebra, the left column, and by algebraic geometry, the right column.

Table 7: The two interpretations facing each other

Interpretation by premodern algebra	Sketch of the solution	Interpretation by algebraic geometry
$a^2 + b^2 = c \rightrightarrows A^2 + B^2$ *Generic enunciation*	To divide a given number, which is composed of two squares, into two other squares.	$a^2 + b^2 = c = x^2 + y^2$ *Generic enunciation*
$2^2 + 3^2 = 13 \rightrightarrows A^2 + B^2$ *Instantiated enunciation*	Let the <number> 13, which is composed of the squares 4 and 9, to be divided anew into two other squares.	$2^2 + (-3)^2 = 13 = x^2 + y^2$ *Instantiated enunciation*
$A \coloneqq 1x \,\&\, 2,\ B \coloneqq 2x \,/\, 3$ *Naming the unknowns*	Let the sides of the requested squares be assigned to be 1 *arithmos*, 2 units, the one, ... 2 *arithmoi* less 3 units, the other.	$x = 2 + t,\ y = -3 + ut\ (u = 2)$ *Substitutions*, defining a rational straight line passing through the point (2, −3) of the circle defined by the equation $x^2 + y^2 = 13$
$5x^2 \,\&\, 13 \,/\, 8x \coloneqq 13$ *Equation* (to which the problem is converted)	The two added together make 5 *dynameis*, 13 units less 8 *arithmoi*. These are equal to 13 units.	$t[(u^2 + 1)t + 2(2 - 3u)] = 0\ (u = 2)$ *Resulting equation*
$x = \frac{8}{5}$. *Solution of the equation*	And the *arithmos* becomes $\frac{8}{5}$.	$t = \frac{2(3u-2)}{u^2+1}$, which for $u = 2$ gives $t = \frac{8}{5}$. *Solution of the resulting equation*
$A = \frac{18}{5},\ B = \frac{1}{5}$ *Answer to the problem*	To the numerical values.	$x = \frac{18}{5},\ y = \frac{1}{5}$ $\left(\frac{18}{5}, \frac{1}{5}\right)$ is a *new rational point* of the circle

The table above shows that the two interpretations are structurally isomorphic; yet, conceptually they are completely different. In the interpretation *via* algebraic geometry, the problem is conceived directly as an equation; the names assigned to the numbers the finding of which the enunciation of the problem calls for have become 'substitutions' or 'changes of variable' in the equation; the equation to which the problem has been converted has become 'resulting equation'. By virtue of the above, the task of "bringing a question to an equation," to use Isaac Newton's phrase, is for the solution *via* algebraic geometry empty of meaning. Nevertheless, this is the most important stage of the solution of indeterminate problems not only in Diophantus but also in Abū Kāmil and the premodern algebra in general. Therefore, the interpretation by algebraic geometry distorts the very core of premodern algebraic problem solving, which is *the passage from problem to equation.*

Besides, the interpretation by algebraic geometry has other drawbacks. For example, in the problem under discussion, it does not explain the difference in the methods by which Diophantus and Abū Kāmil assign names to the sought-after numbers. Strictly speaking only the naming of Diophantus, in which the two names are assigned instantaneously, is explained by means of the cord method. In the case of the solution of Abū Kāmil the two names are assigned at different times, therefore different methods of naming are applied. The interpretation by algebraic geometry disregards this difference in the methods of naming used by Diophantus and Abū Kāmil,[40] thus making the solution of Abū Kāmil identical to the slightest details with that of Diophantus, which of course is wrong.[41]

This is only one example showing how the interpretation by algebraic geometry barbarously distorts premodern algebra, not only conceptually, but also at the level of the internal layout of the texts. Having today in our disposal the historiographical category of premodern algebra, in comparison with which the interpretative capacity of modern algebraic geometry when applied in premodern texts has only drawbacks to present and no advantages, any attempt of expulsion of the *Arithmetica* of Diophantus from the history of algebra, under the pretext of an alleged genealogy that connects it with algebraic geometry, must be rejected unequivocally.

V Conclusion

Let me summarize the main findings of this study. In part I, the recently elaborated notion of 'premodern algebra' was introduced, and its fundamental features were outlined, with the aim to investigate whether it could be adopted as a historically sane basis for situating and understanding Diophantus' mathematical practice. This examination was performed in part II, where a concrete example, problem 33 of the second book of the *Arithmetica*, was used as a case study to illustrate that Diophantus' solution does meet all the basic presuppositions underlying problem solving within premodern algebra. In part III, the differences that separate premodern and modern algebra, but also their common elements, were briefly discussed. Lastly, in part IV, by analyzing another concrete example solved by two premodern algebraists, Diophantus and Abū Kāmil, it was demonstrated that the analysis by means of tools

40 For example, this is how Rashed interprets the solution of Abū Kāmil *via* algebraic geometry: "Cette fois encore, Abū Kāmil procède par la méthode de la corde. Il pose $x = a + t$ et $y = b - ut$ et, en substituant dans l'équation, il obtient $t = \frac{2(bu-a)}{u^2+1}$"; Rashed 2012, 152.

41 The difference is pointed out by Sesiano, although not in the context of 'naming' as the present study maintains; cf. Sesiano 1977, 95.

and techniques of the modern algebraic geometry that a few modern historians of mathematics advocate, besides of being blatantly ahistorical, it does not offer any comparative advantage over the historically sensitive and sympathetic analysis through the tools and the techniques of premodern algebra.

The notion of premodern algebra was introduced and elaborated primarily by Jeffrey A. Oaks, with the aim to portray the algebraic activity of the medieval Arabic, Latin and Italian mathematicians. This study has shown that the same notion is also suitable to account for Diophantus' method of solving arithmetical problems. Such an interpretation authorizes us to contextualize Diophantus' mathematical activity within the broad tradition of premodern algebra, by avoiding, at the same time, the danger of anachronism embodied in the general term 'algebra' when applied to historical material like the one discussed here.

Bibliography

Texts quoted

Abū Kāmil: Rashed, R. 2012. *Abū Kāmil. Algèbre et Analyse Diophantienne. Édition, traduction et commentaire by Roshdi Rashed.* Berlin.

Al-Khwārizmī: Rashed, R. 2007. *Le commencement de l'algèbre. Texte établi, traduit et commenté par Roshdi Rashed.* Paris.

Al-Nadīm: Dodge, B. 1970. *The Fihrist of al-Nadīm: A Tenth Century AD Survey of Islamic Culture. Bayard Dodge editor and translator.* 2 vols. New York.

Diophante: Rashed, R. 1984. *Diophante: Les Arithmétiques. Texte établi et traduit par Roshdi Rashed.* Tome III: *Livre IV*; Tome IV: *Livres V–VII.* Paris.

Diophantus, *Arithmetica*: Tannery, P. 1893–1895. *Diophanti Alexandrini Opera Omnia cum Graeciis Commentariis.* 2 vols. Leipzig (repr. Stuttgart: 1974).

Diophantus: Stamatis, E. (ed.) 1963. *Διοφάντου Αριθμητικά. Η άλγεβρα των αρχαίων Ελλήνων.* Athens.

Euler, L. 1972. *Elements of Algebra* (orig. St. Petersburg 1767). Transl. J. Hewlett. New York.

Lagrange, J. L. 1877. *Œuvres de Lagrange.* Vol. 7. Paris.

Newton, I. 1720. *Universal Arithmetick: or, a Treatise of Arithmetical Composition and Resolution* (orig. 1707). Transl. J. Raphson. London.

Works quoted

Abdeljaouad, M. 2002. "Le manuscrit mathématique de Jerba: Une pratique des symboles algébriques maghrébins en pleine maturité." In: *Quaderni de Ricerca in Didattica* 11, 110–173.

Bachmakova, I. G. 1966. "Diophante et Fermat." In: *Revue d'Histoire des Sciences et de Leurs Applications* 19, 289–306.

Bashmakova, I. G. 1981. "Arithmetic of Algebraic Curves From Diophantus to Poincaré." In: *Historia Mathematica* 8, 393–416.

Bašmakova, I. G. 1974. *Diophant und diophantische Gleichungen.* Basel.

Bernard, A. & J. Christianidis 2012. "A New Analytical Framework for the Understanding of Diophantus's Arithmetica I–III." In: *Archive for History of Exact Sciences* 66, 1–69.

Byrne, J. S. 2006. "A Humanist History of Mathematics. Regiomontanus's Padua Oration in Context." In: *Journal of the History of Ideas* 67, 41–61.

Christianidis, J. 2007. "The Way of Diophantus: Some Clarifications on Diophantus' Method of Solution." In: *Historia Mathematica* 34, 289–305.

Christianidis, J. 2015. "The Meaning of Hypostasis in Diophantus' Arithmetica." In: T. Arabatzis *et al.* (eds.), *Relocating the History of Science: Essays in Honor of Kostas Gavroglu*. New York, 315–327.

Christianidis, J. & J. Oaks 2013. "Practicing Algebra in Late Antiquity: The Problem-Solving of Diophantus of Alexandria." In: *Historia Mathematica* 40, 127–163.

Christianidis, J. & I. Skoura 2013. "Solving Problems by Algebra in Late Antiquity: New Evidence From an Unpublished Fragment of Theon's Commentary on the *Almagest*." In: *SCIAMVS* 14, 41–57.

Cifoletti, G. 1996. "The Creation of the History of Algebra in the Sixteenth Century." In: C. Goldstein *et al.* (eds.), *L'Europe Mathématique – Mathematical Europe*. Paris, 121–142.

Corry, L. 2013. "Geometry and Arithmetic in the Medieval Traditions of Euclid's Elements: A View From Book II." In: *Archive for History of Exact Sciences* 67, 637–705.

Corry, L. 2004. *Modern Algebra and the Rise of Mathematical Structures* (orig. 1996). 2nd ed. Basel.

Fauvel, J. & J. Gray 1987. *The History of Mathematics: An Open University Course Reader*. London.

Hankel, H. 1965. *Zur Geschichte der Mathematik in Alterthum und Mittelalter* (orig. Leipzig 1874). Hildesheim.

Heath, T. L. 1921. *A History of Greek Mathematics*. 2 vols. Oxford.

Heath, T. L. (ed.) 1964. *Diophantus of Alexandria: a Study in the History of Greek Algebra* (orig. Cambridge 1910). New York.

Høyrup, J. 1998. "A New Art in Ancient Clothes. Itineraries Chosen between Scholasticism and Baroque in Order to Make Algebra Appear Legitimate, and Their Impact on the Substance of the Discipline." In: *Physis* 35, 11–50.

Jaouiche, K. 1988. Review of Diophante 1984 (Rashed 1984). In: *Annals of Science* 44, 308–311.

Jaouiche, K. 1990. Review of Rashed 1984 (Rashed 1994). In: *Isis* 81, 337–338.

Klein, J. 1981. "The World of Physics and the 'Natural' World." In: *The St. John's Review* 33, 22–34.

Klein, J. 1992. *Greek Mathematical Thought and the Origin of Algebra* (orig. Berlin 1934–1936). New York.

Nesselmann, G. H. F. 1842. *Versuch einer kritischen Geschichte der Algebra. Nach den Quellen bearbeitet. Erster Theil: Die Algebra der Griechen*. Berlin.

Oaks, J. 2007. "Medieval Arabic Algebra as an Artificial Language." In: *Journal of Indian Philosophy* 35, 543–575.

Oaks, J. A. 2009. "Polynomials and Equations in Arabic algebra." In: *Archive for History of Exact Sciences* 63, 169–203.

Oaks, J. 2010a. "Equations and Equating in Arabic mathematics." In: *Archives Internationales d'Histoire des Sciences* 60, 265–298.

Oaks, J. 2010b. "Polynomials and Equations in Medieval Italian Algebra." In: *Bollettino di Storia delle Scienze Matematiche* 30, 23–60.

Oaks, J. A. 2011a. "Geometry and Proof in Abū Kāmil's Algebra." In: Association Tunisienne des Sciences Mathématiques Arabes (ed.), *Actes du Dixième Colloque Maghrébin sur l'Histoire des Mathématiques Arabes*. Tunis, 234–256.

Oaks, J. A. 2011b. Review of Rashed 2007. In: *BSHM Bulletin: Journal of the British Society of the History of Mathematics* 26, 68–70.

Oaks, J. 2012. "Algebraic Symbolism in Medieval Arabic algebra." In: *Philosophica* 87, 27–83.

Oaks, J. A. 2014. Review of Rashed 2012. In: *Aestimatio* 11, 24–49.

Oaks, J. A. 2017. "Irrational 'Coefficients' in Renaissance Algebra." In: *Science in Context* 30, 141–172.

Oaks, J. A. forthcoming. "Algebra According to Early Arabic Authors." In: Chemla, K. & T. Miao (eds.), *A Worldwide Approach to the Early History of Algebra*. Heidelberg.

Oaks, J. A. & H. M. Alkhateeb 2005. "Māl, Enunciations, and the Prehistory of Arabic Algebra." In: *Historia Mathematica* 32, 400–425.

Oaks, J. A. & H. M. Alkhateeb 2007. "Simplifying Equations in Arabic Algebra." In: *Historia Mathematica* 34, 45–61.

Rashed, R. 1974. "Les Travaux Perdus de Diophante (I)." In: *Revue d'Histoire des Sciences et de Leurs Applications* 27, 97–122.

Rashed, R. 1975. "Les Travaux Perdus de Diophante (II)." In: *Revue d'Histoire des Sciences et de Leurs Applications* 28, 3–30.

Rashed, R. 1994. *The Development of Arabic Mathematics: Between Arithmetic and Algebra* (orig. Paris 1984). Transl. A. F. W. Armstrong. Dordrecht.

Rashed, R. & C. Houzel 2013. *Les 'Arithmétiques' de Diophante: Lecture Historique et Mathématique*. Berlin.

Sesiano, J. 1977. "Les Méthodes d'Analyse Indéterminée Chez Abū Kāmil." In: *Centaurus* 21, 89–105.

Sesiano, J. (ed.) 1982. *Books IV to VII of Diophantus' Arithmetica in the Arabic Tradition Attributed to Qusṭā ibn Lūcā*. New York.

Sialaros, M. & J. Christianidis 2016. "Situating the Debate on 'Geometrical Algebra' Within the Framework of Premodern Algebra." In: *Science in Context* 29, 129–150.

Skoura, I. & J. Christianidis 2014. "As Diophantus Says': Theon of Alexandria, an Erudite Reader of Diophantus." In: *Neusis: Journal for the History and Philosophy of Science and Technology* 22, 5–52. (In Greek)

Steinschneider, M. 1897. *Die arabischen Uebersetzungen aus dem Griechischen*. Leipzig.

Suter, H. 1900. *Die Mathematiker und Astronomen der Araber und ihre Werke*. Leipzig.

Van Egmont, W. 1996. "Types and Traditions of Mathematical Problems: A Challenge for Historians of Mathematics." In: M. Folkerts (ed.), *Mathematische Probleme im Mittelalter*. Wiesbaden, 379–428.

Courtney Roby

Geometer, in a Landscape: Embodied Mathematics in Hero's *Dioptra*

Abstract: In this paper, I will use recent work on embodied cognition to analyze the techniques Hero uses to embody the imposition of mathematical constructions on the physical world in his *Dioptra*, a text on surveying. Hero shifts geometrical objects, and techniques for manipulating them, into the real world by invoking material attributes that purely geometrical objects do not possess. This shift, in turn, demands an innovative restructuring of conceptual categories, triggered by processes that strongly recall Kuhn's work on conceptual incommensurability. Hero rarely refers directly to surveying personnel; rather the surveyor appears as a focalizer of the surveying action: orienting the measurement of the distance from his own side of a river to the opposite bank; precisely sighting the top of a rod to gauge the pitch of a planned tunnel; and so on. The surveyor may be equipped with hands as well as eyes, to discover a screw's pitch haptically, adjust a water-level, or delicately trace the path of a tunnel without disturbing the weighted cords hung along it. Such embodied actors represent a striking alternative to the disembodied 'Helping Hand' described by Fowler and Taisbak as characteristic of mathematics in the Euclidean tradition and shifts into this material domain are often linguistically marked. At the same time, Hero also emphasizes the geometrical techniques his actors employ, maintaining a continuous connection with the mathematical tradition in which he inscribes himself.

Keywords: Incommensurability, Land-surveying, Embodied Cognition, Distributed Cognition.

I Introduction

It is not always easy to find a home for 'applied mathematics' in the Greek tradition. The notional division of Greek mathematics into one tradition of 'pure' geometry and one of 'applied' algorithmic numerical methods is certainly reductive and artificial.[1]

1 Barriers between 'pure' and 'applied' mathematics in antiquity have been drawn in many ways and for many purposes. Here I have in mind in particular allegations like those of Kline, that in mathematical practices outside those of classical Greece, "the arithmetic and algebraic processes and the geometrical rules were the end result of physical evidence, trial and error, and insight", in contradistinction to the ordered rationality of classical Greek geometry (Kline 1972, 14, discussed with counterexamples in Høyrup 2012). Here the difference between 'classical' geometrical methodologies on

https://doi.org/10.1515/9783110565959-004

Among other drawbacks, this approach does not account well for applications of geometrical methods to solving practical problems, particularly in cases where material instruments and physical activities are involved. Shifting geometrical objects, and techniques for manipulating them, into the real world demands an innovative restructuring of conceptual categories, triggered by processes that strongly recall Kuhn's work on conceptual incommensurability.

Kuhn traces his theory of incommensurability back to his work with Feyerabend on how old scientific theories are supplanted by new ones when the language of one theory becomes unintelligible to the other.[2] The most intractable kind of unintelligibility emerges from shifts in meanings, and in particular in contingently linked systems of 'interrelated or interdefined' signifiers.[3] While Kuhn's theory of conceptual incommensurability has since undergone considerable refinement—at his own hands and those of others—, Kuhn maintains the central idea that the transition from one paradigm to another involves work of interpretation, not just translation, to create a discourse suitable for shifting the referents of the old concepts into the new.[4] Andersen, Barker, and Chen observe that although Kuhn's theory of concepts fell into disuse by scholars who focused on other aspects of his work, his ideas nevertheless resonate with those of cognitive psychologists, like Rosch, who focused

the one hand and algebraic, arithmetical, and quasi-geometrical 'rules' methodologies on the other is mapped uncritically onto a distinction between a pure mathematics capable of producing proofs and a devalued set of less rigorous quasi-mathematical practices. In her prologue to a volume containing a wealth of countervailing analyses of algorithmic and other varieties of proof in Greek mathematical practice as well as that of other cultures, Chemla (2012, 1) thoroughly discusses and dismantles this 'disturbingly simple' identification of mathematical proof with a method of geometrical proof assigned to a moment in classical Greek culture. As Chemla points out, the simplistic assertion of the 'incontrovertibility' of a certain style of mathematical proof as the only truly rigorous form of mathematical activity (and the concomitant exclusion of mathematical activities that do not exhibit this style from the 'standard early history of mathematics') continues to shape current mathematical discourse on the prerequisites and capabilities of proofs (ibid., 12–19). Høyrup's distinction between 'scientific' and 'subscientific' traditions in 'mathematical cultures inspiring or inspired by ancient Greek mathematics' offers a more socially and historically nuanced model that values both approaches. 'Scientific knowledge' is in this model defined as "knowledge that is pursued *systematically* and *for its own sake*", whereas 'subscientific knowledge' is "*specialists' knowledge*... acquired and transmitted *in view of its applicability*" (Høyrup 1994, 25; emphasis his). Here there is no devaluation of either side, but rather a distinction between the overarching aims, rhetoric, and sociohistorical roles of differing and possibly parallel traditions. Indeed, Høyrup argues that 'subscientific' knowledge "may well be much more refined than 'scientific' knowledge" (ibid.).

2 Kuhn 1982a, 669.

3 Ibid., 676.

4 Kuhn's theory of incommensurability was immediately put into dialogue with competing ideas in the same issue of *PSA*: see Kitcher 1982; Hesse 1982; Kuhn 1982b.

on the construction of categories that structure the 'phenomenal' world.[5] That cognitive-psychological work developed in its early stages roughly contemporaneously with Kuhn's work on conceptual theories, and would eventually lead to the development of prototype and framing theory.[6]

The objects of our 'real world' experience are not fully commensurate in attributes with the objects of 'pure' Greek geometry. To be sure, they have some common properties: they can have extension in length, breadth, and width; they can be represented in diagrammatic form; and they may conform at least approximately to certain definitions and manipulations adopted from geometrical texts. At the same time, they also have attributes with no analogues in the schema of the purely geometrical object: they can be made of concrete materials; they are seen and defined from an eye-level perspective; and they are often reified on the ground by a team of embodied humans. The conceptual transformations that allow adapted versions of geometrical objects to operate in the world highlight the incommensurabilities between abstract and material objects.

Andersen, Barker, and Chen argue that piecewise shifts like this can indeed be construed as constituting a Kuhnian conceptual revolution: conceptual revolutions do not need to destabilize the whole conceptual structure of Greek mathematics, but only replace the incommensurable attributes within their domain of operation.[7] At the same time, Kuhn insists that the newly reinterpreted system must maintain internally consistent conceptual taxonomies; rather than swapping out just one object or attribute for one by another name, whole sets of objects and the conceptual schemas into which they fit must be altered (Kuhn 1982a, 683). A conceptual revolution might thus be highly visible locally without toppling the reformed system on a global level.

The array of 'mathlike' objects that can be realized in the phenomenal world through transformative interpretations of their geometrical antecedents is, of course, vast. Here, I focus on Hero of Alexandria's application of geometrical techniques to surveying problems in his *Dioptra*. In this text, Hero shifts geometrical objects to a new earthbound domain, activating attributes 'incommensurable' with a purely geometrical schema. I will examine in particular the linguistic alterations to the familiar attributes of geometrical prose that signal those techniques' being put to work by human hands in a physical environment. The resulting changes are localized but linguistically quite visible, a re-interpretive quake that leaves much terminology in place but shifts the linguistic matrix in which it is deployed in sometimes dramatic ways.

---

5 I intend this 'phenomenal' world along the lines sketched by Andersen, Barker, and Chen (2006, 170): "to emphasize that there is not one unique subdivision of the world into entities, but that the world may be perceived as consisting of different entities dependent upon which features are considered important, we shall call the perceptually and conceptually divided world a *phenomenal world*".

6 Ibid., 5.

7 Ibid., 164–165.

The conventions of Greek mathematical prose require no lengthy review in this context, but one feature is worth particular notice here: namely, what Fowler and Taisbak refer to as the 'Helping Hand':[8]

> a well-known factotum in Greek geometry who takes care that lines are drawn, circles are described, points are taken, perpendiculars are dropped, etc. The perfect imperative passive is its verbal mask: "Let a circle have been described (γεγράφθω) with center A and radius AB."

As Fowler and Taisbak observe, the 'Helping Hand' is simply the agent of operations previously validated by postulates and propositions, but the peculiar sense of detachment with which it carries out those operations makes it linguistically remarkable. Asper too observes the striking impersonality of the linguistic structures characteristic of Greek geometry, where the only personality that usually intrudes is the author's λέγω ὅτι.[9] The use of third-person imperatives (usually in the passive or middle voice) is indeed largely true of Greek geometrical prose, and it is quite peculiarly true of Greek geometrical prose.[10]

Fowler and Taisbak make a jesting reference to the Helping Hand's ability to defend the geometrical construction against 'contamination', and a more serious suggestion that it represents a stance, developed along Platonic lines, that "the stage on which mathematics is played out is the Realm of Intelligence."[11] Asper situates the Platonic development of this line of thinking within a longer lineage including Speusippus and Proclus.[12] From this viewpoint, Asper observes, the impersonal imperatives so often used for these operations strongly encourage the view of the geometrical tradition as geared toward contemplation rather than operation.

At the same time, Greek mathematical texts must not be treated as though they conformed universally to these linguistic norms. In mathematical texts of the

8 Fowler and Taisbak 1999, 362.

9 Asper 2007, 125–127. Netz (2009, 98) calls the 'λέγω ὅτι' an "invitation to a demonstration", as it is a preliminary guarantee that the desired result will be produced. The same authorial voice may of course adopt other locutions, like the 'ὁμοίως δὴ δείξομεν' particularly featured by Apollonius, noted at Acerbi 2010, 38.

10 These assertions are based on statistics from the TLG. Their use in mechanical texts often parallels their use in mathematical texts, including the use of letter labels and geometrical objects, so I consider this to be essentially the same textual phenomenon. Likewise, Asper (2007, 127) observes that the distinctive directive forms of geometry are used as well by authors on harmonics like Aristoxenus, who is intent on downplaying the role of individual perceptual experience in determining the quantitative relationships between harmonic intervals. Creese (2010, 68–72) correlates the use or avoidance of those forms to the relationships between diagrams and instruments, and to differing degrees of rigorous proof. They are also sometimes used, alongside quite different types of objects, in philosophical and medical texts, and they catch on strongly in Byzantine theological works.

11 Fowler and Taisbak 1999, 362.

12 Proclus, *In Eucl.* 77.17-78.8.

algorithmic problem-solving style Asper labels the 'Rezeptstil', first- and second-person verbs are the rule rather than the exception.[13] Even within the sphere of geometrical texts, the impersonal style of the 'Helping Hand' is not a universal rule. Indeed, Netz warns that such totalizing characterizations yield a caricature of Greek mathematics as a rigidly formulaic zone, where no authorial 'modulation of style' is tolerated.[14] The real picture is, of course, more complicated; the 'Helping Hand's' grip on Greek geometry is far from total. Netz, for example, notes Archimedes' *Sphere and Cylinder* I.3, where 'we bisect' an angle. The use of the active voice is marked here, and, as Netz notes, there must be a reason for it: the active voice here "signifies the real action of bisecting," while the passive voice in the next proposition "signifies the virtual action of contemplating the possibility of an action."[15] First-person verbs of construction like this are scattered, though sparsely, throughout Archimedes' corpus. The more numerous first-person instances of verbs of discourse represent, like λέγω ὅτι, a different kind of activity: there the first person is part of the author's work *qua* constructor of text, not of mathematical objects.

Texts that invite geometrical objects into the real world push back even more forcefully against a monolithic view of Greek mathematics. The surveying problems of Hero's *Dioptra* are for the most part presented in the 'impersonal' language typical of Greek geometrical texts. However, under select circumstances, Hero departs from those conventions, using personal verbs and referencing individual actors in the landscape. These actors have hands, which they use to perform the concrete manipulations the 'Helping Hand' elides, and eyes, which give them a particular vantage point on their surveying work rather than the geometer's bird's-eye view. They operate with instruments that require manual adjustments based on sensory and proprioceptive feedback. That information, in turn, depends sensitively on the characteristics of an environment that is peculiar to each problem, and cannot be specified within the text. So, the surveyor, with his individual ergonomic situation in a particular topography, equipped with his instruments, is enmeshed within a problem-solving complex.

The work done within this complex is *embodied*: physical interaction with objects in the world is required, and those interactions will be shaped by all the peculiar factors mentioned above.[16] Embodied problem-solving within such a complex is, as Clark observes, a fundamentally different kind of work than abstract, purely propositional problem-solving.[17] Within the text of the *Dioptra* the reader, of course, finds only generalizations of the experience of embodied problem-solving. However, they

13 Asper 2007, 198–210. For the narrative of 'algorithmic' mathematics as a second-rate practice and some welcome attempts to change it, see, for example, Høyrup 1990; Høyrup 1997; Høyrup and Damerow 2001; Mau and Müller 2009.

14 Netz 2009, 102–103.

15 Netz 2004, 52.

16 Varela, Thompson, and Rosch 1991, 173; Langacker 2008, 524; Shapiro 2011, 52–56.

17 Clark 1997, 5–6.

are expressed in such a way that they trigger something like what Langacker calls a 'simulation', that is an imagined perception of at least a subset of attributes of an embodied activity that can produce 'epistemic immediacy' even in the absence of the relevant sensory input.[18]

It may be useful to begin with an example. *Dioptra* 21 provides a concise glimpse of the many varieties of embodied surveying work Hero addresses in this text:[19]

> Διὰ διόπτρας ἀπολαβεῖν ἀπὸ ἡμῶν διάστημα ἐπὶ τῆς δοθείσης εὐθείας, ἴσον τῷ δοθέντι διαστήματι. ἔστω ἡ δοθεῖσα εὐθεῖα ἐφ' ἧς δεῖ ἀπολαβεῖν <ἡ ΑΒ· τὸ δὲ δοθὲν διάστημα ὃ δεῖ ἀπολαβεῖν> ἔστω τὸ ΑΒ· ἀφ' οὗ δὲ δεῖ σημείου ἀπολαβεῖν, ἔστω τοῦ Α. ἐλθὼν ἐπί τινος ἀκλινοῦς ἐπιπέδου τόπου οἷον τοῦ ΓΔ, τίθημι τὴν διόπτραν τὴν ΕΖ· καὶ ταύτης ἔμπροσθεν κανόνα ὀρθόν, μήκους ὡς πηχῶν ι, τὸν ΗΘ, ἀπέχοντα ἀπὸ τῆς διόπτρας, τουτέστιν ἀπὸ τοῦ Ε σημείου, ὃ βούλομαι διάστημα, ἔστω δὴ πηχῶν γ. ἀπέλαβον οὖν ἀπὸ τοῦ Ε ἐν ἐπιπέδῳ εὐθεῖαν τὴν ΕΔ πηχῶν ὅσων ἐὰν βούλωμαι, ἔστω δὴ πηχῶν φ, καὶ καταλείψας σημεῖον πρὸς τῷ Δ, ἐγκλίνω τὸν ἐν τῇ διόπτρᾳ κανόνα, ἄχρις ἂν φανῇ δι' αὐτοῦ τὸ Δ σημεῖον. καὶ μένοντος αὐτοῦ ἀκινήτου, ἀντιπεριστὰς ἔλαβον δι' αὐτοῦ σημεῖον ἐπὶ τοῦ ΗΘ κανόνος τὸ Μ, καὶ ἐπέγραψα πηχῶν φ.
>
> Using the dioptra, to lay out a distance from us along a given line, equal to a given distance. Let there be given a line on which it is necessary to lay out <the line AB; and as for the given distance which it is necessary to lay out> let it be [the distance] AB; as for the point from which it is necessary to lay it out, let it be from A. Coming to some level plane on the ground, e.g. ΓΔ, I take the dioptra EZ, and in front of this an upright rod, in length say ten cubits, HΘ, extending from the dioptra, that is from the point E, at the distance I want—let it be three cubits. Then I took the line EΔ from the point E to the line on the plane, of however many cubits I want—let it be 500 cubits. Having left a marking-point at Δ, I tilt the bar [alidade] on the dioptra, until the point Δ appears through it. And with this remaining stationary, I took a corresponding point M across it on the rod HΘ, and I wrote on it '500 cubits'. (21.1-16)

The enunciation of this passage hybridizes the 'problem'-style infinitive enunciation (['the problem is] to...') with a first-person actor possessing a definite perspective – the problem is not to lay out a line 'from a given point,' or 'from any point,' but a distance 'from us'.[20] Hero proceeds to an ekthesis that would not look out of place in a purely mathematical text, as it is framed exclusively in third-person imperatives: while they are enveloped in the ekthesis the objects of the proof are not signaled as incommensurable with the framework of 'normal' geometry. Once outside that envelope, however, the domain-shift once again becomes prominently marked.

When the surveyor's work on the ground begins, the mathematical impersonality of the text really subsides in favor of active verbs. It is 'I' who take up the dioptra and the sighting-rods and 'I' who angle the dioptra to align my individual sightline with the target. 'I' make my marks upon the ground and on the sighting-rod,

---

18 Langacker 2009, 199.

19 The text of the *Dioptra* is from Schöne 1903. Translations are my own.

20 For the infinitive as typical of the enunciation of the 'problem', see Acerbi 2010, 89.

changing the notional 'whatever distance I want' to a particular length. Sighting from my specific ground-level perspective, adjusting the equipment to line up my sight, marking down defined distances: as we shall see, all of these and more are tasks that invoke the surveyor as a physical actor in a physical landscape, blending manual work with geometrical constructions to create a hybrid of mathematical and material. Hero chooses linguistic structures that mirror this hybridization of 'hands-on' and 'hands-off' work; in what follows, I will examine the nuanced pattern of these shifts alongside recent work in cognitive linguistics that may help to explain them.

II Manual Engagement

The surveyor brings a panoply of equipment into the landscape that will be his canvas. Available tools ranged from the complex dioptra (a precision instrument that seems to have used stepped-down gear ratios to enable very fine adjustments to both horizontal and vertical viewing angle) to the simple groma (essentially a vertical pole topped by a cross-bar bracket and a cross, from the vertices of which weighted strings depend).[21] The dioptra is, unsurprisingly, at the center of Hero's eponymous treatise; often its presence is only implicit, but when the task involves manipulating the instrument in the field it makes a sudden appearance, accompanied by its operator.

This occurs, for example, in *Dioptra* 6, where the problem is to determine the height difference between two points separated by an arbitrary distance, with the ultimate aim of constructing a water-course between them. The problem is posed using an infinitive in the enunciation: "Given two points at an arbitrary distance apart, to sight which of them is higher or lower, and by how much..." (6.1-3). Hero continues in the same characteristically geometrical style, with an ekthesis in third-person imperatives: "let the given places, that is the points, be A and B" and so forth (6.7-11). But things change when the equipment appears; for this intrusion of the material into the construction, evidently the 'Helping Hand' does not suffice. Suddenly an authorial 'I' appears to do the job:

> ἕνα οὖν τῶν εἰρημένων κανόνων ἵστημι πρὸς τῷ Α, καὶ ἔστω ὁ ΑΓ. εἶτα ἀποστήσας τὴν διόπτραν ἀπὸ τοῦ Α τοσοῦτον, ἐφ' ὅσον δυνάμεθα ὁρᾶν τὸν ΑΓ κανόνα, ἐπὶ τὰ αὐτὰ μέρη τῷ Β, ἐπιστρέφω τὸν ἐπ' ἄκρῳ τῷ στυλίσκῳ, ἐν ᾧ ἐστὶ τὰ ὑάλινα κυλίνδρια, ἄχρις ἂν ἐπ' εὐθείας γένηται ὁ πλάγιος κανὼν τῷ ΑΓ. εἶτα ἐπιστρέψας τὰ κοχλίδια ἐν τῷ κανόνι ἀνάγω τὰς λεπίδας, ἄχρις ἂν αἱ ἐν αὐταῖς ἀνατομαὶ γένωνται κατὰ τὰς ἐν τοῖς ὑαλίνοις γραμμάς, ἃς ποιεῖ ἡ τοῦ ὕδατος ἐν αὐτοῖς ἐπιφάνεια.

21 On these instruments, especially the dioptra, see Drachmann 1969; Argoud and Guillaumin 2000; Lewis 2001; Coulton 2002; Lewis 2012.

> Then I stand one of the aforementioned rods at A, and let this be ΑΓ. So, having erected the dioptra at such a distance from A as we can see the rod ΑΓ, directed at B, I turn the knob on the top of the stand, where the glass cylinders are, up to where the sideways bar appears in a straight line with ΑΓ. Then, having turned the screws on the bar, I bring up the plates until their slits appear on the lines on the glass [tubes] that the surface of the water in them makes. (6.12-21)

The operations described in this passage involve some quite complex and delicate manual operations predicated on careful visual observation, alongside simpler physical activities like setting up the sighting-rod ΑΓ. First, the sighting line is given a physical length, not merely by placing a letter label on a point, but by placing a rod on the earth. The length is itself ergonomically constrained by both physiology and topography: the dioptra must be placed at such a distance that the sighting-rod remains visible. The surveyor's eyes are then engaged in another way, adjusting the height of the dioptra's stand, and then aligning the height of the sighting-targets with the water-level on the dioptra.

Hero could have expressed all these relationships as a purely geometrical construction, representing them using parallel lines and points of intersection. Indeed, this would have been a natural approach, given that the passage began with impersonal verb forms and immaterial objects that would have been quite at home in a geometrical text. Up to the point where the instruments are mentioned, these were only points separated by an unspecified amount of distance, even though one of them is notionally located near a water-source (which will become more important later in the problem). But Hero does not choose this approach; instead, he introduces a first-person figure into the text, quite different from the formulaic 'λέγω ὅτι' of the author who appears in mathematical texts to stake an epistemological claim. The first-person agent here has a particular perspective, a line of sight, and engages manually with the device, making fine adjustments to its height to align everything along that line of sight. The process as a whole is underdetermined in the text; the point of correct alignment is to be determined on the basis of the sensory and proprioceptive feedback that guides the work of embodied agents.

Other passages of the *Dioptra* exhibit comparable linguistic shifts when a process of manual manipulation or embodied sighting is invoked. *Dioptra* 20, which describes the construction of a maintenance passage for a tunnel, involves a series of transitions between the 'hands-off' and 'hands-on' domains. A shadowy first-person actor appears almost immediately; the problem is to find a place "from which, when a passage is created at the given tunnel, we will reach that place" so as to clear out rubble and carry out repairs. However, after this introduction, Hero reverts to the 'impersonal' diction of geometry for the ekthesis: "Let there be given the tunnel ΑΒΓΔΕ, and passages leading to it ΗΘ and ΚΛ," and so on. Why should this be? Material *realia* take a back seat to spatial relationships in this passage, and moreover the tunnel's features remain still: the tunnel predates the construction of the maintenance passage, so nothing in it is being moved or manipulated.

In this sense, the tunnel is closer to the eternally extant objects of a mathematical construction.

After the tunnel has been laid out, manipulation of material objects once again becomes the focus, and the first-person figure re-enters the text. The first task is to measure out some distances using a cord.

> καὶ λαβὼν σχοινίον εὖ ἐκτεταμένον καὶ προβεβασανισμένον, ὥστε μηκέτι ἐπεκτείνεσθαι ἢ συστέλλεσθαι, τὴν μὲν ἀρχὴν αὐτοῦ τίθημι πρὸς τῷ Σ. λαβὼν δέ τι σημεῖον ἐπὶ τοῦ ΑΒΓ τοίχου τὸ Τ, ἐπεκτείνω τί σχοινίον ἐπὶ τὸ Τ, καὶ ὁμοίως ἐπὶ τὸ Π, καὶ σημειωσάμενος τὰ μήκη τῶν ΤΣ, ΤΠ ἐφαρμόζω αὐτὰ ἐν τῷ ἐπάνω ἐδάφει, ὥστε γενέσθαι τρίγωνον τὸ ΡΥΟ, τὴν μὲν ΡΥ ἴσην ἔχον τῇ ΤΣ, τὴν δὲ ΥΟ τῇ ΤΠ.
>
> And taking a cord that has been well stretched and tested in advance, so that it will no longer stretch or contract, I place its beginning at Σ. Then taking some point on the wall ΑΒΓ, say Τ, I stretch a cord to Τ, and similarly to Π, and when I have marked out the lengths ΤΣ and ΤΠ I join them on the ground above, so that the triangle ΡΥΟ is created, having (the side) ΡΥ equal to (the side) ΤΣ, and ΥΟ to ΤΠ. (20.16-24)

The process is then repeated, continually measuring out triangles with the cord until point M is reached. The cords in question can be represented in text by sets of two letter labels, just like a line segment in a geometrical proof. However, they remain cords (σχοινία), not lines, and they have the properties of cords, such as elasticity that has to be eliminated in advance.

But the presence of cords with peculiar material properties is not the only factor here that differentiates the action here from a geometrical construction. There is also the first-person figure who takes up the cord, pins down one of its ends at Σ and stretches the other to T, and transforms the line into marks on the tunnel walls. The whole process of manipulating the cord is focalized through him; he moves from one vertex to another, creating a kind of materially reified diagram as he goes. Once this process is over and the triangles have been constructed, Hero reverts back to 'impersonal' third-person imperatives. The bridge between these sections is one final first-person construction, an apologetic "so that we don't go into too elaborate detail (ἵνα μὴ ποικιλογραφῶμεν)." This figure is the first person of the λέγω ὅτι, the author who intrudes to shape the text rather than to shape the earth, in this case moving from the material particularities of constructing the maintenance passage, as seen through the eyes of someone inside the tunnel, to the geometrical generalities of the solution.

Hero thus shifts back and forth within the problem's text between 'impersonal' and 'personal' modes, depending on whether his focus is on objects that lend themselves well, *for the role they play in the problem*, to description in terms of purely geometrical attributes, or those whose role relies on attributes unavailable from the geometrical schema. This *ad hoc* domain shifting should not suggest that Hero's approach is haphazard; rather, it reflects the kind of flexible, iterative category construction Rosch observed in her experiments. Rosch found that (contrary to her initial expectations) her experiments on taxonomies of "basic level objects" like animals and furniture indicated that as subjects develop and refine conceptual categories, "both the constraint of real-world factors and the construction and reconstruction of attributes are continually present."[22] Hero's textual rendition of the surveyor's problem-solving process likewise presents the reader with comparable but incommensurable sets of objects, flexibly deployed by drawing on each set for the parts of the problem that match up best with their available attributes.

Near the end of the *Dioptra*, Hero provides instructions for building an odometer, consisting of a box full of interconnected gears, which are themselves connected to a screw that advances them with each turn of the wagon-axle to which it is in turn attached. Hero introduces the initial components of the odometer through the usual third-person imperatives: "let there be a frame (γεγονέτω οὖν πῆγμα)" and so forth, as he instructs the reader in sizing gears so that the odometer will be able to measure a reasonable day's journey. His approach changes, however, when he introduces the screw. Though it is numbered among the simple machines, the screw is far from simple to craft precisely. Hero acknowledges this difficulty explicitly, and suggests manual testing as a solution. As the work of building the odometer shifts into the hands-on domain of craft, Hero's language shifts into the first person:

> ἀλλ' ἐπεὶ ἡ ἑκάστου κοχλίου στροφὴ οὐκ ἀκριβῶς οὐδὲ μεμετρημένως τοὺς παρακειμένους ὀδόντας στρέφει, ἡμεῖς τῇ πείρᾳ ἐπιστρέφομεν τὸν πρῶτον κοχλίαν, ἕως οὗ τὸ παρακείμενον αὐτῷ ὀδοντωτὸν τύμπανον μίαν ἀποκατάστασιν λάβῃ, μετροῦντες ὁσάκις αὐτὸς ἐπιστρέφεται.
>
> But since the turn of every screw does not turn the engaging teeth exactly or precisely, we turn the first screw as a test, until the disk with teeth engaged with it has made one turn, while counting how many times the screw is turned. (34.68-73)

Likewise, he advises later that the gears in the odometer might not mesh correctly at a first attempt: another problem whose solution cannot be precisely specified in geometrical terms. Again, a first-person figure is invoked to determine their proper placement manually: "we must move them further apart so that they do not get in the way" (34.109-114).

Vignettes of manual manipulation, of active placement and displacement of surveying tools like the dioptra and sighting rods, thus seem to provoke Hero to choose different grammatical forms than the more purely mathematical manipulations that characterize the hands-off components of the surveyor's work. Why does he contravene the conventions of geometrical prose in precisely this way? The shift might be explained by cognitive concepts now often collected under the rubrics of 'embodied cognition' and 'distributed cognition.' The first of these originally gained prominence largely through the work of Varela, Thompson, and Rosch. They defined cognition

22 Rosch 1978, 42.

as 'embodied action', where 'embodied' indicates that cognition depends crucially on sensorimotor experience and feedback, and 'action' that "sensory and motor processes, perception and action, are fundamentally inseparable in lived cognition."[23] Though some of the particulars of their model have been extended or adjusted by later work, the central idea—that cognition is influenced by the sensorimotor feedback loop in which we are constantly enmeshed—has provided a much-needed escape from dualist views of cognition. By shining a spotlight on the body's involvement in cognition, through perceptions and actions that are all bodily mediated, embodied cognition seems to get at just the kind of distinction Hero manifests grammatically. Here are problems whose solution requires collaboration between the surveyor's body and mind – the 'Helping Hand' alone will not do.

Andy Clark's work in distributed cognition is particularly relevant to the problems Hero tackles in the *Dioptra*. Clark's 'extended mind' theory gives the brain a minimally privileged role in problem-solving, particularly in cases where the body's interaction with its environment is concerned—such cases can be comparatively exotic problems of robotic control, but the same principles apply to activities as familiar as walking. Clark argues that "[i]n the natural context of body and world, the ways brains solve problems is fundamentally transformed", and that by thinking of cognition purely as abstract problem-solving we lose sight of the ways many problems are actually solved in the physical world.[24]

In the passages of the *Dioptra* where problems are solved through manual manipulation, hand-eye feedback loops, and other such methods, something subtly but crucially different is happening compared to those where the problem remains located in a purely geometrical domain, where the intervention of human hands is not required (or, perhaps, even possible). Out in the field, the surveyor is confronted by problems that can only be solved bodily, using manual manipulations that can only be approximated in text. These depend upon the interworking of sensorimotor complexes like the eyes that sight through the glass levels while the hands adjust the screw until the two coincide—precisely the sort of sensorimotor loop Varela, Thompson, and Rosch invoke, and a kind of problem-solving that requires thinking about the whole body's involvement, as Clark suggests.

III Working in the world

Many of the *Dioptra*'s problems are constrained by what a particular practitioner would actually be able to do or see in the field. Problem 21 supplies perhaps the

23 Varela, Thompson, and Rosch 1991, 173.
24 Clark 1997, 5–6.

simplest formula expressing the ground-level perspective surveying problems involve: 'Using the dioptra, to lay out a line from us directed at a given point, extending a given distance.' This formulation of the problem is a reminder that every surveying problem starts from somewhere. Maybe 'somewhere' is a place with defined topographical features, maybe not; but, in any case, the problem-solving landscape is altered from the geometrical bird's-eye view just by defining the extension of objects from a specific perspective. The diagrams of the *Dioptra* are often presented from above or as cutaway side views (for problems set on a hill), but the surveyor's work in the world is carried out on the ground, from a series of points each with its own perspective. The surveyor walks the paths that the geometer constructs from above.

Let us return to the water-course construction of *Dioptra* 6. The sighting operations do not end with the leveling adjustment seen earlier, but continue as follows:

> καὶ κατασταθέντων οὕτως τῶν λεπιδίων διὰ τῶν ἐν αὐτοῖς ἀνατομῶν διοπτεύω θεωρῶν τὸν ΑΓ κανόνα, τῆς ἀσπιδίσκης μετεωριζομένης ἢ ταπεινουμένης, ἄχρις ἂν φανῇ ἡ μέση τοῦ λευκοῦ καὶ μέλανος χρώματος γραμμή. καὶ μενούσης τῆς διόπτρας ἀκινήτου μεταβὰς ἐκ τοῦ ἑτέρου μέρους διοπτεύω διὰ τῶν ἀνατομῶν, ἀποστήσας ἀπὸ τῆς διόπτρας τὸν ἕτερον κανόνα τοσοῦτον ὥστε βλέπεσθαι.
>
> And when the plates have been placed thus, I sight through the slits in them, observing the rod ΑΓ, while the small disk is raised or lowered to the point where the line between the white and black color appears. And with the dioptra remaining stationary, I shift and sight from the other side through the slits, having stood the other rod far enough away from the dioptra so as to be able to see it. (6.21-29)

The dioptra is designed for use with sighting targets painted to facilitate telling from afar when the instrument has been accurately aligned. At this point in the problem, a purely geometrical, disembodied description of the operation would not be easily produced. While some features could be described in this way (for example, by stipulating a bisecting line on the target that could intersect with a line that corresponded to the notional line of sight across the dioptra), other parts of the construction (like standing the rod far enough away to be able to see it) demand the experientially and ergonomically based knowledge of an operator on the scene. In order to switch from forward to backward sighting, the surveyor leaves the dioptra unmoved, while moving to change his own perspective.

With grounded location and perspective come ergonomic limitations. So, Langacker argues that "[a]n observer's experience is enabled, shaped, and ineluctably constrained by its biological endowment and developmental history (the products—phylogenetic and ontogenic—of interaction with a structured environment)."[25] Certainly, the surveyor's problem-solving experience is reshaped as he iteratively interacts with the landscape to gradually impose a roughly geometrical structure upon it. In placing the sighting-rods

25 Langacker 2000, 203–204.

and their targets, he will be constrained by the distance he can see, which will be in part determined by his own height relative to the local topography. Hero cannot specify this in the text, of course; this is problem-solving that has to happen in the field.

Dioptra 9, which describes how to find the perpendicular width of a river and, hence, its narrowest crossing, offers a glimpse of spatial construction through the eyes of a specific observer. The passage begins by stating that the problem is "for those who are on one bank to find the least width of a river," and specifying the banks AB and ΓΔ.

> στήσας οὖν τὴν διόπτραν πρὸς τῇ ΓΔ ὄχθῃ, ὡς ἐπὶ τὸ Ε, ἐπέστρεψα τὸν κανόνα, ἄχρις ἂν φανῇ δι' αὐτοῦ σημεῖον ἐπὶ τῆς ΓΔ ὄχθης τὸ Δ. καὶ τῇ ΕΔ διὰ τῆς διόπτρας πρὸς ὀρθὰς ἤγαγον τὴν ΕΖ ἐπιστρέψας τὸν κανόνα. εἶτα ἐγκλίνω τὸ ἡμικύκλιον ἄχρις ἂν ἐπὶ τῆς ΑΒ ὄχθης φανῇ τι σημεῖον διὰ τοῦ κανόνος. πεφηνέτω τὸ Ζ· ἔσται δὴ τὸ ἐλάχιστον πλάτος τοῦ ποταμοῦ τὸ ΕΖ· ἡ γὰρ ΕΖ ὡσανεὶ κάθετός ἐστιν ἐπ' ἀμφοτέρας τὰς ὄχθας, εἴπερ παραλλήλους αὐτὰς ἐννοοίμεθα..

> Then, having stood the dioptra on the bank ΓΔ, say at E, I turned the bar until the point Δ on the bank ΓΔ appeared through it. And I extended [the line] EZ perpendicular to EΔ through the dioptra, having turned the bar. Then I tilt the semicircle in, until some point on [the line] AB appears through the bar; let Z have appeared. The least width of the river will be EZ, for [the line] EZ is as though perpendicular to both banks, if we imagine them to be parallel. (9.1-14)

The perspective of the operator is emphasized early in the problem: he is located on one bank, looking to the other, rather than viewing the problem from above, as the accompanying diagram does. In order to solve the problem, the operator will once again rely on his particular perspective and lines of sight. He physically manipulates the dioptra, tilting the disk and alidade toward or away from him (so highlighting the operator's defined, embodied location) and adjusting them until the stipulated points appear.

Hero's nod to the fact that the banks of the river can only be approximated as parallel is far from the *Dioptra*'s only concession to the irregularities of the landscape in which the surveyor operates. The water-course problem of *Dioptra* 6 acknowledges that the path sighted between the starting-point and the water-source will not necessarily be constructed; the solution so far has only established an idealized sketch of the water's path. Determining the most suitable actual route for the water demands close local scrutiny by a surveyor on the lookout for common practical problems: obstacles too hard to tunnel through and too high to surmount, terrain that is likely to contaminate the water with sulfur, and so on (6.120-122). Hero makes a list of these, and then instructs that "whenever we encounter such as these, we will turn away, so as not to harm the water-course (τοιούτοις ὅταν περιτύχωμεν, ἐκνεύσομεν, ὥστε κατὰ μηδὲν βλάπτεσθαι τὴν τοῦ ὕδατος ἀγωγήν)" (6.123-124). That is, when the water-course shifts domains from an idealized geometrical construction to a physical construction that accounts for terrestrial irregularities, 'we' are again at the helm to focalize the transition. The solutions to these problems and others like them are facilitated by a figure portrayed as moving through the problem's space, viewing it through a particular set of eyes rather than from an artificial top-down distance.

IV Instantiating and inscribing

Hero's use of active verb forms often centers on another, related type of engagement with the material world: manipulating the written records that were a vital element of the surveyor's work.[26] Hero shows us the part of this process that happens in the field: marking down the particular length measurements of a problem as they are made. For example, solving the water-course problem of *Dioptra* 6 requires a chain of sightings; at each point, a sighting will be taken backward to the last point and forward to the next, to determine the overall slope of the course.[27] Since the lengths of the reverse sights add up to the distance to the place where the water is meant to end up, if this number is the greater (as it is here), the water will be carried downward. Otherwise, a water-lifting device may be necessary, and Hero provides a brief list of the mechanical means available for this task.

Though the points and lines representing the sighting-rods, dioptra, positioning points etc. are first introduced in an 'impersonal' style, Hero resumes the use of personal verbs when it comes time to put numbers to the measurements, and put those numbers down on paper:

> ἐμέτρησα οὖν ἑκατέραν τῶν ΑΓ, ΔΕ καὶ ἔστω ἡ μὲν ΑΓ ηὑρημένη πηχῶν ϛ, ἡ δὲ ΔΕ πηχῶν β. ἀπεγραψάμην οὖν δύο στίχους, ἐν μὲν τῷ ἑνὶ ἐπιγράψας καταβάσεως, <ἐν δὲ τῷ ἑτέρῳ ἀναβάσεως>, ὡς ὑπογέγραπται· καὶ τοὺς μὲν ἓξ πήχεις ἐν τῷ τῆς καταβάσεω στίχῳ σημειοῦμαι, τοὺς δὲ δύο ἐν τῷ τῆς ἀναβάσεως.
>
> Then I measured each of ΑΓ and ΔΕ, and let [the line] ΑΓ have been found to be six cubits, and ΔΕ two cubits. I then wrote out a list of two columns, in the first of which I put the title of 'reverse [sights]', and in the second of 'forward sights', as written below. And I indicate the six cubits in the column of forward sights, and the two in that of reverse sights. (6.36-40)

Look closely at how Hero treats this notional six-cubit measurement. When it is the imagined distance of line segment ΑΓ, it remains in the hands-off domain of the third-person imperative. But when it comes time to mark it down on the chart, Hero shifts to the first person. Perhaps this marks a further material manipulation in the surveyor's toolbox: making a new symbol on the chart to represent a distance that was always there in the landscape.

---

26 Most surviving evidence for such inscriptions comes from the world of Roman surveying rather than the quite different system of practice Hero seems to describe. Still, some particularly useful information for comparison is at Nicolet 1991, 149–169; Moatti 1993, chapters 2–3; Chouquer 2010, 246–261; Mayer and Olesti Vila 2001.

27 On this procedure, see Lewis 2001, 89–97.

Having established the layout of the number table (two columns with headers), Hero continues the instructions for manipulating the sighting-rods and the dioptra itself. Once again, active verbs govern his account of moving the dioptra from place to place, turning the bar on top of it to catch sight of the rods, raising the sighting-disks into view, moving the rods into place, and so forth. Then he returns to the problem of writing down the measurements thus made:

> πάλιν οὖν τὸ μὲν ἀπὸ τοῦ Η διάστημα ἄχρι τοῦ ἐδάφους σημειοῦμαι εἰς τὸν τῆς καταβάσεως στίχον, τὸ δὲ ἀπὸ τοῦ Θ εἰς τὸν τῆς ἀναβάσεως· καὶ ἔστωσαν μὲν καταβάσεως πήχεις τέσσαρες, ἀναβάσεως δὲ πήχεις δύο. καὶ πάλιν μένοντος τοῦ πρὸς τῷ Θ κανόνος μετατίθημι τὴν διόπτραν καὶ τὸν ἕτερον κανόνα <καὶ> καταστήσας, ὡς προείρηται, ἐπ' εὐθείας τάς τε ἀσπιδίσκας καὶ τὰς ἀνατομὰς λαμβάνω [καὶ] ἐπὶ τῶν κανόνων σημεῖα τὰ Λ, Μ.
>
> Then once again I indicate in the column of the reverse sights the distance from H to the ground, and that from Θ in that of the forward sights. And let [the distance of] the reverse sight be four cubits, that of the forward sight be two cubits. And with the rod in turn remaining at Θ I move the dioptra and the other rod; having positioned it as described previously I take in, on a straight line, the disks and the slits and the points Λ and M of the rods. (6.49-57)

Hero then works through a sample of the procedure all the way to the surveyor's arrival at point B (6.57-74). Throughout this process, the sample distances are suggested using third-person imperatives: "let there be four cubits to the reverse sight at Ξ (πρὸς μὲν τῷ Ξ καταβάσεως ἔστωσαν πήχεις τέσσαρες)" and so on. He follows this up with a sequential list of the two sets of sample numbers, for the reverse and forward sights, concluding with their sums (6.75-85).

After the numerical example, Hero brings the reader back to the initial problem, reattaching the problem of measuring lengths to the problem of sighting the path of a watercourse. He resumes the problem and applies letter labels to the points of interest with a smattering of third-person imperatives. When he returns to the two-column inscription, this time to add up the numbers and add the final sum, active verbs appear again:

> Τῶν οὖν ἀριθμῶν σεσημειωμένων ἐν τοῖς εἰρημένοις στίχοις συντίθημι πάντας τοὺς τῆς καταβάσεως ἀριθμούς· εἰσὶ δὲ λγ· ὁμοίως καὶ τοὺς τῆς ἀναβάσεως· εἰσὶ δὲ κγ· ὥστε ὑπεροχὴ πήχεις ι.
>
> With the numbers having been indicated in the aforementioned columns, I add together all the numbers of the reverse sight; they are 33; similarly those of the forward sight; they are 23, so the difference is 10 cubits. (6.92-95)

It is useful to compare and contrast the patterns of arithmetical problem-solving here to the pattern Asper labels the 'Rezeptstil'.[28] Asper lists markers of the 'Rezeptstil' that include an eisagogic style, references to accompanying diagrams, guidance through

28 Asper 2007, 198–210.

imperative verb forms, well-marked steps of solution, and intermediate summary statements. Problems of this style are presented and solved algorithmically; the reader is meant to proceed through the steps as though following a recipe. First—or second—person forms of address bridge the textual distance between author and reader.[29]

Is the 'Rezeptstil' at work in these passages of the *Dioptra*? Yes and no. There is no doubt that the Heronian corpus contains examples of the style, though the clearest examples come from the texts least likely to be Hero's own work.[30] Asper further observes Hero's ability to merge the formal markers of geometry's general theorizing with numerically specified algorithmic problem-solving procedures.[31] But some caveats must be observed in comparing the *Dioptra*'s use of personal verbs to the 'Rezeptstil' texts. First, Hero does not usually proceed through the solution to a surveying problem using the numbers he gives; *Dioptra* 6 is something of an outlier in this regard, as is the numerical example of setting the odometer's gear ratios in *Dioptra* 34.[32] Second, Hero uses a linguistic pattern essentially different from that of the 'Rezeptstil', since the numbers for the examples are originally proposed using ἔστω, calling attention to their arbitrariness.[33] Active verbs are only deployed for the processes of adding them up (which lands squarely in the territory of the 'Rezeptstil') and explicitly writing them on a chart (which does not, and which occurs often in the *Dioptra* even absent such arithmetical maneuvers). Third, as we have seen, examples of numerical procedures are just a small subset of the passages in the *Dioptra* where active verbs are involved, most of which are firmly located in 'geometrical' rather than 'algorithmic' territory. So, while the influence of the 'Rezeptstil' may certainly have played a part in determining the linguistic patterns Hero selects here, it does not tell the whole story. All this is part and parcel of Hero's approach to pre-existing mathematical

---

29 Ibid., 198.

30 For example, the motley collections of *Stereometrica* mostly present problems structured along the lines of the 'Rezeptstil'. So, for example, *Stereometrica* 1.2 in Heiberg's edition commences 'Σφαῖρα, ἧς ἡ περίμετρος ποδῶν κβ• εὑρεῖν αὐτῆς τὸ στερεόν. ποίει οὕτως• λαβὲ ἀπὸ τῆς περιμέτρου τὴν διάμετρον ἀπὸ τοῦ ὑποκειμένου ὑποδείγματος τῶν κύκλων• καὶ ἔσται ἡ διάμετρος ποδῶν ζ. ταῦτα ἐφ' ἑαυτά• γίνονται μθ', and concludes with the solution 'τοσούτων ἔσται ποδῶν ἡ ἐπιφάνεια τῆς σφαίρας'. Note in particular the use of imperatives like ποίει and λαβέ, and how the problem guides the reader all the way through the numerical solution. On the further problems raised by Heiberg's edition of the *Stereometrica*, and the authorship issues that plague the Heronian metrological corpus more generally, see Vitrac 2010, 1–5, 42.

31 Asper 2007, 203.

32 ἐὰν ἄρα καὶ τοῦτο τὸ τύμπανον ἔχῃ ὀδόντας λ, ὅπερ εἶναι εἰκὸς καὶ πλείονας γίνεσθαι, ἅπαξ στραφέντος αὐτοῦ, στροφαὶ τοῦ τροχοῦ δηλωθήσονται ͵ζσ• ἂν [δὲ] ἄρα ὁ τροχὸς ἔχῃ τὴν περίμετρον πηχῶν ι, ἔσονται πήχεις μζ ͵β. ἔστιν στάδια ρπ. καὶ ταῦτα μὲν ἐπὶ τοῦ β' τυμπανίου εὕρηται• πλειόνων δὲ ὄντων καὶ τῶν ὀδόντων κατὰ τὸ πλῆθος αὐξομένων πολλοστὸ<ν> τῆς ὁδοῦ μέγεθος <εὑρεθ>ήσεται μετρούμενον. (*Dioptra* 34.55-64)

33 A better match in this respect than the 'Rezeptstil' might be the approach taken by Diophantus in the Greek books of the *Arithmetica*, where the emphasis is on setting up the structure of an equation exemplified by a certain number rather than escorting the reader through a particular numerical solution in detail. On these attributes of Diophantus's approach, see Christianidis 2007, 296–299.

traditions; he creates a fluid transition between different approaches, in this case 'geometrical' and 'algorithmic' styles.

Let us return to the task of inscribing measurements. In *Dioptra* 7, Hero proposes a method for sighting from a given point to another which cannot be seen, to produce a straight line between two points at an arbitrary distance from one another. Most of the action of solving this problem, which involves sighting a chain of line segments, is handled through third-person imperatives. Likewise, he proposes a list of notional sample distances using this 'impersonal' form. However, acts of inscription again call for a shift in voice, as 'we' write down the results of those measurements in the first-person plural indicative:

> ἅμα δὴ διοπτεύοντες γράψομεν ἐν χάρτῃ ἢ δέλτῳ τό τε σχῆμα τοῦ διοπτρισμοῦ, τουτέστιν τὰς κλάσεις τῶν εὐθειῶν, καὶ ἔτι τὰ μεγέθη ἑκάστης αὐτῶν ἐπιγράψομεν.
>
> At the same time as we are sighting, we will write on papyrus or tablet the form of the sightings, that is the breaks of the lines, and moreover we will inscribe the length of each of them. (7.21-24)

The list of measurements takes on additional concreteness here with Hero's stipulation that it be written on papyrus or tablet—a new material element for the surveyor to engage with.[34] So, the same pattern occurs as in the acts of inscription in *Dioptra* 6: notional distances in the world are proposed through third-person imperatives, but actually writing them down, instantiating figure on paper or tablet, is a concrete engagement with the world that requires an actor. As we saw earlier, the same pattern obtains in *Dioptra* 21.

There is a cognitive difference between imagining or describing a general type and a specific instance of something, and this might be what the linguistic shifts in these passages reflect. Langacker connects the distinction between types and instances to cognitive-grammatical 'grounding', which accounts for "the speech event, its participants (speaker and hearer), their interaction, and the immediate circumstances (notably, the time and place of speaking)."[35] For Langacker, looking at a speech event together with its 'ground' introduces an asymmetry between subject and object: if the event has to do with describing something seen from someone's particular perspective, the eyes are maximally subjective, while the thing seen is maximally objective. For the particular case of instantiation, "an instance (as opposed to a type) is thought of as having a particular location in the domain of instantiation."[36] In Hero's case, inscribing the distance of a particular sighting into the table seems to be an act of reification, turning an arbitrary 6-cubit distance into the measured result of a sighting by a surveyor from his own particular perspective.

---

34 On the use of papyrus, wooden boards, and wax tablets in mathematical reckoning and record-keeping, see Fowler 1999, 204–217.

35 Langacker 2008, 259–260.

36 Ibid., 268.

Hero also offers active instructions for creating another type of inscription, now in the visual rather than the verbal domain. Usually, diagrams are introduced as geometrical constructions, governed by the usual third-person imperatives. However, in *Dioptra* 17 Hero gives the diagram a newly concrete life, instantiated in the physical world as a template. The problem at hand is, at first, to create a harbor bounded by any given section of a circle; in order to do this the dioptra is set up with its (circular) drum horizontal, and the sighting rods are used to identify a path ΓΔΕ similar to the circular segment desired for the harbor, but at a smaller scale. The rays from the dioptra to these points are then further extended to generate the harbor's actual outline:

καὶ οὕτως κατασταθέντων προσβεβλήσθω ἀπὸ τοῦ Ζ ἀκτὶς πρὸς τὴν ΓΔ εὐθεῖαν, ἕως οὗ συμπέσῃ τῷ ἐδάφει κατὰ τὸ Θ· ἔσται δὴ τὸ Θ ἐπὶ τῆς περιγραφομένης ἐν τῷ λιμένι γραμμῆς. ὁμοίως δὲ καὶ ἕτερα λαμβάνοντες τῷ Θ περιγράψομεν τὴν ΒΘΑ γραμμήν.

And when they have been set up in this way let a ray have been extended from Z to the line ΓΔ, up to the point where it intersects the ground, Θ. Θ will be on the line circumscribed on the harbor. Taking the others in the same way as for Θ, we will circumscribe the line ΒΘΑ. (17.13-17)

This pattern is by now familiar: a geometrical construction, made using third-person imperatives, serves as a kind of lemma that helps to produce the actual desired result of a construction on the ground, which is itself described using first-person active verbs.

The particular textual innovation of *Dioptra* 17 appears in the next passage, where Hero proceeds beyond the circular outline that can be made with materials conveniently at hand using the circular drum of the dioptra. Here, he promises that the harbor can, in fact, be bounded by any conic section or indeed any other line, by constructing a template in the desired shape and using the same ray-extension method as before:

ὁμοίως δὲ ἐὰν βουλώμεθα τὴν περιγραφομένην μὴ εἶναι κύκλου περιφέρειαν, ἀλλὰ ἐλλείψεως, ἢ καὶ ὅλην ἔλλειψιν ἢ καὶ παραβολὴν ἢ ὑπερβολὴν ἢ ἄλλην τινὰ γραμμήν, ποιήσομεν ὁμοίαν αὐτῇ ἐκ σανίδος· καὶ ἐφαρμόσαντες ἐπὶ τὸ ΓΔ τύμπανον, ὥστε συμφυὲς αὐτῷ γενέσθαι, ὑπερέχειν <δὲ> εἰς τὸ ἐκτὸς τοῦ τυμπάνου τὴν ἐκ τῆς σανίδος περιτμηθεῖσαν γραμμήν, τὰ αὐτὰ ποιήσομεν τοῖς ἐπὶ τῆς ΓΔΕ περιφερείας εἰρημένοις. οὕτως οὖν πάσῃ τῇ δοθείσῃ γραμμῇ ὁμοίαν περιγράψομεν.

Similarly, if we want the circumscribed line not to be an arc of a circle, but that of an ellipse, or a whole ellipse, or a parabola or hyperbola or some other line, we will make [a line] similar to it from a plank. And attaching it to the drum ΓΔ, so that it becomes fused with it and a line is cut around the outside of the plank on top of the drum toward the outside, we will do the same things as mentioned before on the circumference ΓΔΕ. In this way we will circumscribe a line similar to the given line. (17.29-37)

Since the object similar to the harbor's desired form is not immediately available as was the dioptra's circular drum, 'we' must make it from a plank of wood.[37] 'We' then

37 Hero likewise recommends the use of a board (σανίς) as a template for using a diagram to shape components of a catapult at *Belopoeica* 26.18.

affix the plank to the dioptra and carry out the same sighting exercise as before, but this time the whole procedure is presented through active verbs rather than including third-person imperatives as in the initial circular case. It is as though the template, when first invoked as a material object crafted by a first-person figure, shifts the whole problem into the domain of hands-on, active engagement. The template changes from a geometrical construction to another piece of equipment for the surveyor to pick up and manually engage with.

V Conclusion

Hero's *Dioptra* presents the reader with a hybrid landscape, part mathematical and part material, that is itself a reminder of the varied activities encompassed and enabled by Greek mathematics. Applied geometry of the kind the *Dioptra* gives us can use previously established mathematical results to generate new linguistic, textual, and material objects, a new technical 'phenotype'. The hybrid discipline of surveying as represented in Hero's *Dioptra* has to be able to prompt a vivid imagined 'simulation' of hands-on, problem-specific surveying work that can never be fully described in the text. It also has to be able to represent problem-solving from the point of view of a person standing on the ground, not just the synoptic bird's-eye view of geometry. Moreover, this will be a particular person with a particular bodily disposition including height and so forth, located within a landscape that has its own peculiar characteristics: each problem in the field will be an 'instantiation' rather than a 'type'.

The surveyor's work in that landscape is far from solitary; though Hero does not mention them (as the Roman texts do), the surveyor's world is populated by practitioners and assistants, perhaps also by apprentice learners, together with all the instruments and written artifacts that serve as tools of their trade. It is, thus, home to something along the lines of the 'distributed cognition' described by Clark and others. Hero's text itself, as well as the inscriptions it guides the reader to construct, should be counted among the 'external representations' that are intrinsic to the cognitive processes of scientific and technical work. Giere identifies among these the instruments, charts, and diagrams which connect the different actors engaging in the work, accessible by virtue of the fact that they "are not localized in a person's brain or in a computer, but somewhere external to these locations."[38] The dioptra and the *Dioptra* are both engaged in the cognition of land-surveying as Hero describes it, and he constructs the text so as to highlight the collectivity of this

38 Giere 2007, 314. Giere is here comparing 'distributed' cognition and 'collective' cognition, associated respectively with Clark (1997) and Hutchins (1995).

technical practice, even as he effaces figures like assistants from the landscape he depicts.

Each act of embodied problem-solving happens within a human-environment complex that engages and changes the problem's terrain, often in unpredictable ways. The shift from the 'blank slate' diagrammed environment in which a geometrical problem is solved to this kind of complex is signaled in the text by a linguistic shift from an 'impersonal' to a 'personal' mode. This shift is nuanced, a localized revolution such as Andersen, Barker, and Chen describe; nevertheless Hero's strategy is quite consistent, and strikingly visible once the pattern has once been discerned. He fuses the personal with the impersonal, the material with the geometrical, a thoughtful textual response to the hybrid objects the surveyor works on. He maintains a continuous connection to the characteristic language of mathematical texts while shifting geometrical objects and techniques into the applied domain, his strategic departures into vignettes of concretely embodied work highlighting the domain shift rather than effacing it.

This strategy recalls Kuhn's emphasis on the work of re-interpretation that must occur in order for an audience used to one conceptual structure to understand a new one incommensurable with the old. He warns, "where structure is different, the world is different, language is private, and communication ceases until one party acquires the language of the other."[39] Hero's repeated shifts in the *Dioptra* between familiar geometrical objects and their novel extensions into the domain of the surveyor's embodied actions foster the work of guided reinterpretation Kuhn demands. In creating a conceptual domain that melds geometrical objects with the surveyor's interventions in the landscape, Hero builds a new schematic structure for the objects within that hybrid domain. He thus carries out a conceptual revolution: highly specialized and localized, but a radical transformation nonetheless.

Bibliography

Texts quoted

Hero, *Dioptra*: Schöne, H. 1903. *Heronis Alexandrini opera quae supersunt omnia*. Vol. 3, 188–314. Leipzig.

Hero, *Stereometrica*: Heiberg, J.L. 1914. *Heronis Alexandrini opera quae supersunt omnia*. Vol. 5, 2–162, Leipzig.

Proclus, *In Eucl.*: Friedlein, G. 1873. *Commentarii in Primum Euclidis Elementorum Librum*. Leipzig.

39 Kuhn 1982a, 683.

Works quoted

Acerbi, F. 2010. *Il Silenzio delle Sirene: la Matematica Greca Antica*. Roma.

Andersen, H. *et al.* 2006. *The Cognitive Structure of Scientific Revolutions*. Cambridge.

Argoud, G., & Guillaumin, J.-Y. (eds.). 2000. *Autour de « La Dioptre » d'Héron d'Alexandrie : Actes du Colloque International de Saint-Étienne : (17, 18, 19 juin 1999) / Textes Réunis et éd. par Gilbert Argoud, Jean-Yves Guillaumin*. Saint-Étienne.

Asper, M. 2007. *Griechische Wissenschaftstexte: Formen, Funktionen, Differenzierungsgeschichten*. Stuttgart.

Chemla, K. (ed.). 2012. *The History of Mathematical proof in Ancient Traditions*. Cambridge.

Chouquer, G. 2010. *La Terre dans le Monde Romain: Anthropologie, Droit, Géographie*. Paris.

Christianidis, J. 2007. The Way of Diophantus: Some Clarifications on Diophantus' Method of Solution. In: *Historia Mathematica*, 34(3), 289–305.

Clark, A. 1997. *Being There: Putting Brain, Body, and World Together Again*. Cambridge, Mass.

Coulton, J. J. 2002. "The Dioptra of Hero of Alexandria." In: Tuplin, C. J. & T. E. Rihll (eds.), *Science and Mathematics in Ancient Greek Culture*. Oxford, 150–164.

Creese, D. E. 2010. *The Monochord in Ancient Greek Harmonic Science*. Cambridge.

Drachmann, A. G. 1969. "A Detail of Heron's Dioptra." In: *Centaurus* 13, 241–247.

Fowler, D. 1999. *The Mathematics of Plato's Academy: A New Reconstruction* (2nd ed). Oxford.

Fowler, D. & C. M. Taisbak 1999. "Did Euclid's Circles Have Two Kinds of Radius?" In: *Historia Mathematica* 26(4), 361–364.

Giere, R. N. 2007. "Distributed Cognition without Distributed Knowing." In: *Social Epistemology* 21, 313–320.

Hesse, M. 1982. "Comment on Kuhn's 'Commensurability, Comparability, Communicability'." In: *PSA: Proceedings of the Biennial Meeting of the Philosophy of Science Association* 1982, 704–711.

Høyrup, J. 1990. "Sub-scientific Mathematics: Observations on a Pre-Modern Phenomenon." In: *History of Science* 28, 63–87.

Høyrup, J. 1994. *In Measure, Number, and Weight: Studies in Mathematics and Culture*. Albany, NY.

Høyrup, J. 1997. "Hero, ps.-Hero, and Near Eastern Practical Geometry: An Investigation of Metrica, Geometrica, and Other Treatises." In: *Antike Naturwissenschaft und Ihre Rezeption* 7, 67–93.

Høyrup, J. 2012. "Mathematical Justification as Non-Conceptualized Practice: The Babylonian Example." In: K. Chemla (ed.), *The History of Mathematical Proof in Ancient Traditions*. Cambridge, 362–383.

Høyrup, J. & P. Damerow (eds.) 2001. *Changing Views on Ancient Near Eastern Mathematics*. Berlin.

Hutchins, E. 1995. *Cognition in the Wild*. Cambridge, MA.

Kitcher, P. 1982. "Implications of Incommensurability." In: *PSA: Proceedings of the Biennial Meeting of the Philosophy of Science Association* 1982, 689–703.

Kline, M. 1972. *Mathematical Thought from Ancient to Modern Times*. New York.

Kuhn, T. S. 1982a. "Commensurability, Comparability, Communicability." In: *PSA: Proceedings of the Biennial Meeting of the Philosophy of Science Association* 1982, 669–688.

Kuhn, T. S. 1982b. "Response to Commentaries [by Kitcher and Hesse]." In: *PSA: Proceedings of the Biennial Meeting of the Philosophy of Science Association* 1982, 712–716.

Langacker, R. W. 2000. *Grammar and Conceptualization*. Berlin.

Langacker, R. W. 2008. *Cognitive Grammar: a Basic Introduction*. Oxford.

Langacker, R. W. 2009. *Cognitive Linguistics Research: Investigations in Cognitive Grammar*. Berlin.

Lewis, M. J. T. 2001. *Surveying Instruments of Greece and Rome*. Cambridge.

Lewis, M. J. T. 2012. "Greek and Roman Surveying and Surveying Instruments." In: R. J. A. Talbert (ed.), *Ancient Perspectives: Maps and their Place in Mesopotamia, Egypt, Greece, and Rome*. Chicago, 129–162.

Mau, J. & W. Müller 1960. "Mathematische Ostraka aus der Berliner Sammlung." In: *Archiv für Papyrusforschung und Verwandte Gebiete* 17, 1–10.

Mayer, M. & O. Olesti 2001. "La Sortitio de Ilici. Del Documento Epigráfico al Paisaje Histórico." In: *Dialogues d'Histoire Ancienne* 27, 109–130.

Moatti, C. 1993. *Archives et Partage de la Terre dans le Monde Romain (IIe siècle avant-Ier siècle après J.-C.)*. Rome.

Netz, R. (ed.) 2004. *The Works of Archimedes: Translation and Commentary. Volume I: The Two Books On the Sphere and the Cylinder*. Cambridge.

Netz, R. 2009. *Ludic Proof: Greek Mathematics and the Alexandrian Aesthetic*. Cambridge.

Nicolet, C. 1991. *Space, Geography, and Politics in the Early Roman Empire*. Ann Arbor, MI.

Rosch, E. 1978. "Principles of Categorization." In: Rosch, E. & B. B. Lloyd (eds.), *Cognition and Categorization*. Hillsdale, NJ, 27–48.

Shapiro, L. A. 2011. *Embodied Cognition*. New York.

Varela, F. J. *et al.* 1991. *The Embodied Mind: Cognitive Science and Human Experience*. Cambridge, MA.

Vitrac, B. 2010. Héron d'Alexandrie et le Corpus Métrologique: État des Lieux. In: *"Géométrie(s), Pratiques d'Arpentage et Enseignement: Quels Liens et dans Quel Contexte?"*. Paris.

Michalis Sialaros

How Much Does a Theorem Cost?[1]

Abstract: Euclid is phenomenally famous as the author of the *Elements*. Indeed, no mathematical book has ever been read, edited, praised or criticized more than his *magnum opus*. However, very little is known about his life. For centuries, scholars have tried to discover new evidence with not much success. Thus, a reasonable question arises: 'Is there anything left to say?' This paper provides evidence that the answer to this question is positive. But what is crucial here is that this particular evidence is revealed, not via a discovery of some new material, but as a result of a revolutionary practice: the digitalization of the ancient Greek literature. As a case study, this paper focuses on the extant anecdotes associated with Euclid; one recorded by the philosopher Proclus Lycaeus and the other by the anthologist Joannes Stobaeus. In an attempt to shed some light on the matter of the origins of the latter anecdote, I draw parallels between this and a hitherto largely ignored Pythagorean *symbol*. This approach, I argue, can offer valuable insight into the textual tradition of the *symbol* and the conceptual background of the anecdote in question.

Keywords: Euclid, Anecdotes, Three-obol, Stobaeus, Pythagoras, Symbol, Aenigma.

I Introduction: The Euclid Anecdotes and Their Sources

a Proclus' Anecdote

The first anecdote associated with Euclid of Alexandria is recorded by Proclus Lycaeus (ca. fifth century AD), a philosopher and head of the Neoplatonic Academy, in his *Commentary on Book I of Euclid's Elements*:[2]

1 I would like to express my sincerest appreciation to readers of earlier versions of this paper: Serafina Cuomo, Jean Christianidis, Alain Bernard, Peter Riedlberger, and Vassilis Karasmanis. Special thanks go to the PhD students of the Research Training Group 'Philosophy, Science and the Sciences' of Humboldt University of Berlin, who provided valuable feedback: Rosa Matera, Jan Gerhold, Gonzalo Gamarra, Benjamin Wilck, Georgia-Maria Korobili, Mikhail Silian, Nicholas Allan Aubin, and Robert Roreitner. Above all, I am indebted to Danae Karydaki, who provided unconditionally—as always—her wise counsel and sympathetic ear. This work was generously supported by the RTG 'Philosophy, Science and the Sciences', at Humboldt University of Berlin.

2 Proclus, *In Eucl.* 68.13-17. My translations, here and henceforth, unless otherwise noted.

https://doi.org/10.1515/9783110565959-005

(Φ)ασιν ὅτι Πτολεμαῖος ἤρετό ποτε αὐτόν, εἴ τίς ἐστιν περὶ γεωμετρίαν ὁδὸς συντομωτέρα τῆς στοιχειώσεως· ὁ δὲ ἀπεκρίνατο, μὴ εἶναι βασιλικὴν ἀτραπὸν ἐπὶ γεωμετρίαν.

(T)hey say that once Ptolemy[3] asked him if there is in geometry some way shorter than that of the *Elements*. And he replied that there is no royal road to geometry.

Proclus provides numerous hints in his *Commentary* from which we can safely assume that he draws heavily on previous works.[4] Regarding, however, the source from which he obtained the Euclid anecdote, Proclus, unfortunately, remains silent.[5] In fact, the situation is more frustrating: there is no reference to the Euclid-Ptolemy anecdote (henceforth abbreviated as the 'EP-anecdote') in the extant Greek literature. Within this framework, it seems very difficult to cross-examine the validity of Proclus' account.

Interestingly enough, an almost identical version of the anecdote appears in Joannes Stobaeus' *Anthologium* (ca. fifth century AD). In this case, the dialogue takes place between Menaechmus and Alexander the Great (henceforth abbreviated as the 'MA-anecdote'):[6]

Μέναιχμον τὸν γεωμέτρην Ἀλέξανδρος ἠξίου συντόμως αὐτῷ παραδοῦναι τὴν γεωμετρίαν· ὁ δέ "ὦ βασιλεῦ", εἶπε, "κατὰ μὲν τὴν χώραν ὁδοί εἰσιν ἰδιωτικαὶ καὶ βασιλικαί, ἐν δὲ τῇ γεωμετρίᾳ πᾶσίν ἐστιν ὁδὸς μία".

Alexander requested from the geometer Menaechmus to teach him geometry concisely. And, "O King", he said, "through the country there are private and royal roads; but, in geometry, there is one road for all".

3 It has correctly been noted that Proclus does not explicitly state here that the meeting took place between Euclid and Ptolemy the First (cf. Acerbi 2007, 183; Vitrac and Caveing 1990–2001, I 16; Heath 1956, I 2). It must, however, also be observed that Proclus invokes the anecdote only to use it as a premise in order to establish Euclid's dating, not vice versa. This crucial remark paves the way for the hypothesis that Proclus must have somehow known that the story referred to Ptolemy the First—perhaps from the unidentified source of the anecdote; otherwise, his deduction would not stand to reason. One, of course, cannot completely exclude the latter possibility.

4 For a representative, yet not exhaustive, list, see Morrow 1970, xlviii–lvi. For a more thorough analysis, see Vitrac and Caveing 1990–2001, I 38–44. Mueller's 1987 comparative study identifies strong similarities between the first part of Proclus' *Commentary* and Iamblichus' *De Communi Mathematica Scientia*; on this issue, also see O'Meara 2014, 415.

5 A few lines before presenting the anecdote (*In Eucl.* 65.7-70.18), Proclus cites his famous 'catalogue of geometers' in connection with "those who recorded histories" (*In Eucl.* 68.4). Much ink has been spilled on the identity of these authors. Although several questions remain open, there seems to be an agreement among scholars that the primary source of the catalogue was Eudemus' perished *History of Geometry*. Nonetheless, the catalogue's final form—which includes the added information on Euclid—is most likely a compilation, either by Proclus himself, or, most probably, by an earlier scholar. For a thorough literature review on this issue, see Zhmud 2006, 179–190.

6 Stobaeus, *Anth.* 2.31.115.

Based on the striking similarities between the two anecdotes, some scholars cast doubt on the authenticity of the EP-anecdote, arguing that (a) only one of these stories could reflect a historical incident and that (b) the MA-anecdote is most likely to be that story. This is how this view is expressed by Thomas Heath: '(T)here is a temptation to transfer such a story at a later date to the most famous mathematicians or rulers'; and by Ivor Bulmer-Thomas: '(B)ut it would be natural to transfer it to the most famous geometer, and the attribution to Menaechmus is to be preferred'.[7] The core idea of this argument—namely, that there is a tendency to transfer stories to the most famous people—seems right. However, it fails to be conclusive for two reasons: First, the idea that stories are transferred at later dates is not always right. In other words, people (in several cases and for a variety of reasons) may attribute contemporary actions or quotes to famous people of the past. Second, one should not ignore that what we consider as 'fame' is a context-related evaluation: Euclid was arguably a more famous mathematician that Menaechmus;[8] however, on the other hand, Alexander the Great was a much more famous political leader than Ptolemy. So, through this prism, which of the two 'famous' figures is the anecdotes' pole of attraction, Euclid or Ptolemy? This deceivingly simple question is potentially paralyzing, as it casts doubt on the very possibility of approaching these anecdotes as sources of historical information. And, of course, it cannot be answered, if one relies only on arguments like Heath's and Bulmer-Thomas's. Thus, we need to turn to other arguments.

Let us begin by noting that, in both anecdotes, the concept of 'royal road' is employed for rhetorical effect; namely, as a metaphor to denote effortless learning. An audience, however, in order to comprehend a metaphor, requires an actual point of reference. The question that must, therefore, be asked is 'what is, literally speaking, a royal road?' Fortunately, the answer to this question is quite straightforward, as we have historical knowledge of only one actual highway with this name: this is the Royal Road mentioned by Herodotus, which ran through the Persian Empire, from Asia Minor all the way up to Babylon.[9] Within this framework, the mathematician's reply to his ruler that "through the country, there are private and royal roads" seems to make sense only in the context of an empire that actually contained such a road. And, if this is correct, that empire could only be Alexander's empire, given that Ptolemy never reigned a kingdom that contained the Royal Road.

Let us now have a closer look at the MA-anecdote. According to the text, Menaechmus appears to divide the existing roads in the country into two categories, 'royal' (βασιλικαί) and 'private' (ἰδιωτικαὶ); the former to be used by royals and the latter by

7 Heath 1956, I 252; Bulmer-Thomas 1974, 269.

8 This claim is not made without hesitation. Menaechmus, the alleged discoverer of the conic sections, was probably one of the most famous mathematicians of his time, and the leading member of a mathematical community: Proclus (*In Eucl.* 78.9, 254.5) speaks twice about "the mathematicians around Menaechmus" and, in some cases (*In Eucl.* 67.2-12), he also provides some names. Cf. Bulmer-Thomas 1974; Heath 1956, I 251–253; Suda, s.v.

9 Herodotus, *Hist.* 5.52-54, 8.98; cf. Calder 1925.

citizens. Then, he compares these roads with the road of geometry, remarking that the latter is 'for all' (πᾶσίν). This is an indication that the unknown author of the anecdote understands the concept of a 'royal road' as a road used by royals[10]—which is wrong, of course, as the historical Royal Road was public.[11] Based on this mistake, one could assume that the anecdote was not written at the time of the described event, but much later, when the Royal Road became more a metaphor than an actual point of reference. In other words, this is an indication that one should also cast doubt on the authenticity of the MA-anecdote.

Finally, one should also take into consideration that, despite their strong similarities, the two anecdotes are not identical: (a) There is a crucial difference in the way the two rulers request a shorter way to geometrical learning. While Alexander just seeks to learn geometry concisely (συντόμως), Ptolemy requests a way shorter than that of the *Elements* (ὁδὸς συντομωτέρα τῆς στοιχειώσεως). *Le bon Dieu est dans les détails.* This phraseology reveals that Ptolemy—the hero of the anecdote, not the historical Ptolemy—is well aware that he is addressing the author of the *Elements*—probably a renowned geometer at the time. (b) Although Menaechmus is very specific as to where the private and royal roads are to be found (i.e. 'in the country' (κατὰ μὲν τὴν χώραν)), Ptolemy's question is without spatial specification (i.e. 'if there is in geometry some way' (εἴ τίς ἐστιν περὶ γεωμετρίαν ὁδὸς)). These differences must be seen as significant elements of historical localisation. That is to say, the information provided here is strong enough to support the hypothesis that the anecdotes have emerged from and shaped by distinct historical settings.[12]

It should be by now evident that the available textual sources related to the EP-anecdote, do not let us draw conclusions on whether it reflects a historical fact. On the other hand, its close similarity to the MA-anecdote allows us to identify a core pattern, a *topos*, which remained unchanged despite the fact that the anecdotes seem to have emerged from different periods. I shall return to the concept of topos in Euclid anecdotes in Section II. For now, let us have a look at the second anecdote related to Euclid.

b Stobaeus' Anecdote and His Sources

The second Euclid anecdote, which is described by Heath (1956, I 3) as one of those stories "one would like to believe they are true", is recorded by the anthologist Joannes Stobaeus, who is, in turn, drawing on some Serenus:[13]

---

10 I would like to thank Peter Riedlberger for pointing this out.

11 Cf. fn. 9. For a recent study, see Graf 1994.

12 Two additional, rather unlikely scenarios must be recorded: (a) that both anecdotes might represent historical events, which took place independent from one another. (b) that both anecdotes might represent historical events, but, in this case, Euclid was aware of Menaechmus' reply to Alexander.

13 Stobaeus, *Anth.* 2.31.114.

Ἐκ τῶν Σερήνου. Παρ' Εὐκλείδῃ τις ἀρξάμενος γεωμετρεῖν, ὡς τὸ πρῶτον θεώρημα ἔμαθεν, ἤρετο τὸν Εὐκλείδην· "τί δέ μοι πλέον ἔσται ταῦτα μαθόντι;", καὶ ὁ Εὐκλείδης τὸν παῖδα καλέσας "δός", ἔφη, "αὐτῷ τριώβολον, ἐπειδὴ δεῖ αὐτῷ ἐξ ὧν μανθάνει κερδαίνειν".

From the <*Records*> of Serenus.[14] Someone who had begun to learn geometry with Euclid, when he learned the first theorem, asked Euclid "what shall I earn by having learnt these?" And Euclid called the servant and said: "Give him three obols because he must gain out of whatever he learns".

A first question naturally emerges here: 'Who is Stobaeus' Serenus?' With the exception of a reference by the Ecumenical Patriarch of Constantinople Photius (ca. ninth century AD), who describes this Serenus as a 'philosopher', his identity is shrouded in mystery.[15] One possibility[16] is that Stobaeus refers to Serenus of Antinoupolis, a third/fourth-century mathematician and philosopher, who wrote a perished commentary on Apollonius' *Conics*.[17] Another candidate is Aelius Serenus, a grammarian of uncertain date (ca. second-fourth century AD), mentioned in the *Suda*:[18]

Σερῆνος· ὁ καὶ Ἀθηναῖος, ὁ Αἴλιος χρηματίσας· γραμματικός. Ἐπιτομὴν τῆς Φίλωνος πραγματείας περὶ πόλεων καὶ τίνες ἐφ' ἑκάστης ἔνδοξοι, βιβλία γ', Ἐπιτομὴν τῶν Φιλοξένου εἰς Ὅμηρον α'.

Serenus (or Athenaeus), also called 'Aelius', grammarian. <He wrote> an Epitome of Philo's treatise *On Cities and Their Famous Citizens*, three books; an epitome of Philoxenus' <comments> on Homer, one <book>.

Finally, Photius mentions Serenus 'the Grammarian', a compiler of dramas (δράματα) in various meters.[19]

The time of action of the last two Sereni (Aelius Serenus and Serenus the Grammarian), as well as the fact that they are both described as grammarians, let Paul von Rohden to the hypothesis that they might be the same person.[20] Robert Kaster,

---

14 The omitted word here is clearly '*Records*' (Ἀπομνημονευμάτων). In the *Anthologium*, Stobaeus quotes Serenus eighteen times in total. In six of these cases, we have the complete phrase 'Ἐκ τῶν Σερήνου Ἀπομνημονευμάτων'; cf. *Anth.* 2.1.22, 2.2.17, 3.11.23, 4.6.20, 4.19.48, 4.24a.11.

15 Photius, *Bibl.* 167.114b18. It is not clear, however, whether Photius deduces this description from his own interpretation of the contents of the *Records* as they appear in Stobaeus' *Anthologium*, or if he has some (unknown to us) additional information on Serenus.

16 The proposed list does not include those Sereni who wrote in Latin (e.g. Vibius Serenus, Annaeus Serenus, Septimius Serenus), or some completely unknown Sereni (like the two martyrs who belonged to Origenes' circle of disciplines).

17 Theon of Smyrna attributes a certain proposition to Serenus 'the Philosopher', who is probably the same person; cf. Heiberg 1896, xviii–xix. For the latest discussion on Serenus of Antinoupolis' dating, see Decorps-Foulquier 1992. Two extant treatises of Serenus, entitled *On the Section of a Cylinder* and *On the Section of a Cone*, were edited by Heiberg; cf. Heiberg 1896.

18 Suda, s.v. For Aelius Serenus' dating, see Theodoridis 1976, 11.

19 Photius, *Bibl.* 279.536a.8-15.

20 See Von Rohden 1893, 532; cf. Hammerstaedt 1997, 108 fn. 32.

on the other hand, cast doubt on this idea, on the basis of the following argument:[21] Photius mentions Serenus the Grammarian together with four poets (Hermias of Hermopolis, Andronicus of Hermopolis, Horapollon, and Cyrus of Antaeopolis), all of whom lived in Egypt around the fourth century AD. Thus, Kaster's argument goes, Serenus was probably an Egyptian who lived around that period. As a result, he could not be the same person as Aelius Serenus, because the latter was an Athenian. In my view, however, there is a serious objection to this argument. The idea that Aelius Serenus lived in Athens is based on an arguably false reading of the corresponding entry in the *Suda* 'Σερῆνος· ὁ καὶ Ἀθηναῖος'. This phrase should not be translated, following Kaster, as 'Serenus of Athens', but either as (a) 'Serenus (or Athenaeus)' or (b) 'Serenus, also called the Athenian'. In the first case, 'Athenaeus' is a proper name and not a description of his place of action; thus, Kaster's argument is refuted. Now, in the second case, 'the Athenian' stands for Serenus' nickname and not a description of his place of action. If this reading is right, Serenus' nickname is a strong indication that he did not live in Athens, but in a different place: since nicknames are employed in order to differentiate people from one another, no Athenian gets the nickname 'Athenian' in Athens. So, again, Kaster's objection is refuted. Therefore, Aelius Serenus, the grammarian, may have lived in Egypt, and thus it could be the case that he was the same person as Serenus the Grammarian, mentioned by Photius.

Returning now to the *Anthologium*, we notice that Stobaeus cites at least[22] seventeen anecdotes from Serenus' *Records*, besides the one about Euclid: on Thales (2.1.22); on Arcesilaus (2.2.17); on Heraclitus and Anaxarchus (2.31.116); on Euripides and Plato (3.5.36); on Diogenes (3.6.17); on Alexander of Pherae and Thebe (3.7.62); on Plato and Polemon (3.11.23); on Antipater (3.13.48); on Philip (3.13.49); on a woman from Sparta (3.13.58, 4.24a.11); on the Spartans (3.39.27); on a man from Sybaris (3.29.96); on the Persian Kings (4.2.26); on Croesus (4.6.20); on Gellias of Acragas (4.19.48); and on a man from Sparta (4.22f.134). No strong interest is shown in these passages in mathematics.[23] Based on this observation, I am more inclined to accept that Serenus' *Records* was the work of Aelius Serenus, a scholar trained in preparing compilations. This is, of course, due to lack of further evidence, just a hypothesis. Indeed, one should also not ignore the possibility that none of the aforementioned Sereni is identical with Stobaeus' Serenus. If this is the case, the scarcity of sources does not, unfortunately, allow us to say much about the identity of this man.

---

21 See Kaster 1988, 354–355. For the same point of view, see Hammerstaedt 1997.

22 I use 'at least' since one cannot exclude the possibility that some of the numerous anecdotes that appear in the *Anthologium* without citation might also be taken from Serenus' *Records*.

23 The sole exception is the passage referring to Thales (*Anth.* 2.1.22), a version of the famous story Plato mentions in the *Theaetetus* 174a in which Thales falls into a well while observing the sky.

II Theorems and Three-obols

Let us now have a closer look at Stobaeus' anecdote (henceforth abbreviated as the 'S-anecdote'). The story goes that a new student of Euclid, after having learned his first theorem, asked his teacher what his benefit would be from the knowledge he had just acquired. Euclid then ordered his servant to give the boy some money and criticized his wish to gain something out of his achievement. The fundamental narrative elements of the anecdote can be summarized as follows:

(a) *Characters*: There are two protagonists (namely, a mathematician and a new student of geometry), as well as an auxiliary character (namely, a servant).
(b) *Staging*: The setting is not explicitly described. It could be any place where a lesson of geometry could be held.
(c) *Plot*: The sequence of events is divided into three phases: first, a lesson in geometry takes place; second, the student raises a question regarding the value of his new piece of knowledge; and, finally, the teacher—obviously in a sarcastic mood—orders the servant to offer a financial reward to the student.
(d) *Moral Lesson:* Besides the aforementioned narrative elements, the S-anecdote contains a clearly implied moral lesson: one should not approach mathematical (or even scientific, in general) knowledge only for the sake of a financial reward.

A number of interesting questions regarding the historical value of the anecdote can be raised: does this story provide evidence of Euclid's institutional setting, his social status, or even his character? If, on the one hand, the S-anecdote depicts a historical incident, one could deduce from it some historical information: for example, that Euclid was not only an author but a teacher as well; that he was not poor (since he could support at least one servant); and that he had a particular approach towards the value of mathematical knowledge. If, on the other hand, the anecdote is fictional, is there any historical information left? And, if yes, considering the apparent unique appearance of the anecdote in the Greek tradition, how can one approach this knowledge? I propose to address these questions in what follows.

a A Figure and a Step

Unlike the case of the EP/MA-anecdotes, the S-anecdote is believed to have no parallel in the extant Greek literature. I am convinced, however, that the following passage, with which Proclus concludes his introduction to his *Commentary on book I of the Elements* proves this idea wrong:[24]

24 Proclus, *In Eucl.* 84.8-23; my emphasis. Proclus' commentary is accompanied by two sets of introductory notes, usually described as the 'first' and the 'second' prologue, respectively; cf. Morrow 1970, 39.

Ἀρχὴν δὲ ποιούμενοι τῆς τῶν καθ' ἕκαστα ζητήσεως προαγορεύομεν τοῖς ἐντευξομένοις, μὴ ταῦτα παρ' ἡμῶν ἀπαιτεῖν ὅσα διατεθρύληται τοῖς πρὸ ἡμῶν λημμάτια καὶ πτώσεις καὶ εἴ τι τοιοῦτο. τούτων μὲν γὰρ διακορεῖς ἐσμὲν καὶ σπανίως αὐτῶν ἐφαψόμεθα. ὅσα δὲ πραγματειωδεστέραν ἔχει θεωρίαν καὶ συντελεῖ πρὸς τὴν ὅλην φιλοσοφίαν, τούτων προηγουμένην ποιησόμεθα τὴν ὑπόμνησιν, ζηλοῦντες τοὺς <Πυθαγορείους>, οἷς πρόχειρον ἦν καὶ τοῦτο σύμβολον "σχᾶμα καὶ βᾶμα, ἀλλ' οὐ σχᾶμα καὶ τριώβολον" ἐνδεικνυμένων, ὡς ἄρα δεῖ τὴν γεωμετρίαν ἐκείνην μεταδιώκειν, ἣ καθ' ἕκαστον θεώρημα βῆμα τίθησιν εἰς ἄνοδον καὶ ἀπαίρει τὴν ψυχὴν εἰς ὕψος, ἀλλ' οὐκ ἐν τοῖς αἰσθητοῖς καταβαίνειν ἀφίησιν καὶ τὴν σύνοικον τοῖς θνητοῖς χρείαν ἀποπληροῦν καὶ ταύτης στοχαζομένην τῆς ἐντεῦθεν περιαγωγῆς καταμελεῖν.

As we are about to make a beginning to the inquiry of the particular <propositions>, we declare to the readers that they should not demand from us <an examination> of those <subjects> that have been discussed over and over by our predecessors, such as lemmas, cases, and the likes. For we are surfeited with these and shall rarely touch on them. But we shall make our principal concern to mention those <subjects> that contain substantial theory and contribute to the whole philosophy, emulating in this way the Pythagoreans, amongst whom also this symbol was common: *"a figure and a step, but not a figure and a three-obol"*. By this they wanted to show that one must pursue that kind of geometry, in which each and every theorem places a step upwards and raises the soul to height, instead of letting it descend amongst the sensibles to satisfy the needs of the mortals and, aiming at these, neglect to turn around hence.

This testimony is rich in information. In the beginning, Proclus records that Euclid's *Elements* had already been the subject of detailed commentating by previous scholars (ὅσα διατεθρύληται τοῖς πρὸ ἡμῶν). He distinguishes, however, his own agenda, by explicitly stating that he is not going to provide an account of all these discussions, but he is, instead, only going to select and consider those subjects that "contain substantial theory and contribute to the whole philosophy" (πραγματειωδεστέραν ἔχει θεωρίαν καὶ συντελεῖ πρὸς τὴν ὅλην φιλοσοφίαν). Immediately afterward, he draws parallels between his approach and a *symbol* which he attributes to the Pythagoreans (henceforth abbreviated as the 'P-*symbol*'): 'a figure and a step, but not a figure and a three-obol' (σχᾶμα καὶ βᾶμα, ἀλλ' οὐ σχᾶμα καὶ τριώβολον).[25] Finally, he concludes by providing an interpretation of the *symbol*, according to which one must pursue that special kind of geometry in which every theory helps the soul to raise and not to descent amongst trivial human needs (τίθησιν εἰς ἄνοδον καὶ ἀπαίρει τὴν ψυχὴν εἰς ὕψος).

25 *Symbols* (also known as *akousmata* or *ainigmata*) are sayings or proverbs thought to represent the oldest form of transmission of Pythagoras' teachings; for a bibliography on collections of *symbols*, see Thom 2013, 77. A definite description of these sayings is difficult to provide, given their diversity in terms of form and content. Iamblichus (*VP* 82), probably based on Aristotle, provides a three-fold classifications: (a) *symbols* expressing what something is (τί ἔστι); (b) *symbols* expressing what the highest form of something is (τί μάλιστα); and (c) *symbols* suggesting what one must do (τί πρακτέον). For a discussion on the connection between Iamblichus and Aristotle, see Burkert 1972, 167. Zhmud (2012, 173–174) questions Burkert's reading. For a study on the third kind of *akousmata*, which is most relevant to the *symbol* Proclus refers to, see Horky 2013, 3–35.

The pivotal role of the *symbol* in the interpretation of this passage (and, subsequently, in the interpretation of Proclus' whole agenda) is profound. Unfortunately, far too little attention has been paid to it.[26] The following note by Thomas Taylor found in Glenn Morrow's commentary on Proclus *Commentary on Book I of the Elements* is indicative of this: "I do not find this aenigma among the Pythagorean symbols that are extant; so that it is probably no where mentioned but in the present work."[27] We shall see that Taylor is wrong here. However, before turning to the other testimonies, let us first examine the *symbol* in some detail.

Regarding grammar, the first thing one notices in the *symbol* 'σχᾶμα καὶ βᾶμα, ἀλλ' οὐ σχᾶμα καὶ τριώβολον' is the presence of the long α instead of the expected η, in the words σχᾶμα (rather than σχῆμα) and βᾶμα (in place of βῆμα). This is a clear indication that the *symbol* is written in Doric dialect. Byzantine scholars shared the same view. For example, Eustathius, a twelfth-century bishop and Homer commentator, noticed this peculiarity and explicitly linked the *symbol* with Doric Greek.[28] Additionally, a number of Byzantine lexica established the same connection: (a) the *Etymologicum Genuinum* (ca. tenth century AD) writes: "and 'βᾶμα καὶ σχᾶμα' is said, according to the voice of the Dorians, instead of 'σχῆμα καὶ βῆμα'"; (b) the *Etymologicum Magnum* (ca. twelfth century AD) "and 'βᾶμα καὶ σχᾶμα' is said, according to the voice of the Dorians, instead of 'σχῆμα καὶ βῆμα'"; and, (c) the *Etymologicum Symeonis* (ca. twelfth century AD) "and 'βᾶμα καὶ σχᾶμα' is said, according to the voice of <the> Dorians".[29]

Interestingly enough, the *symbol*'s key Doric words βᾶμα and σχᾶμα are extremely rare in the Greek corpus, and they usually appear only in relation to the *symbol*.[30] Only in some rare cases these words appear in different contexts. So, βᾶμα is an entry to the lexicon of unusual Greek words compiled by Hesychius of Alexandria (ca. fifth/sixth century AD) and explained as 'basis, step';[31] moreover, the word appears in a fragment attributed to the lyrical poetess Sappho.[32] Now, as regards σχᾶμα, the word appears three times in Archimedes,[33] three times in [Ps.] Archytas' fragments,[34] once in a fragment attributed to [Ps.] Timaeus,[35] and once in Iamblichus.[36] All these appearances frame the Doric origin of the *symbol*.

26 For a recent interest in this passage, cf. Harari 2015, 164 fn. 4.
27 Morrow 1970, 69 fn. 81. Vitrac and Caveing (1990–2001, I 16), on the other hand, are more careful.
28 Eustathius, *In Il.* I, 782, 13–14.
29 *E Gen*, s.v..; *EM* 196.37; *EtSym* I, 432.23.
30 Τριώβολον, the third key word, is very usual and appears more than 500 times in the extant corpus.
31 Hesychius, s.v. Cf. Theognostus, *Can.* 433.6.
32 Sappho, fr.16.
33 Archimedes, *Stom.* III.70.9-15
34 [Ps.] Archytas, 39,13 and 7,6-7; cf. Thesleff 1965, 8–11.
35 [Ps.] Timaeus 208,7; Thesleff 1965, 23.
36 Iamblichus, *De Comm.* 8.121. For the Pythagorean pseudepigrapha, see fn. 43.

Now, as regards the meaning of the P-*symbol*, some difficulties arise from the way one should interpret the word σχᾶμα. Its literal translation, *i.e.* 'form/shape/figure', does not offer much help as it does not lead to a meaning. We should, instead, read σχᾶμα as a metonym for a mathematical proposition.[37] In this way, the reading of the *symbol* is clear: the knowledge of a mathematical theorem should be seen as a step towards a higher goal, not as a source of a material—or even financial—profit. Through this prism, it is manifest that the P-*symbol* presents significant structural similarities to the S-anecdote, the most striking of which is, of course, the financial value of each theorem: a three-obol. The table below summarizes these elements:

	S-anecdote	P-*symbol*
Lesson of geometry	Mentioned	Clearly implied
Protagonists: A teacher and a student	Mentioned	Clearly implied
Financial value of a theorem: a three-obol	Mentioned	Mentioned
Moral lesson	Clearly Implied	Mentioned

It should be by now evident that the table reveals a striking relation between the S-anecdote and the P-*symbol*: the latter contains all the conceptual elements of the former. In other words, if a certain dramatic setting were added to the P-*symbol*, the S-anecdote would result.

b The P-*symbol* in the Greek Literature

Returning now to the issue of the apparent literary isolation of the P-*symbol* mentioned in Section IIa, let us recall that Morrow, following Thomas, admitted that he could not find the P-*symbol* among the extant Pythagorean *symbols*.[38] The rest of this section will show that Morrow and Thomas's view is simply wrong.[39]

(1) The earliest reference to the P-*symbol* is provided by Iamblichus (ca. third century AD), who compiled a list of Pythagorean *symbols* (σύμβολα) in his

37 For a similar interpretation, see Lloyd 2012, 395–396; Netz 1999, 35–38; Karasmanis 1987, 22; cf. Aristotle, *Met.* 998a.25-28; *APr.* 46a.8-9; Plato, *Phaedrus*. 73b1; *Crat.* 436d2. This also appears in the Islamic mathematical tradition; cf., for example, Lo Bello 2003, 105.

38 Cf. fn. 27.

39 This statement is not meant to undermine the undisputed scholarly value of Morrow or Thomas. It does show, however, how the digitalization of the ancient Greek literature, an effort which started in the 1970s, has allowed us to have a more accurate overview of the whole corpus. This relatively new tool is especially useful for historians of Greek mathematics, who usually struggle with the scarcity of their sources, as it promotes the idea to revisit and check older historians' statements such as the above. For the purposes of the paper, I relied heavily on TLG.

Protrepticus.[40] The *symbol* under number 36 goes: "Prefer the 'a figure and a step' than the 'a figure and a three-obol'" (λς'. Προτίμα τὸ σχῆμα καὶ βῆμα τοῦ σχῆμα καὶ τριώβολον).[41] A few lines later, Iamblichus also provides an interpretation of the *symbol*:[42]

Τὸ δὲ <προτίμα τὸ σχῆμα καὶ βῆμα τοῦ σχῆμα καὶ τριώβολον> παραγγέλλει τοῦτο· φιλοσόφει καὶ μαθήματα μέτιθι μὴ παρέργως, καὶ δι' αὐτῶν ὥσπερ διαβάθρας ἐπὶ τὸ προκείμενον χώρει· κατάπτυε δὲ τῶν προτιμωμένων καὶ πρεσβευομένων τοῖς πολλοῖς, καὶ προτίμα τὴν Ἰταλικὴν φιλοσοφίαν τὴν τὰ ἀσώματα καθ' αὑτὰ θεωροῦσαν τῆς Ἰωνικῆς τῆς τὰ σώματα προηγουμένως ἐπισκοπουμένης.

And the ['prefer a figure and a step than a figure and a three-obol'] orders this: pursue philosophy and commit yourself to mathematical sciences, not as a by-work. And through these, as if it were a ladder, proceed to the question under investigation. Detest those preferred and honored by the many, and prefer the Italic philosophy, which contemplates the incorporeal beings themselves, to the Ionic, which makes the corporeal beings the principal subject of investigation.

Curiously enough, Iamblichus does not cite the P-*symbol* in the Doric dialect.[43] He does, nevertheless, clearly link the *symbol* to the Pythagoreans and the so-called "Italic" philosophy. Likewise Proclus, Iamblichus proposes a scheme in which mathematics should be a medium to lead the learner upwards, and not down to the trivial world of corporeal beings.

(2) A second reference to the P-*symbol* is provided by Elias (ca. sixth century AD) in his commentary on Porphyry's *Isagoge*.[44] Expounding on the need to use mathematics as a ladder or a bridge (κλίμαξιν ἐοικότα ἢ γεφύραις) to comprehend

---

40 Some of Iamblichus' *symbols* also appear in other sources; cf. Olympiodorus, *In Phaed.* 1.13; Simplicius, *In Epict.* LXVIII, 18–9; Philoponus, *In de An.* 116, 30–117, 24. We know that Iamblichus was not the first to prepare a collection of *symbols*: Anaximander the Younger (ca. fifth century BC) is reported to have written a treatise entitled 'Explanation of Pythagorean *Symbols*' (cf. *Suda*, s.v.). Also Diodorus of Aspendus, Philochorus, and Androcydes are believed to have published on the matter. Iamblichus himself may have written a book *On the Symbols*, which is, unfortunately, lost. In addition, there exist quotations of *symbols* in several other authors, such as Plutarch, Athenaeus, Clement of Alexandria, Hippolytus of Rome, Diogenes Laertius, and Porphyry; cf. Burkert 1972, 166–191. For a more recent study, see Thom 2013.

41 Iamblichus, *Pr.* 108-11-12.

42 Iamblichus, *Pr.* 125.1-8.

43 One of the most controversial issues in the history of the reception of Pythagoreanism relates to the so-called 'pseudepigrapha', a group of treatises that was written in an artificial Doric to match its apparent Pythagorean origin. For a detailed study, see Thesleff 1965; cf. Centrone 2014 and Bonazzi 2013, 386 fn.3. Despite the fact that there is an ongoing debate on whether *symbols* belong to the so-called 'early' or 'later' Pythagorean tradition, there is little doubt that the P-*symbol* is pre-Euclidean.

44 Elias, *In Porph.* 25.24-29.2. The evidence that Elias even existed is scant. It is possible that his commentary on Porphyry's *Isagoge* was composed as a transcript of Olympiodorus' lectures. On this issue, see. Wildberg 1990, 37; Blumenthal 1981; and Westerink 1967, introduction.

immaterial beings, Elias writes: "For this is what the Pythagoreans also say 'a figure and a step'" (οὕτω γάρ φασι καὶ οἱ Πυθαγόρειοι 'σχᾶμα καὶ βᾶμα'). Strikingly enough, Elias provides here only half of the *symbol*, this time in the Doric dialect. Similar to Iamblichus and Proclus, Elias connects the *symbol* to a cognitive function that leads the soul upwards. Additionally, a few lines after citing the P-*symbol*, Elias notes an interesting connection (28.27-31). Addressing those who wonder what is special about mathematics, he argues that mathematics has two unique characteristics: (a) unlike other subjects, mathematics has the power of accurate knowledge that demonstrations provide (τὸ ἀραρὸς τῶν ἀποδείξεων); and, (b) those who taste mathematical knowledge have the same problem as the Homeric Lotus-eaters (πάσχοντες τὸ τῶν Λωτοφάγων): they cannot put it away. I shall return to this passage in Section IIc.

(3) Let us now come to a letter written by Michael Psellos (ca. eleventh century AD), a renowned Byzantine polymath, to his student Ioannes of Longobardia.[45] Psellos begins his letter by praising Ioannes for his eagerness to advance fast in sciences. (Ἡ περὶ τὰς ἐπιστήμας προθυμία σου πολλοῦ ἂν ἀξία ἦν) (18.3), and continues by expressing criticism of the careless way he approaches geometry (οὕτως ἐπὶ τὴν γεωμετρίαν ἀπεριμερίμνως χωρεῖς) (18.42-3). Immediately afterwards he says: "and you may have forgotten that Pythagorean <*symbol*>, which <goes> 'a figure and a step'" (τὸ δὲ Πυθαγορικὸν ἐκεῖνο ἴσως ἠγνόηκας ὅτι σχᾶμα καὶ βᾶμα) (18.45). He concludes the letter, by arguing that the learning of mathematical sciences must be conceived as a ladder, through which a man will raise to the immaterial beings (πρὸς τὰς πάντῃ ἀΰλους τῶν οὐσιῶν) (18.46-7). Psellus' letter is truly interesting in its own right. For the needs of the current research, however, I will only make some preliminary remarks regarding the way in which the author employs the P-*symbol*. First, it is evident that Psellus knows the *symbol* very well and, most importantly, expects his student Ioannes to be aware of it too: in a critical manner, he says "and you may have forgotten". Second, Psellus attributes the Doric *symbol* to the Pythagoreans. Finally, similar to Elias, Psellus quotes only the first half of the *symbol*. There is little doubt, nevertheless, that he knows its full version: after quoting it, he repeatedly gives advice to his student to avoid learning geometry for the sake of a financial reward.[46]

(4) The last reference to the P-*symbol* is provided by Eustathius—whom I have already mentioned in relation to his comments on the Doric structure of the *symbol* in Section IIa. In his monumental commentary on Homer's *Iliad*, Eustathius notes: "As it is proverbially denoted by the <*symbol*> 'figure and a three-obol'" (ὡς παροιμιακῶς δηλοῖ καὶ τὸ «σχῆμα καὶ τριώβολον»).[47] A few lines later, and in connection to the origins of the *symbol*, he proposes two short stories, one in relation to the tragic poet

45 Psellus, *Oratoria minora* 18.
46 Cf. Psellus, *Oratoria minora* 18.63-5; 18.86-89.
47 Eustathius, *In Il.* III 529.11-12.

Phrynichus (529.12-14) and another related to Pythagoras (529.14-17). For the needs of this paper, I shall present only the latter: Eustathius, admittedly based on Iamblichus, records that Pythagoras had once a skilful (δεξιῷ), yet lazy (ὀκνηρῷ), student. In order to motivate the student, Pythagoras paid him a three-obol for each theorem he would learn. Fortunately for us, the Iamblichus passage which Eustathius mentions is extant.

c Pythagoras' Lesson of Geometry

In his *Life of Pythagoras*, Iamblichus mentions an incredible story (henceforth abbreviated as the 'P-anecdote') about Pythagoras' first student.[48] The story begins when Pythagoras returned to his home island, Samos, after a series of journeys. He was eager to find a student to pass him on the knowledge he had acquired. This was proved to be a difficult task. Eventually, he recruited a young man—also called Pythagoras—and offered to pay him (guess what) a three-obol for every theorem he would learn (καὶ διδάσκων παντὸς σχήματος, ὅ ἐστι διαγράμματος, μισθὸν καὶ ἀντίπονον παρεῖχε τῷ νεανίᾳ τριώβολον) (5.22.4-6). After some time, Pythagoras observed that the student developed a growing passion for geometry. Pretending he was poor, he told him that they should stop the lessons. But the student replied: "From now on, I will provide for you, and cherish in return, in some way. For, in my turn, I will give you a three-obol for each figure" ('ἐγώ σοι λοιπὸνποριῶ καὶ ἀντιπελαργήσω τρόπον τινά· κατὰ γὰρ ἕκαστον σχῆμα τριώβολον καὐτός σοι ἀντιπαρέξω'). Iamblichus ends this story, by mentioning that Pythagoras (the student), alone of all the Samians, migrated with Pythagoras to Croton.

The source of this anecdote remains unknown. Iamblichus says nothing on the matter, and, moreover, the story itself is not found in any of the other extant biographies of Pythagoras. If it is not completely made up, a possible candidate for a source is Aristoxenus, Pythagoras' biographer and Aristotle's pupil, whom Iamblichus mentions four times in his *Life of Pythagoras*.[49] Certainty, nonetheless, on this matter is unattainable.

Let us now examine the fundamental narrative elements of the P-anecdote:

(a) *Characters*: We have two protagonists, a mathematician and his new student of geometry.

---

48 Iamblichus, *VP.* 5.21.1-5.25.3. For the view that Iamblichus' *VP* was the first of nine (or even ten) *logoi* (discourses, texts) by Iamblichus on Pythagoreanism, see O'Meara 2014.

49 Cf. Iamblichus, *VP.* 33.233.14; 33.234.1; 33.237.1; 35.251.9. According to Rohde (1871, 1872), the *VP* is made up of bits taken from a life of Pythagoras by Nicomachus of Gerasa and one by Apollonius of Tyana. In Rohde's view, Nicomachus conveys earlier sources, and most notably Aristoxenus, whereas Apollonius is extremely unreliable. These claims have been contested; cf. Burkert 1972, 98–101, Staab 2002, 217–237. O'Meara (2014, 412–414) argues that Iamblichus probably used more than two sources. Iamblichus himself quotes several texts, amongst which, Pythagoras *Sacred Discourse* (*VP* 82.12), Aristotle's lost books on Pythagorean philosophy (*VP* 18.12-13), and Androcydes *On the Pythagorean Symbols* (*VP* 81.11-2).

(b) *Staging*: The place where the geometry lessons are held is not explicitly described. We only know that all events take place on the island of Samos.
(c) *Plot*: The sequence of events can be divided into three phases. First, a desperate teacher recruits a new student promising to provide him with a three-obol for every new theorem he would learn. Second, the teacher decides to stop the lessons, after noticing that the student had developed a passion for geometry. Finally, the student, unable to stop learning geometry, offers to pay his teacher a three-obol for each new theorem he would teach him.
(d) *Moral lesson*: The anecdote clearly brings out a moral lesson: Even if one approaches mathematical (or, more generally, scientific) knowledge for the sake of a financial (or, more generally, material) reward, eventually one would be seduced by knowledge itself.

It is worth noting that this moral lesson establishes a further connection with the passage where Elias (*In Porph.* 28.27-31) discusses the P-*symbol* (mentioned above in Section IIb). I remind the reader that, after citing the P-*symbol*, Elias mentions two unique characteristics of mathematics: (a) unlike other subjects, mathematics has the power of accurate knowledge that demonstrations provide; and, (b) those who taste mathematical knowledge have the same problem as the Homeric Lotus-eaters: they cannot put it away. This was apparently also the case of Pythagoras' 'unfortunate', Lotus-eater, student.

III Final Remarks

The aim of this paper was two-fold: first, to use Euclid anecdotes as a case study in order to highlight the importance of using new, revolutionary digital tools to study ancient mathematical texts; second, to revisit the extant Euclid anecdotes, in an attempt to grasp their historical development. In addition to introducing the reader to the problems imposed by the scarcity of evidence, I have drawn parallels between the Euclid anecdote provided by Joannes Stobaeus and a hitherto largely ignored Pythagorean *symbol*, in an attempt to shed light on the matter of their origin. Within this framework, I have presented the extant textual tradition related to this *symbol*, as well as its link to a Pythagorean anecdote. This is the detailed list of the new findings:

- A new argument that offers cumulative support to the primacy of the MA-anecdote (Section Ia).
- A counter-argument against Kaster's view that Aelius Serenus and Serenus the Grammarian could not be the same person (Section Ib).
- A comparative study of the P-*symbol* and the S-anecdote (Section IIa).
- A complete list of the Greek sources in which the P-*symbol* appears. As a consequence, Morrow and Taylor's idea that the *symbol* appears only in Proclus is rejected (Section IIb).

- A presentation of Eustathius' testimony, which is the only source that links the *symbol* with Iamblichus' anecdote on Pythagoras first student (Section IIb).
- An establishment of the connection between Iamblichus' story and Elias' discussion on the P-*symbol* (Section III).

It should be by now evident that the S-anecdote, the P-*symbol*, and the P-anecdote, all share a common topos, a coherent cognitive structure, and most probably a common origin. And to reply to the paper's title question: 'How much does a theorem cost?', the answer is the same no matter if we examine the 6th century BC, the Hellenistic period, or the Byzantine times: 'a three-obol'.[50] We have also seen that a rich textual tradition attests that the P-*symbol* was most probably in circulation before Euclid's time. The P-*symbol* itself might have emerged from another anecdote (the P-anecdote), or *vice-versa*, but this hypothesis requires a separate study. Through this prism, our original question re-emerges: 'Is it even possible to trace the S-anecdote's origins?' Unfortunately, due to the scarcity and complexity of our sources, no definite answer can be provided. I am more inclined to reject the idea that the S-anecdote might depict a historical incident, only on the basis of the 'topos' it shares with the P-*symbol*. No one, however, can be absolute on this matter. Perhaps, one way to untangle this problem is to rephrase the question into 'What is truly at stake, in case the S-anecdote does not depict a historical dialogue?' My answer to this question is 'not much'. If one wishes to support the idea that Euclid was a teacher of mathematics, one may use other sources. For example, Pappus explicitly mentions that Apollonius spent much time studying with the pupils of Euclid in Alexandria;[51] most of Euclid's books seem to be structured with an eye towards education;[52] and, Proclus, describing the now lost *Pseudaria*, records that it contained a collection of false proofs and methods to help beginners avoid fallacies in geometry.[53]

To conclude, this paper demonstrates that the extant Euclid anecdotes are much more than brief, entertaining stories about a famous mathematician. They are the outcome of a multilevel interaction between several traditions (among which, early and Late Pythagoreanism, Neopythagoreanism, Aristotelian and Hellenistic editorial practices), emerging in different contexts and responding to different problems. As such, they compose a complex and multifaceted literary phaenomenon, worthy of further examination.

---

50 I have tried to trace the actual value of a three-obol (half a drachma) from 600 BC to 200 BC, and I have also taken into consideration variations in terms of place. Unfortunately, there is not a change great enough to allow us to establish conclusions on the primacy of one of the anecdotes.

51 Pappus, *Coll.* 7.678.10-12.

52 Later sources are definite on this: Marinus, for example, hints at the pedagogic usefulness of Euclid's *Data* and Proclus praises the educational value of the *Elements*; cf. Marinus, *In Dt.* 256.22-23; Proclus, *In Elem.* 71.5-24, 82.3-83.6.

53 Proclus, *In Elem.* 70.1-18. For more information on the *Pseudaria*, see Acerbi 2008.

Bibliography

Texts quoted

Archimedes, *Stom.*: Mugler, C. 1971. *Archimède*. Vol. 3. Paris.
Archytas [Ps.]: Thesleff, H. 1965. *The Pythagorean Texts of the Hellenistic Period*. Åbo.
Aristotle, *APr.*: Ross, W. D. 1964. *Aristotle's Prior and Posterior Analytics*. Oxford (repr. 1968).
Aristotle, *Met.*: Ross, W. D. 1924. *Aristotle's Metaphysics*. 2 vols. Oxford (repr. 1970).
Elias [Ps.], *In Porph.*: Busse, A. 1900. *Eliae in Porphyrii isagogen et Aristotelis Categorias Commentaria*. Commentaria in Aristotelem Graeca 18.1. Berlin.
Etymologicum Genuinum, *E Gen*: Lasserre, F. & N. Livadaras. 1992. *Etymologicum Magnum Genuinum. Symeonis Etymologicum una cum Magna Grammatica. Etymologicum Magnum Auctum*. Vol. 2. Athens.
Etymologicum Magnum, *EM*: Gaisford, T. 1848. *Etymologicum Magnum*. Oxford (repr. Amsterdam: 1967).
Etymologicum Symeonis, *EtSym*: Lasserre, F. & N. Livadaras. 1992. *Etymologicum Magnum Genuinum. Symeonis Etymologicum una cum Magna Grammatica. Etymologicum Magnum Auctum*. Vol. 2. Athens.
Eustathius, *In Il.*: van der Valk, M. 1971–1987. *Eustathii Archiepiscopi Thessalonicensis Commentarii ad Homeri Iliadem Pertinentes*. 4 vols. Leiden.
Herodotus, *Hist.*: Legrand, Ph.-E. 1932–1954. *Hérodote: Histoires*. 9 vols. Paris (repr.1963–1970).
Hesychius: Latte, K. 1953–1966. *Hesychii Alexandrini Lexicon*. Vols 1–2. Copenhagen.
Iamblichus, *De Comm.*: Klein, U. (post N. Festa) 1891. *Iamblichi de Communi Mathematica Scientia Liber*. Leipzig (repr. Stuttgart: 1975).
Iamblichus, *Pr.*: Pistelli, H. 1888. *Iamblichi Protrepticus ad Fidem Codicis Florentini*. Leipzig (repr. Stuttgart: 1967).
Iamblichus, *VP.*: Klein, U. (post L. Deubner) 1937. *Iamblichi de Vita Pythagorica Liber*. Leipzig (repr. Stuttgart: 1975).
Marinus, *In Dt.*: Heiberg, J. L. & H. Menge. 1896–1916. *Euclidis Opera Omnia*. Vol. 6. Leipzig.
Olympiodorus, *In Phaed.*: Westerink, L. G. 1976. *The Greek Commentaries on Plato's Phaedo*. Vol. 1. Amsterdam.
Pappus, *Coll.*: Hultsch, F. 1876–1878. *Pappi Alexandrini Collectionis quae supersunt*. Berlin.
Philoponus, *In de An.*: Hayduck, M. 1897. *Ioannis Philoponi in Aristotelis de Anima Libros Commentaria*. Commentaria in Aristotelem Graeca 15. Berlin.
Photius, *Bibl.*: Henry, R. 1959–1977. *Photius: Bibliothèque*. 8 vols. Paris.
Plato, *Crat.*: Burnet, J. 1900–1907. *Platonis Opera*. Vol. 1. Oxford (repr.1967).
Plato, *Phaedrus*: Burnet, J. 1900–1907. *Platonis Opera*. Vol. 2. Oxford (repr. 1967).
Plato, *Theaet.*: Burnet, J. 1900–1907. *Platonis Opera*. Vol. 1. Oxford (repr.1967).
Proclus, *In Eucl.*: Friedlein, G. 1873. *Commentarii in Primum Euclidis Elementorum Librum*. Leipzig.
Psellus: Littlewood, A. R. 1985. *Michaelis Pselli Oratoria Minora*. Leipzig.
Sappho: Lobel, E. & D. L. Page. 1955. *Poetarum Lesbiorum Fragmenta*. Oxford (repr.1968).
Serenus: Heiberg, J. L. 1896. *Sereni Antinoensis Opuscula*. Leipzig.
Simplicius, *In Epict.*: Dübner, F. 1842. *Theophrasti Characteres*. Paris.
Stobaeus, *Anth.*: Wachsmuth, C. & O. Hense. 1884–1912. *Ioannis Stobaei Anthologium*. 5 vols. Berlin (repr. 1958).
Suda: Bekkeri, A. 1854. *Suidae Lexicon*. Berlin.
Theognostus, *Can.*: Cramer, J. A. 1835. *Anecdota Graeca e codd. Manuscriptis Bibliothecarum Oxoniensium*. Vol. 2. Oxford (repr. Amsterdam: 1963).
Timaeus [Ps.]: Thesleff, H. 1965. *The Pythagorean Texts of the Hellenistic Period*. Åbo.

Works quoted

Acerbi, F. (ed.) 2007. *Euclide: Tutte le Opere*. Milan.

Acerbi, F. 2008. "Euclid's Pseudaria." In: *Archive for History of Exact Sciences* 62, 511–551.

Blumenthal, H. J. 1981. "Pseudo-Elias and the Isagoge Commentaries Again." In: *Rheinisches Museum* 124, 188–192.

Bulmer-Thomas, I. 1974. "Menaechmus." In: C. C. Gillispie (ed.), *Dictionary of Scientific Biography*. Vol. 9. New York, 268–277.

Bonazzi, M. 2013. "Eudorus of Alexandria and the 'Pythagorean' Pseudepigrapha." In: G. Cornelli *et al.* (eds.), *On Pythagoreanism*. Berlin, 385–404.

Burkert, W. 1972. *Lore and Science in Ancient Pythagoreanism*. Cambridge, MA.

Calder, W. M. 1925. "The Royal Road in Herodotus." In: *The Classical Review* 39, 7–11.

Centrone, B. 2014. "The pseudo-Pythagorean Writings." In: C. A. Huffman (ed.), *A History of Pythagoreanism*. Cambridge, 315–340.

Decorps-Foulquier, M. 1992. "L'époque où vécut le géomètre Sérénus d'Antinoé." In: J.-Y. Guillaumin (ed.), *Mathématiques dans l'Antiquité*. Saint-Etienne, 51–58.

Graf, F. D. 1994. "The Persian Royal Road System." In: A. Kuhrt *et al.* (eds.), *Achaemenid History VIII: Continuity and Change*. Leiden, 167–189.

Hammerstaedt, J. 1997. "Photios über einen verlorenen Codex mit Autoren des vierten Jahrhunderts n.Chr. aus Mittel- bzw. Oberägypten." In: *Zeitschrift für Papyrologie und Epigraphik* 115, 105–116.

Harari, O. 2015. Essay review of Riedlberger 2013. In: *Mnemosyne* 68, 162–165.

Heath, T. L. (ed.) 1956. *The Thirteen Books of Euclid's Elements* (orig. Cambridge 1908). 3 vols. New York.

Horky, P. S. 2013. *Plato and Pythagoreanism*. Oxford.

Karasmanis, V. 1987. *The Hypothetical Method in Plato's Middle Dialogues*. Diss. Oxford.

Kaster, R. A. 1988. *Guardians of Language: The Grammarian and Society in Late Antiquity*. Berkeley.

Lloyd, G. E. R. 2012. "Mathematics and Narrative: An Aristotelian Perspective." In: Doxiadis, A. & B. Mazur (eds.), *Circles Disturbed: The Interplay of Mathematics and Narrative*. Princeton, 389–406.

Lo Bello, A. (ed.) 2003. *The Commentary of al-Nayrizi on Book I of Euclid's Elements of Geometry*. Boston.

Morrow, G. R. (ed.) 1970. *Proclus: A Commentary on the First Book of Euclid's Elements*. Princeton.

Mueller, I. 1987. "Iamblichus and Proclus' Euclid Commentary." In: *Hermes* 115, 334–348.

Netz, R. 1999. *The Shaping of Deduction in Greek Mathematics: A Study in Cognitive History*. Cambridge.

O'Meara, D. J. 2014. "Iamblichus' On the Pythagorean Life in context." In: C. A. Huffman (ed.), *A History of Pythagoreanism*. Cambridge, 399–415.

Riedlberger, P. (ed.) 2013. *Domninus of Larissa, Encheiridion and Spurious Works*. Pisa.

Rohde, E. 1871. "Die Quellen des Jamblichus in seiner Biographie des Pythagoras." In: *Rheinisches Museum für Philologie* 26, 554–576.

Rohde, E. 1872. "Die Quellen des Jamblichus in seiner Biographie des Pythagoras." In: *Rheinisches Museum für Philologie* 27, 23–61.

von Rohden, P. 1893. "Aelius (137)." In: Pauly A. & G. Wissowa (eds.), *Realencyclopädie der classischen Altertumswissenschaft*. Vol. 1. Stuttgart, 532.

Staab, G. 2002. *Pythagoras in der Spätantike. Studien zu 'De Vita Pythagorica' des Iamblichos von Chalkis*. Leipzig.

Theodoridis, C. 1976. *Die Fragmente des Grammatikers Philoxenos*. Berlin.

Thesleff, H. 1965. *The Pythagorean Texts of the Hellenistic Period*. Åbo.

Thom, J. C. 2013. "The Pythagorean Akousmata and Early Pythagoreanism." In: G. Cornelli *et al.* (eds.), *On Pythagoreanism*. Berlin, 77–102.

Vitrac, B. & M. Caveing (eds.) 1990–2001. *Euclid: Les Éléments*. 4 vols. Paris.

Westerink, L. G. 1967. *Pseudo-Elias (Pseudo-David), Lectures on Porphyry's Isagoge*. Amsterdam.
Wildberg, C. 1990. "Three Neoplatonic Introductions to Philosophy: Ammonius, David, Elias." In: *Hermathena* 149, 33–51.
Zhmud, L. 2006. *The Origin of the History of Science in Classical Antiquity*. Berlin.
Zhmud, L. 2012. *Pythagoras and the Early Pythagoreans*. Oxford.

Claas Lattmann

Diagrammatizing Mathematics: Some Remarks on a Revolutionary Aspect of Ancient Greek Mathematics

Abstract: This paper takes a fresh look at the ancient mathematical diagram from the viewpoint of semiotic model theory. It discusses, first, the ontological nature and epistemic function of diagrams; second, the characteristics of their interrelation and interaction with the textual part of mathematical proofs; third, their epistemic validity; and, fourth, their 'revolutionary' nature as per Thomas S. Kuhn's theory of scientific revolutions. The analysis is based on an understanding of ancient mathematical diagrams as 'diagrams' as defined by Charles S. Peirce. Such a 'diagram' is a specific sort of iconic sign, and thus 'model,' that represents the relational structure of its object by way of its own relational quality. Against this backdrop, this paper shows: (1) that the diagram stands as a holistic model for a general mathematical structure (2) that consists of 'diagrammatical atoms' (3) that represent the general entities of mathematical theory; (4) and that this abstract model is being constructed and subsequently analyzed in the textual part of the proof, (5) the purpose of which is, accordingly, to 'show' that this model has the specific sought-for, yet unknown mathematical properties. (6) Given this, Greek mathematical proof turns out to be an analysis of static general models, rather than an either implicit or failed (mathematically) inductive argumentation concerned with particular cases.

Keywords: Greek Mathematics, Diagrams, Models, Icons, Thomas S. Kuhn, Euclid, Plato, Charles S. Peirce.

I Introduction

One of the most striking features of ancient Greek mathematics is the ubiquitous use of diagrams: for example, there is not a single proof in Euclid's *Elements* that is not accompanied by a diagram. Surprisingly, this is the case not only with regard to the geometrical propositions of the *Elements*, but also with regard to the propositions concerned with arithmetic and proportion theory (books VII–IX and V, respectively).[1]

1 Cf. Netz 1999a, 40–43; esp. at 42: "That diagrams were considered essential for mathematics is proved by books V, VII–IX of Euclid's *Elements*. There, all the propositions are accompanied by diagrams, as individual—and as far as the situations allow—as elaborate as any geometrical diagram. Yet, in a sense, they are redundant, for they no longer represent the situations discussed."

https://doi.org/10.1515/9783110565959-006

Despite its apparent methodological centrality, the ancient Greek mathematical diagram has, until recently, either been ignored or seen as some sort of superfluous embellishment—or, to the contrary, as something that, in principle, invalidates the mathematical proof it is supposed to illustrate. If one assumes that the heart of ancient mathematics is its 'objective mathematical content,' as expressed in modern symbolic notation,[2] diagrams not only add nothing to, but, by introducing and referring to visual elements, pose a severe threat to its logical validity.[3] Nonetheless, from a history of science point of view, it is doubtful (and, actually, has been doubted) whether it is appropriate to regard ancient mathematics as a more-or-less deficient form of modern mathematics, as such a critique would presuppose.[4] It might have been a phenomenon *sui generis*, separated from modern mathematics by a scientific 'revolution' as defined by Thomas S. Kuhn.[5] This would make both forms of mathematics incommensurably different, and these differences might also relate to the specific nature and function of the diagrams. Therefore, if we found a way of describing ancient mathematics not from the perspective of modern mathematics, there might be a possibility of better understanding and appreciating the ontological and epistemological status of ancient mathematical diagrams.[6]

In this vein, this paper proposes a new reading of the ancient mathematical diagram from the perspective of semiotic model theory. On the basis of understanding

2 Cf. Netz 1999a, 12: "That diagrams play a crucial role in Greek mathematics is a fact often alluded to in the modern literature, but little discussed. The focus of the literature is on the verbal aspect of mathematics." Cf., e.g., Heath 1921, and Mueller 1981. It is telling that in Heath's 1926 edition of Euclid's *Elements*, the index does not contain the lemma 'diagram'. In modern mathematics, diagrams are superfluous indeed; see Giaquinto 2008, and cf. Hilbert 1899, and Tarski 1959.

3 See, e.g., Manders 2008a, 65–68 and cf. section II below.

4 Cf. the controversy with regard to 'geometric algebra,' in particular Unguru's 1975 seminal article and, recently, Sialaros & Christianidis 2016, esp. at 129–133. For the broader picture, see Netz 2003.

5 See Kuhn 1996.

6 The ever-growing interest in the ancient mathematical diagram is evidenced by an ever-increasing number of studies on this topic in recent times from a history and / or philosophy of science perspective. Often they, too, abandon the perspective of modern mathematics and, thus, adhere to the methodological principle underlying this paper; however, no consensus as to what a more adequate perspective might be has been reached yet. Cf., e.g., Manders 2008b, Norman 2006, Miller 2007, Mumma 2006, Mumma 2008a, Mumma 2008b, Avigad et al. 2009, Mahr & Robering 2009, Macbeth 2010, Mumma 2010, Coliva 2012, Mumma 2012, Mumma & Panza 2012. One of the starting-points of this discussion is Netz's 1999a study. Cf. the following note.

diagrams as signs that represent their objects as iconic relational 'models,'[7] I will argue, it is possible to adequately recognize, first, their nature and, second, their specific function in the process of ancient mathematical reasoning. As a first step, I will briefly expound why diagrams have been regarded as problematic, followed by, second, an explication of ancient mathematical diagrams as a specific form of iconic signs as defined in the framework of Charles S. Peirce's theory of signs. Third, I will discuss how diagrams relate to the corresponding textual part of the mathematical propositions. Finally, some suggestions will be made as to how classical ancient Greek mathematics had 'revolutionary' features that made it incommensurably different from all the preceding ancient mathematical approaches (including those of early Greece and Egypt / Mesopotamia) as well as from modern mathematics.

II The Problematic Ontological and Epistemological Status of the Ancient Mathematical Diagram

Ancient mathematical diagrams have often been regarded as problematic. A cursory look at a representative proposition makes it transparent why this is the case. To keep it simple, consider proposition I.1 in Euclid's *Elements:*[8]

> Ἐπὶ τῆς δοθείσης εὐθείας πεπερασμένης τρίγωνον ἰσόπλευρον συστήσασθαι.
> Ἔστω ἡ δοθεῖσα εὐθεῖα πεπερασμένη ἡ AB.
> Δεῖ δὴ ἐπὶ τῆς AB εὐθείας τρίγωνον ἰσόπλευρον συστήσασθαι.
> Κέντρῳ μὲν τῷ A διαστήματι δὲ τῷ AB κύκλος γεγράφθω ὁ ΒΓΔ, καὶ πάλιν κέντρῳ μὲν τῷ B διαστήματι δὲ τῷ BA

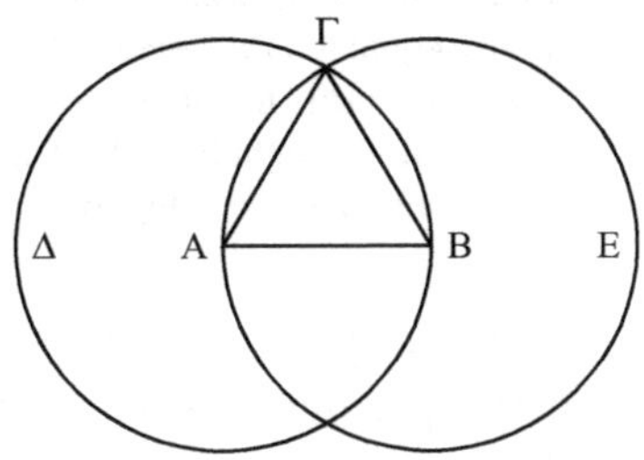

7 Such a proposal has not been put forward yet. Though Netz 1999a also discusses the ancient diagram from a semiotic perspective, his results are quite different from the ones that I propose here; *mutatis mutandis* this is also the case with regard to Mahr & Robering 2009. This being said, Macbeth 2010 has pointed to some of the semiotic aspects explicated here. Nonetheless, as will be shown, her views differ significantly from those this paper advocates, especially with regard to the whole diagrammatic structure. The semiotic model-theoretic approach used for the following analysis has been put forward in Kralemann & Lattmann 2013b; for an overview on the contemporary theory of models, see, e.g., Frigg & Hartmann 2017, and Bailer-Jones 2009.

8 In this paper, Euclid is quoted from Heiberg & Stamatis's edition, and the translations follow Heath 1926 (unless otherwise noted). However, in the example, I have added a translation of the athetized sentence and marked off the six parts of the proposition: cf. section IV below (also for a more adequate translation of the *ekthesis* and, by implication, the *kataskeue*).

κύκλος γεγράφθω ὁ ΑΓΕ, καὶ ἀπὸ τοῦ Γ σημείου, καθ' ὃ τέμνουσιν ἀλλήλους οἱ κύκλοι, ἐπὶ τὰ Α, Β σημεῖα ἐπεζεύχθωσαν εὐθεῖαι αἱ ΓΑ, ΓΒ.
Καὶ ἐπεὶ τὸ Α σημεῖον κέντρον ἐστὶ τοῦ ΓΔΒ κύκλου, ἴση ἐστὶν ἡ ΑΓ τῇ ΑΒ· πάλιν, ἐπεὶ τὸ Β σημεῖον κέντρον ἐστὶ τοῦ ΓΑΕ κύκλου, ἴση ἐστὶν ἡ ΒΓ τῇ ΒΑ. ἐδείχθη δὲ καὶ ἡ ΓΑ τῇ ΑΒ ἴση· ἑκατέρα ἄρα τῶν ΓΑ, ΓΒ τῇ ΑΒ ἐστιν ἴση. τὰ δὲ τῷ αὐτῷ ἴσα καὶ ἀλλήλοις ἐστὶν ἴσα· καὶ ἡ ΓΑ ἄρα τῇ ΓΒ ἐστιν ἴση· αἱ τρεῖς ἄρα αἱ ΓΑ, ΑΒ, ΒΓ ἴσαι ἀλλήλαις εἰσίν.
Ἰσόπλευρον ἄρα ἐστὶ τὸ ΑΒΓ τρίγωνον, καὶ συνέσταται ἐπὶ τῆς δοθείσης εὐθείας πεπερασμένης τῆς ΑΒ.
[Ἐπὶ τῆς δοθείσης ἄρα εὐθείας πεπερασμένης τρίγωνον ἰσόπλευρον συνέσταται]· ὅπερ ἔδει ποιῆσαι.

[*Protasis:*] On a given finite straight line to construct an equilateral triangle.
[*Ekthesis:*] Let AB be the given finite straight line.
[*Dihorismos:*] Thus it is required to construct an equilateral triangle on the straight line AB.
[*Kataskeue:*] With centre A and distance AB let the circle BCD be described; again, with centre B and distance BA let the circle ACE be described; and from the point C, in which the circles cut one another, to the points A, B let the straight lines CA, CB be joined.

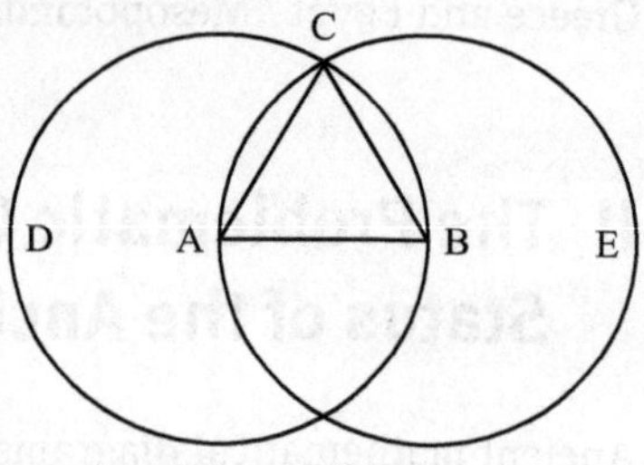

[*Apodeixis:*] Now, since the point A is the centre of the circle CDB, AC is equal to AB. Again, since the point B is the centre of the circle CAE, BC is equal to BA. But CA was also proved equal to AB; therefore each of the straight lines CA, CB is equal to AB. And things which are equal to the same thing are also equal to one another; therefore CA is also equal to CB. Therefore the three straight lines CA, AB, BC are equal to one another.
[*Symperasma:*] Therefore the triangle ABC is equilateral; and it has been constructed on the given finite straight line AB. [Therefore, on a given finite straight line an equilateral triangle has been constructed.] (Being) what it was required to do.

From the viewpoint of modern logic and mathematics, the specific use of the diagram in this proposition seems to be problematic.[9] Two points are especially worrisome:

(1) Whereas the mathematical task whose possibility the proposition is supposed to prove is given in general terms at the beginning (the *protasis*), the rest of the

9 The diagram shown here is a reproduction of the one printed in Heiberg & Stamatis's edition. However, this edition cannot, as modern editions of Euclid's *Elements* in general, be regarded as a critical edition with regard to the diagrams: see Saito 2009, esp. at 817–825, and Saito & Sidoli 2012. This fact, however, will not prove problematic for the approach put forward here, for it operates on a more fundamental, *semantic* level. On the general editorial problems connected with ancient mathematical diagrams, see, e.g., Netz 2004, 8–10 with regard to Archimedes (cf. Netz 2012, esp. at 164–176); see also, more comprehensively, Saito 2006, and, specifically with regard to the variants of the diagrams found in the Arabic manuscript tradition of Euclid's *Elements*, see De Young 2012. One of the main goals scholars are attempting to achieve with regard to ancient mathematical diagrams is to provide a critical edition of them. The following discussion will suggest, however, that such an edition should aim not simply at reproducing the (variation in the) outer appearance of the diagrams, but at giving an appropriate representation of their (unique) *semantic* content, which consists in the specific *abstract* mathematical relations they exhibit.

proposition (and especially the proof proper, the *apodeixis*) apparently deals with particular objects that are instantiated in the accompanying diagram.[10] Although the mathematical reasoning validly leads to the conclusion (the *symperasma*), such a conclusion alone, which only refers to the particular elements of the diagram, would not allow for the generalizing step taken from there to the universal *protasis*. This situation prompts the crucial question of what was thought to guarantee the epistemic validity of the proof procedure in Greek mathematics in principle[11]—and whether it *actually* did what it was supposed to do.

(2) A second worry is that for accomplishing the proof, both the text of the proposition and the diagram seem to be relevant, but either one alone is, apparently, insufficient: on the one hand, the diagram relies on the text, for it evidently cannot constitute a proof in the proper sense on its own, at least with regard to what counts as a mathematical proof in the context of Greek mathematics;[12] on the other hand, the text refers to, and thus intricately relies on, the diagram, in particular insofar as some of the proof-relevant mathematical objects are identifiable only with the visual help of the diagram, such as the triangle ABC in proposition I.1.[13] Given this, it seems to follow that "the diagram is not directly recoverable from the text,"[14] so that the latter is, at least in part, dependent upon the former. However, it is not easy to understand and explicate how exactly diagram and text interact.[15]

Taking these two worries together, the use of diagrams in ancient Greek mathematics seems to severely undermine its logical validity. Diagrams appear to have been treated as some sort of black box, which somehow miraculously provided the desired results, without its being clear (or even of interest) to mathematicians how they managed to do this.[16] From the perspective of the history of science, this is a very curious situation, not the least in view of the obvious methodological centrality diagrams had.

---

10 On this problem, see Mueller 1981, 12–14, and Netz 1999a, 240–270.

11 Cf. Netz 1999a, 142.

12 See, e.g., Netz 1999a, 12–88.

13 In a different conceptual framework, an exclusively diagrammatical proof could, of course, be both possible and sufficient: cf. Miller 2007 (also Mahr & Robering 2009, 292–294, and Coliva 2012) and the studies referred to in n. 6 above.

14 Netz 1999a, 19.

15 Netz 1999a, 19. However, see also Patterson 2007, 11–25.

16 Cf. Netz 1999a, 56–57. On the whole issue of mathematical proof in antiquity, cf. Chemla 2012a, in particular her 2012b introduction.

III The Ancient Mathematical Diagram from a Semiotic Model-Theoretic Perspective

However severe they may seem, the problems connected with the ancient mathematical diagram can be solved. The conceptual basis for such a solution is to regard diagrams as signs, and specifically as *icons,* that is 'models.' This opens up the possibility of looking at them from the perspective of contemporary semiotic model theory—and this, quite surprisingly, in a way very similar to how Plato speaks about them in the context of the 'Divided Line.'[17] This suggests that the proposed solution may not only be adequate from a *philosophy* of science, but also from a *history* of science point of view, for it would be *(mutatis mutandis)* compatible with (at least some of) the ancients' views on the practice of mathematics itself. This being said, such a convergence of ancient and modern views is, in principle, not a necessary precondition for the historical validity of the following analysis. Even if it were the case that the ancient mathematicians had a different understanding of diagrams and their use, whether in theory or in practice, it is, in principle, the case that ancient diagrams can be investigated via semiotics. The only precondition for doing so is to accept that diagrams can be regarded as signs. That this is possible, and arguably even appropriate, is evident. Therefore, semiotics, as the theory of signs in general, is also *a fortiori* the theory of diagrams in particular.

This methodological clarification leads to the introduction of the model-theoretic approach itself: every diagram is a representation, for it stands for something other than itself. For example, the lines making up a triangle in a diagram are not only lines that are combined in a specific spatial configuration, but they also represent (and, thus, stand for) the sides of a mathematical 'triangle.' Every representation, in turn, is by definition a sign, at least from a semiotic perspective, and in particular from the point of view of Charles S. Peirce's theory of signs. *Qua* sign a diagram therefore is "something which stands to somebody for something in some respect or capacity,"[18] implying (*inter alia*) that it has an 'object' (which it represents) and a 'meaning' (more or less what Peirce calls the 'interpretant').

A diagram, however, is not simply a sign. Given Peirce's general exhaustive classification of all signs into *symbols, indices,* and *icons,* it necessarily follows that diagrams must belong to the class of *icons.*[19] A *symbol,* first, is a sign that represents its object on the basis of habituation or convention, such as ordinary words. An *index,*

17 Given the complexity of the 'Divided Line,' this similarity cannot be explicated here in detail; however, cf. below with n. 28.

18 So the general definition of a sign in Peirce CP 2.228. Cf. Kralemann & Lattmann 2013b, 3400–3404, and see, comprehensively, Short 2007.

19 On this classification of signs, see Peirce CP 2.247–249; cf. Lattmann 2012, 537–539, Kralemann & Lattmann 2013b, 3404–3410. This classification is only one among many, though a very important one.

second, is a sign that represents its object on the basis of an existential, that is actual relation to its object; an example is a weather vane, insofar as it represents the direction of the wind in a given moment in time by being made to point in a certain direction by the wind itself. Finally, an *icon* is a sign that represents its object on the basis of a similarity between its own perceivable properties and the properties of its object, such as portraits or photographs—where 'similarity,' *nota bene*, does not imply a relation of identity or isomorphism. It suffices that the relevant properties of sign and object may (at least potentially) be regarded as similar, in whatever (in principle, subjective) way someone deems appropriate. In the framework of this exhaustive classification, it is evident that diagrams are to be classified as *icons*, for a diagram (*qua* diagram) neither represents primarily on the basis of habituation (such as a *symbol*), nor on the basis of an existential relation (such as an *index*). Therefore, diagrams must be *icons* and, thus, they represent on the basis and by way of a similarity to the object they represent.

The insight that diagrams are *icons* can be further refined. First, it has to be recognized that *icons* are nothing but models, for the two notions are identical. As will be discussed below in detail, the recognition of this identity will allow us to understand the specific manner of representation diagrams exhibit, including the nature of their characteristic iconic similarity. Second, it is possible to determine the specific kind of *icon* diagrams are if one takes into account Peirce's further exhaustive classification of all *icons* into (as he calls them) *images, diagrams,* and *metaphors.*[20] An *image* exhibits a direct, simple similarity to its object, such as a portrait when it is interpreted as some form of 'copy' of the human being it represents; a *diagram* exhibits a similarity to its object with regard to its inner structure, which is given by the direct, pairwise relations among its constitutive elements; and a *metaphor* implies a similarity between the habitual object of one sign to the object of another sign by representing a corresponding semiotic quality connected to the first subordinated sign.[21] Obviously, Peirce's notion of *icon* allows for more sophisticated and complex forms of similarity than those commonly termed 'iconic' (which, in turn, would pertain only to the class of *images*).

In the framework of this exhaustive classification, ancient mathematical diagrams must be assigned to that class of *icons* which Peirce calls 'diagrams,' for they evidently possess a directly relational quality; that is, a quality that is expressible by way of one or many two-term relations. The core of the capability of these signs

---

20 On both points, see Kralemann & Lattmann 2013b; cf. Kralemann & Lattmann 2013a.

21 With regard to the philosophical basis of this classification (i.e., Peirce's theory of categories, which is based on the notion of relation) and with regard to metaphors in particular, see Lattmann 2012. An example of a *metaphor* is the sentence 'Achilles is a lion,' which implies a novel semantic quality of the word 'lion' for what, habitually, the word 'Achilles' represents, and this on the basis of an (implied) similarity between a lion and Achilles. Cf. Kralemann & Lattmann 2013b.

to represent is that they, *qua* diagram, exhibit an inner structure that consists of the mathematical relations connected with their constitutive elements. Our example proposition from Euclid's *Elements* (I.1) makes clear what this means. In this proposition, the 'diagram' represents specific direct relations among the points A, B, the lines AB, AC, BC, the circles ACE, BCD, and the triangle ABC, and this it does within the framework of an encompassing holistic structure, that is, itself.[22] Thus, the diagram is a representation based on, and by way of, specific relations it possesses *qua* diagram, and its perceivable qualitative character with regard to these relations corresponds to certain 'mathematical' relations that are supposed to be the semantic core of what the diagram expresses.

Accordingly, ancient mathematical diagrams are to be classified as 'diagrammatical *icons*.' However, the very same holds true for the individual parts of the diagram, too, that is, for the points, lines, circles etc. that make up the whole diagrammatical structure. On the one hand, they represent the basic notions of mathematical theory, as defined, for example, by the definitions at the beginning of Euclid's *Elements*.[23] On the other hand, they do not represent these basic notions in the form of words (that is, semiotically, *symbols*), such as 'circle' in English or κύκλος in Greek, but, evidently, as signs that directly exhibit the essential mathematical relations that are characteristic of these basic notions. For example, Euclid defines a circle as "a plane figure contained by one line such that all the straight lines falling upon it from one point among those lying within the figure are equal to one another,"[24] and it is primarily mathematical relations such as these that the diagram of a circle is meant to exhibit and, thus, make accessible to perception.

In view of this, the individual parts of the whole diagram are iconic 'diagrams,' too. They can be regarded as the basic building blocks, or 'atoms,' of any given ancient mathematical diagram—which, accordingly, is nothing but the combination

22 This explains Mander's observation that nowhere in Euclid's *Elements* are exact properties relevant, but only (as he calls them somewhat misleadingly) "co-exact" properties (see Manders 2008b, esp. at 91–94). The latter are properties "that are insensitive to the effects of a range of variation in diagram entries" (Manders 2008a, 69) or, in the terms of modern mathematics, they are properties of which one might say that "they express the topology of the diagram" (Manders 2008b, 93). Using the modern concept of topology is, however, *historically* problematic, not so much because it is a *modern* theoretical approach, but because it introduces categories that cannot be adequately mapped onto the specific features of the ancient mathematical situation.

23 Of course, from a contemporary point of view, these definitions are clearly not sufficient for proving what propositions are meant to prove: Euclidean proofs "appear more full of holes than a Swiss cheese" (Patterson 2007, 18; cf. Mumma 2010, and Neuenschwander 1973, esp. at 326). See, in particular, Patterson 2007, 14–18 on how it is especially diagrams that make accessible and use the implicit assumptions etc. left out by the definitions.

24 Euclid, *Elem.* I. df. 15: Κύκλος ἐστὶ σχῆμα ἐπίπεδον ὑπὸ μιᾶς γραμμῆς περιεχόμενον [ἣ καλεῖται περιφέρεια], πρὸς ἣν ἀφ' ἑνὸς σημείου τῶν ἐντὸς τοῦ σχήματος κειμένων πᾶσαι αἱ προσπίπτουσαι εὐθεῖαι [πρὸς τὴν τοῦ κύκλου περιφέρειαν] ἴσαι ἀλλήλαις εἰσίν.

of such atoms into a specific structure. This structure, in turn, is a specific mathematical situation that is intentionally generated for the purpose of proving a mathematical assertion or the possibility of a mathematical task, as expressed by the *protasis* of a proposition (that is, the former in the case of a so-called *theorema,* the latter in the case of a so-called *problema:* cf. section IV below).

However, from the general definition of a sign it follows that the perceivable diagram is not identical to its semiotic object, for the thing that represents, in principle, is not the same as the thing that is represented. In consequence, one has to further distinguish between the printed diagram—that is, the perceivable sign itself—and its semiotic object.[25] Given this, it is primarily not the (printed) *representamen* ('sign'), but only its object that possesses the universal mathematical properties that are represented by the sign. This it does without possessing additional non-mathematical properties, which would not be relevant in the context of mathematics, in general, or any given mathematical proposition, in particular. Therefore it is this semiotic object that serves as the mathematical diagram proper, whereas the printed diagram, though *qua icon* by definition being similar to the abstract diagram in all the mathematically relevant aspects, displays certain additional properties that are not relevant from a mathematical point of view and might even interfere with the mathematical properties proper. For example, the printed diagram also possesses attributes that relate to its locatedness in space and time, such as a specific width and a specific color, and, even worse, its points have an extension and its straight lines are not mathematically straight. Actually, the printed diagram can depart very far from (so-to-speak) exactly 'copying' the 'mathematical' diagram. However, there is, in principle, a mapping between each other's attributes and relations, irrespective of how different they are; and it is exactly this mapping that defines iconic similarity in general and, thus, forms the semantic core of any iconic representational relation.[26]

Therefore, only the abstract diagram the printed diagram stands for has to be regarded as the mathematical diagram proper.[27] This is in general accordance, as mentioned above, with the interpretation of the ancient mathematical diagram

25 This is one of the main differences to Netz's 1999a analysis: "I have argued that letters are primarily indices, so that representations employing them cannot but refer to the concrete diagram" (51). This leads him to assuming that "the option that the diagram points towards an ideal mathematical object can be disposed of" (51). However, according to Peirce (on which Netz, too, bases his analysis), *indices* can, of course, have a generic character. One may only think of the demonstrative pronoun 'this,' which Peirce often adduces as an example of *indices*. In general, the definition of *indices* only implies that there is an existential relation between sign and object; however, this existential relation can have a particular or a generic (or a potential) nature (which fact implies, in accordance with Peirce's three fundamental categories, yet another exhaustive threefold classification of all signs).

26 See Kralemann & Lattmann 2013b.

27 The 'printed diagram' may be regarded as an 'example' of this more abstract diagram. On this epistemic function of diagrams in general, see Patterson 2007, 11–13.

given by as eminent an ancient philosopher as Plato, at least in the 'Divided Line' (esp. at *Resp.* 510c1–511b2). In this passage, Plato distinguishes between the diagram seen with the eyes and the diagram seen with the faculty of *dianoia*, or, put another way, between a (visually) perceivable and an intelligible, but in all the relevant respects equivalent, diagram.[28] Of course, some of the shortcomings of the perceivable (printed) diagram had probably already been noted before Plato, most notoriously by Protagoras (cf. DK80 B7). Plato, however, was the first to posit the existence of an intelligible representational object of the visible diagram and, in the framework of his theory of Forms, to thus attempt to propose a novel and original solution to this fundamental problem.

To sum up, the preceding analysis has argued for an interpretation of ancient mathematical diagrams as signs and, specifically, as diagrammatical *icons* as defined by Charles S. Peirce's theory of signs. This has led to the following insights: diagrams represent a holistic relational structure consisting of simple diagrammatical *icons*, which represent the basic notions of mathematical theory, which themselves are (constituted by) nothing but (direct) relations; accordingly, the whole diagram represents a specific mathematical situation consisting of mathematical relations; and, finally, the printed diagram is not identical to its abstract semiotic object, the mathematical diagram proper, though, in principle and by definition, the one is similar to the other with regard to at least the mathematically relevant relations.

IV The Diagram and the Parts of a Proposition

The semiotic philosophy of science analysis given in the preceding section has yielded valuable insights into the ontological nature of the ancient mathematical diagram. A subsequent step will use these insights to investigate its epistemic dimension. In this vein, our example proposition (Euclid, *Elem.* I.1) will be analyzed with a view on how the diagram and the text relate to each other.[29]

Following the, at least according to Proclus, traditional ancient view, a mathematical proposition has, at most, six parts: the *protasis* (πρότασις, 'enunciation'), the *ekthesis* (ἔκθεσις, 'setting-out'), the *dihorismos* (διορισμός, 'definition'), the *kataskeue* (κατασκευή, 'preparation'), the *apodeixis* (ἀπόδειξις, 'proof'), and, finally,

28 Cf. the (difficult) passage Plato, *Resp.* 510d5–511a1. These sweeping remarks must suffice for the present purpose; I will propose an in-depth interpretation of this model elsewhere.
29 On the potential problems resulting from using Euclid, *Elem.* I.1 as an example to discuss the parts of a proposition, see Netz 1999b, 289. However, cf. the following note and the remarks on the *symperasma* below.

the *symperasma* (συμπέρασμα, 'conclusion').[30] The model-theoretic framework put forward here can show that each of these parts has a specific function that relates in a unique way to the diagram of the proposition. Interestingly enough, the following results have, *mutatis mutandis*, a surprising similarity to the description given by Proclus himself. In view of this convergence, the analysis explicated here, though originating from a decidedly modern perspective, may make a claim of being *historically* adequate and, thus, of contributing to a better understanding of the *ancient* perspective on mathematics, namely by making transparent what the factual basis of Proclus's assessment of the mathematical situation might have been.

(1) The *protasis* states the mathematical claim the whole proposition is going to prove in general terms. In the given example ("On a given finite straight line to construct an equilateral triangle"), the goal is to prove the possibility of constructing a specific mathematical object, that is an isosceles triangle, in such a way that it has a specific relation to a certain given mathematical object, here a straight line of finite length.

(2) In the example proposition, the *ekthesis* consists of the sentence Ἔστω ἡ δοθεῖσα εὐθεῖα πεπερασμένη ἡ AB. Contrary to the traditional view—compare Heath's translation "Let AB be the given finite straight line"—, this sentence should be translated as "Let there be the given finite straight line, that is, AB"; that is, the predicate ἔστω should not be understood as a copula connecting the subject 'AB' with the predicate noun 'the given finite straight line' (or *vice versa*), but as having an existential meaning relating to the subject 'the given finite straight

30 Cf. the extensive discussion in Proclus, *In Eucl.* 203.1–205.12, esp. at 203.5–15; cf. also 205.13–210.16 for the application of this schema to Euclid, *Elem.* I.1; I render the terms in accordance with (or, in the case of *dihorismos* and *kataskeue*, on the basis of) Netz 1999b, 294–302. The schema can also be found (*inter alia*) in Hero, *Def.* 137.1. On some of the problems that scholars have seen with regard to the division of the proposition as per Proclus's analysis, see Netz 1999a, 253, and, more comprehensively, Netz 1999b. One of the main points of criticism is that "the structure of Greek propositions is not as rigid as Proclus seems to imply. There is much variability, which shows that this conventionalisation [...] could not have been the result of an explicit codification" (Netz 1999a, 253). This problem, however, can be solved by understanding Proclus's division primarily in a functional way. This means that we should expect to find (more or less) all of what Proclus describes in any proposition (though with certain exceptions and / or modifications, depending on the specific case), but not necessarily in the same order and as strictly separated as in Euclid's *Elements*. Put differently, the evident variability we encounter should not be understood as revealing (at least, *per se*) the inadequacy of Proclus's analysis, but rather the possibility of different modes of presentation in different mathematical authors and / or works and / or situations, while in any case the specific mathematical functions of the different parts as enumerated by Proclus are being preserved. If so, the variability would, in principle, not be problematic at all. This, in particular, would, for obvious reasons, hold true for those propositions in which mathematical objects are also constructed within the proof proper *(apodeixis)* (cf. Netz 1999b, 290–294). Similarly, the validity of Proclus's description does not depend on the terminology he uses: on this problem, cf. Netz 1999b, 294–302. On the ancient (especially mathematical) *stoicheia* literature in general, see Asper 2007, 94–212.

light,' which is supplemented by the specifying apposition 'AB.'[31] Given this, the *ekthesis* does nothing but produce a finite straight line and, in the process of doing this, assigns the name 'AB' to it. In effect, the *ekthesis* generates something that instantiates the mathematical object '(finite) straight line' and, thus, has the character of an iconic model that represents this peculiar mathematical object, the mathematically relevant properties of which inhere in it. In addition, we can see that, as this finite straight line is the only mathematical object referred to in the *protasis* as being given, the goal of the *ekthesis* is to produce the material starting-point, and thus 'model' the basis, for proving the *protasis*.

(3) As the wording in the example proposition implies ("Thus it is required to construct an equilateral triangle on the straight line AB"), the *dihorismos* reformulates the abstract *protasis* in the terms of the concrete diagrammatical atoms that have been produced in the *ekthesis*.

(4) Following upon the *dihorismos,* the *kataskeue* produces additional diagrammatical atoms, in the example proposition the circles BCD, ACE and the lines CA, CB. In principle, therefore, the *kataskeue* does the same as the *ekthesis*. However, there is one significant difference. The models produced in the *kataskeue* have not been referred to in the *protasis* and, accordingly, are not regarded as being 'given' for the whole proposition (that is, in the general sense of being the material starting-point for proving the *protasis*). They rather have the status of supplements to the initial configuration of the diagram and 'prepare' it for the subsequent proof proper, that is, insofar as they are necessary for accomplishing it. Taken together, *ekthesis* and *kataskeue* interdependently produce the sufficient basis (in the form of diagrammatical models, which combined are nothing but 'the' diagram) needed for the *apodeixis*.

(5) The *apodeixis,* the proof proper, proceeds by referring to specific relational properties that the whole diagrammatical model possesses with regard to its parts, that is, the diagrammatical atoms. It is, thus, aiming at laying the basis for recognizing that the mathematical task formulated in the *protasis* is actually possible (or, in the case of a *theorema,* that the mathematical assertion formulated in the *protasis* is true). In consequence, the *apodeixis* is not some sort of logical

31 Cf. Federspiel 1992, 15–17. For a more in-depth analysis, see Netz 1999a, 43–56, esp. at 44: "First, what such clauses do not assert: they do not assert a relation between a symbol and an object. Rather, they assert an action—in the case above, the taking for granted of a certain line—and they proceed to localise that action in the diagram, on the basis of an independently established reference of the letters." It has to be noted that the syntactical situation in the given example is obscured by the articulation of both δοθεῖσα εὐθεῖα πεπερασμένη and AB; this articulation, however, represents only a relative determination. Accordingly, we do not find an articulation in comparable contexts in, e.g., Euclid's *Elements* when the proposition does not deal with 'given' objects. Cf., in general, Kühner & Gerth § 461 n. 4 and, with regard to the (specifying) articulation of only the second element of such a phrase (such as in Ἔστω δύο τρίγωνα τὰ ΑΒΓ, ΔΕΖ in Euclid, *Elem.* I.4), Kühner & Gerth § 463, 3 A.

deduction (at least in the common sense), but a 'demonstration'; namely, a quasi-deictic demonstration that certain relations hold true in a specific mathematical setting that has been deliberately created for this very purpose.[32] Proving, therefore, is 'showing' (*apodeiknynai* or, alternatively, the simplex *deiknynai*) in a very literal sense. In view of this, Greek mathematical proof does not have a synthetic, but a genuinely analytic character: it is nothing but an explication of specific relational properties of a static diagram.[33]

(6) Finally, the *symperasma* reformulates the abstract *protasis* in terms of the model (that is, diagram) that has been produced and analyzed up to this point. This implies, at least in the example proposition, that that subpart of the diagram, the production of which was the goal of the proposition, is eventually identified as such—and thus shown to have been produced actually. Accordingly, the *symperasma* states that the triangle ABC, which has not been referred to before, is equilateral and stands on the given finite straight line AB. In hindsight, the triangle ABC is identified as a model of the searched-for mathematical object 'equilateral triangle,' which turns out to be nothing but a not yet recognized emergent relational property of the whole diagram. To provide such novel insights in general, however, is, to quote Peirce, "a great distinguishing property of the icon," and so *a fortiori* also of the diagram: "By the direct observation of it other truths concerning its object can be discovered than those which suffice to determine its construction" (Peirce CP 1.179). In other words, ancient mathematical diagrams are deliberately constructed in such a way as to possess and display emergent properties that expand the body of mathematical knowledge.

32 Cf. Mueller 1974 on the lack of logical deduction in Greek mathematics, though from a different viewpoint; cf. section V below. Macbeth 2010 introduces the notion of *icon* into the scholarly debate, but she identifies these *icons* (only) with the subordinate sub-models, i.e., the 'diagrammatical atoms' (as they are called here), such as the circle ACE in Euclid, *Elem.* I.1. This, however, does not allow one to recognize and adequately deal with the fact that the whole diagrammatical structure, too, possesses mathematical attributes and relations of its own (cf. 255–258). Obviously, this insight has far-reaching consequences, in particular with regard to an adequate understanding of the nature of the (as she calls them) 'pop up objects,' such as the triangle ABC in our example proposition.

33 Cf. Patterson 2007, 9, though from a different perspective: "The reader of Euclid will in fact often have just this sort of experience: the drawing of the diagram prior to the portion of the proof labeled 'Demonstration' (*apodeixis*) can make it immediately obvious that a theorem must be true and even suggest a proof strategy; the *apodeixis* then presents the worked-out proof." Evidently, the proof proper has a textual nature, for only in the text the relevant attributes and relations can be explicitly addressed and dealt with. Therefore, it is misleading to say that "it is the diagram that is the site of reasoning in Euclid, not the text. The text, on this account, is merely a script to guide one's words, and thereby one's thoughts, as one walks oneself through the demonstration" (Macbeth 2010, 260). However, such a view on the ancient mathematical proof could be the result of an (ahistorical) implicit transformation of the single Euclidean diagram into a sequence of diagrams that each separately represent one single step of the proof (for which sequence such a description might be adequate indeed, i.e., depending on the concrete circumstances): cf. Macbeth 2010, 259–265 (she adduces Euclid, *Elem.* I.31 and III.1 as examples).

To sum up, the analysis of the relation between the textual part of the proposition and its diagram has yielded the following results (which, to be sure, hold true not only for *problemata,* such as Euclid, *Elem.* I.1, but *mutatis mutandis* also for *theoremata*):[34]

(1) The possibility of the mathematical task (or the truth of the assertion) formulated in the *protasis* is proved by constructing a specifically devised and generated diagrammatical, that is, relational iconic model, in the framework of which the mathematical relations pertaining to this task (or assertion) are shown to exist (or hold).
(2) This model is produced in the *ekthesis* and the *kataskeue*: in the *ekthesis,* insofar as the respective mathematical entities are supposed to be the given material basis for the proof; in the *kataskeue,* insofar as they are not given, but their addition to the diagram is deemed necessary and / or helpful for accomplishing the proof.
(3) The *apodeixis* operates on the basis of exactly those relations that are implied by the diagrammatical atoms and / or their combination, that is, the whole diagrammatical structure. The *apodeixis* has the aim of identifying emergent relations (either of, or in the form of, mathematical objects) that are relevant for proving that the *protasis* is possible (or true, respectively), but that have not been brought into the diagram by way of explicitly and directly constructing them.
(4) Whereas the text of the proposition produces an abstract diagrammatical model, this abstract model is, in its final configuration, represented in the form of the printed 'diagram.' Evidently, this diagram is a concrete diagrammatical model, too (though secondary to and ontologically as well as epistemologically dependent on the abstract model).

34 There are two important differences that are of interest for the given purpose (cf., e.g., the *theorema* Euclid, *Elem.* I.4): (1) In the *dihorismos,* the *problema* states the goal of the construction by using a generic identifier, whereas in the *theorema* all mathematical objects that are being constructed are explicitly identified. The reason is obvious: in the *problema,* the construction of the sought-for object is the end-point of the proposition, and, therefore, it cannot be (shown as) constructed at an earlier point. (2) In the *symperasma,* the *problema* refers to the concrete mathematical objects, whereas the *theorema* refers to the abstract mathematical entities, resulting in the *symperasma's* being almost identical with the *protasis* (except for the transparent addition of the concluding particle ἄρα). The reason is, more or less, the same as in the case of the *dihorismos:* the goal of a *problema* is the construction of a specific generic mathematical object, and this object has to be constructed within and / or as an extension to the concrete diagrammatical structure that has been constructed for this very purpose—and, in the concluding *symperasma,* it subsequently has also to be *identified* as such. This, however, makes it necessary that there be an explicit reference to concrete diagrammatical atoms.

(5) The whole diagram stands for a specific mathematical situation, which is implicitly seen as being capable of providing a solution to the given task or assertion; namely, of bringing forth, and to light, additional, yet unknown mathematical objects and / or relations that are of mathematical interest.[35]

V The Epistemological Status of the Ancient Mathematical Proof

The model-theoretic reading of the ancient mathematical diagram explicated in the preceding two sections leads to a new perspective on the epistemological status of proofs in ancient mathematics that will help answer the question as to why ancient mathematics may have been revolutionary in the terms of Thomas S. Kuhn's theory of scientific revolutions. As we have seen in the second section, the validity of ancient mathematical proofs has often been called into question, especially with regard to the ubiquitous use of diagrams. One of the main problems, it has been maintained, is the universal applicability of the proof:

> How can one move from an argument based upon a particular example to a general conclusion, from an argument about the straight line AB to a conclusion about any straight line? I do not believe that the Greeks ever answered this question satisfactorily [...]. [...]. [T]here is no reason to suppose the Greeks to have had anything like modern logic to represent actual mathematical argument, and the Euclidean style makes it look as though a proof is thought of as being carried out with respect to a particular object, but in a way assumed to be generalizable. In the absence of something like the rules of logic there is no uniform procedure for checking the correctness of this assumption in individual cases. Rather one must rely on general mathematical intelligence.[36]

Put slightly differently, Greek mathematics supposes that "to prove a particular case is to count as proving a general proposition," where "generality is the repeatability of necessity."[37] The validity of the proof, therefore, is only insecurely implied, but not

---

35 Cf. Netz 1999a, 35–43. In this regard, there are several similarities to Mahr & Robering's 2009 approach, which describes Euclid's mathematical 'diagrams as images that represent models' (the article's title is "Diagramme als Bilder, die Modelle repräsentieren"). However, there are significant differences, most of which result from the use of another model-theoretic framework (Mahr 2009) that limits the insights into the nature of diagrams as semiotic entities. For example, the most specific result of Mahr & Robering's analysis is the assertion that 'all diagrams illustrate the proof of a proposition' ("alle Diagramme veranschaulichen den Beweisgang einer Proposition"). However, this general assertion does not explain sufficiently enough what ancient mathematical diagrams are and what they do.

36 Mueller 1981, 13 and 14, respectively. See also Klein 1968, 117–125, Mueller 1981, 12–14, Netz 1999a, 240–270, and Manders 2008a, 72–75.

37 Mueller 1981, 13, and Netz 1999a, 270, respectively.

in principle guaranteed by the application of some specific methodological tool or formalism.

Against the backdrop of the semiotic approach explicated here, a different answer lies at hand. Its basis is to recognize that ancient mathematics operated in a way quite distinct from modern mathematics, from which it is separated by a fundamental methodological revolution. Though seldom spelled out explicitly, it is commonly presupposed that in ancient mathematics, diagrams and their parts had (*mutatis mutandis*) an ontological status equivalent to variables in modern mathematics. Compare, for example, Heath's translation of the *ekthesis* of Euclid, *Elem.* I.1, which reads, evidently after the model of modern mathematics, as "Let AB be the given finite straight line."[38] According to the common view, the purpose of this sentence is to identify, and thus to single out, the object AB as a member of the complete set of 'finite straight lines'; as such, the line AB is a particular mathematical object, which, in principle, is different from all the other members of the respective set, such as the natural number 2 is different from all the other natural numbers.

The above analysis has shown that ancient mathematics instead used abstract diagrammatical models that represent the essential relations that are characteristic of the respective mathematical objects they represent, put into a specific mathematical structure, that is, the whole diagram. This diagram possesses emergent mathematical relations of its own, that is, relations that are, on the one hand, implied by the specific nature of the diagrammatical atoms it consists of, but that, on the other hand, only come to light when these objects are being combined into this specific structure. Accordingly, the ultimate goal of each ancient mathematical proposition is the analysis of this structure with regard to its relational qualities. Proofs, therefore, have a genuinely static character—*nota bene* in contrast to modern mathematics, which is rather interested in investigating the (generic) *dynamic* interaction of particular mathematical objects, for example, by way of functions.

In consequence, ancient mathematics conceives of its objects not as particular variables, but as universal models that, unlike variables, fully represent the relevant essential relational properties of the respective general basic mathematical entities. Given this, ancient mathematics is not deficient with regard to the epistemic validity of its proofs, at least not in the way it is commonly believed. For example, the triangle in the proposition I.1 of Euclid's *Elements* is not 'any' triangle (equivalently to, e.g., $n \in N$), but it is *the* triangle—that is, at least insofar as, and to the extent that, this triangle does not differ from any other triangle with regard to any

38 For a more adequate translation of this sentence see above with n. 31. Though, in particular, Netz rightly rejects Heath's translation and gives a more adequate translation, he interprets the sentence in a rather different way: see Netz 1999a, 43–56.

essential property it has *qua* mathematical triangle.[39] Accordingly, there is, in principle, no jump from the particular to the universal, at least insofar as the mathematically relevant essential properties are concerned that ancient mathematics is interested in. This is quite different from an approach in terms of variables: such as, e.g., n is *per definitionem* not $n + 1$, so is any particular triangle, when conceived of as a variable, not identical with any other particular triangle with regard to an essential part of its being a mathematical triangle. Just in this difference, however, lies the necessity of having to generalize the results in the context of modern mathematics, for example, by way of a formalism such as mathematical induction, a requirement that, as we can see now, is not a requirement classical Greek mathematics had to (or even could) make.[40]

In effect, classical ancient Greek and modern mathematics turn out to be incommensurably different with regard to methodology and aims—and this holds true also, as can be tentatively suggested, but not discussed here in detail, with regard to all earlier ancient mathematical approaches, including (1) Egyptian and Babylonian mathematics as well as (2) Greek mathematics prior to (roughly) the fourth century BCE.

(1) Apparently, both Egyptian and Babylonian mathematics, irrespective of their sophistication, were only interested in discussing and solving specific *particular* problems. They were not concerned with general mathematical models that, in the form of diagrammatical *icons*, represent, and thereby bring to light, the essential universal relations inherent in mathematical objects, at least not for their own sake.[41]

(2) The same seems to have been the case for early Greek mathematics. Though we barely have any authentic firsthand testimonies for Greek mathematics prior to the fourth century BCE (i.e., that have not been transformed by later, 'Euclidean' mathematics), there is, nonetheless, sufficient evidence to provide at least some

---

39 At some places, Euclid speaks about arbitrary points etc.; cf., e.g., *Elem.* I.5 and see the overview given by Netz 1999, 246–252. While it is evident that such mathematical objects do have a particular nature, it is equally evident that there is a significant difference to the fundamental methodological phenomenon that is discussed here. It is transparent that in such cases, Euclid does not investigate the essential invariant properties of the universal mathematical objects *per se*, but explicitly singles out and focuses on only one arbitrary instantiation of an indefinite set of similar objects, and this is reflected in the invariable addition of modifiers such as τις oder τυχών (which are *nota bene* never used in the default case). This, however, leads indeed to the epistemological problems sketched above.

40 On the question of mathematical induction in antiquity, cf. Unguru 1991, Fowler 1994, and Netz 1999a, 269.

41 Cf. Unguru's paper in this volume and, as a telling example, Waschkies 1989 for the relation between oriental and Greek arithmetic; see also Høyrup 2002, Waschkies 2004, and Asper 2007, 94–96 and 377–381. For diagrams especially in Egyptian mathematics, see De Young's 2009 comprehensive survey; cf. also, e.g., the *Rhind Papyrus*, which clearly demonstrates the primarily particular and practical nature of Egyptian mathematics (see Robins & Shute 1987).

plausibility for such a view.[42] Apart from early testimonies coming from practical mathematics such as the diagrams in the 'Tunnel of Eupalinus,'[43] which serve as a mathematical representation of a specific particular case, we also have relevant testimonies originating from the time when Greek mathematics seems to have taken on its Euclidean shape, that is, the first half of the fourth century BCE.

One of them is the notorious passage in Plato's *Theatetus* on Theodorus's attempt to show the incommensurability of some of the square roots of the natural numbers up to 17 with the number 1 (147d–148b). Irrespective of the fact that this testimonium evidently has, to a certain degree, a fictional setting (which fact does, of course, have a bearing on several aspects of its interpretation), it definitely belongs to the very few authentic and firsthand testimonies for pre-Euclidean mathematics and, at any rate, it can give insights into the mathematics of Plato's own times. Though there are, in part grave, interpretative problems connected with this passage,[44] what it shows without doubt is that Theodorus did not have a general, universal proof of the incommensurability of the square root with the number 1. Evidently, Plato represents Theodorus as being concerned with discussing only *particular* cases of square roots, which were, we may suppose, represented by diagrams, too, but then only by *particular* diagrams. If this is the case, however, no respective general proof of incommensurability could have existed—for if it did, it would have been useless to discuss particular cases of square roots separately. All 'proofs' would have been exactly identical. To be sure, this would square well with what we know of the early history of mathematics at large, for the first general theory of incommensurability seems to have required Eudoxus's theory of proportion as a necessary precondition; that is, it could only have been put forward at some point in the first half of the fourth century BCE.[45]

In addition to the *Theaetetus* passage, there is another interesting passage in Plato's dialogues that is relevant for the given purpose, namely the passage on the

42 The controversial details cannot be discussed here. For the broader picture, however, see Asper 2007, 197–212, Asper 2009, Cuomo 2001, 4–61, Waschkies 2004. See also Hahn 2017, especially the introduction.

43 See Käppel 1999. Cf. the similar case of geometrical drawings found on tiles used for building the temple of Artemis in Ephesos dating back even to the seventh century BCE; see Schädler 2001.

44 Cf., e.g., Knorr 1975, esp. at 62–108, and Burnyeat 1978 for an in-depth discussion.

45 This holds irrespective of when exactly incommensurability was 'found'; see Cuomo 2001, 30 for a brief overview and cf. Lehman & Weinman's paper in this volume. As I will show elsewhere, the recognition that such a phenomenon as incommensurability exists (in contrast to simply finding certain pairs of magnitudes for which it was *practically* difficult to determine the common measure, but which only *later* were to be described as being in principle incommensurable, that is, after a scientific revolution) historically presupposes both the invention of the universal mathematical diagram and Eudoxus's theory of proportion.

duplication of the square in Plato's *Meno* (82a–85b). Not only does this passage, too, reveal the methodological shift from an early form of Greek mathematics, which was concerned with particulars, to a later form of Greek mathematics, which was concerned with universal facts, it is also one of the first authentic attestations (or even *the* first authentic attestation) of a proof by means of a universal diagrammatical *icon*. The approach chosen by 'Socrates' not only exactly conforms to the Euclidean approach, as it has been explicated above, but at the same time also exhibits, on the surface level of the conversation between 'Socrates' and the slave, the earlier approach of mathematics focused on particulars.[46]

All this, however, cannot (and need not) be discussed here in detail. It suffices to see that the Euclidean approach is separated by a veritable scientific revolution at least from Egyptian and Babylonian mathematics—and that, if the observations concerning the two Plato passages point in the right direction, this scientific revolution must have taken place at some point in the first half of the fourth century BCE, when, finally, Greek mathematics began a scientific life of its own and became 'classical,' 'Euclidean' mathematics.

VI Conclusion

This paper has taken a fresh look at the ancient mathematical diagram from the viewpoint of contemporary semiotic model theory. This has led to new insights into, first, the ontological nature and epistemic function of diagrams; second, the characteristics of their interrelation and interaction with the textual part of the mathematical proposition; third, the question of their general epistemic validity; and, fourth, the history of ancient and, in particular, Greek mathematics at large.

It has turned out that the key to an adequate understanding of the ancient mathematical diagram is the recognition that it is nothing but a 'diagram' as defined by Charles S. Peirce, that is, a specific sort of iconic sign (and thus model) that displays the relational structure of its object by way of its own relational quality. This has led to the insight that the mathematical diagram has, on the one hand, a concrete manifestation, that is, the printed diagram; and, second, an abstract counterpart, that is, the semiotic object of the printed diagram—and that it is this abstract semiotic object only that is to be regarded as the mathematically relevant 'diagram' proper.

46 The *Meno* passage will be discussed elsewhere, too. For a reading of the passage in the context of 'Euclidean' mathematics see also Patterson 2007, who first discussed the similarity of this passage to (some sort of) Euclidean methodology; cf. Patterson 2007, 7: "Note that, although Socrates' version is idiosyncratic, it bears significant resemblance to a Euclidian production proof [...]."

Against this backdrop, it could be shown (1) that the diagram stands as a holistic model for a general mathematical structure (2) that consists of 'diagrammatical atoms' (3) that represent the general entities of mathematical theory; (4) and that this abstract model is being constructed and subsequently analyzed in the textual part of the proof, (5) the purpose of which is, accordingly, to 'show' that this model has the specific sought-for, yet unknown mathematical properties. (6) Given this, Greek mathematical proof turns out to be an analysis of static general models, rather than an either implicit or failed (mathematically) inductive argumentation concerned with particular cases.

In all, the argument this paper makes is primarily a historical one: it does not only claim that the interpretation of the ancient mathematical diagram it puts forward is an interpretation that explicates its ontological nature and epistemic function from a contemporary *philosophy* of science point of view; but it also claims, from a *history* of science perspective, that this interpretation *(mutatis mutandis)* conforms to, and can explain, how ancient mathematics itself interpreted diagrams, at least implicitly as shown by its own practice.[47]

The most important upshot of this analysis, however, is the insight that the classical Greek approach to mathematics is incommensurably opposed both to Egyptian and Babylonian mathematics as well as to early Greek mathematics, all of which show an orientation toward particular cases; and, on the other hand, to the generality of modern mathematics, which is, in a vital aspect, tied to the concept of a variable and the use of mathematical induction. This insight allows us to recognize that Greek mathematics is a paradigm of mathematics set apart from ancient as well as modern paradigms of mathematics by two fundamental scientific revolutions.*

---

47 This is one of the significant differences of the approach explicated here from topological approaches to explaining the ancient mathematical diagram; cf., e.g., Manders 2008b, Miller 2007, and Mumma 2006. Though these analyses can explain many aspects of the use of diagrams from a modern theoretical perspective, they are not sufficiently grounded in the *historical* practice and theory of classical Greek mathematics.

* I am grateful to the Alexander von Humboldt Foundation for awarding me a Feodor Lynen Research Fellowship, which has given me the wonderful opportunity to work on this project at the Department of Classics at Emory University. My academic host Peter Bing and his colleagues have made this research stay a most enriching experience. Special thanks go to Richard Patterson for stimulating and enlightening discussions about Plato and ancient mathematics as well as for his generous help in improving both the argument and the English of this paper. I also feel obliged to express my gratitude to the American Friends of the Alexander von Humboldt Foundation for awarding me the Calder Fellowships 2013, 2014, and 2015, not only for the reason that they have allowed me to present my thoughts on the ancient diagram at this wonderful conference, for the organization of which I thank Michalis Sialaros. Finally, heartfelt thanks to Lutz Käppel for his never-ending support, and to the late Hans-Joachim Waschkies for his friendship and for all that I had the honor of learning from him.

Bibliography

Texts quoted

Euclid, *Elem.*: Heiberg, J. & E. S. Stamatis 1969–1977. *Euclidis Elementa.* Vols. 1–5. 2nd ed. Leipzig.

Hero, *Def.*: Heiberg, J. 1903. *Definitiones, Opera quae supersunt omnia.* Vol. 4. Leipzig.

Plato, *Meno*: Burnet, J. 1900–1907. *Platonis Opera.* Vol. 3. Oxford.

Plato, *Resp.*: Slings, S. R. 2003. *Platonis Respublica.* Oxford.

Plato, *Theaet.*: Duke, E. A. *et al.* 1995. *Platonis Opera.* Vol. 1. Oxford.

Proclus, *In Eucl.*: Friedlein, G. 1873. *Commentarii in Primum Euclidis Elementorum Librum.* Leipzig.

Works quoted

Asper, M. 2007. *Griechische Wissenschaftstexte. Formen, Funktionen, Differenzierungsgeschichten.* Stuttgart.

Asper, M. 2009. “The Two Cultures of Mathematics in Ancient Greece.” In: E. Robson & J. Stedall (eds.), *The Oxford Handbook of the History of Mathematics.* Oxford, 107–132.

Avigad, J. *et al.* 2009. “A Formal System for Euclid’s Elements.” In: *The Review of Symbolic Logic* 2, 700–768.

Bailer-Jones, D. M. 2009. *Scientific Models in Philosophy of Science.* Pittsburgh.

Balzer, W. 1997. *Die Wissenschaft und ihre Methoden: Grundsätze der Wissenschaftstheorie.* Freiburg.

Burnyeat, M. 1978. “The Philosophical Sense of Theaetetus’ Mathematics.” In: *Isis* 69, 489–513.

Chemla, K. (ed.) 2012a. *The History of Mathematical Proof in Ancient Traditions.* Cambridge.

Chemla, K. 2012b. “Prologue. Historiography and History of Mathematical Proof: A Research Programme.” In: K. Chemla (ed.), *The History of Mathematical Proof in Ancient Traditions.* Cambridge, 1–68.

Coliva, A. 2012. “Human Diagrammatic Reasoning and Seeing-as.” In: *Synthese* 186, 121–148.

De Young, G. 2009. “Diagrams in Ancient Egyptian Geometry: Survey and Assessment.” In: *Historia Mathematica* 36, 321–373.

De Young, G. 2012. “Mathematical Diagrams from Manuscript to Print: Examples from the Arabic Euclidean Transmission.” In: *Synthese* 186, 21–54.

Federspiel, M. 1992. “Sur la locution ἐφ’ οὗ / ἐφ’ ᾧ servant à désigner des êtres géométriques par des lettres. Avec des remarques sur l’art de traduire et sur la langue mathématique d’Hippocrate de Chio.” In: J.-Y. Guillaumin (ed.), *Mathématiques dans l’Antiquité.* Saint-Etienne, 9–25.

Fowler, D. H. 1994. “Could the Greeks Have Used Mathematical Induction? Did They Use It?” In: *Physis* 31, 253–265.

Frigg, R. & S. Hartmann 2017. “Models in Science ” In: N. Zalta (ed.), *The Stanford Encyclopedia of Philosophy* (Spring 2017 edition). URL = https://plato.stanford.edu/archives/spr2017/entries/models-science/.

Giaquinto, M. 2008. “Visualizing in Mathematics.” In: P. Mancosu (ed.), *Philosophy of Mathematical Practice.* Oxford, 22–42.

Hahn, R. 2017. *The Metaphysics of the Pythagorean Theorem.* Albany.

Hartshorne, C. *et al.* (eds.) 1931–1958. *Collected Papers of Charles Sanders Peirce.* Cambridge, MA.

Heath, T. L. 1921. *A History of Greek Mathematics.* 2 vols. Oxford.

Heath, T. L. (ed.) 1926. *The Thirteen Books of Euclid’s Elements* (orig. Cambridge 1908). 3 vols. New York.

Hilbert, D. 1899. *Grundlagen der Geometrie.* Leipzig.

Høyrup, J. 2002. *Lengths, Widths, Surfaces. A Portrait of Old Babylonian Algebra and Its Kin.* New York.

Käppel, L. 1999. “Die PARADEGMA-Inschrift im Tunnel des Eupalinos auf Samos.” In: *Antike & Abendland* 45, 75–100.

Klein, J. 1968. *Greek Mathematical Thought and the Origin of Algebra* (orig. Berlin 1934–1936). New York.

Knorr, W. R. 1975. *The Evolution of the Euclidean Elements: A Study of the Theory of Incommensurable Magnitudes and Its Significance for Early Greek Geometry*. Dordrecht.

Kralemann, B. & C. Lattmann 2013a. “The Semantics of Models: a Semiotic Philosophy of Science Approach.” In: K.-D. Schewe & B. Thalheim (eds.), *Semantics in Data and Knowledge Bases: 5th International Workshop SDKB 2011, Zürich, Switzerland, July 3, 2011. Revised Selected Papers*. Berlin, 50–69.

Kralemann, B. & C. Lattmann 2013b. “Models as Icons: Modeling Models in the Semiotic Framework of Peirce’s Theory of Signs.” In: *Synthese* 190, 3397–3420.

Kuhn, T. S. 1996. *The Structure of Scientific Revolutions* (orig. 1962). Chicago.

Kühner, R. & B. Gerth 1898–1904. *Ausführliche Grammatik der griechischen Sprache. Zweiter Teil: Satzlehre*. 3rd ed. 2 vols. Hannover.

Lattmann, C. 2012. “Icons of Novel Thought. A New Perspective on Peirce’s Definition of Metaphor (CP 2.277).” In: *Semiotica* 192, 535–556.

Macbeth, D. 2010. “Diagrammatic Reasoning in Euclid’s Elements.” In: B. van Kerkhove *et al.* (eds.), *Philosophical Perspectives on Mathematical Practice*. London, 235–267.

Mahr, B. 2009. “Information Science and the Logic of Models.” In: *Software and System Modeling* 8, 365–383.

Mahr, B. & K. Robering 2009. “Diagramme als Bilder, die Modelle repräsentieren: Diagrammgebrauch in der Elementargeometrie.” In: *Zeitschrift für Semiotik* 31, 275–309.

Manders, K. 2008a. “Diagram-Based Geometric Practice.” In: P. Mancosu (ed.), *Philosophy of Mathematical Practice*. Oxford, 65–79.

Manders, K. 2008b. “The Euclidean Diagram (1995).” In: P. Mancosu (ed.), *Philosophy of Mathematical Practice*. Oxford, 80–133.

Miller, N. 2007. *Euclid and His Twentieth Century Rivals. Diagrams in the Logic of Euclidean Geometry*. Stanford.

Mueller, I. 1974. “Greek Mathematics and Greek Logic.” In: J. Corcoran (ed.), *Ancient Logic and its Modern Interpretations*. Dordrecht, 35–70.

Mueller, I. 1981. *Philosophy of Mathematics and Deductive Structure in Euclid’s Elements*. Cambridge, MA.

Mumma, J. 2006. *Intuition Formalized: Ancient and Modern Methods of Proof in Elementary Euclidean Geometry*. Diss. Pittsburgh.

Mumma, J. 2008a. Review of Miller 2007. In: *Philosophia Mathematica* 16, 256–264.

Mumma, J. 2008b. “Ensuring Generality in Euclid’s Diagrammatic Arguments.” In: G. Stapleton *et al.* (eds.), *Diagrammatic Representation and Inference. 5th International Conference, Diagrams 2008, Herrsching, Germany, September 19–21, 2008. Proceedings*. Berlin, 188–205.

Mumma, J. 2010. “Proofs, Pictures, and Euclid.” In: *Synthese* 175, 255–287.

Mumma, J. 2012. “Constructive Geometrical Reasoning and Diagrams.” In: *Synthese* 186, 103–119.

Mumma, J. & M. Panza 2012. “Diagrams in Mathematics: History and Philosophy. Introduction.” In: *Synthese* 186, 1–5.

Netz, R. 1999a. *The Shaping of Deduction in Greek Mathematics: A Study in Cognitive History*. Cambridge.

Netz, R. 1999b. “Proclus’ Division of the Mathematical Proposition into Parts: How and Why Was It Formulated?” In: *Classical Quarterly* 49, 282–303.

Netz, R. 2003. “Introduction: The History of Early Mathematics—Ways of Re-Writing.” In: *Science in Context* 16, 275–286.

Netz, R. (ed.) 2004. *The Works of Archimedes: Translation and Commentary. Volume I: The Two Books On the Sphere and the Cylinder*. Cambridge.

Netz, R. 2012. "The Texture of Archimedes' Writings: Through Heiberg's Veil." In: K. Chemla (ed.), *The History of Mathematical Proof in Ancient Traditions*. Cambridge, 163–205.

Neuenschwander, E. 1973. "Die ersten vier Bücher der Elemente Euklids. Untersuchungen über den mathematischen Aufbau, die Zitierweise und die Entstehungsgeschichte." In: *Archive for History of Exact Sciences* 9, 325–380.

Norman, J. 2006. *After Euclid. Visual Reasoning & the Epistemology of Diagrams*. Stanford.

Patterson, R. 2007. "Diagrams, Dialectic, and Mathematical Foundations in Plato." In: *Apeiron* 40, 1–33.

Robins, G. & C. Shute 1987. *The Rhind Mathematical Papyrus*. London.

Saito, K. 2006. "A Preliminary Study in the Critical Assessment of Diagrams in Greek Mathematical Works." In: *SCIAMVS* 7, 81–144.

Saito, K. 2009. "Reading Ancient Greek Mathematics." In: E. Robson & J. Stedall (eds.), *The Oxford Handbook of the History of Mathematics*. Oxford, 801–826.

Saito, K. & N. Sidoli 2012. "Diagrams and Arguments in Ancient Greek Mathematics: Lessons Drawn from Comparisons of the Manuscript Diagrams With Those in Modern Critical Editions." In: K. Chemla (ed.), *The History of Mathematical Proof in Ancient Traditions*. Cambridge, 135–162.

Schädler, U. 2001. "Der Kosmos der Artemis von Ephesos." In: U. Muss (ed.), *Griechische Geometrie im Artemision von Ephesos*. Vienna, 279–287.

Short, T. L. 2007. *Peirce's Theory of Signs*. Cambridge.

Sialaros, M. & J. Christianidis 2016. "Situating the Debate on 'Geometrical Algebra' Within the Framework of Premodern Algebra." In: *Science in Context* 29, 129–150.

Tarski, A. 1959. "What is Elementary Geometry?" In: L. Henkin *et al.* (eds.), *The Axiomatic Method, With Special Reference to Geometry and Physics*. Amsterdam, 16–29.

Unguru, S. 1975. "On the Need to Rewrite the History of Greek Mathematics." In: *Archive for History of Exact Sciences* 15, 67–114.

Unguru, S. 1991. "Greek Mathematics and Mathematical Induction." In: *Physis* 28, 273–289.

Waschkies, H.-J. 1989. *Anfänge der Arithmetik im alten Orient und bei den Griechen*. Amsterdam.

Waschkies, H.-J. 2004. "Introduction." In: J. Christianidis (ed.), Classics in the History of Greek Mathematics. Dordrecht, 3–18.

Fabio Acerbi

Composition and Removal of Ratios in Geometric and Logistic Texts from the Hellenistic to the Byzantine Period

Abstract: The historical path of the operations of 'composition' and 'removal' of ratios from Greek antiquity to Byzantine times is followed by presenting and briefly describing the extant historical evidence.

Keywords: Greek mathematics; Byzantine mathematics; Ratio.

I Introduction

What follows is a long essay on the monolithic inertia of the ancient Greek approach to mathematics. One might speculate about the social or cultural origins of such a formidable inertia, but what I would like to offer in this paper is simply a series of facts and documents attesting to its existence. Of course, the only reasonable way to do this is to pick up a specific conception or tool, and then check whether its pattern of evolution, from Greek antiquity to Byzantine times, is continuous or discontinuous. It so happens that concepts or techniques suitable to serve as examples are quite difficult to find: to the best of my knowledge, the notion discussed in the present paper is the most suitable one—if not the only available one over a long time span; in this sense, my analysis is tendentious. Whether from this example one may infer that *no* revolution *could* take place in the field of logistic (or even in Greek mathematics as a whole), and, if so, why, I leave it up to the reader to decide.

To be more precise, the present study has two particular aims: (a) to follow the historical path of the operations of 'composition'[1] and 'removal' of ratios from Greek antiquity to Byzantine times; (b) to present and briefly describe the extant historical evidence. These operations are both first attested in the context of geometric proofs, since they are two species of a small constellation of standard manipulations of ratios and proportions. In particular, composition of ratios bears a misleading resemblance to the operation named *ex aequali* (δι' ἴσου), whose theoretical foundations are laid in *Elem.* V.20 and 22. After a puzzling first appearance in the *Elements* and a handful of applications in the Archimedean corpus, compounded ratios were used quite

1 As it will be shown below, 'composition' was never regarded as an operation—with a couple of noteworthy exceptions. I employ the term here for the sake of being concise.

https://doi.org/10.1515/9783110565959-007

frequently—and apparently with systemic aims—by such authors as Apollonius and Pappus.

The golden age of composition and removal of ratios came along with the proliferation of 'logistic'[2] texts that characterized the Late Antiquity and the Byzantine period. These texts have the form of computational primers to the *Almagest*. Expositions of composition and removal of ratios were included in them because the operation of removal is a crucial step in all computations involving the Sector Theorem, which was, in its turn, the key mathematical tool of the entire *Almagest*.

In order to be transferred from proportion theory to logistic, composition and removal of ratios—and, in fact, the notion of ratio itself—were adapted and modified, by introducing the crucial notion of πηλικότης; namely, the 'value' or 'size' of a ratio, i.e. a numerical object.

Two historiographic points of some interest result from an analysis of the sources I am going to present. First (and this is a unique example in ancient Greek mathematics), the evolution of a concept can be followed closely from its original, proportion-theoretical context to a strictly computational framework—in which, most notably, all ratios are numerical objects. Second, a standard procedure of removal of a numerical ratio from another did not exist: on the contrary, several procedures are attested, more or less markedly different from one another. Such a variety originated from three facts: first, all sources insisted on viewing removal as an operation on ratios provided in compounded form; second, the applications of the Sector Theorem introduce the additional constraint that one of the terms of the ratio resulting from the removal is assigned; third, and most importantly, the idea that removal may be reduced to taking a suitable fourth proportional of three assigned numbers can be implemented in several ways. The latter feature shows the extent to which the conception of a ratio as a relation—and not as a numerical object—was deeply entrenched in mathematical practice, despite the introduction of the πηλικότης. As a matter of fact, the idea that the πηλικότητες can be effectively used in order to shortcut and unify all these procedures—the result of a removal of one ratio from another being simply the ratio of

---

2 According to Eutocius, dividing the unit does not pertain to arithmetic but to logistic (*AOO* III, 120.28–30: ὥστ' ἐπ' ἐκείνων [*scil.* superparticular and superpartient ratios] διαιρετέον τὴν μονάδα, ὃ εἰ καὶ μὴ κατὰ τὸ προσῆκον τῇ ἀριθμητικῇ ἀλλὰ τῇ λογιστικῇ τυγχάνει). An earlier definition of logistic—which almost certainly can be ascribed to Geminus (a first-century BC mathematically-minded philosopher and polymath, maybe a pupil of Posidonius)—does not allow dividing the unit. This definition can be found in pseudo-Hero, *Def.* 135.5–6 (*HOO* IV, 98.12–100.3), and is also preserved, by a different line of tradition, as a scholium to Plato's *Chrm.* 165E6 (Scholium 27 in Cufalo 2007, 173). It is possible that the domain of logistic was expanded to include fractional parts as a consequence of the adoption of the sexagesimal system in Greek mathematical astronomy. This happened some time around Hipparchus' life span, which certainly included the interval [147-127] BC. The best introduction to Greek logistic is still Vogel 1936.

the 'values' of the two ratios—is stated and proved only in Barlaam's treatise, whose brilliant exposition remained, however, isolated and unparalleled.

This study is a first attempt to expound in a compact way some results coming from my own editorial work on sources relevant to the issue, in particular the *Almagest* scholia, both *vetera* and Byzantine, and the Ur-text of the Pseudo-Domninus there discovered (cf. Acerbi & Riedlberger 2014); the *Prolegomena to the Almagest* (henceforth '*Prolegomena*'; cf. Acerbi *et al.* 2010); the *Isagoge* of George of Trebizond;[3] the treatises of Theodoros Metochites and John Chortasmenos; a short text of George Gemistus Plethon (cf. Acerbi *et al.* 2016).

The presentation of the material is divided into two broad sections, each one dealing, respectively, with the geometric and the logistic approach to composition and removal of ratios. The first section contains the evidence found in Euclid (third century BC), Archimedes (287–212 BC), Apollonius (early second century BC), Ptolemy (ca. 150 AD), Pappus (ca. 320 AD), and Theon of Alexandria (ca. 360 AD). I also discuss some relevant features of this evidence. The second section deals in detail with general expositions and worked-out examples, found in late logistic texts. The authors whose contributions are presented and discussed in this section are: Nicomachus (second half of the second century AD); again Theon of Alexandria; Eutocius (early sixth century AD); the scholiast of the *Almagest* whose text on removal of ratios was anonymously reworked and came to be ascribed to Domninus of Larissa (middle fifth century AD, but the scholium comes from the same scholarly circles as the *Prolegomena*); the anonymous author of the *Prolegomena* (early sixth century AD); again an anonymous scholiast of the *Almagest*; Leo the Mathematician (early ninth century AD); some scholiasts to the *Elements*, among whom Maximus Planudes (second half of the thirteenth century AD); George Pachymeres (second half of the thirteenth century AD); Manuel Bryennios (early fourteenth century AD); Theodoros Metochites (early fourteenth century AD); Barlaam of Seminara (middle fourteenth century AD); Demetrios Cydones (second half of the fourteenth century AD); John Chortasmenos (early fifteenth century AD); George Gemistus Plethon (early fifteenth century AD); George of Trebizond (middle fifteenth century AD). Quite surprisingly, the issue of composition and removal of ratios does not at all feature in the computational primer to the *Almagest* written by Theodoros Meliteniotes, a distinguished scholar contemporary with Demetrios Cydones (ed. Leurquin 1990). Less surprisingly, the same omission appears in an anonymous computational primer that proves to be a compilation of extracts from some of the authors just mentioned (ed. Moll 1965), and in Maximus Planudes' *Great Calculation According to the Indians* (Allard 1981) and in

3 Critical edition, with an introduction and a commentary, of the relevant *Almagest* scholia in Acerbi 2017. I am also preparing a critical edition of the *Isagoge* of George of Trebizond, which is attested in two, quite different, versions in Latin and in Greek.

its anonymous source (Allard 1977): these two treatises bear no connections with the *Almagest*.

The choice of the presented authors was made on the basis of mere linguistic continuity: they all write in ancient Greek. Still, it is worth mentioning the sophisticated general exposition of composition of ratios, characterized by a markedly combinatorial approach, redacted by Thābit ibn Qurra (Lorch 2001).

II Geometric Approach

II.1 Euclid's *Elements* and the *Sectio Canonis*

The standard definition of 'compounded ratio' is *Elem.* VI.df.5: "A ratio is said to be compounded of ratios when the values of the ratios, multiplied by one another, make some <value of a ratio>".[4] Heiberg, on good grounds, though the definition spurious:[5] it is never used in the rest of the *Elements*; it is placed after definition 2 in the Theonine manuscripts Bodl. Dorv. 301 and Par. gr. 2466; it is only transcribed, with a sign referring to the last line of definition 4, in the lower margin of f. 86r of the unique non-Theonine manuscript Vat. gr. 190. Some remarks are in order:[6]

- The compounding ratios are not 'multiplied': what is 'multiplied' are their πηλικότητες.[7] For this reason, when writing a compounded ratio in symbolic form, I shall put the standard mathematical sign of composition '∘' between the two compounding ratios, thus: $a:b = (c:d)\circ(e:f)$.[8] As we shall see below, the formula stands for the expression "the ratio of a to b is compounded both of the ratio of c to d and of that of e to f".
- A ratio is said to be 'compounded' of two ratios: it neither 'is' the two compounding ratios, nor it is 'equal' or 'identical' to them. The sign '=' in the above formula is thus misleading and I have introduced it *faute de mieux*. I shall call 'left-hand side' and 'right-hand side' what appears on the left or right side of the sign '=', respectively.

4 Λόγος ἐκ λόγων συγκεῖσθαι λέγεται, ὅταν αἱ τῶν λόγων πηλικότητες ἐφ' ἑαυτὰς πολλαπλασιασθεῖσαι ποιῶσί τινα; *EOO* II, 72.13–15. My translation of *Elem.* VI.df.5 includes a final integration based on Theon, *in Alm. I.13*, *iA*, 533.1–2, who is the only independent source from the Late Antiquity that completes the final τινα of *Elem.* VI.df.5 with πηλικότητα λόγου.

5 See *EOO* II, 72.7 and 73.13–15 *app.*; see also 73 n. 2.

6 See Saito 1986 and Sylla 1984, 18–19 and 22–23, for earlier discussions.

7 As we shall see, Greek arithmetic theory employs an 'additive' lexicon to conceptualize and formulate these operations on ratios.

8 I shall write a ratio in the form $a:b$ and the corresponding πηλικότης in the form a/b.

- Despite the fact that VI.df.5 reduces 'composition' to a multiplication, a compounded ratio was never conceived of as the result of an *operation* on ratios. With reference to the formula above, the two ratios *c*:*d* and *e*:*f* are not 'compounded' to produce *a*:*b*; what is 'compounded' is *a*:*b* itself, while *c*:*d* and *e*:*f* are the 'compounding' ratios. In fact, if an operation is at issue here, this is not a 'composition', but a 'de-composition' of *a*:*b* into component ratios. This conceptualization is strictly adhered to in all our sources and had momentous consequences from a mathematical point of view. Only in later sources the noun 'composition' came to be used, but only as a convenient shortcut, as I have done in the introductory section.

In the *Elements*, compounded ratios only feature in propositions VI.23 and VIII.5, whose enunciations and proofs are strictly parallel. These two propositions, one must stress, are deductive dead ends. Let us read their enunciations:[9]

> VI.23. Equiangular parallelograms have to one another the <ratio> compounded of <those of> their sides.
> VIII.5. Plane numbers have to one another the <ratio> compounded of <those of> their sides.

Before discussing the proofs of these propositions, it is better to have a look at the standard designation of a compounded ratio; this will also allow us to fix some conventions. Let us read the first occurrence of the formula, that I shall call 'canonical', in the *Elements*:[10]

> ὁ τῆς Κ πρὸς Μ λόγος σύγκειται ἔκ τε τοῦ τῆς Κ πρὸς Λ λόγου καὶ τοῦ τῆς Λ πρὸς Μ
> the ratio of K to M is compounded both of the ratio of K to Λ and of that of Λ to M.

This expression has the following, noteworthy features:
- The predicate is a form of 'to be compounded' (συγκεῖσθαι), usually used as the passive of συντιθέναι. It is often replaced by the synonym 'to be conjoined' (συνάπτεσθαι).
- The correlative structure 'both ... and' (τε ... καί) links the two compounding ratios more tightly than a simple 'and' (καί).
- The term 'ratio' (λόγος) is often understood, especially in the designations of the compounding ratios.
- The article that determines λόγος is always present, even if the noun is understood. The double article τοῦ, which would precede any of the compounding ratios when numbers (masc.) are at issue, is frequently simplified. For instance, we read: ἔκ τε τοῦ Κ πρὸς Λ καὶ τοῦ Λ πρὸς Μ, and not ἔκ τε τοῦ τοῦ Κ πρὸς τὸν Λ καὶ τοῦ τοῦ Λ

---

9 *EOO* II, 146.9–10 and 286.2–3, respectively.
10 *Elem.* VI.23, *EOO* II, 146.24–25.

πρὸς τὸν Μ. Here, the first τοῦ in both occurrences of τοῦ τοῦ stands for τοῦ λόγου, while the second is an integral part of the designation of the first term of the ratio.[11]

- Quite unexpectedly if one has *Elem.* VI.df.5 in mind, there is a common term Λ to the two compounding ratios. This common term was called 'middle' (μέσος). This name is unfortunate since it obviously interferes with the standard meaning of 'mean <proportional>' (μέσος) C of two terms A and B, namely, the term that realizes the continuous proportion A:C::C:B. Of course, a mean proportional can always be used as the middle term in a compounded ratio, but this does not need to always be the case: any term homogeneous with the terms of a given ratio can serve as a middle in order to write the ratio in compounded form. I shall say that a compounded ratio is 'written in standard form' when there is a common term to the two compounding ratios.

Anticipating on what we shall see in what follows, here are some further terminological points.

- I shall call 'replacement of identical ratios' in a compounded ratio $a{:}b = (c{:}d)\circ(e{:}f)$ any substitution of a ratio with another identical to it: if $a{:}b = (c{:}d)\circ(e{:}f)$ and $c{:}d{::}m{:}n$, then $a{:}b = (m{:}n)\circ(e{:}f)$. No proof in ancient Greek sources validates this inference, that amounts to requiring that any definition of 'compounded ratio' is well-founded (that is, it only depends on equivalence classes of ratios). Note that we cannot apply V.11 (transitivity of identity of ratios) in order to validate this inference, simply because we do not know what kind of relation is the one (if any) holding between a ratio and any rewriting of it in compounded form—this is the relation I have written '='. Nor can we apply the δι' ἴσου proposition V.22, for the same reason as for V.11 and for the additional reason that V.22 allows operating on compounded ratios 'written in standard form' only.
- If a ratio is in common between two compounded ratios $a{:}b = (c{:}d)\circ(e{:}f)$ and $m{:}n = (c{:}d)\circ(p{:}q)$, as here $c{:}d$, it will be called the 'shared ratio'. Of course, it is impossible for two *different* ratios $a{:}b$ and $m{:}n$ written in compounded standard form to have a shared ratio among their compounding ratios. If two identical compounded ratios admit of a shared ratio, then the other compounding ratios will also be identical: if $(c{:}d)\circ(e{:}f) = a{:}b{::}m{:}n = (c{:}d)\circ(p{:}q)$, then $e{:}f{::}p{:}q$. Operationally, this amounts to 'removing' the shared ratio (see sections II.2–4). No proof in ancient Greek sources validates this inference.

The proof of *Elem.* VI.23 runs as follows, where $p(a,b)$ is the parallelogram whose sides are a and b:

1) *Setting out* and *determination*: the equiangular parallelograms $p(a,b)$, $p(x,y)$ are assigned; to prove that $p(a,b){:}p(x,y) = (a{:}x)\circ(b{:}y)$.

---

11 Similarly, in the enunciations of *Elem.* VI.23 and VIII.5 we read λόγον ἔχει τὸν συγκείμενον ἐκ <u>τῶν</u> πλευρῶν and not λόγον ἔχει τὸν συγκείμενον ἐκ <u>τῶν τῶν</u> πλευρῶν; my emphasis.

2) *Construction*: set sides a and x adjacent and on a straight line. As a consequence, sides b and y are also adjacent and on a straight line; complete parallelogram $p(b,x)$.
3) *Construction*: introduce auxiliary straight lines K, Λ, M such that K:Λ::a:x and Λ:M::b:y. Since a, b, x, and y are given, this can be done by taking a chance straight line K and by applying twice the construction of fourth proportional of two straight lines (VI.12) to find Λ and M.
4) *Proof*: Since a:x and b:y are the ratios of the sides of the assigned parallelograms, and since K:M = (K:Λ)∘(Λ:M), it follows that K:M is compounded of the ratios of the sides of the parallelograms. This involves (i) a replacement of identical ratios applied to the two compounding ratios: K:Λ → a:x and Λ:M → b:y; (ii) a change of designation from denotative letters to a definite description and a related shift in the wording of the formula: ὁ X λόγος σύγκειται ἔκ τε τοῦ Y καὶ τοῦ Z → X λόγον ἔχει τὸν συγκείμενον ἐκ + {definite description of Y&Z}. As we have seen, no proof in ancient Greek sources validates either of these inferences. For obvious reasons, I shall call the complex of steps (i)–(ii) 'transfer of denomination' on the right-hand side of a compounded ratio.
5) Two applications of VI.1 and V.11 give $p(a,b)$:$p(b,x)$::a:x::K:Λ and $p(b,x)$:$p(x,y)$:: b:y::Λ:M.
6) Application of the δι' ἴσου proposition V.22 to K:Λ::$p(a,b)$:$p(b,x)$ and Λ:M::$p(b,x)$: $p(x,y)$ gives K:M::$p(a,b)$:$p(x,y)$.
7) Transfer of denomination on the left-hand side (it involves the compounded ratio): since K:M is compounded of the ratios of the sides of the parallelograms, also $p(a,b)$:$p(x,y)$ is. Again, nothing validates this inference, that amounts to assuming that, if two ratios are identical, they can be said to be compounded of identical ratios. This is a form of replacement of identical ratios, applied to a discursive identification of the compounded ratio.

Here is a tabular representation of the proof; the numbers correspond to the steps just listed.

	(a:x)∘(b:y)	=	(K:Λ)∘(Λ:M)	=	K:M			::	$p(a,b)$:$p(x,y)$	
K:Λ::a:x						::	$p(a,b)$:$p(b,x)$			
Λ:M::b:y						::	$p(b,x)$:$p(x,y)$			
3		4				5		6		7

Proposition VIII.5 is parallel to VI.23. The corresponding steps in the two propositions are listed in the following table:[12]

12 *Elem.* VIII.5 also implicitly uses commutativity of multiplication, namely, *Elem.* VII.16. This has no counterpart in VI.23, since the proposition holds for 'parallelograms' (+ two-letter designation) and not simply for 'rectangles contained by' (+ three-letter designation). In fact, the geometric 'counterpart' of commutativity of multiplication simply amounts to a permutation of the letters denoting the parallelogram.

VI.23	completion of a parallelogram	VI.12	transf. denom.	VI.1	V.11	V.22 δι' ἴσου	transf. denom.
VIII.5	multiplication of sides	VIII.4	/	VII.17	/	VII.14 δι' ἴσου	transf. denom.

The completion of a parallelogram in VI.23 corresponds, in VIII.5, to multiplying one side coming from any of the plane numbers by one side coming from the other. Taking a fourth proportional of two straight lines—VI.12, but note that this step is not explicit in the construction—is made to correspond, in VIII.5, to finding the least three numbers containing two given ratios (VIII.4). The last proposition even deserves an instantiated citation in VIII.5: this shows that VIII.5 fits better the deductive structure of Book VIII than its counterpart VI.23 fits the structure of Book VI. Transitivity of identity of ratios (proposition V.11) is not proved in the arithmetic books of the *Elements*, but its validity is obvious, given the definition of numbers in proportion at VII.df.21.[13] It is important to note that, in VIII.5, the entities having the same function the straight lines K, Λ, M have in VI.23 come out more 'naturally' than in the geometric case, since they are the least three numbers containing the ratios of the sides.

Both propositions were modified by Theon: he completed the setting-out of VI.23, in order to make the reference to the names of the sides explicit;[14] and he completed the proof of VIII.5, by adding the transfer of denomination that corresponds exactly to step 4 of the proof of VI.23.[15]

The drawbacks of VI.23 are obvious: first, the introduction of the straight lines K, Λ, M—and hence the application of the δι' ἴσου—is unnecessary; second, the replacement of identical ratios, both on the left and right-hand side of the compounded ratio, is, as we have seen, an operation that is left unproven. Add to this that VI.df.5 is not applied at all in the proof.

As for the first point, it is enough to note that the proof works equally well if one directly introduces parallelogram $p(b,x)$ as a middle term in the ratio of parallelograms $p(a,b):p(x,y)$ and then argues by means of VI.1, replacement of identical ratios, and transfer of denomination. It suffices to read the Euclidean proof backwards to realize this.[16] On the other hand, if one wants to keep the deductive arrow going from

13 The definition is a disjunction of conditions and only involves the relational operators 'to be equal to' and 'to be identical to'.

14 See *EOO* II, 146.14 *app.* Theon's supplement is unnecessary, since the names of the sides are univocally determined by the names of the parallelograms and by the names of their equal angles, and these names are provided in the setting-out. Moreover, Theon's supplement shows that he misunderstood the logic of the proof, since there are two perfectly equivalent possibilities as to which sides are to be coupled in the compounding ratios, and this symmetry is spontaneously broken only in the construction. Note that the same overspecification is forced by my, mildly symbolic, formulation of the proof.

15 The deduction at issue in VI.23 is at *EOO* II, 146.21–148.1, to be compared with *EOO* II, 286.13 *app.*

16 Reasons for introducing the lines K, Λ, M are adduced in Mueller 1981, 154–155 (where the above alternative proof is also proposed), and Saito 1986, 32. The solution of this interpretive problem can only be a matter of guesswork.

sides to parallelograms, and assuming that a compounded ratio must be written in standard form, the introduction of K, Λ, M is necessary. Note that, contrary to what happens in all other geometric authors we shall be dealing with, in the proofs of VI.23 and VIII.5 no compounded ratio features that is not written in standard form:[17] Euclid never operates 'inside' a compounded ratio; the transfer of denomination can be taken as a trick to avoid doing so.

The second point is more delicate and shows, either that the concept of a compounded ratio is not well-defined or that using it is a sheer verbalism. The way the subject is handled in the *Elements*, in fact, hides a trivial substratum, that will show up in the purely geometric exposition we shall read in Theon: it is simply obvious that any middle term can make a ratio a compounded ratio, and there is nothing to prove about this. If, however, one wants to perform operations on compounded ratios, like replacing identical ratios or removing shared ratios (as Euclid, Archimedes, or Apollonius do), the concept of compounded ratio must be grounded more firmly compared to how it is treated in the *Elements*. The introduction of πηλικότης and of VI.df.5 associated to it, apparently had the function of de-trivializing the whole affair. As a by-product—and since two ratios are identical if and only if their πηλικότητες are equal—, this move will make any (misleading) use of δι' ἴσου unnecessary, as we shall see in Barlaam's exposition (see section III.1.8). On the other hand, VI.df.5 will obviously give a decidedly logistic turn to the subject of compounded ratios and of the operations on them. Still, the subject will allow being founded on purely number-theoretical grounds, as Eutocius will do (see section III.1.2).

In the *Sectio Canonis*, a treatise ascribed to Euclid, compounded ratios are handled perfunctorily (cf. *Document 1* in the Appendix). The concept is involved in almost every proposition of the theoretical part of the treatise. It is always assumed that a compounded ratio (which is called 'interval' (διάστημα) in the language of harmonic theory adhered to in the *Sectio Canonis*) is written in standard form, and such that the middle term is also a mean proportional. In proposition 7, it is shown that a triple interval is compounded of a double and of a sesquialter interval; namely, $3:1 = (2:1)\circ(3:2)$. In proposition 8, it is shown that, if one removes a sesquitertian interval from a sesquialter, the remainder is a sesquioctave interval, namely, the assumed compounded ratio is $3:2 = (9:8)\circ(4:3)$. Both proofs argue by partition of the antecedent of a ratio once the consequent of the same ratio is taken as the unit, and by counting the 'units' resulting from taking a suitable multiple of both; for instance, in a sesquitertian ratio, the antecedent contains the consequent once and a third of it ($^4/_3 = 1 + {}^1/_3$), and, therefore, 12 consequents are equal to 9 antecedents. A similar argument is used by Theon in his commentary on the *Almagest* (see section III.1.1).

17 Actually, in VIII.5, no canonical formula is present at all.

II.2 The Archimedean corpus

Compounded ratios in the Archimedean corpus can be found in a handful of propositions.[18] In *Fluit.* II.10, a compounded ratio is first introduced; after a replacement of identical ratios, it is reduced to the ratio compounded of 2:5 and 5:1, which is stated to be identical to 2:1.[19] In *Con. sph.* 10 (but this simply consists in an enunciation), 23, 24, and 31, a property analogous to one that follows from combining two theorems proved in the *Elements* is formulated in the language of compounded ratios; any of such ratios is subsequently transformed by replacement of identical ratios.[20] In *Sph. cyl.* II.4, a compounded ratio written in standard form is introduced, repeatedly transformed by replacement of identical ratios, and finally 'simplified' by 'removal' of a ratio shared with another compounded ratio.[21] Only II.8 *aliter*, a proposition that, however, is almost certainly spurious, depends in a decisive way on a clever application of compounded ratios: at the beginning of the proof, a multiple compounded ratio between solids, and written in standard form with two middle terms, is introduced and subsequently transformed by replacement of identical ratios; one of these replacements involves an application of the theorem we read as *Elem.* VI.23.[22] In the sequel of the proof, however, in order to express ratios of solids as compounded of the ratios of suitable surfaces and linear elements of the solids themselves, compounded ratios are replaced by expressions involving a 'multiplication' of these magnitudes.[23]

When Archimedes introduces compounded ratios outside applications of previously stated theorems, he does insert a middle term in *Sph. cyl.* II.4 and II.8 *aliter*.

---

18 As for the verbal forms employed in the canonical formula, the Archimedean corpus offers 12 occurrences of forms of συγκεῖσθαι (they all are participial forms); 6 of forms of συνάπτεσθαι—of these, 5 are of the form συνῆπται (in *Sph. cyl.* II.4 (*quater*) and in II.8 *aliter*), and 1 of the participle συνημμένος (in *Sph. cyl.* II.8 *aliter*).

19 See *AOO* II, 388.12–14 and 390.1–4, respectively.

20 See *AOO* I, 304.13–18; 364.10–18; 364.22–26 (rightly bracketed by Heiberg) and 366.6–9; 368.25–370.3 and 370.4–8; 430.25–432.4 and 432.10–14, respectively. The property states that cones and their segments have to one another the ratio compounded of that of their bases and of that of their heights; it combines *Elem.* XII.11 and 14. As Pappus also does, and very much in line with a 'solid' counterpart of *Elem.* VI.23, in *Con. sph.* 10, 23, 24, and 31, compounded ratios are thus used by Archimedes to express ratios of solids as compounded of ratios of suitable surfaces and linear elements of the solids themselves.

21 See *AOO* I, 190.4–6, 8–10, 14–18.

22 See *AOO* I, 216.14–20 and 216.24–26 (application of what we read as *Elem.* VI.23). To justify the insertion of several middle terms, Eutocius (*AOO* III, 203.2–16) refers back to his own general treatment of compounded ratios, which we shall discuss in sect. III.1.2.

23 On this thorny issue, see Netz 2004 and Acerbi 2005. Eutocius provides a very long commentary on *Sph. cyl.* II.8 *aliter*, in which he completes the attested proof at several points, adding, in particular, the synthesis (*AOO* III, 202.1–216.12). Before this, he proves an interesting lemma, linking the formulation in terms of compounded ratios with that in terms of 'multiplication' of magnitudes; see the first footnote in section III.1.2.

He does not do so in *Fluit.* II.10. The involved property, however, requires a lengthy proof, that Archimedes does not provide—the proof suggested by Heiberg[24] introduces the compounded ratio written in standard form. On the other hand, in *Con. sph.* 31 Archimedes transforms the original compounded ratio in order to obtain one with a middle term, which is eliminated in the subsequent step to get a simple ratio between the extremes.

In *Sph. cyl.* II.4, shared ratios are 'removed' from identical compounded ratios to produce identical ratios. The removal is expressed by the canonical clause κοινὸς ἀφῃρήσθω + 'object to be removed' in the nominative, a formula that is usually applied to removal of magnitudes common to the two sides of an equality.[25] Heiberg bracketed this clause because Eutocius repeats the entire argument by explicitly stating the identity of the compounded ratios and by performing the removal.[26]

The occurrence in *Sph. cyl.* II.4 triggered Eutocius' exposition on compounded ratios which will be discussed in section III.1.2.

II.3 Apollonius' *Conica*

In Apollonius' *Conica*, compounded ratios appear in 22 propositions, and in 5 instances already in the enunciation (with an asterisk *). These are: *Con.* I.11, 12, 13, 38, 39*, 40* (restatement of 39), 41*, 43 (instantiated quotation of 39), 45 (instantiated quotation of 39), 54, 55, 56, 58; II.11, 20; III.14 (instantiated quotation of I.40), 15 (instantiated quotation of I.41), 24, 53, 54*, 55, 56*.[27] Propositions I.11, 12, 13, 54, 55, 56, 58; II.11, 20; III.24, 53, 54, 55, 56 use—but do not quote—*Elem.* VI.23, often in combination with VI.1, in order to convert ratios of surfaces to ratios of straight lines and vice versa.

When Apollonius introduces compounded ratios outside applications of *Elem.* VI.23 or of previous theorems in the *Conica*, he always writes them in standard form. This happens in *Con.* I.55, 58; III.15; IV.54, 56. Inside compounded ratios, Apollonius freely replaces ratios with ratios identical to them.

In I.41, 54, and II.20, shared ratios are 'removed' from identical compounded ratios to produce identical ratios. In I.41 and 54, the removal is expressed by the canonical clause κοινὸς ἀφῃρήσθω + 'object to be removed' in the nominative.

Our data suggest that it is Apollonius who first systematically employed compounded rations. Apollonius' use of this tool is so deeply entrenched in crucial

24 In *AOO* II, 389 n. 4.

25 There are 21 occurrences of this formula in the *Elements*.

26 We read Eutocius' argument in *AOO* III, 128.20–130.2.

27 As for the verbal forms employed in the canonical formula, the *Conica* offers 79 occurrences of συγκεῖσθαι (59 of participial forms, 20 of σύγκειται); 4 of συνάπτεσθαι (1 of the form συνῆπται (in I.38), 3 of the participle συνημμένον (in III.14 and 15 (*bis*)).

features of his approach to the theory of conic sections that we may safely assume that it was not introduced by Eutocius, when the latter produced the only recension of the *Conica* we have access to.

The occurrence in *Con.* I.11 spurred Eutocius himself to reprise the exposition on compounded ratios he had previously inserted in his commentary on *Sph. cyl.* II.4. We shall discuss them in section III.1.2.

II.4 Ptolemy's *Almagest*

The most celebrated result of Greek spherical trigonometry is the Sector Theorem, also known as 'Menelaus' Theorem' because its being attested in the *Sphaerica*—it appears as proposition III.1 in Abū Naṣr's redaction. The Sector Theorem is a powerful mathematical tool, suited to determine arcs of a great circle on the surface of a sphere. It is the keystone of some of the most important technical results in the *Almagest*, where it is applied seventeen times in total.[28] It comes as no surprise then that the Sector Theorem is also proved in *Alm.* I.13 and, with many more cases on offer, in Theon, *in Alm. I.13*, *iA*, 535.10–570.12 (see section II.6).[29]

The configuration adopted by Ptolemy in his proof is the following: From the endpoints B, Γ of two mutually intersecting arcs AB, AΓ of great circles on the surface of a sphere, two arcs BE, ΓΔ are drawn across, meeting at Z and intersecting arcs AΓ, AB at E, Δ, respectively. All these arcs must be less than a semicircle. Then the following relations hold:

$$ch(2\Gamma E):ch(2EA) = [ch(2\Gamma Z):ch(2Z\Delta)]\circ[ch(2\Delta B):ch(2BA)]$$
$$ch(2\Gamma A):ch(2AE) = [ch(2\Gamma\Delta):ch(2\Delta Z)]\circ[ch(2ZB):ch(2BE)],$$

where, for instance, *ch*(2ΓE) is the chord subtended by twice the arc ΓE. Ptolemy provides a detailed proof that the first relation holds, leaving the proof of the second to the reader. The Sector Theorem is proved by Ptolemy last of a series of seven propositions. The first of these propositions is a version of the Theorem in the plane, involving compounded ratios of line segments.[30] The argument runs as follows (see Fig. 1): from the endpoints B, Γ of two mutually intersecting straight lines AB, AΓ, two lines BE, ΓΔ are drawn across, meeting at Z and intersecting straight lines AΓ, AB at E, Δ, respectively. It is required to show that ΓA:AE = (ΓΔ:ΔZ)∘(ZB:BE). The proof sets out the 'obvious' compounded ratio written in standard form ΓΔ:HE = (ΓΔ:ΔZ)∘(ΔZ:HE),

28 This happens in *Alm* I.14, 16, II.2, 3 (*ter*), 7 (*bis*), 10, 11, 12 (*bis*), VIII.5 (*ter*), 6 (*bis*); see *iA*, 555 n. 1.
29 For a clear exposition of the mathematics involved, see Neugebauer 1975, 26–30.
30 The plane case is proved in *POO* I.1, 68.23–70.16, and the Sector Theorem at 74.9–76.9.

draws a suitable parallel HE to one of the assigned straight lines, and readily argues by similar triangles and replacements of identical ratios in a compounded ratio.

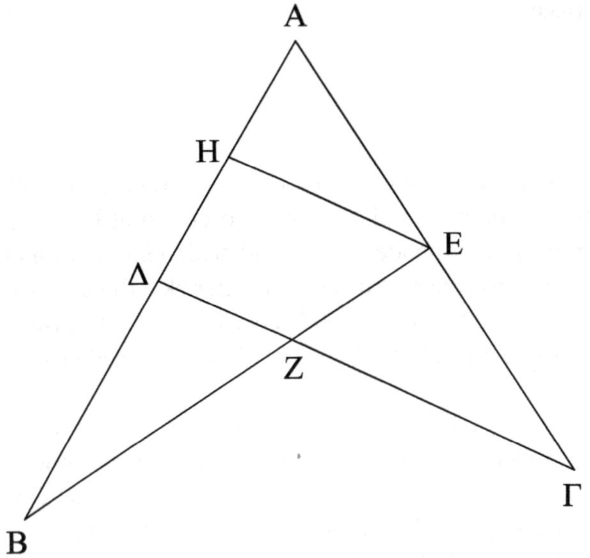

Fig. 1

In the proof of the plane case of the Theorem, the compounded ratio is written in standard form when introduced, and then transformed by replacement of identical ratios; the middle term is said to be taken 'from outside' (ἔξωθεν). In the proof of the spherical case, the compounded ratio arising in a suitably related plane case is simply transformed by replacement of identical ratios.

In all applications of the Sector Theorem in the *Almagest*, the relevant compounded ratio features only once, and a relation of the kind written above is simply stated to hold; five of the six arcs are numerically given, and one must find the sixth. The procedure, which was never worked out in detail by Ptolemy, amounts to using the Table of Chords to calculate chords from arcs and vice versa, and to performing the operation of ἀφαίρεσις 'removal' of a ratio from a ratio.[31]

Since Ptolemy never explains how to remove a ratio from a ratio, the gap was filled in Pappus' and Theon's commentaries on the *Almagest*, and by all subsequent generations of (Late-Antique and Byzantine) scholiasts. Filling the gap entailed either producing general expositions that tried to unify the several cases of the operation of

31 As for the verbal forms employed in the canonical formula, the *Almagest* has 6 occurrences of συγκεῖσθαι (1 of the form συγκείμενος, 1 of συγκεῖσθαι, 4 of σύγκειται); 24 occurrences of συνάπτεσθαι (6 of συνημμένος, 16 of συνῆπται, 1 each of συνῆφθαι and συναφθήσεται).

removal under a unitary treatment, or working out the calculations involved in a specific application of the Sector Theorem. The major role played by the Sector Theorem in the *Almagest* explains even the mere existence of most of the general expositions I will summarize in the following pages.

II.5 Pappus' *Collectio*

Pappus, in his *Collectio*, is the first mathematician who uses compounded ratios with explicit metamathematical goals: he offers several alternative proofs that hinge on compounded ratios, clearly suggesting that the idea of a grand-scale exploitation of the method was his own and that he was proud of it (with an asterisk * the applications of VI.23): see *Coll.* VII.37–40*, 42*, 68*, 74*, 75*, 84*, 86*, 194, 197*, 210, 246*, 253*, 255*, 272*, to which one must also add VII.240 (which will be discussed later in this section) and VIII.9–10.[32]

When Pappus introduces compounded ratios (this may happen several times within the same proposition), he rarely writes them in standard form. The exceptions are in *Coll.* VII.68, 84, 86, 197, 210, 240, VIII.9–10. In VIII.10, the middle term is twice said to be taken 'from outside' (ἔξωθεν). Inside compounded ratios, Pappus freely replaces ratios with ratios identical to them. In VII.194 there is no application of VI.23, nor are the compounded ratios written in standard form when introduced. However, the passage is obviously corrupt. In VII.210, one finds a unique instance of a replacement, within a compounded ratio identical to another, of a compounding ratio with a compounded form of it, and the subsequent removal of shared ratios is applied, on the one side, to a compounded ratio with one middle term and, on the other side, to a multiple compounded ratio with two middle terms.

In VII.194, 210 (*bis*), shared ratios are 'removed' from identical compounded ratios to produce identical ratios. The removal is expressed by the clause κοινὸς ἐκκεκρύσθω + 'object to be removed' in the nominative, a totally un-canonical *variatio* of the standard formula that is usually applied to removal of magnitudes common to the two sides of an equality.

A striking application of the method can be found in *Coll.* VII.37–40,[33] where the problem is to provide a generalization of the enunciation of the 'locus on three and four lines'. Pappus first remarks that the enunciation immediately generalizes to five or six lines: it is enough to replace the rectangles with parallelepipeds. The higher loci of this type raise problems, "since there is nothing contained by more than three

---

32 As for the verbal forms employed in the canonical formula, the *Collectio* has 20 occurrences of συγκεῖσθαι (10 of participial forms, 10 of σύγκειται); 67 of συνάπτεσθαι (20 of the form συνῆπται, 47 of forms of the participle συνημμένος).

33 See also the enunciation of what is known as 'Gouldin's theorem', in *Coll.* VII.42.

dimensions".[34] Pappus claims that his predecessors did not employ a language adequate to the formulation of such general enunciations, since "it was possible to enunciate and generally to prove these things by means of compounded ratios".[35] Pappus provides an example of an adequate formulation:

> ἐὰν ἀπό τινος σημείου ἐπὶ θέσει δεδομένας εὐθείας καταχθῶσιν εὐθεῖαι ἐν δεδομέναις γωνίαις, καὶ δεδομένος ᾖ λόγος ὁ συνημμένος ἐξ οὗ ἔχει μία κατηγμένη πρὸς μίαν καὶ ἑτέρα πρὸς ἑτέραν καὶ ἄλλη πρὸς ἄλλην καὶ ἡ λοιπὴ πρὸς δοθεῖσαν (ἐὰν ὦσιν ζ′· ἐὰν δὲ η′, καὶ ἡ λοιπὴ πρὸς λοιπήν), τὸ σημεῖον ἅψεται θέσει δεδομένης γραμμῆς.

> If straight lines are drawn from some point at given angles onto straight lines given in position, and there is given the ratio compounded of that which one drawn line has to one, and another to another, and a different one to a different one, and the remaining one to a given (if there are seven—but if eight, the remaining to the remaining one), the point will touch a curve given in position.

From a theoretical point of view, the most relevant passage is *Coll.* VII.240, a lemma to Apollonius, *Con.* I.39 or 41 (cf. *Document 2* in the Appendix). The proof does not make use of the πηλικότης, and it proves that, if $a{:}b = (c{:}d)\circ(e{:}f)$, then $c{:}d = (a{:}b)\circ(f{:}e)$. The proof runs as follows:

- Let h be such that $d{:}h{::}e{:}f$.
- Since $a{:}b = (c{:}d)\circ(e{:}f) = (c{:}d)\circ(d{:}h)$ by assumption, and $(c{:}d)\circ(d{:}h) = c{:}h$ by inserting the middle term d in ratio $c{:}h$, then $a{:}b{::}c{:}h$. This amounts to assuming that, if two ratios are compounded of identical ratios, they are identical.
- Since $c{:}d = (c{:}h)\circ(h{:}d)$ by inserting the middle term h in ratio $c{:}d$, and $c{:}h{::}a{:}b$ by the previous deduction, and $h{:}d{::}f{:}e$ by inversion of the construction for h, then $c{:}d = (a{:}b)\circ(f{:}e)$.

Thus, the proof argues by replacement of ratios both in the assigned compounded ratio and in suitable compounded ratios written in standard form.

II.6 Theon's *Commentary on the Almagest*

In this section, I shall only deal with the long proof of the Sector Theorem in *in Alm. I.13*, *iA*, 535.10–570.12 (see sections III.1.1 for Theon's general expositions on compounded ratios and removal, and III.2.1 for the procedures of removal he employs in the applications).

Theon's long proof of the Sector Theorem is just an expansion of that provided by Ptolemy in *Alm.* I.13. The well-known habit of mathematical commentators to imitate

34 Here and henceforth, I follow Jones' 1986 translation.

35 The language deemed inadequate by Pappus resorted to 'multiplication' in Heronian style of squares and rectangles; this is the approach of *Sph. cyl.* II.8 *aliter*, as we have seen.

the style and lexicon of the model makes Theon's text a poorly significant witness of linguistic changes. At any rate, in the proof of the plane case, the compounded ratio is written in standard form when introduced, and then transformed by replacement of identical ratios; the middle term is said to be taken 'from outside' (ἔξωθεν).[36] In the several partial proofs of the spherical case, either the compounded ratio arising in a suitable plane case is simply transformed by replacement of identical ratios, or compounded ratios arising in other spherical cases are transformed by replacement of identical ratios. Theon also provides complete enunciations of all the theorems that Ptolemy had instead enunciated in instantiated form. In the first of these enunciations, Theon implicitly introduces the reader, by means of the disjunction "is conjoined, that is, is compounded" (συνῆπται ἤτοι σύγκειται), to the major verbal *variatio* in the canonical formula of a compounded ratio.[37]

III Logistic Approach

The data collected in section II have already given prominence to the two actors that will play a prime role in what follows: the Sector Theorem and *Elem*. VI.df.5. In principle, logistic is only involved in explanations of the former, but we shall see that extensive register-crossing took place. For this reason, and because of their context, I have included in this section the expositions 'in geometric style' (γραμμικῶς)—that is, in deductive style—by Theon and by Eutocius. Also notice—since it belongs to the same category—the very short argument formulated in the 'language of the givens' which appears in section III.1.3.

Before proceeding, I recall the main terminological conventions pertaining to our subject and typical of the logistic domain.

- A 'ratio' (λόγος) is a relation between two 'terms' (ὅροι), called in the order 'antecedent' (πρόλογος) and 'consequent' (ὑπόλογος), and much less frequently ἡγούμενος and ἑπόμενος. To avoid ambiguities, I shall often use the noun 'extreme' as a synonym of 'term'. In specific numerical examples, recall that the terms of a ratio are usually given in the form 'greater-to-lesser'.
- In the operation of 'removal' (ἀφαίρεσις) of a ratio from a συγκείμενος λόγος 'compounded ratio', the basic elements are the ratio 'from which the removal is going to occur' (ἀφ' οὗ ἡ ἀφαίρεσις γίνεται), the ratio 'to be removed' (ἀφαιρούμενος),

36 In *iA*, 543.6, ἔξωθεν, in the expression ἔξωθεν λαμβανομένης, is replaced by μέσης.

37 As for verbal forms employed in the canonical formula, books I–IV of Theon's commentary on the *Almagest* have 62 occurrences of συγκεῖσθαι (8 of forms of συγκείμενος, 1 of συγκεῖσθαι and of συγκείσθω, 52 of σύγκειται); 18 occurrences of συνάπτεσθαι (17 of συνῆπται, 1 of συνῆφθαι). I have excluded the occurrences in quotations from the *Almagest*. In *iA*, 542.6–7 and 545.2–3, we read citations of *Elem*. VI.23.

and 'the remainder' (ὁ λοιπός). I shall translate the verb καταλείπειν by 'to leave out', even in the case in which participial forms are involved. We shall also see that a type of the procedure of removal will require 'to fit' (ἐναρμόζειν) a ratio to another.

- The adjective 'paronymous' (παρώνυμος) is applied to the number that expresses the πηλικότης 'value' of a ratio: 'six' is the πηλικότης of the sextuple ratio and is paronymous to it; there cannot be strict homonymy.[38] Likewise, a part is paronymous to a number if it bears the same name as the number, only transformed into an ordinal, like 'three' and '<one> third'.

We have already seen that there is no operation of 'composition' of ratios. This is quite paradoxical, since *Elem.* VI.df.5 defines a compounded ratio in terms of an operation performed on the values of the ratios themselves. Even more paradoxical is the fact that the removal of a ratio from a compounded ratio was indeed conceived of as an operation. Moreover, the implicit assumption that one most appropriately remove a ratio from a compounded ratio, and not simply from a ratio *tout court*, had as a consequence that the operation of removal more or less entirely consisted in a prescription for writing an assigned ratio in a suitable compounded form. It is such a prescription that quite naturally—that is, according to the 'suitable' compounded form required by the context—branches in a variety of cases.[39] Accordingly, the operation of removal was almost always performed on ratios that were already provided in compounded form. Therefore, if from $a{:}b = (a{:}d)\circ(d{:}b)$ we want to remove ratio $d{:}b$, it suffices to literally 'remove' (ἀφαιρεῖν) $d{:}b$ from the right-hand side; ratio $a{:}d$ is, in a most concrete sense, the 'remainder'.

The translation of the verb ἀφαιρεῖν by 'to remove' and of the associated *nomen actionis* ἀφαίρεσις by 'removal' calls for a few words of justification.[40] On the one hand, ἀφαιρεῖν (and the substantives related to it) have a well-defined technical meaning in the Greek mathematical lexicon: 'to subtract', either a magnitude from a magnitude or a number from a number. Moreover, as we have already seen, the Greek

38 Still, 'homonymous' (ὁμώνυμος) is the adjective we read, with identical meaning, in *Elem.* VII.37–9 and in some of the sources we shall meet in the following pages.

39 In some sources, this division into cases is also motivated by the fact—*a priori* irrelevant if one only sets out to remove ratios from compounded ratios—that no applications of the Sector Theorem simply require to remove a ratio from another, but also to calculate one of the terms of the remainder, the other term being given.

40 Heath (1921 II, 538), and Bulmer-Thomas (1971, 160) following him, translate the verb as 'to take out'; Knorr (1989, 202–204) has 'to subtract'; Ruelle (1883) is more inconsistent by using 'retrancher' in the introduction, and 'ôter' or 'enlever' in the translation; Tannery (1885, 221 of the reprint) 'supprimer'; Riedlberger (2013) 'to remove'. The problem of this translation is discussed by O. Riemann in note 1 of Ruelle 1883, 85; cf. Riedlberger, 2013, 201–202; Acerbi & Riedlberger 2014, 406; and Acerbi & Pérez Martín 2015, 123 (the last two articles use 'rimuovere').

mathematical tradition formulates within the lexicon of adding the operations on ratios we express within the lexicon of multiplying. Both conceptually and operationally, an ἀφαίρεσις λόγων does not correspond to the modern conceptualization of the 'same' operation as a division of fractions. On the other hand, if Greek mathematical style has two lexical families corresponding to what we would call 'adding' (namely, the verbs προστιθέναι and συντιθέναι and the substantives related to them),[41] it has just one lexical family corresponding to what we would call 'subtracting' (namely, ἀφαιρεῖν and the substantives related to it). Such considerations lead to regard as *a priori* perfectly legitimate to translate ἀφαιρεῖν by 'to subtract'. Still, employing 'to subtract' would call for 'to add' as a natural correlative, and not 'to compound', which is currently used for συντιθέναι. Further, such renderings would clash with well-entrenched modern mathematical habits, since the clause 'to subtract a ratio from another' has a well-defined and technical meaning also in modern mathematics, and this meaning is at variance with the meaning of the Greek technical expression the above clause is intended to translate.

III.1 General Expositions

III.1.0 Nicomachus' *Introductio Arithmetica*

In *Ar.* II.2.5, Nicomachus briefly deals with compounded ratios. He introduces and perfunctorily employs the lexicon of 'compounding' and 'composition' (substantive σύνθεσις, adjective σύνθετος 'compound', compounding ratios that are said to be 'compounded' to give a ratio), but does not explain what such an 'operation' might amount to. Moreover, his lexicon is far from fixed, as he also uses the verbs συλλαμβάνεσθαι 'to be taken together', ἀναλύεσθαι and διαλύεσθαι 'to be analyzed', συνίστασθαι 'to be constructed', and the noun σύστημα 'complex' to denote the same operation or its results, either from the viewpoint of the compounded ratio, or from the viewpoint of the compounding ratios. In particular, the lexicon of 'analysis' is employed when a given ratio is shown, by inserting a middle term, to be compounded of two ratios. In this way—and these are Nicomachus' examples—, a double ratio is shown to be analyzed into a sesquialter and a sesquitertian ratio (terms 4 3 2 and 6 4 3). Conversely, from a sesquialter and a sesquitertian ratio, if compounded, a double ratio is in any case constructed. Nicomachus then applies the lexicon of composition to reformulate a property expounded in *Ar.* II.2.3–4; namely, that any generic multiple ratio, compounded with a suitable superparticular ratio, gives the immediately

41 The former verb as a prevalent arithmetical connotation, while the latter as a prevalent geometrical connotation, even if register-crossing may occur. As we have seen, only συντιθέναι—and most frequently its passive συγκεῖσθαι—occurs in the sectorial lexicon of compounded ratios.

subsequent multiple ratio (or, conversely, that the latter can be analyzed into the former two). For instance, a double ratio, compounded with a sesquialter ratio, gives a triple ratio (terms 18 12 6 and 18 9 6, and again 6 3 2); a triple ratio, compounded with a sesquitertian ratio, gives a quadruple ratio, and so on.

III.1.1 Theon on Compounded Ratios in his *Commentary on the Almagest*

Theon has two general expositions that can be summarized as follows (cf. *Document 3* in the Appendix):[42]

1) The opening passage of *in Alm. I.13*, *iA*, 532.1–535.9, which appears just before Theon's extended proof of the Sector Theorem.

The argument is of the kind later deemed 'inductive' by Eutocius (see section III.1.2). After a transcription of *Elem.* VI.df.5 (notice that Theon is the only independent source in the Late Antiquity who completes the final τινα with πηλικότητα λόγου), Theon claims that a ratio can be written in a compounded form using any possible middle term. This means that the middle terms can be either (a) lesser that the greater extreme of the ratio but greater than the lesser extreme, or (b) greater than the two extremes, or (c) lesser than the two extremes. Theon gives two different 'proofs' of his claim. The first proof operates by multiplication of the denominations of the ratios. It is a circular argument, as Theon himself implicitly admits by resorting to the 'composition' of multiple ratios in order to justify the assertion that double times triple gives sextuple.[43] The second proof argues by partition of the antecedent of a ratio once the consequent of the same ratio is taken as unit, and by counting the 'units' resulting from taking a suitable multiple of both. The argument is similar to the one used in the proof of *Sectio can.* 7 and 8 (see section II.1). Theon's examples do not identify the ratios by setting out their terms, as Eutocius does, but their denominations: double, triple, sextuple; triple, half, sesquialter; subdouble (*sic*), sesquitertian, subsesquialter. The presence of the middle term in Theon's exposition—and, as we shall see, in Eutocius'—shows that the goal of his arguments, that seemingly deal with compounded ratios, is, in fact, to set the stage for operating a removal. Theon ends his exposition by pointing out that (*iA*, 535.7–9) "it is obvious that, if from a compounded ratio one whatever of the compounding <ratios> is removed, one of the extremes being made to disappear, the remaining compounding <ratio> will be left out".

42 Cf. *iA*, 532 n. 1, for a discussion on the authenticity of the first passage and of its location in the manuscripts.
43 Cf. *iA*, 534.9.

2) *In Alm. II.11*, *iA*, 759.8–762.2.

This is a sketchy justification of the operation of removal in deductive style, only applied to a case of the Sector Theorem that can be deduced, by simple inversion of a ratio, from a case that had been treated in the commentary on Book I of the *Almagest*: if $a{:}c = (a{:}b)\circ(b{:}c)$, then $a{:}b = (a{:}c)\circ(c{:}b)$. The proof is said to be immediately evident and simply amounts to point out that it is enough to 'deplace' the term *b*.[44] Theon's trivial proof is, in fact, all that is needed if one limits itself to compounded ratios written in standard form. We have seen a proof in deductive style of the same statement, but applied to compounded ratios not written in standard form, in Pappus, *Coll.* VII.240 (cf. section II.5 and *Document 2* in the Appendix). It is thus possible that Theon compiles here an analogous argument in Pappus *in Alm. II.11*, now lost.

III.1.2 Eutocius on Compounded Ratios in his Commentaries on Archimedes' *De Sphaera et Cylindro* and on Apollonius' *Conica*.

The early sixth-century Neoplatonic philosopher Eutocius redacted two expositions on compounded ratios.[45] These are, as Eutocius himself says and as the following parallel examination will confirm, the one a (summary) reprise of the other. I shall refer to the two expositions in their order of redaction: *in Sph. cyl.* first, *in Con.* second; references to the texts are separated by the sign '/'.

Eutocius starts with an historico-doctinal *excursus*, which justifies the introduction of his digression, and boasts about his own contribution; namely, that this is the first abstract exposition ever written (*in Sph. cyl.*, *AOO* III, 120.3–11: he complains about the fact that only 'inductive' (ἐπαγωγῇ) expositions can be found, among others, in Pappus, Theon, Arcadius / *in Con.*, *AGE* II, 218.6–15: he mentions his own previous exposition in the commentary *in Sph. cyl.*, and another in σχόλια to the first book of the *Almagest*).[46]

44 The text is probably corrupt; see Rome's discussion at *iA*, 762 n. 2. In the enunciation of this theorem, one must athetize ἀνάπαλιν at 761.1, *contra* Rome.

45 Cf. *In Sph. cyl. II.4*, *AOO* III, 120.3–126.20 and *in Con.*, *AGE* II, 218.6–220.25. The texts that triggered Eutocius' digressions are Archimedes, *Sph. cyl.* II.4, *AOO* I, 190.4–6, and Apollonius, *Con.* I.11, *AGE* I, 40.8–10, respectively. Add to these the proof that, if there are four terms A, Γ, Δ, B, "the ratio compounded of that of the <rectangle contained> by A, B to the <square described> on Γ with the ratio of B to Δ is identical to that of the <rectangle contained> by A, B times B to the <square described> on Γ times Δ" (*in Sph. cyl. II.8*, *AOO* III, 198.19–200.31). The proof, that applies the δι' ἴσου proposition, is translated as *Document 6* in the Appendix.

46 The text reads as follows (*in Con.*, *AGE* II, 218.6–15): "since this was treated by the commentators in a most inductive way, and not according to cogent methods, I had some thoughts on that and I have written them down in my own published <notes> to the fourth theorem of the second book of Archimedes' *On the sphere and the cylinder*, and in the annotations to the first book of Ptolemy's *Syntaxis*. It will do no harm that this be also written here, because it is by no means obvious that

The definition of compounded ratio as appears in *Elem.* VI.df.5 is then transcribed, accompanied by remarks on the πηλικότης of a ratio, the key notion in the abstract proof Eutocius is going to offer (the text at 120.12–122.9 is richer: it has in addition a definition of πηλικότης, according to Nicomachus and his commentator Heronas, as "the number to which the given ratio is paronymous [παρώνυμος]"; this is stated to be equivalent to the definition "the number that, multiplied by the consequent term of a ratio, also gives the antecedent"; note also the characterization, at 120.28–30, of logistic as the discipline in which one divides the unit / 218.16–26).

After this, Eutocius provides the non-inductive proof that ratio $a{:}c$ is compounded of $a{:}b$ and $b{:}c$. He does this by showing that, if d and e are the πηλικότητες of $a{:}b$ and of $b{:}c$, respectively, and if $de = z$, then z is the πηλικότης of $a{:}c$ (122.10–124.7 / 218.27–220.16; they are nearly identical, the latter being more concise; cf. *Document 7* in the Appendix).

Only in the commentary *in Sph. cyl.* (124.8–126.3), we read an 'inductive' proof that a ratio can be written in compounded form using any possible middle term, namely, whether this is (a) lesser that the greater extreme of the ratio but greater than the lesser extreme, (b) greater than the two extremes, or (c) lesser than the two extremes. This is the same argument as that outlined in the first passage in Theon, summarized in section III.1.1. In his examples, Eutocius identifies the ratios by setting out their terms and by providing a middle term that satisfies each of the stated conditions: the middle term lies between the two extremes (12, 4, 2), is greater than both (9, 12, 6), or is lesser than both (9, 4, 6).

Again in *in Sph. cyl.* only (126.4–20), we read an instantiated enunciation that generalizes the main result to several ratios and middle terms: $a{:}b = (a{:}c)\circ(c{:}d)\circ(d{:}b)$. The proof amounts to taking for granted that, within a compounded ratio, one can replace ratios with compounded ratios: since $a{:}b = (a{:}d)\circ(d{:}b)$ and $a{:}d = (a{:}c)\circ(c{:}d)$, then $a{:}b = (a{:}c)\circ(c{:}d)\circ(d{:}b)$.

---

the present readers will also read those expositions, and since almost the entire treatise of the *Conics* makes use of it" (ἐπεὶ δὲ ἐπακτικώτερον μᾶλλον καὶ οὐ κατὰ τὸν ἀναγκαῖον τρόπον ὑπὸ τῶν ὑπομνηματιστῶν ἐλέγετο, ἐζητήσαμεν αὐτὸ καὶ γέγραπται ἐν τοῖς ἐκδεδομένοις ἡμῖν εἰς τὸ τέταρτον θεώρημα τοῦ δευτέρου βιβλίου τῶν Ἀρχιμήδους περὶ σφαίρας καὶ κυλίνδρου καὶ ἐν τοῖς σχολίοις τοῦ πρώτου βιβλίου τῆς Πτολεμαίου συντάξεως· οὐ χεῖρον δὲ καὶ ἐνταῦθα τοῦτο γραφῆναι διὰ τὸ μὴ πάντως τοὺς ἀναγινώσκοντας κἀκείνοις ἐντυγχάνειν, καὶ ὅτι σχεδὸν τὸ ὅλον σύνταγμα τῶν κωνικῶν κέχρηται αὐτῷ). Thus, quite clearly Eutocius asserts that what we are going to read is more or less identical to two previous expositions of his. One must infer that the now lost 'annotations to the first book of Ptolemy's *Syntaxis*' contained the same material as we find in the two surviving commentaries. Moreover, Arcadius cannot be the author of the part of the *Prolegomena* we shall deal with in section III.1.4: on the one hand, he is mentioned by Eutocius because he adhered to the 'inductive' mode of presentation; on the other hand, the *Prolegomena* does not resort to this mode but expressly refers to Pappus for a general exposition and otherwise only provides specific examples (which *do not* amount to an 'inductive' proof). These remarks disprove both Mogenet's contention (1956, 13–44) that Eutocius was the author of the *Prolegomena* and Knorr's (wrong) inference from his own (correct) rebuttal of Mogenet's thesis that Arcadius himself, a character we simply know nothing about, authored the *Prolegomena* (1989, 155–177).

Only in the commentary *in Con.* (220.17–25), Eutocius adds a final remark about the commonalities between logistic and arithmetic (that is, number theory, characterized by proofs γραμμικῶς); in this way, he intends to justify his own choice of using language and methods of arithmetic in order to prove a result belonging to logistic.

III.1.3 A Scholium on Ptolemy's *Almagest* and the Treatise on Removal of Ratios Ascribed to Domninus of Larissa

As said in section II.4, in any application of the Sector Theorem in the *Almagest*, a relation involving a compounded ratio whose terms are chords of suitable arcs is stated to hold, five of the six arcs are numerically given, and one must find the sixth. The procedure, whose details are never worked out by Ptolemy and which admits of a very rich combinatorics of cases and subcases, requires performing the operation of removal of a ratio from a ratio. Since Ptolemy never explains how to perform such a removal, the gap was filled by commentators on the entire *Almagest*, such as Pappus and Theon, as well as by a number of well-known or anonymous scholiasts in the Late Antiquity and in the Byzantine period. The unitary and general exposition, just enriched with a few numerical examples, we find in one of the *Almagest* scholia (cf. *Document 4* in the Appendix; see below in this section for a summary of the actual content of the scholium) can be expounded in a most 'inductive' way as follows.

The problem that has bewildered all commentators is that, in the applications, no given compounded ratio is ever written in standard form. Two ratios are instead assigned, one of which must be removed from the other, the latter being assigned as a compounded ratio not written in standard form. As a consequence, the ratio from which the removal is going to occur must preliminary be written as a compounded ratio in standard form. This is done by operating a 'fitting' inside it (the related verb is ἐναρμόζειν), that automatically takes into account the form of the ratio to be removed. A first ratio is 'fitted' to a second ratio by transforming it into a ratio which is equivalent to it and having the antecedent or the consequent equal to the antecedent or to the consequent of the second; therefore, one has only to calculate the remaining term of the first ratio, and this can be done by taking a suitable fourth proportional.

For instance, let it be required to remove 4:3 from 12:6. Now, 'to fit' 4:3 to 12:6 amounts to find a ratio equivalent to 4:3 with the following, alternative, constraints:

a. either 12 is the new antecedent; thus one must calculate the new consequent as a fourth proportional $4:3::12:x$,[47] yielding as a result $x = (12\times3)/4 = 9$;
b. or 6 is the new consequent; thus one must calculate the new antecedent as a fourth proportional $4:3::x:6$, yielding as a result $x = (4\times6)/3 = 8$.

47 Contrary to what is suggested by the qualifier 'fourth', the position in the proportion of the number to be determined is immaterial, if we allow for a rearrangement of the given terms.

In case (a), 4:3 'fitted' to 12:6 gives 12:9, which is a ratio equivalent to 4:3 and whose antecedent is identical to the antecedent of 12:6. In case (b), instead, the ratio obtained by 'fitting' 4:3 to 12:6 is 8:6, whose consequent is identical to that of 12:6.

Conversely, 'to fit' 12:6 to 4:3 amounts to find a ratio equivalent to 12:6 with the following, alternative, constraints:

c. either 4 is the new antecedent; thus one must calculate the new consequent as a fourth proportional 12:6::4:x, yielding as a result $x = (6\times4)/12 = 2$;
d. or 3 is the new consequent; thus one must calculate the new antecedent as a fourth proportional 12:6::x:3, yielding as a result $x = (12\times3)/6 = 6$.

In case (c), 12:6 'fitted' to 4:3 gives 4:2; in case (d), it gives 6:3.

If we want to remove 4:3 from 12:6, let us choose for instance procedure (a): 'fit' 4:3 to 12:6 to give 12:9; insert the middle term 9 between the terms of ratio 12:6 in order to write it as the compounded ratio (12:9)∘(9:6);[48] literally 'remove' 12:9 and the remainder is 9:6. Since the ratio has been removed that contains the antecedent of the compounded ratio 12:6, the operation of removal is said to be performed 'with respect to the antecedent' (πρὸς τὸν πρόλογον); otherwise, it is said to be performed 'with respect to the consequent' (πρὸς τὸν ὑπόλογον). Procedures (a) and (c) can only give rise to the first type of removal, (b) and (d) to the second.

Finally, it always happens that one is interested in calculating one of the terms of the remainder, the other being given. This is the reason why a 'fitting' has four cases and not two (a, b or c, d): the position of the term to be calculated in the remainder can make it necessary to 'fit' either the compounded ratio to the ratio to be removed, or vice versa. If the given term does not match the homologous term of the remainder calculated by any of the above procedures, a further taking of a fourth proportional must be performed. Suppose, as above, that we want to remove 4:3 from 12:6 in order to calculate the antecedent of the remainder, its consequent being given as 8. The operation of removal according to procedure (a) yields 9:6, that must be reduced to a ratio with consequent 8. This is done by setting out the proportion 9:6::x:8, yielding as a result $x = (9\times8)/6 = 12$.

This scholium was cut, pasted, and enriched with a complete series of examples by some anonymous revisor. The accidents of textual transmission made the result to be attributed to Domninus of Larissa, a fifth-century Neoplatonic philosopher contemporary of Proclus.[49] The text ascribed to Domninus is organized as follows (the section numbers are canonical since the first edition).

48 And, in fact, passing to πηλικότητες *iuxta Elem.* VI.df.5 and inserting number 9, we get $^{12}/_{6} = (^{12}/_{9})(^{9}/_{6})$, where we are entitled to write the πηλικότητες of the ratios as fractions.

49 The standard edition of Domninus' works is now Riedlberger 2013, which also contains the best available discussion of his life and works. The scholium is edited and thoroughly discussed in Acerbi & Riedlberger 2014.

- Introduction to the problem; definition of compounded ratio (always written in standard form) according to *Elem.* VI.df.5; two examples: 4 and 40 as middle terms of 2 and 8 (sections 1–5).
- Explanation of the principles of removal of ratios, the removed ratio being always less than the one from which the removal is going to occur: (i) 'fitting' of the lesser ratio to the greater ratio or vice versa; (ii) removal made with respect to the antecedent or with respect to the consequent of the ratio to which the fitting of point (i) is made; as a consequence, there are four possible cases (section 6).
- Examples, always with 12:3 and 6:4 (these correspond to my set of examples above). *a)* The first ratio is written as compounded of 12:8 and 8:3 (removal performed with respect to the antecedent of the greater ratio: 12:8 is identical to the ratio to be removed 6:4; the remainder is 8:3) or as compounded of 12:4½ and 4½:3 (removal with respect to the consequent of the greater ratio: 4½:3 is identical to the ratio to be removed 6:4; the remainder is 12:4½) (sections 7–15). *b)* Either ratio 6:1½ (that is, 12:3) is written as compounded of 6:4 and 4:1½ (removal with respect to the antecedent of the lesser ratio; the remainder is 4:1½), or ratio 16:4 (that is, again 12:3) is written as compounded of 16:6 and 6:4 (removal with respect to the consequent of the lesser ratio; the remainder is 16:6) (sections 16–18).
- Things become simpler if the two ratios have the same antecedent or the same consequent. Example: to remove 12:4 from 12:3; the remainder is 4:3, taken by simply forming the ratio of the extremes that are not identical (sections 19–21).
- Again, the definition of compounded ratio (section 22).
- Validation of the operation of removal of ratios using the 'language of the givens': if a given ratio is removed from a given ratio, the remainder is also given (section 23).

The original scholium contains only sections 1–5, 22–23, 6, 19–21, in this order. In particular, the presence of sections 22–23 is fully justified only in the original position; the revisor that produced the text ascribed to Domninus also added the long, central portion featuring the numerical examples of removal. Both the reworked text that came to be ascribed to Domninus and the original scholium repeatedly make the conceptual mistake of multiplying ratios and not the related πηλικότητες.

III.1.4 The *Prolegomena ad Almagestum*

The *Prolegomena* is a computational primer to the *Almagest*: it contains a tightly organized 'handbook of logistic', featuring as its main themes: an introduction to the sexagesimal system; a description of computational algorithms for multiplication, division, and extraction of approximate square root; a presentation of interpolation techniques; an exposition about compounded ratios and removal of a ratio from an assigned ratio. According to the author, no comprehensive previous exposition of this kind existed.

The *Prolegomena* deals with compounded ratios as a preliminary to removal.[50] The author starts by quoting *Elem.* VI.df.5. He then gives a numerical example: the ratio of 100 to 5 makes the twentyfold ratio. This can be seen as a compounded ratio by interpolating one or several middle terms. No general exposition is provided, since "the geometer Pappus has shown that, if there are two magnitudes, and <magnitudes> are set between them, the ratio <contained> by the extreme magnitudes is equal to that <compounded> of all intermediate ratios". A numerical example follows: 20 as a middle term of 100 and 5; therefore, the twentyfold ratio is compounded of the quintuple and of the quadruple ratio. The case of three ratios with two middle terms is also treated on the basis of the same example, inserting 10 as an additional term. That's all. The exposition is not inductive (as, for instance, Theon's is), for no general result is stated and allegedly proved by an example.

Removal of ratios is treated in the *Prolegomena* by means of paradigmatic examples; in order to summarize the procedure underlying them I resort to a general exposition with dummy letters. The similarity of this exposition and of the one seen in the previous section is obvious.

One is required to remove $c:d$ from $a:b$. To this end, one must (i) write the ratio $a:b$ in compounded form in such a way that one of the compounding ratios is $c:d$, say $a:b = (x:y)\circ(c:d)$, where $x:y$ will obviously be the remainder, but (ii) have the compounded ratio written in standard form: $a:b = (a:m)\circ(m:b)$, where m is a middle term. One, then, has to find such a middle term m. This can be done in a general way according to two procedures:

- *With respect to the antecedent.* In $a:b = (a:m)\circ(m:b)$, one identifies $a:m$ as the ratio to be removed (it has the same *antecedent* as the ratio from which the removal is going to occur), and therefore $m:b$ as the remainder. But the ratio to be removed was assigned as $c:d$. Therefore, m can be calculated as a fourth proportional: $c:d::a:m$. By *Elem.* VII.19, $m = ad/c$ is the required middle term.
- *With respect to the consequent.* In $a:b = (a:m)\circ(m:b)$, one identifies $m:b$ as the ratio to be removed (it has the same *consequent* as the ratio from which the removal is going to occur), and therefore $a:m$ as the remainder. But the ratio to be removed was assigned as $c:d$. Therefore, m can be calculated as a fourth proportional: $c:d::m:b$. By *Elem.* VII.19, $m = bc/d$ is the required middle term.

It is obvious that the conditions identifying the two procedures are exhaustive and exclusive. These can be summarized by means of one single prescription: 'find the fourth proportional of the two extremes of the ratio to be removed and of the extreme of the ratio from which the removal is going to occur not featuring in the name of the

---

50 I use the critical text of the *Prolegomena* that I have established in collaboration with N. Vinel and B. Vitrac. A debased edition (and a bad translation of the section of the *Prolegomena* on compounded ratios and removal) appears in Knorr 1989, 185–201.

procedure, each keeping its original position as antecedent or as consequent; this fourth proportional replaces, in the ratio from which the removal is going to occur, the extreme not featuring in the name of the procedure'. The *Prolegomena*, and other sources as we shall see, make the game more complicated by introducing a further parameter. This comes from the fact, in principle without relevance, that one of the extremes of the ratio to be removed might be identical to one of the extremes of the ratio from which the removal is going to occur. If one excludes, as the author of the *Prolegomena* does, the case in which each of the terms of the ratio to be removed is identical to one or the other of the extremes of the ratio from which the removal is going to occur, five possibilities remain, whose full list is provided: 1) $c = a$ (and therefore $d \neq b$); 2) $c = b$ (and therefore $d \neq a$); 3) $d = a$ (and therefore $c \neq b$); 4) $d = b$ (and therefore $c \neq a$); 5) $c \neq a$ and $d \neq b$. For no apparent reasons, in the sequel the author only works out the details of cases 1, 3 and 5, always in the order 'with respect to the antecedent' / 'with respect to the consequent'. He adds at the end of the exposition, as a 'postponed case', the one in which the ratio to be removed is the inverse of the ratio from which the removal is going to occur ($c = b$ and $d = a$)—a case that had just been excluded at the beginning.

Finally, if the consequent or the antecedent of the remainder do not coincide with the given consequent or the given antecedent, the author of the *Prolegomena* shows, on the basis of an example, how to transform the remainder to a ratio having the given term; this is done, as we have seen in the previous section, by means of a further taking of a fourth proportional. We shall see that a number of commentators will consider this final transformation of the remainder as an integral part of the operation of removal.

III.1.5 Leo the Mathematician on Composition and Removal of Ratios

A long scholium on *Elem.* VI, entitled ὑπόμνημα σχόλιον εἰς τὰς τῶν λόγων σύνθεσίν τε καὶ ἀφαίρεσιν, Λέοντος, is ascribed to the early ninth-century scholar Leo the Mathematician. The scholium is transcribed by the hand of the main copyist at the end of Book VI, and together with other exegetic material, at ff. 120–121 of Bodl. Dorv. 301.[51]

As the title suggests, the scholium deals with 'composition and removal of ratios'. As we have seen, this subject is only marginally dealt with in *Elem.* VI, even if df.5 provides a definition of compounded ratio. It is, therefore, misleading to assert, as it is commonly done, that the scholium is related to *Elem.* VI.df.5.[52] The structure of this text

---

51 The scholium is edited in *EOO* V, 714.17–718.22, as a part of *sch.* 1 of the *Appendix Scholiorum* III.
52 On the basis of a misinterpretation of the alphabetic letters employed there as ordinals, the scholium was also absurdly held to contain germs of algebraic notation; cf. Vogel 1960, 661. To employ ordinals as dummy letters was a common and widespread practice in Greek mathematics; it is enough to get a look at *Elem.* V. These alphabetic signs are treated in the scholium as real denotative letters, since they are marked by a superimposed bar and not by an apex, as ordinal numbers usually are.

is as follows: a series of four arithmetic 'lemmas', followed by the proof of the main result on composition of ratios. Lemma 2 is followed by an 'arithmetical clarification' of both lemmas 1 and 2; the proof of the main result is followed by an 'example' (numerical ὑπόδειγμα), in which it is shown that ratio 7:5 is compounded of ratios 7:11 and 11:5, by resorting to the πηλικότητες of the ratios. The scholium concludes with a short prescription on how to remove a ratio from another. Let us read the enunciations of the lemmas and of the main result, and see the numerical examples following them:

- <Lemma 1>: "Let there be numbers A, B, Γ, and let Δ be the <number contained> by A, B, and E by B, Γ, and again Z by A, Γ, and once more let H be the <number contained> by A, E, and Θ by B, Z and again K by Γ, Δ. I say that numbers H, Θ, K are equal to one another". In short, the product of three numbers is the same, whichever is the order of the factors (recall that multiplication is a binary operator, and that its commutativity is proved in *Elem.* VII.16; the proof of the lemma is a straightforward adaptation of that of VII.16).
- Lemma 2: "Let a number A be multiple of B according to Γ. I say that B is also multiple of A according to the part homonymous to Γ". Arithmetical clarification of lemma 1: 4×5×6 is equal to 5×6×4, that is, to 120. Clarification of lemma 2: 100 is multiple of 20 according to 5, while 20 is multiple of 100 according to $^1/_5$.
- Lemma 3: "Let A be superparticular of B according to Γ. I say that, alternately, B is also superparticular of A according to the part homonymous to Γ". Example: the superparticular ratio 4:3 and the superparticular (in fact, subparticular) ratio 3:4.
- Lemma 4: "The same also happens with epimeric <ratios> [...] one must also conceive the same for compound ratios, such as multiple-superparticular and multiple-superpartient".[53] Examples: 7:5 and 5:7; 7:3 and 3:7; 13:5 and 5:13.
- Main result: "Let there be a magnitude A having a ratio to B, let the value of this ratio be Γ, and between A, B let a chance magnitude Δ fall. I say that the ratio of A to B is conjoined both of that which A has to Δ and Δ to B". The proof presents the mistake of compounding πηλικότητες, that by definition can only be multiplied. (For this and the subsequent example, cf. *Document 8* in the Appendix.)
- Example of removal of ratios by means the procedure 'with respect to the antecedent' (see sections III.1.3 and III.1.4; the connection is not made explicit by Leo).

Only the proofs of lemma 1 and of the main result are fully-fledged demonstrations (the main result ends with a sentence that can be read as a hint as to how to iterate the proof in order to include nested compounded ratios). The proofs of lemmas 2–4 and of the final prescription are nothing but explanations carried out on paradigmatic examples; in particular, the procedure of removal of ratios remains quite obscure. Only lemmas 2–4 are identified as such in the text, which begins directly with the

53 A superparticular ratio has the form $(n + 1):n$; a superpartient ratio has the form $(n + k):n$, with $n > k$.

enunciation of lemma 1. Lemma 1 is expressly referred to (expression 'by the first lemma' (διὰ τὸ α' λῆμμα)) within the arithmetical clarification of lemma 2 and within the main proof; lemma 2 is expressly referred to again within its own arithmetical clarification.

As is clear from the enunciations listed above, the scholium displays a deductive problem: lemmas 2–4 have nothing to do with the rest of the text, and in fact they are not used in the proof of the main result. If we were presented with such a state of affairs in a more structured exposition, there is no doubt that we would regard lemmas 2–4 as interpolated.

If we skip lemmas 2–4, the combination 'lemma 1 + main result' adopts a deductive progression that is found, exactly in this form, nowhere in ancient sources, even if it can be seen as a rewriting (longer and more cumbersome) of the proof presented by Eutocius in his two expositions (see section III.1.2). The terminology of ratios adopted in lemmas 2–4 is at best misleading, owing to the fact that the standard prefix ὑπο– for ratios lesser than unit is systematically omitted: see the example above, where it is said that '20 is multiple of 100' instead of '20 is submultiple of 100'.

There is no doubt, then, that the scholium is badly organized, and sometimes inappropriately worded. My impression is that it is the result of the stratification of at least two layers: a core text comprising lemma 1 and the proof of the main result—maybe extracted from an ancient exposition to which we have no longer access—and the rest: lemmas 2–4, all numerical examples and the final prescription. Add to this that lemmas 1 and 2 are in their turn annotated (recall that the text is written on the full page in the manuscript) by three references to appropriate propositions of *Elem.* VII, which amounts to a logical *hysteron proteron* once the scholium is appended to *Elem.* VI; a short comment scholium to the effect of clarifying the notion of 'homonymous part' is instead contained in an indentation of the text at the beginning of lemma 2. The two- or multi-layered structure of the resulting text could explain its seemingly bizarre designation as ὑπόμνημα σχόλιον.[54]

The hypothesis about the composition of Leo's scholium is confirmed by what precedes it at ff. 118v–119v of Bodl. Dorv. 301, and edited by Heiberg at *EOO* V, 711.6–714.16. Here is a synopsis.

- At 711.6–12. Abbreviated quotation of the definition of compounded ratio; example: the twelvefold ratio is compound both of the triple ratio and the quadruple, or of the double and the sextuple.
- At 711.12–18. Quotation of the enunciation of *Elem.* VI.19, with numerical example (if sides are in a triple ratio, triangles are in a ninefold ratio).
- At 711.19–24. List of the standard names of numerical ratios: multiple, superparticular, superpartient, etc.

54 The second word is given in an abbreviated form. Heiberg, tentatively—and unnecessarily in my opinion—proposed the reading σχολικόν.

- At 712.1–713.7. (Wrong) examples of πηλικότητες: 3:1 is the πηλικότης of the triple ratio, 4:1 that of the quadruple ratio, 3:2 that of the sesquialter ratio (712.1–3; these are ratios, not values, indeed). Examples of composition, with associated diagrams: the text appears to be a description of the diagrams (712.4–9). Example of removal, with two associated diagrams: the remainder of the removal of a triple ratio from a double is sesquialter (712.9–713.7).
- At 713.8–11. Statement that "whenever the three terms of the proportion are not in identity of ratios, then we do not say that the first has to the third a duplicate ratio than to the second".
- At 713.12–714.16. Texts that coincide almost verbatim with Leo's lemma 1 and main proof. The main proof of this scholium contains the reference to the previous result in the form 'by the lemma before itself' (διὰ τὸ πρὸ ἑαυτοῦ λῆμμα). A quick glance at the texts is enough to show that Leo's scholium is beyond doubt a reworking of the anonymous scholium. In particular, the references to appropriate propositions or principles of *Elem.* VII one read in the margin of Leo's lemma 1 are a subset of those annotated in the margin of the first proof in the anonymous scholium.

The typical diagrams associated with compounded ratios consist of three numbers arranged in a row, connected by arcs to which the denominations of the ratios are attached; we shall see an example of them in section III.2.2.

III.1.6 Some *Elements* Scholia on Compounded Ratios

Scanty information comes from the scholia n. 2–11 on Book VI of the *Elements* (*EOO* V, 320.6–331.4), all referring to df.5. Scholium 4 (*ibid.*, 324.10–326.7) is identical to the first exposition by Theon, summarized in section III.1.1. The other scholia mainly insist on providing long series of numerical examples, sometimes within the sexagesimal system in order to ease calculations (n. 3), thereby showing the dependency on a logistic context. One also finds short clarifications of the concept of πηλικότης (n. 2—with mention of Diophantus and a lexicon coming directly from the Platonic *Apologia Socratis* (verb ταλαιπωρεῖν)—3, 6 and 10), including a short discussion of the extension of this concept to non-exprimable ratios (again n. 2). The most articulated exposition is ascribed to Maximus Planudes (n. 6, cf. *Document 10* in the Appendix). It contains: examples; (implicit) distinction between the number paronymous to a ratio and the ratio itself; assertion that "once the <numbers> paronymous to the compounded ratios (συντιθεμένοις λόγοις) are taken and multiplied by one another, it results a number paronymous to the compounded ratio"; examples. Finally, scholium n. 9 (*ibid.*, 330.6–8) is worth a transcription: σημείωσαι τὸ λόγος ἐκ λόγων· ἐν τῷ πέμπτῳ τοῦ ὀγδόου ἡ σύνθεσις εὕρηται καὶ ἡ διαίρεσις ἐν τῇ ἀρχῇ τοῦ θ′ ("*nota bene* 'ratio from ratios': the composition can be found in the fifth of the eight, the division at the beginning of the ninth"). These indications are exact, even if the subject of interpolation of middle proportionals is one of the *leitmotive* of the entire books VIII–IX of the *Elements*.

III.1.7 George Pachymeres' *Quadrivium*

The first Byzantine treatise containing some elements of logistic is George Pachymeres' *Quadrivium*. These elements can be found in the first six chapters of Book IV, devoted to astronomy. Compounded ratios and removal of a ratio from another are expounded in chapters IV.3 and 4, respectively (cf. Tannery 1940, 352.14–357.30).

Pachymeres' rule for compounding ratios is simple: take the πηλικότητες of the ratios to be compounded and multiply them by one another; the result is a number, that can always be taken to be the πηλικότης of a ratio in lowest terms; this is the ratio compounded of the given ratios. As a consequence, to Pachymeres 'composition' is an operation. A series of numerical examples follows.

Things become more complicated with removal. Pachymeres first gives a series of instructions for removing a given ratio from a given ratio, the lesser from the greater, by means of a diagrammatic set-up. The removal may take place either with respect to the antecedent or with respect to the consequent of the ratio from which the removal is going to occur. Pachymeres introduces three unnecessary complications: First, the choice of removing with respect to the antecedent or to the consequent is made to depend from the 'position', in the compounded ratio from which the removal is going to occur, of the ratio to be removed (that is, a double ratio can be compounded either from a sesquitertian and a sequialter or from a sesquialter and a sesquitertian, in this order; numerical examples: sequialter 3:2 to be removed from double 8:4 or 12:6, which give the series of terms 8 6 4 and 12 8 6, respectively). Second, Pachymeres gives his instructions twice, the first exposition requiring in fact that the removal occurs from a ratio already provided in compounded form, the second expounding "by means of a certain rule which ratio is left out after the removal of the lesser ratio from the greater", so that the remainder is not known *a priori* but must be calculated (numerical examples in the second exposition: again sequialter 3:2 to be removed from double 8:4; sesquiquintum 6:5 to be removed from double 20:10; Pachymeres also provides a rule for calculating the πηλικότης of the remainder). Third, he requires, as many authors did before him and will do after him, that the ratio to be removed is less than the ratio from which the removal is going to occur.

When formulating his general rules, Pachymeres always uses the denominations of the ratios (that is, he refers to whole equivalence classes of ratios); he sets out specific terms of them only in the examples. For this reason, his exposition could not be devised to fit the practice of the *Almagest*, in which specific ratios are always set out—and in fact, Pachymeres offers no applications to this effect of the rules he states: his treatise is a compilation intended to remain on a theoretical level, not a computational primer to the *Almagest*. For Pachymeres' rule and instructions, cf. *Document 9* in the Appendix.

The distinguished Byzantine scholar Theodoros Metochites wrote a computational primer on the *Almagest* in the lines of the *Prolegomena*. He included three chapters on compounded ratios and removal of a ratio from another (chapters 8–10 of Book II of his *Abridged Astronomical Elements*; we read them in Vat. gr. 181, ff. 26r–35r): this is nothing but a plagiarism, with enormous expansions (Metochites' verbosity is legendary), of what we read in Pachymeres' *Quadrivium*.

III.1.8 Barlaam of Seminara's *Treatise of Logistic*

The distinguished Middle-Palaiologan theologian and polemicist Barlaam of Seminara was actively engaged in the study of astronomy and harmonics, and redacted a high-level compendium of number theory, the *Treatise of Logistic*.

Here is a list of the definitions and of the enunciations of Book V (see Carelos 1996, 69–93).

1. A value of a ratio is a number that, multiplied by the consequent term of the ratio, makes the antecedent.
2. A ratio is said to be compounded of ratios when the values of the ratios, multiplied by one another, make the value of the ratio.
3. When I remove a ratio from a ratio, I call 'antecedent [ἡγούμενον] ratio' that from which I make the removal, 'consequent' [ἑπόμενον] that <ratio> which I remove. The <ratio> that with the consequent <ratio> makes the antecedent <ratio>, this is said 'remainder' or 'left out' ratio.
4. When I prescribe to remove a ratio from a ratio, then I prescribe to find the remainder.
5. When someone, being prescribed to remove a ratio from a ratio, finds the <ratio> left out, the prescription turns out to be accomplished.

1. To find the value of a given ratio.
2. To find the mean proportional of two given numbers.
3. To find the third proportional of two given numbers.
4. To find the fourth proportional of three given numbers.
5. If there be three numbers in continued <proportion>, the ratio of the first to the third is compounded both of the ratio which the first has to the second and of the ratio which the second has to the third.
6. If there be several numbers in continued <proportion>, the ratio of the first to the last is compounded of the ratios of the numbers lying in continued <proportion>, from the first as far as the last.
7. If a ratio be compounded of several ratios, the inverse <ratio> is also compounded of the inverse ratios.
8. If a ratio be compounded of two ratios and their values be taken, the same ratio is also compounded of the two ratios which its [*scil.* of the compounded ratio] value has to each of the remaining values.
9. The values of identical ratios are equal, and the ratios whose values are equal are identical.
10. If a ratio be compounded of two ratios, this same <ratio> is also compounded of two ratios, both the one that the antecedent of either ratio has to the consequent of the other, and the one that the antecedent of the latter has to the consequent of the former.

11. If a ratio be compounded of two ratios, any of these is compounded of two ratios, both the one that the antecedent of the compounded <ratio> has to the antecedent of the other ratio, and the one that the consequent of the latter has to the consequent of the compounded <ratio>.
12. If a ratio be compounded of two ratios and again of two other <ratios>, and either of the former two be identical to either of the latter two, the remainder will also be identical to the remainder.
13. If a ratio be compounded of several ratios, each of the compounding ratios is compounded both of the ratio that turned out to be compounded and of the inverse of all the remaining ones.
14. If the values of two ratios are taken, any of them is compounded both of the other and of the ratio which its value has to the value of the former.
15. As many ratios as one pleases and a number being given, to find which is the ratio compounded of them whose antecedent will be the given number.
16. As many ratios as one pleases and a number being given, to find which is the ratio compounded of them whose consequent will be the given number.
17. A given <ratio> being removed from a given ratio, to find which is the ratio left out.
18. To remove from a given ratio the inverse ratio in the same terms, so that the <ratio> left out is left out with respect to a given term.
19. To remove from a given ratio a given <ratio> having the same antecedent, so that the <ratio> left out is left out with respect to a given term.
20. To remove from a given ratio a given <ratio> having the same consequent, so that the <ratio> left out is left out with respect to a given term.
21. To remove from a given ratio a given <ratio> having the antecedent of the antecedent ratio as consequent, so that the <ratio> left out is left out with respect to a given term.
22. To remove from a given ratio a given <ratio> having the consequent of the antecedent ratio as antecedent, so that the <ratio> left out is left out with respect to a given term.
23. To remove from a given ratio a given <ratio> in specific terms, so that the <ratio> left out is left out with respect to a given term.

A few remarks are in order.

- Barlaam regards the compounded ratios written in standard form just as a species of the genus ‘compounded ratio’. He never speaks of an operation of ‘composition’ of ratios. He only deals with numerical ratios. He never uses the numerical δι’ ἴσου *Elem.* VII.14 in his proofs. He introduces a lexicon which is unique with his exposition (the ‘antecedent’ and ‘consequent’ ratio).
- Proposition 1 is a problem and explains how to find the πηλικότης of a given ratio: just divide the antecedent by the consequent.
- The proof of proposition 5 is identical to Eutocius’ corresponding proposition. Barlaam also provides an alternative proof.
- In proposition 6, and very much like Eutocius in his exposition at *in Sph. cyl. II.4*, Barlaam takes for granted that, within a compounded ratio, one can replace ratios with compounded ratios: if $a{:}b = (a{:}d)\circ(d{:}b)$ and $a{:}d = (a{:}c)\circ(c{:}d)$, then $a{:}b = (a{:}c)\circ(c{:}d)\circ(d{:}b)$.
- Proposition 8 shows that a compounded ratio can be written in a canonical form that involves only the πηλικότητες of the compounded and of the compounding

ratios, and hence puts the operation of 'replacement of identical ratios' on safe grounds.

- Proposition 9 shows that ratios with the same πηλικότης form an equivalence class.
- Proposition 10 shows that the antecedents or the consequents of the compounding ratios can be interchanged.
- Propositions 11 and 13 show that any ratio compounding a compounded ratio can be written as a ratio compounded of the original compounded ratio and of the inverses of the other compounding ratios, possibly with antecedents or consequents of different ratios interchanged.
- Proposition 12 shows that the remainder of a removal of a ratio from a compounded ratio only depends on the πηλικότης of the removed ratio, that is, in modern parlance, on its equivalence class.
- Proposition 14 shows that either of any two ratios can be written as the ratio compounded of the other and of the ratio of their πηλικότητες.
- Propositions 15–23 are problems: 15–16 explains how to find the ratio compounded of a number of given ratios, whose antecedent or consequent is a given number; 17 shows, as an immediate consequence of proposition 14, that the result of a removal of one ratio from another is simply the ratio of the πηλικότητες of the two ratios. Propositions 18–23 require to find, in a plurality of cases of removal, a remainder having a given extreme; they operate first the removal and then, by taking a suitable fourth proportional, transform the remainder to a ratio having the given extreme.

For Propositions 1, 8, 9, 14, and 17 cf. *Document 12* in the Appendix.

The distinguished Byzantine scholar John Chortasmenos wrote a tract on compounded ratios and removal of a ratio from another (we read it in the autograph Vindob. suppl. gr. 75, ff. 234r-256v, a copy of which is Ambros. C 263 inf., ff. 195r-212r): this is nothing but a plagiarism of a part of Book V of Barlaam's treatise, in particular propositions 18–23.

III.1.9 Demetrios Cydones on Removal of Ratios

The Middle-Palaiologan scholar Demetrios Cydones was actively engaged in the study of astronomy. He personally owned an important manuscript of Ptolemy's *Almagest*, Laur. Plut. 28.1, in which we read many scholia penned by his own hand,[55]

55 For a first assessment, see Acerbi & Pérez Martín 2015. A complete edition of Cydones' scholia—which have never been studied—will follow.

as well as Manuel Bryennios' scholia we shall deal with in section III.2.3. In one of his own scholia, Cydones asserts that Ptolemy did not prove all cases of removal he actually uses in his treatise, and sets out to prove all of them. He does so by considering the compounded ratios arising in a plane case of the Sector Theorem actually proved by Ptolemy (see sections II.4 and III.1.3). He refers to the actual configuration of such a plane case, that provides many instances of identical ratios between line segments, in order to replace identical ratios in compounded ratios written in standard form. Cydones treats in this way most of the combinatorics allowed by the Sector Theorem.

Cydones states some of his results in general terms, and calls them 'porisms' (πορίσματα); cf. *Document 13* in the Appendix. Since his arguments refer to an actual configuration of the plane case of the Sector Theorem, they cannot be considered as valid proofs of the general statements he sets out. For instance, the proof leading to his first result has the following argument scheme.

– In a suitable plane case of the Sector Theorem, Ptolemy proves that $a{:}b = (c{:}d)\circ(e{:}f)$.
– To show that $b{:}a = (d{:}c)\circ(f{:}e)$.
– In fact, the actual configuration of the plane case envisaged by Ptolemy gives that, for a suitable line segment h, $b{:}a{::}h{:}c$ and $h{:}d{::}f{:}e$.
– Therefore, $b{:}a{::}h{:}c = (h{:}d)\circ(d{:}c) = (f{:}e)\circ(d{:}c)$.
– Therefore, $b{:}a = (d{:}c)\circ(f{:}e)$.
– Therefore, if a ratio be compounded of two ratios, the inverse is also compounded of the inverse ratios.

III.1.10 George Gemistus Plethon on Composition and Removal of Ratios and on Taking a Fourth Proportional

The short text of the distinguished Renaissance philosopher George Gemistus Plethon (cf. *Document 14* in the Appendix) was edited by Heiberg as an anonymous scholium to the *Elements*.[56] Only very recently the autograph, allowing the ascription to Plethon, has been discovered and published (Acerbi *et al.* 2016). The text contains concise descriptions of the operations of composition of ratios, of removal of a ratio from another, and of taking of a fourth proportional in the most generic case. The most interesting feature of Plethon's approach is that he introduces composition as an operation that removes the middle term between two extremes, all of them as a whole being conceived as a triad of terms: $(a,m,b) \equiv (a{:}m)\circ(m{:}b) \rightarrow a{:}b$. Removal is described along the same lines, the aim to keep the parallel between the two descriptions being evident: $(a,b,c) \equiv a{:}b \wedge c{:}b \rightarrow (a{:}c)\circ(c{:}b) \rightarrow a{:}c$. Finally, the operation of

56 *EOO* V, 720.1–721.2, as *sch.* 2 of the *Appendix scholiorum* III.

taking of a fourth proportional is described as follows (note that Plethon's formulation in natural language is far clearer):

$(a,b,c) \rightarrow [ac/b \wedge (a{:}b{::}m{:}c \vee a{:}m{::}b{:}c \vee c{:}b{::}m{:}a \vee c{:}m{::}b{:}a)] \vee$
$\vee\ [ab/c \wedge (c{:}a{::}b{:}m \vee m{:}a{::}b{:}c \vee c{:}b{::}a{:}m \vee m{:}b{::}a{:}c)] \vee$
$\vee\ [bc/a \wedge (a{:}b{::}c{:}m \vee m{:}b{::}c{:}a \vee a{:}c{::}b{:}m \vee m{:}c{::}b{:}a)].$

III.2 Worked-out Examples of Removal of Ratios: Tabular Set-ups

III.2.1 Pappus' and Theon's *Commentaries* on the *Almagest*

Pappus, in the portion of his commentary on the *Almagest* that has been transmitted (namely, Books V–VI), applies the Sector Theorem and performs a removal of ratio from a ratio a number of times. This happens in *iA*, 81.15–83.3 and 83.4–19 (*Alm.* V.13), 101.16–102.15 (*Alm.* V.14), 185.7–16 (*Alm.* VI.5), 228.5–21 (*Alm.* VI.6), 243.14–244.5 (*Alm.* VI.7) (I will refer to these passages as numbers from 1 to 6). Surprisingly enough, an application of the Sector Theorem does not correspond in the *Almagest* to any of these passages. Pappus applies three different procedures: it is always required to remove $c{:}d$ from $a{:}b$, where $a{:}b$ is a compounded ratio not written in standard form, and to find one of the terms of the remainder the other being given (cf. section III.1.4). Of course, a, b, c, d are dummy letters standing for the specific numbers set out by Pappus.

a) Passages 3 and 4. Case $d \neq a$ and $c \neq b$. In passage 3, Pappus computes $m = (bc)/d$—that is, he takes a fourth proportional m according to the proportion $m{:}b{::}c{:}d$—stating that "we reduce [ὑποβάλωμεν] ratio $c{:}d$ to term b by multiplying the second by the third and taking one d^{th} of the result" (*iA*, 102.12–14). He then directly gives $a{:}m$ as the remainder. The intended procedure (see Theon's item i just below) inserts m as a middle term between a and b: $a{:}b = (a{:}m)\circ(m{:}b)$, removes $m{:}b{::}c{:}d$ and gets the remainder $a{:}m$. No explanatory calculation at all is provided in passage 4: Pappus just gives $a{:}m$ as the remainder.
b) Passages 1, 2, and 6. Case $d \neq a$ and $c \neq b$, with the exception of 2, in which $c = b$ and $d \neq a$. Pappus computes $m = (ad)/b$—that is, he takes a fourth proportional m according to the proportion $a{:}b{::}m{:}d$. He then sets (the verb is a passive participial form of τάσσειν) c as a middle term between m and d: $a{:}b{::}m{:}d = (m{:}c)\circ(c{:}d)$, so that $m{:}c$ obviously is the result of the removal of $c{:}d$ from $a{:}b$. In passage 1, the ratio $m{:}c$ is immediately stated to be identical to a suitable one having the given consequent of 120 degrees; in passage 2, $c = 120$ degrees and no such additional step is required. No explanatory calculation at all is provided in passage 6: Pappus just gives the ratio transformed of $m{:}c$ to the given consequent of 120 degrees as the remainder.
c) Passage 5. Case $d = b$ and $c \neq a$. Without stating it as a general rule, Pappus directly gives $a{:}c$ (namely, the ratio of the homologous terms which are not identical) as the remainder.

In the portion of his commentary on the *Almagest* that has been edited by Rome, namely, Books I–IV, Theon repeatedly has the occasion of working out the details of the removals of ratio from a ratio that Ptolemy only alludes to in his treatise. This happens in *iA*, 575.8–578.5 (*Alm.* I.14), 578.17–579.10 (*Alm.* I.14), 591.5–594.7 (*Alm.* I.16), 595.18–596.4 (*Alm.* I.16), 619.14–620.7 (*Alm.* II.2), 622.5–623.7 (*Alm.* II.3), 624.3–13 (*Alm.* II.3) (I will refer to these passages as numbers from 1 to 7). Theon applies three different procedures; it is always required to remove $c{:}d$ from $a{:}b$, where $a{:}b$ is a compounded ratio not written in standard form, and to find one of the terms of the remainder the other being given (cf. again section III.1.4).

i) Passages 1 and 2. Case $d = a$ and $c \neq b$. Theon takes a fourth proportional m according to the proportion $d{:}c{::}b{:}m$. He then takes (in all passages, the verb is a passive participial form of λαμβάνειν) m as a middle term between a and b: $a{:}b = (a{:}m)\circ(m{:}b)$, removes $m{:}b{::}c{:}d$ and gets the remainder $a{:}m$. In passage 1, he also provides a recipe to identify which is the remainder and which is the ratio to be removed between the two compounding ratios that result from inserting m as a middle term: one must always remove the second compounding ratio, that is, the one containing the consequent of the ratio from which the removal is going to occur (*iA*, 577.1–4).

ii) Passages 3 to 6. Case $d \neq a$ and $c \neq b$, with the exception of 5, in which $c = b$ and $d \neq a$. Theon computes a fourth proportional m according to the proportion $a{:}b{::}m{:}d$, but takes c as a middle term between x and d: $a{:}b{::}m{:}d = (m{:}c)\circ(c{:}d)$, so that $m{:}c$ obviously is the result of the removal of $c{:}d$ from $a{:}b$. Passage 3 is enriched by a remark to the effect of introducing an alternative procedure, identical to that which we have called 'with respect to the antecedent' (cf. again section III.1.4). Theon also points out that, in this case, the middle term does not yield the result with respect to the antecedent of the ratio from which the removal is going to occur, but with respect to the consequent. Therefore one must always remove the first compounding ratio, and keep the second as the remainder. At the end of this very passage, he comments on the variety of cases involved in the operation of removal of ratios (*iA*, 592.19–594.6).

iii) Passage 7. Again case $d \neq a$ and $c \neq b$. Theon takes a fourth proportional m according to the proportion $c{:}d{::}a{:}m$. He then takes m as a middle term between a and b: $a{:}b = (a{:}m)\circ(m{:}b)$, removes $a{:}m{::}c{:}d$ and gets the remainder $m{:}b$. Without stating it as a general rule, he removes the first compounding ratio, that is, the one containing the antecedent of the ratio from which the removal is going to occur.

If the consequent or the antecedent of the remainder do not coincide with the given consequent or the given antecedent (passages 3, 4, 6, 7: the given value is always 120 degrees), Theon shows how to transform the remainder to a ratio having the given term; this is done by means of a further taking of a fourth proportional. We have seen (and we shall see again) that a number of commentators will consider this final 'fitting' as an integral part of the operation of removal.

III.2.2 Two Ancient Scholia to Ptolemy's *Almagest*

The context is the same as that outlined in section III.1.3. The scholia described here (cf. *Document 5* in the Appendix) almost immediately follow, in the margins of our manuscripts of the *Almagest*, the scholium described in section III.1.3 (cf. *Document 4* in the Appendix).

The first scholium presented in this section works out the details of the procedure of removal of ratios.[57] It does so by outlining first a general prescription. A tabular set-up allowing to perform the removal in an orderly way is then described. The tabular set-up so described coincides with a second scholium, to be seen below. In the general prescription, the case in which the antecedent (consequent) of the compounded ratio coincides with the antecedent (consequent) of the removed ratio is treated first: the terms different from those which coincide will form the ratio resulting from the removal. This case is exemplified by setting out two triples of numbers, 12 6 4 and 12 4 3, joined by suitable arcs, so as to make clear that the compounded ratio coincides with 12:4 or 12:3, respectively, the removed ratio with 12:6 or 4:3, respectively (that is: no arc joins 6 and 4 in the sequence 12 6 4, no arc joins 12 and 4 in the sequence 12 4 3). The triple 12 4 3 coincides with the one set out to the same effect in the scholium described in section III.1.3, but there the removed ratio is 12:4.

The general case is dealt with by taking a suitable fourth proportional; the numerical examples, set out in a standard *chi*-shaped scheme of calculation, are heavily corrupt and cannot be reconstructed without introducing arbitrary corrections. The general prescription runs as follows, taking again $a:b = (c:d)\circ(e:f)$ as a reference: (i) if you want to remove the first compounding ratio, make $c:d::a:m$ and the remainder is $m:b$; (ii) if you want to remove the second compounding ratio, make $d:c::b:m$ and the remainder is $a:m$.

After this, the scholium describes the tabular set-up associated to the removal alluded to in the *Almagest* passage to which the scholium itself refers. This amounts to a modification of the standard *chi*-shaped scheme of calculation of a fourth proportional; the gist of the description resides in the general indications about the places to be assigned, in the tabular set-up, to the relevant terms of the given ratios, namely, the compounded ratio and the removed ratio. After performing the calculation and identifying its result with the relevant chord, the scholiast points out that the case at issue is particular, since the antecedent of the remainder (that is, of the compounded ratio: the removal is made according to procedure (ii) above) is equal to the given

57 Its *relatum* in Ptolemy's treatise is *Alm.* I.14, *POO* I.1, 78.3–7: "therefore, if from the ratio of 120 to 48;31,55 we remove that of 60 to 120, the ratio is left out of the <chord> under the double of <arc> ZΘ to that under the double of ΘH, the one of 120 to 24;15,57" (ἐὰν ἄρα ἀπὸ τοῦ τῶν ρκ πρὸς τὰ μη λα′ νε″ λόγου ἀφέλωμεν τὸν τῶν ξ πρὸς τὰ ρκ, καταλείπεται ὁ λόγος τῆς ὑπὸ τὴν διπλῆν τῆς ZΘ πρὸς τὴν ὑπὸ τὴν διπλῆν τῆς ΘH ὁ τῶν ρκ πρὸς τὰ κδ ιε′ νζ″).

antecedent of the sought ratio. If they were different, one simply would have to take again a fourth proportional between the terms of the remainder and the given antecedent of the sought ratio (note that the scholiast does not state this in general terms). On the other hand, the scholiast rightly omits the case in which the removed ratio comes first in order: as he himself claims, one has to "operate in a similar way", the only difference being that one has to put the first number in order of the removed ratio under the first number in order of the compounded ratio. All in all, this is the simplest and most concise exposition of the operation of removal of a ratio.

The tabular set-up described by the main scholium, and attested as a further scholium, can be summarized as follows: one has to remove the compounding ratio 60:120 from the compounded ratio 120:48;31,55. One sets first in a row the terms of the compounded ratio, then (in case the assigned compounding ratio comes second in order) the consequent 120 of the ratio to be removed under the consequent of the compounded ratio, and its antecedent on the side of the former consequent, 'in the intermediate space'. A standard *chi*-shaped scheme of calculation of a fourth proportional gives the result 24;15,57, to be written between the terms of the compounded ratio, with '4th' marked above. Now there are three terms in a row: 120 and 24;15,57 and 48;31,55; in order to find the remainder of the removal of ratio 24;15,57:48;31,55 (which is identical to ratio 60:120) form ratio 120:48;31,55, it is enough to look at the first two terms in the row. Here is the tabular scholium, where arcs also connect the terms of the compounded ratio and the terms of the remainder.

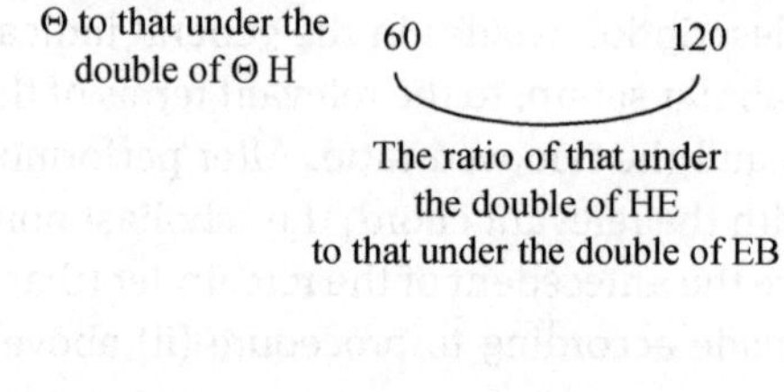

III.2.3 Manuel Bryennios' Scholia to Ptolemy's *Almagest*

Manuel Bryennios' autograph scholia on Ptolemy's *Almagest* have been edited and studied only very recently (Acerbi & Pérez Martín 2015). Four of the five scholia perform in all details some removals only alluded to by Ptolemy. The fifth scholium (cf. *Document 11* in the Appendix) describes a tabular set-up of the calculations involved in the related removal. The procedures at work in the first four scholia can be summarized as follows.

The compounded ratio $a{:}b = (c{:}d)\circ(e{:}f)$ is given and it is required to find e. Take a fourth proportional m such that $a{:}b{::}m{:}d$. Write a 'fictitious' compounded ratio $(m{:}c)\circ(c{:}d) = m{:}d{::}a{:}b = (c{:}d)\circ(e{:}f)$. This entails that $e{:}f{::}m{:}c$. Now, if $c = f$, one immediately has $m = e$ (first and third scholium). If, however, $c \neq f$, by taking again a fourth proportional e of antecedent and consequent of the remainder $m{:}c$ of the 'fictitious' compounded ratio and of term f—that is, by identifying the two remainders above—namely, $m{:}c{::}e{:}f$, one gets the required term as e (second and fourth scholium). Bryennios' procedure, which is an elaboration (with obvious lexical loans) of Theon's prescriptions seen in section III.2.1, can be viewed as an application of the method of simple false position.

The fifth scholium runs as follows.[58] The Sector Theorem gives rise to the relation $a{:}b = (c{:}d)\circ(e{:}f)$, where a = 70;32,3, b = 97;4,56, c = 117;31,15, d = 24;15,57, f = 120; to find e. Ratio $a{:}b$ is given and we have first to find $e{:}f$ by removal of $c{:}d$. Let us trace a diagram in form of letter *chi* and place (i) at its two upper extremes from left to right, the consequent and the antecedent of the compounded ratio, respectively; (ii) at its lower left extreme, the consequent of the removed ratio. Write at the lower right extreme of the *chi* the result m = 17;37,45 of taking the fourth proportional of the terms set out according to this very diagram, namely, by multiplying the two terms coupled by a stroke of the *chi* and by dividing the result by the uncoupled term: $b{:}a{::}d{:}m$. By the side of this result, between the two lower extremes of the *chi*, place the antecedent of the removed ratio. As explained above, if one writes a 'fictitious' compounded ratio $(c{:}d)\circ(e{:}f) = a{:}b{::}m{:}d = (m{:}c)\circ(c{:}d)$, and if one has that $c = f$, one immediately gets $e = m$ = 17;37,45. But $c \neq f$, and by taking again a fourth proportional by means of a *chi*-diagram, applied to antecedent and consequent of the remainder $m{:}c$ of the 'fictitious' compounded ratio and to the term f = 120 that had not yet been employed—that is, by identifying the two remainders above—namely, by setting $m{:}c{::}e{:}f$, one gets e = 18;0,15.

III.2.4 George of Trebizond's *Isagoge* to the *Almagest*

The renowned Cretan scholar and translator George of Trebizond operated in early-Renaissance Italy. Among his several activities, in 1451 he both translated Ptolemy's *Almagest* from Greek into Latin and redacted a monumental commentary thereon, still lying unedited. He had a complex personality which eventually drove him to quarrel with almost all scholars of his time. His controversy with cardinal

58 Its *relatum* in Ptolemy's treatise is *Alm.* II.7, *POO* I.1, 121.21–122.21.

Bessarion, whom he also thought to be the Antichrist, had as a consequence that Regiomontanus, an associate of Bessarion, wrote an even more monumental counter-commentary on the *Almagest*, explaining in detail where his rival was wrong, and where Theon of Alexandria, in his turn attacked by George of Trebizond, was right.[59] Since George of Trebizond's skills as a translator of scientific texts were also questioned by Jacobus of Cremona, a translator of Archimedes from Greek into Latin, the fate of George of Trebizond's scientific translations was decided already at the end of the fifteenth century.

The first section of George of Trebizond's commentary is the so-called *Isagoge*, a computational primer on the *Almagest*, very much in the style of the *Prolegomena* and of its Byzantine avatars. I have studied this text in detail in view of a forthcoming edition. Here is a summary of the sections on compounded ratios and removal.

The operation of composition of ratios is first introduced: "a composition of ratios is that obtained from two ratios, and a compound ratio is that composed of two ratios, either rational or irrational" (σύνθεσις λόγων ἐστὶν ἡ ἐκ δύο λόγων συναγομένη, καὶ σύνθετος λόγος ὁ ἐκ δύο συνιστάμενος λόγων ῥητῶν ἢ ἀρρήτων).[60] After a long series of examples, partly intended to clarify the difficulties involved in compounding irrational ratios, George of Trebizond shows, again by means of examples, that "the denomination of compound ratios can be found [...] once the numbers denominating the compounding ratios are multiplied" (ἡ δὲ παρονομασία τῶν συνθέτων λόγων εὕρηται [...] πολλαπλασιαζομένων τῶν παρονομαζόντων τοὺς συντιθέντας λόγους ἀριθμῶν). After a further series of examples, a final remark points out that any numerical ratio can be thought of as compounded of the ratios of the square roots of its terms.

As for removal, it is defined as follows: "a removal of a ratio from a ratio is that by means of which, one of the compounding <ratios> being thrown away, the other is left over" (ἀφαίρεσις λόγου ἀπὸ λόγου ἐστὶ δι' ἧς τοῦ ἑτέρου τῶν συντιθέντων ἀπορριφθέντος ὁ ἕτερος ἀπολείπεται). The computational tool of 'chiasmus' (χιασμός), or 'crosswise arrangement' (cf. sections III.2.2 and III.2.3) is introduced, in order to perform the removal (cf. *Document 15* in the Appendix). If you are given a compounded ratio written in standard form, for instance 2:8 = (2:4)∘(4:8), whose terms are usually written as the sequence 2 4 8, put the terms of the compounded ratio 2 and 8 at the upper extremes of a *chi*-shaped scheme, the terms of one of the compounding ratios, say 2 and 4, at the lower extremes of the same *chi*-shaped scheme, perform crosswise multiplication and write the results below the corresponding term of the compounded ratio, thus:

59 For George of Trebizond's translation of the *Almagest*, see http://www.wilbourhall.org/index.html#ptolemy; Regiomontanus' *Defensio Theonis* is transcribed and studied in http://regio.dartmouth.edu/about/about-project.html. On George of Trebizond, see Monfasani 1976 and 1984.

60 An 'irrational' ratio is such that its terms contain fractional parts; a 'rational' ratio has only integer numbers as its terms. This was standard terminology in Byzantine authors.

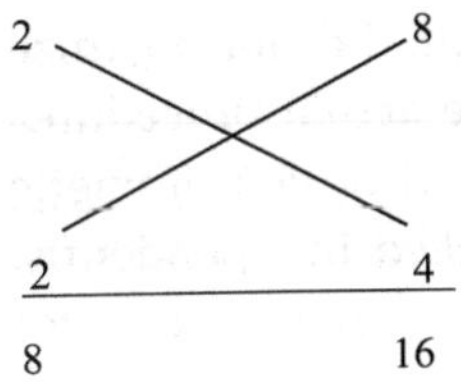

The ratio of the results is the remainder, as is easy to check: if $a{:}b = (a{:}c)\circ(c{:}b)$, the operation performed on the *chi*-shaped scheme amounts to taking the ratio of ac to ab, which is obviously identical to $c{:}b$. If we put the terms of the other compounding ratio, namely, c and b, below the terms a and b of the compounded ratio, we get ab to cb, which give rise to the other compounding ratio $a{:}c$.

If the given compounded ratio is not written in standard form, the same *chi*-shaped scheme provides again the two terms of the reminder, according to the following set-up:

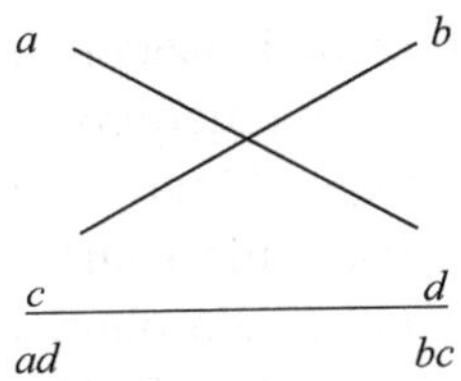

Since $ad{:}bc$ is the remainder of the removal of ratio $c{:}d$ from ratio $a{:}b$, George of Trebizond's prescription is correct and a most general one. He backs up his prescription by a further rule, allegedly conforming to Ptolemy's practice: put the four terms of the compounded and of the removed ratio in a row, in such a way that those of the compounded ratio are the first and the third term, thus: $a\ c\ b\ d$. Then the remainder can be found in four different ways. Form the product of the middle terms c, b or that of the extremes a, d; divide by either of the extremes or of the middle terms, respectively; form the ratio which has the result of the division as the consequent (resp. antecedent) and the term not yet employed among $a\ c\ b\ d$ as the antecedent (resp. consequent): this ratio is the remainder. This amounts to write the remainder $ad{:}bc$ in one of the following four ways: $d{:}(a/bc)$, $a{:}(d/bc)$, $(ad/b){:}c$, $(ad/c){:}b$.

IV Conclusion

The evidence collected above shows that compounded ratios and removal, as a theoretical theme and not as a tool, were exegetical *topoi*, related to two very specific *loci* of the Greek mathematical corpus: the Sector Theorem in the *Almagest* (more specifically related to removal) and VI.df.5 of the *Elements* (more specifically related to compounded ratios). Still, the expositions of the topic did not organize in exegetic chains. On the contrary, the slight variations in the actual procedures adopted to implement an algorithm that involves just a couple of basic arithmetic operations show that, with some exceptions, our sources worked independently. One of the reasons must have been that, on the logistic side, no general recipe for removing ratios was provided in Theon's commentary on the *Almagest*, and that, on the number-theoretic side, no general exposition was provided before the Late Antiquity—at least if we are to believe what Eutocius says, and if we concede that he had access to all relevant sources.

As for the exceptions, we may safely assert: that the exposition of removal in the *Almagest* scholium that eventually came to be ascribed to Domninus (section III.1.3) and the one in the *Prolegomena* (section III.1.4) have the same origin, whereas a different, and simpler approach is suggested in another *Almagest* scholium (section III.2.2); that Leo (section III.1.5) simply appropriated an anonymous scholium that provided a proof, different from the one we read in Eutocius (section III.1.2), of the basic result about compounded ratios written in standard form; that Bryennios (section III.2.3) obviously followed Theon (sections III.1.1 and III.2.1). Two authors stand in isolation: Barlaam (section III.1.8), whose treatise on the number-theoretical foundations of logistic I regard as the solitary achievement of brilliant mind; George of Trebizond (section III.2.4), whose approach may however depend on medieval Latin sources to an extent that future researches will determine. Other authors either contributed trivial material to the issue (this is the case of Maximus Planudes: section III.1.6), stated general rules supported however by non-general proofs (Demetrios Cydones: section III.1.9) or by no proofs at all (George Pachymeres: section III.1.7), or just jotted down a few lines trying to catch the gist of the operations of composition and removal (George Gemistus Plethon: section III.1.10).

Compounded ratios were used perfunctorily in the geometric tradition (sections II.1–6), from the *Elements* to Pappus; they allowed Apollonius to treat a number of properties of conic sections in a compact way; they allowed to formulate mathematical problems that would otherwise have been beyond reach of the Greek mathematical lexicon (the loci on more than four lines).

The logistic tradition dealt with the issues of compounded ratios and, more specifically, of removal of ratios in the context of the applications of the Sector Theorem—that is, one must stress, in a *numerical* context. In any such application, a compounded ratio is provided not written in standard form and it is required to find one of the terms, say e in $a{:}b = (c{:}d)\circ(e{:}f)$, the others being given. Is this trivial?—After all, it is enough to calculate $e = adf/bc$: three multiplications and one division. It was not.

There are two facets to the issue. First, the basic operational unit of what we formulate as a sequence of two multiplications and one division was almost invariably conceptualized as a taking of a suitable fourth proportional: three numbers a, b, c being given in this order, the associated fourth proportional $m(abc)$ satisfies the proportion $a{:}b{::}c{:}m$ and, *as a consequence*, can be calculated as $m = bc/a$. A first complication came from the fact that, on the one hand, in order for m to be a fourth proportional, its position in the above proportion is immaterial: it is not a fourth proportional because it is placed in the *fourth* position—but, on the other hand, changing the position of the fourth proportional in the above proportion amounts to calculating a different number. Was this problem settled once and for all by adopting a convention?—It was not.

Second facet of the issue, the calculations leading to the value of e in $a{:}b = (c{:}d)\circ(e{:}f)$ were almost invariably split in two operations: first, to remove ratio $c{:}d$ from the compounded ratio $a{:}b$ in order to single out a ratio $x{:}y$ corresponding to $e{:}f$; second, since both f and the ratio corresponding to $e{:}f$ are given, to determine e, which is given by *Data* 2. The point is that the ratio corresponding to $e{:}f$ is given as the remainder of a removal, so that in general its consequent y is not equal to f: again, one has to take a fourth proportional $m(yxf)$ in order to finally find e. Since removal of ratios almost uniquely occurred in the applications of the Sector Theorem, were these two operations unified under one single standard prescription, possibly admitting branchings in its formulation?—They were not.

Still, the general procedure for solving the above compounded ratio for e would simply entail no more than two nested takings of a fourth proportional: $e = m(cm(bad)f)$—'Simply'?

Appendix

Document 1: Extracts from Euclid, *Sectio Canonis*[61]

1. If a multiple interval compounded twice [δὶς συντεθέν] make some interval, it will also be multiple.

Let there be an interval ΒΓ, and let Β be a multiple of Γ, and let, as Γ is to Β, Β be made to Δ. Then, I assert that Δ is a multiple of Γ.

In fact, since Β is a multiple of Γ, therefore Γ measures Β; and, as Γ is to Β, Β was to Δ, so that Γ also measures Δ; therefore Δ is a multiple of Γ.

3. If an interval compounded twice make the whole multiple, it will also be multiple.

4. If a non-multiple interval be compounded twice, the whole will neither be multiple nor superparticular.

5. If an interval compounded do not make the whole multiple, it will not be multiple either.

6. A double interval turns out to be composed [συνέστηκεν] of the two greatest superparticular intervals, both the sesquialter and the sesquitertian.

---

61 Translated from *EOO* VIII, 160.5–13 and 14–16; 164.1–3 and 15–17; 166.1–3; 168.1–26. Proposition 3 is excluded since it does not deal with compounded ratios.

7. From a double interval and a sesquialter, a triple interval results.

In fact, let A be double of B and B sesquialter of Γ. I say that A is triple of Γ.

In fact, since A is double of B, therefore A is equal to two B. Again, since B is sesquialter of Γ, therefore B contains Γ and half of it; therefore two B are equal to three Γ; and two B are equal to A; therefore A is also equal to three Γ: therefore A is triple of Γ.

8. If from a sesquialter interval a sesquitertian interval be removed, the remainder is left out as sesquioctave.

In fact, let A be sesquialter of B and let Γ be sesquitertian of B. I say that A is sesquioctave of Γ.

In fact, since A is sesquialter of B, therefore A contains B and half of it; therefore eight A are equal to twelve B. Again, since Γ is sesquitertian of B, therefore Γ contains B and a third of it; therefore nine Γ are equal to twelve B; and twelve B are equal to eight A; therefore eight A are equal to nine Γ; therefore A is equal to Γ and an eighth of it; therefore A is sesquioctave of Γ.

Document 2: Pappus, *Collectio* VII.240[62]

Let A to B have the ratio conjoined both of that which Γ has to Δ and of that which E has to Z. That Γ also has to Δ the ratio conjoined both of that which A has to B and Z to E.

In fact, let the <ratio> of Δ to H be made identical to that of E to Z.

Now, since <the ratio> of A to B is conjoined both of that of Γ to Δ and of that of E to Z, that is, of Δ to H, but the <ratio> conjoined both of that which Γ has to Δ and of that which Δ has to H is that of Γ to H, therefore, <as> A is to B, so Γ is to H. And since Γ has to Δ the ratio conjoined both of that which Γ has to H and of that which H has to Δ, but the <ratio> of Γ to H has been proved identical to that of A to B, and by inversion the <ratio> of H to Δ is identical to that of Z to E, therefore Γ also has to Δ the ratio conjoined both of that which A has to B and of that which Z has to E.

Document 3: Theon's Main Arguments[63]

In fact, let AB to ΓΔ and ΓΔ to EZ have a given ratio. I say that the ratio of AB to EZ is compounded both of AB to ΓΔ and of ΓΔ to EZ, that is, that, if the value of the ratio AB to ΓΔ be multiplied by that of ΓΔ to EZ, it makes that of AB to EZ as a value of ratio.

In fact, let first AB be greater than ΔΓ and ΓΔ than EZ, and let AB be double of ΓΔ and ΓΔ triple of EZ. Now, since ΓΔ is triple of EZ and AB double of ΓΔ, therefore AB is sextuple of EZ—since, if we double the triple of something, the sextuple of it also results, for this is a 'composition' in strict sense.

62 Translated from Jones 1986, 301.30–303.12.

63 Translated from *in Alm.* I.13, edited in *iA*, 533.3–535.7; also, *in Alm.* II.11, edited in *iA*, 761.1–762.2.

Or in this way here. Since AB is double of ΓΔ, let AB be divided into AH, HB equal to ΓΔ. And since ΓΔ is triple of EZ, exactly for the same reasons HB is also triple of EZ; therefore AB as a whole is sextuple of EZ; therefore, through the middle term ΓΔ, the ratio of AB to EZ turns out to be deduced to be compounded both of that of AB to ΓΔ and of ΓΔ to EZ.

And similarly, even if ΓΔ be less than any of AB, EZ, the same will be deduced. In fact, let again AB be triple of ΓΔ and ΓΔ half of EZ. And since ΓΔ is half of EZ and AB triple of ΓΔ, therefore AB is sesquialter of EZ—for, if we triple the half of something, it will contain it once and one half time.

Or also in this way. Since AB is triple of ΓΔ and ΓΔ half of EZ, EZ is so many two of which AB is three equal to ΓΔ, so that AB will be sesquialter of EZ; therefore, through the middle term ΓΔ, the ratio of AB to EZ turns out to be deduced to be compounded both of that of AB to ΓΔ and of ΓΔ to EZ.

But now, again, let ΓΔ be greater than any of AB, EZ, and let AB be half of ΓΔ and ΓΔ sesquitertian of EZ. Now, since ΓΔ is so many four of which AB is two and EZ is so many three of which ΓΔ is four, therefore EZ is also so many three of which AB is two; therefore, through the middle term ΓΔ, the ratio of AB to EZ—the one of two to three, subsesquialter—turns out to be deduced again to be compounded both of the subdouble and of the sesquitertian. Exactly in a similar way also for several <terms> and in the other cases. [...]

We shall prove that if[, inversely,] a ratio be compounded of two ratios, one of the compounding <ratios> is compounded both of the remaining compounding <ratio> inversed and of the compounded one.

In fact, let the ratio of A to Γ, once B is taken as a middle, be compounded both of that of A to B and of B to Γ. I say that the ratio of A to B is also compounded of that of A to Γ and of Γ to B.

And it is immediately evident: for, if we place B after Γ, the ratio of A to B, once Γ is taken as a middle, will be compounded both of that of A to Γ and of Γ to B.

Document 4: The *Almagest* Scholium at the Origin of the Treatise Ascribed to Domninus of Larissa[64]

When we undertake to remove a ratio from a ratio, it is clear that this is nothing other than to resolve [διαλῦσαι] the ratio from which the removal is going to occur into the removed <ratio> and the remaining <ratio> after this. And in fact that from which the removal is going to occur is compounded of the removed <ratio> and the remainder after this—for the lesser <ratio> is removed from the greater. Now, how will the resolution come to occur? Or is it clear that <it will> conversely to composition. A ratio is said to be compounded of ratios when the values of the ratios, multiplied by one another, make some. If, in fact, another term is set between two terms,

64 Translated from Acerbi 2017, 228–229.

the ratio of the first to the second, multiplied by that of the second to the third, makes the ratio of the extremes, whatever the middle <term> might be. For instance, let 4 be a middle of 2 and 8. Now, since 2 has to 4 the ratio of a half, and likewise 4 to 8, I multiply a half by a half and I make $^{1}/_{4}$, and I find that 2 is a fourth part of 8. Yet even if I set 40 instead of 4 as a middle term, the <ratio> of 2 to 8 is, again, compounded both of that of 2 to 40 and of 40 to 8, for 2 is a 20th part of 40, 40 is quintuple of 8, and $^{1}/_{20}$ times 5 makes $^{1}/_{4}$.

Now, since a ratio is said to be compounded of ratios exactly when the values of the ratios, multiplied by one another, make some, it is clear that, if a given ratio be removed from a given ratio, the remainder will also be given. For if we have both the value of the ratio from which the removal is going to occur and the <value> of the removed <ratio>, we will also have the remaining value, which, when it is multiplied by the value of the removed <ratio>, makes the one of the compound <ratio>, $^{1}/_{4}$.

Now, if one removes the value of the lesser of the given ratios from the one of the greater, this will come about in two ways, either by the lesser ratio being transposed to the greater and fitted to it, or by the greater being transformed into the lesser and made to contain it; and each of these cases will be accomplished in two ways. For, each of the ratios having an antecedent and a consequent, if the least is going to be fitted into the greater ratio, the removal will occur either with respect to the antecedent (as for 12 4 8) or with respect to the consequent of the greater ratio (as for 12 $4^{1}/_{2}$ 3) from which the removal is going to occur; if the greater <ratio> has been transformed into the lesser, again, the removal will occur either with respect to the antecedent or with respect to the consequent of the lesser.

Now, this is the way it should be done in general. But if we should happen to find in the greater and the lesser ratio the same term in either the antecedent or the consequent, finding <the ratio> will be easier for us, for these <terms> other than the identical terms will contain the remaining ratio. For instance, from the ratio of twelve to 3, let it be required to remove that of 12 to 4; 12 being fitted to 12 and 4 being set as a middle <term>, the ratio of 12 to 4 turns out to be removed, and that of 4 to 3 remains, which is sesquitertian, for sesquitertian multiplied by triple makes quadruple. Here, too, it is similarly possible to carry out the transition from the greater to the lesser or from the lesser to the greater, and the removal can occur with respect to either the antecedent or the consequent.

Document 5: The Tabular Set-up Described in an *Almagest* Scholium[65]

If, removing a given ratio from a given ratio, we want to find the remaining ‹ratio›, if the antecedent of the removed <ratio> is found to be identical to the antecedent of

65 Translated from Acerbi 2017, 233.

the ratio from which one removes, or the consequent identical to the consequent, the terms in the ratios that are different immediately contain the remaining ratio (as for 12 6 4 or 12 4 3), for the remaining term of the removed ratio will come to be a middle <term> of the <terms> containing the ratio from which one removes. If, instead, there is no such relationship among the <terms>, one must make, either (namely, when the removed ratio is first among the compounding <ratios>) as the antecedent of the removed <ratio> to its consequent, so the antecedent of the ratio from which the removal is going to occur to some other <number>, or (namely, when the removed ratio is second among the compounding <ratios>) as the consequent to the antecedent <of the removed ratio>, so the consequent of the ratio from which the removal <is going to occur> to some other <number>—for the fourth proportional, together with the remaining <term> that contains the ratio from which the removal is going to occur, will make the remaining ratio.

16 8^{4th} 4 16 8^{4th} 4
6 10 6 3

In which way must these things be carried out, and in which way must we arrange the numbers, we shall learn as follows. We first arrange those of the compounded ratio, and we take first the first in the ratio, and second the second (namely, first 120 and second 48;31,55). Next, since we have 3 numbers given and one single ratio given among the compounding <ratios>, we arrange next underneath the ratio which is given among the 2 compounding ratios. And, if the given ratio comes last in order (as in the theorem at issue), we arrange the last number in order of the given ratio under the the last number in order of the compounded ratio (namely, 120 under 48;31,55), and we arrange the first <number> of the given ratio (namely, 60) in between, and so to speak in the intermediate space, and multiplying 60 by 48;31,55 and dividing by 120 we find 24;15,57, and we arrange it above 60, and, clearly, we have removed from the ratio of 120 to 48;31,55 the given ratio of 24;15,57 to 48;31,55 (for it is identical to that of 60 to 120) and we have the ratio of 120 to 24;15,57—which we also seek—given as a remainder.

Now, since we have found the sought ratio as that of 120 to 24;15,57, and the sought ratio was that of the <chord> under the double of <arc> ZΘ to that under the double of ΘH, therefore we have the ratio of the <chord> under the double of <arc> ZΘ to that under the double of ΘH, namely, that of 120 to 24;15,57; and that under the double of ZΘ is 120 (which is also that under the double of ZA); therefore that under the double of ΘH is 24;15,57. And in this way we may find how much is it that under the double of ΘH. If instead that under the double of ZΘ had not been 120, but by hypothesis 40, we would have made, as 120 is to 24;15,57, so 40 to some other <number>, and in this way we may find how much is it that under the double of ΘH. We shall operate in a similar way even if the sought ratio happens to be the last in order.

Document 6: Eutocius on *Sph. Cyl.* II.8[66]

Let there be four terms A, Γ, Δ, B. I say that the ratio compounded of that of the <rectangle contained> by A, B to the <square> on Γ with [μετά] the ratio of B to Δ is identical to that of the <rectangle contained> by A, B times [ἐπί] B to the <square> on Γ times Δ.

In fact, let K be equal to the <rectangle contained> by A, B, and Λ to the <square> on Γ, and let, as B is to Δ, so Λ come to be to M; therefore the ratio of K to M is compounded of that of K to Λ, that is, of that of the <rectangle contained> by A, B to the <square> on Γ, and of Λ to M, that is, of B to Δ.

Then, let K multiplying B make N, and Λ multiplying B make Ξ, multiplying Δ make O.

Now, since the <rectangle contained> by A, B is K, and K multiplying B turns out to make N, therefore N is that by A, B times B. Again, since the <square> on Γ is Λ, and Λ multiplying Δ turns out to make O, therefore O is that on Γ times Δ, so that the ratio of that by A, B times B to that on Γ times Δ is identical to that of N to O. Therefore one has to prove that the ratio of K to M is identical to that of N to O.

Now, since each of K, Λ multiplying B turns out to make each of N, Ξ, therefore, as K is to Λ, so N is to Ξ. Again, since Λ multiplying each of B, Δ turns out to make each of Ξ, O, therefore, as B is to Δ, Ξ is to O; but as B is to Δ, Λ is to M; therefore, also, as Λ is to M, Ξ is to O; therefore K, Λ, M are in the same ratio as N, Ξ, O taken two and two; therefore, also, ex aequali [δι' ἴσου], as K is to M, so N is to O.

Document 7: Eutocius' Main Proof[67]

Let there be two given numbers A, B and let a certain middle Γ be taken of them. Then one must prove that the ratio of A to B is conjoined of that which A has to Γ and Γ to B.

In fact, let the value Δ of ratio A, Γ and E of <ratio> Γ, B, be taken[; therefore Γ multiplying Δ makes A and B multiplying E makes Γ]. Then, let Δ multiplying E make Z. I say that Z is the value of the ratio of A to B, that is, that Z multiplying B makes A.

In fact, let B multiplying Z make H.

Now, since B multiplying Z turns out to make H and multiplying E <turns out to make> Γ, therefore, as Z is to E, H is to Γ. Again, since Δ multiplying E turns out to make Z and multiplying Γ turns out to make A, therefore, as E is to Γ, Z is to A; alternately, as

66 Translated from *in Sph. cyl. II.8 aliter*, edited in *AOO* III, 198.20–200.17.

67 Translated from *in Sph. cyl. II.4*, edited in *AOO* III, 122.11–124.7. The clauses omitted in *in Con.*, *AGE* II, 218.27–220.16, are bracketed. For the underlined deduction also see the main text; slight variations in wording are not considered.

E is to Z, Γ is to A; and inversely, as Z is to E, so A is to Γ; but as Z is to E, H was proved to Γ; therefore, also, as H is to Γ, A is to Γ; therefore A is equal to H; [but B multiplying Z turns out to make H;] therefore B multiplying Z also makes A[; therefore Z is the value of the ratio of A to B; and Z is Δ multiplied by E, that is, the value of ratio A, Γ by the value of ratio Γ, B; therefore the ratio of A to B is compounded both of that which A has to Γ and Γ to B].

The proof at *in Con.* argues more effectively, replacing the underlined text with

Now, since Δ multiplying E turns out to make Z and multiplying Γ turns out to make A, therefore, as E is to Γ, Z is to A. Again, since B multiplying E turns out to make Γ and multiplying Z turns out to make H, therefore, as E is to Z, Γ is to H; alternately, as E is to Γ, Z is to H; and, as E is to Γ, Z was to A;

Document 8: Leo's Main Proof and Example of Removal[68]

This being established as a preliminary, let there be a magnitude A having a ratio to B, let the value of this ratio be Γ, and between A, B let a chance magnitude Δ fall. I say that the ratio of A to B is conjoined both of that which A has to Δ and Δ to B.

In fact, it is clear that B, multiplying the value Γ of the ratio, made A. But again, since magnitude B, multiplying the value Z of the ratio of Δ, B, turns out to make Δ, but, also, magnitude Δ, multiplying the value E of the ratio of A, Δ, turns out to make A, therefore by the first lemma, since B, multiplying the <number resulting> from B, Z, exactly turns out to make A, therefore B, multiplying the <number resulting> from E, Z, also turns out to make A. But indeed, both the <number contained> by B, Γ is A, and again that <contained> by B, Z, E is A; therefore that <contained> by B, Γ is equal to that <contained> by B, Z, E; therefore the value Γ of the ratio of magnitudes A, B is equal to that resulting from the values E, Z; therefore the value Γ is compounded of <value> E multiplied by Z. We shall say the same also if a magnitude fall between A, Δ, and again if another fall between B, Δ, for it is the same procedure.

But now, thus, let it be supposed A having to B the ratio that number 17 has to 13, and <let it be supposed> to remove that which has the ratio of 19 to 11 from it. Now, as 19 is to 11, so I do 17 to 187 nineteenths; therefore the ratio of $^{187}/_{19}$ to 13 units stands as a remainder, that is, if we make 13 nineteen times, in smallest numbers, those of 187 to 247.

68 Translated from *EOO* V, 716.24–717.18 and 718.9–22.

Document 9: George Pachymeres' Rules[69]

When we want to find which ratio is compounded of any given different ratios, we first take their values and multiply them by one another, and we write down the number resulting from their multiplication (whether it is integer or it also has a part or parts). Then we search which ratio is denominated by such a number, and, whatever it is, that is the <ratio> compounded from the composition of the given ratios.

One must know that, whenever we set out to remove a ratio from a ratio—in any case the lesser from the greater—we first take the two terms of the greater ratio (the antecedent and the consequent), from which we set out to do the removal, and write them down. Then we take the lowest [πυθμενικούς] terms of the lesser ratio, which we also set out to remove, and similarly also write them down on a straight line with the said terms of the greater ratio. Then we multiply the consequent term of the greater by the antecedent of the lesser, and write down separately the number resulting from the multiplication of such terms. Then we take it and and we divide it by the consequent term of the lesser ratio, and, taking the number resulting from its division, we write it down between the two terms of the greater ratio: for this will also necessarily have, to the consequent term of it, the ratio itself that we also set out to remove from it. Then we remove such a term [*scil.* the said consequent]—which, clearly, also serves as a consequent of the ratio taken by proportion, that we also set out, as we have said above, to remove from it [*scil.* the greater ratio]—so that such a ratio [*scil.* the lesser ratio] would also have been removed at the same time with it: and in fact, once whatever term is removed of those that are considered to one another in some ratio, necessarily the ratio itself, of which this <term> either is the antecedent or the consequent, is also removed at the same time.

Document 10: Maximus Planudes' Scholium[70]

Of the most learned Maximus Planudes, to the definition 'ratio from ratios' of the sixth, namely, that every ratio can be compounded by two, three, or several ratios. For instance, the double <ratio> 12 to 6 is compounded of two ratios: sesquitertian and sesquialter (both 8 to 6 and 12 to 8), but is also compounded of three: sesquitertian 8 to 6, sesquiquartum 10 to 8, and sesquiquintum 12 to 10. Similarly also from several <ratios>. Now, once the <numbers> paronymous to the compounded ratios are taken and multiplied by one another, it results a number paronymous to the compounded ratio. For instance, since, as said, the double is compounded of sesquitertian and sesquialter, and the sesquitertian has once and a third of the whole under it, I take

69 Translated from Tannery 1940, 352.25–31 and 353.31–354.14.
70 Translated from *EOO* V, 327.12–329.17.

one unit instead of once and $\frac{1}{3}$ instead of a third. Again, since the sesquialter has once and a half of the whole under it, I similarly take one unit instead of once and $\frac{1}{2}$ instead of a half. Now, I multiply these numbers—of course, one unit and a third by the other one unit and a half, and it results two units, that are paronymous to the double. The multiplication can be performed as follows: once one, one: here it is, one unit; once a half, a half; and again a-third times the unit, that is, a third of the unit, a third; and a-third times the half, namely, a third of the half, a sixth. A half and a third and a sixth <give> one unit, that, compounded with that before it, gives two.

In this way, also, from a double and a triple it results a sextuple, for I take two units instead of double, three instead of triple, and I multiply these by one another, and it results six.

If instead the double be compounded of three <ratios>, as shown above: sesquitertian, sesquiquartum, and sesquiquintum, I take again one unit and a third instead of sesquitertian, one unit and a fourth instead of sesquiquartum, one unit and a fifth instead of sesquiquintum, and I multiply these by one another, and it results two units. The multiplication can be performed as follows: in the first place, the unit and $\frac{1}{3}$ by the unit and $\frac{1}{4}$: once one, one; once $\frac{1}{4}$, $\frac{1}{4}$; a-third times one, namely, a third of one, a third; a-third times $\frac{1}{4}$, namely, $\frac{1}{3}$ of $\frac{1}{4}$, $\frac{1}{12}$: and here it is, a unit, $\frac{1}{4}$, $\frac{1}{3}$, and $\frac{1}{12}$. Afterward I multiply the unit and $\frac{1}{5}$ by the unit and $\frac{1}{4}$ $\frac{1}{3}$ $\frac{1}{12}$, and I say: once one, one; once a fourth, a fourth; once a third, a third; once a twelfth, a twelfth; again, a-fifth times one, namely, a fifth of the unith, a fifth; a fifth of a fourth, a twentieth; a fifth of a third, $\frac{1}{15}$; a fifth of a twelfth, a sixtieth. All these parts give one unit as a result, that, conjoined with that before it, gives two as a result. You shall know in the following way that these parts give a unit as a result. One must find the number (let it be taken as one unit) having these parts in the first place starting from a unit, and it is sixty. Indeed, a fourth of this fifteen, a third twenty, a twelfth five, a fifth twelve, a twentieth three, a fiftieth four, a sixtieth one—and fifteen, twenty, five, twelve, three, four, and one <give> sixty.

And in this way, too, from a double, a triple, and a quadruple it results a twenty-fourfold, for instance 2 4 12 48. I take two instead of double, three instead of triple, four instead of quadruple, and I multiply two by three, and it results six. Next, four by six, and it results twenty-four, which is paronymous to twenty-fourfold.

Document 11: Bryennios' Fifth Scholium[71]

Since the ratio of KΔ to ΔΓ, as Ptolemy says, is conjoined of the ratio of KΛ to ΛM and of that of ME to EΓ and the ratio of KΛ to ΛM is given whereas that of ME to EΓ is not given, yet it is given by means of the removal of the said given ratio of KΛ to ΛM (as again he himself says), it is necessary to biefly expound how we shall exactly remove it.

71 Translated from Acerbi & Pérez Martín 2015, 136.

So, one must know that, when we are performing the removal of such a ratio so as to find the remaining ratio (that is to say, ME), we first draw through two straight lines equal to one another and meeting one another in the middle, and we make the sign of the letter chi, in order for the procedure of removal of such a ratio to result perspicuous and invariable. Next, we put, at its upper extremes,[72] both the antecedent and the consequent of the conjoinded ratio, from which we remove that which has to be removed (I mean the antecedent on the left and the consequent on the right, as here[73] 97;4,56 e 70;32,3), and again, at its lower left extreme, the consequent of the removed ratio (as here 24;15,57). Next, we multiply the consequent of the removed ratio (namely, 24;15,57) by the consequent of the conjoined ratio (namely, 70;32,3), and we divide (that is to say we apply[74]) the number resulting from the multiplication by one another of both such consequents by the antecedent of the conjoined ratio (namely, 97;4,56), and we put, at its lower right extreme, the fourth proportional resulting from such a division (as here 17;37,45, for, as 70;32,3 is to 97;4,56, so 17;37,45 is also to 24;15,57). Netx, we take the antecedent of the removed ratio (as here 117;31,15) and we put it between its[75] two lower extremes.

And since the ratio of 17;37,45 to 24;15,57 is identical to the ratio of ΚΔ to ΔΓ and the ratio of 17;37,45 to 117;31,15 is again[76] identical to the ratio of ΜΕ to ΕΓ, and the ratio of 117;31,15 to 24;15,57 is again identical to the ratio of ΚΛ to ΛΜ, we remove the ratio of 17;37,45 to 24;15,57 (namely, its consequent 24;15,57[77]) and the ratio of 117;31,15 to 17;37,45 (namely, of ΜΕ to ΕΓ) is left out (that is to say, it is given).

Now if straight line ΚΛ were equal to ΕΓ, immediately we would also exactly have ΜΕ, which would be 17;37,45 parts. Since instead it is unequal (for this one is 117;31,15, that one is 120), we necessarily transform the ratio of 117;31,15 to 17;37,45 into that <having> 120 <as antecedent> and we also have in this way of how many parts exactly is ΜΕ. We transform it handily as follows: we again form first, as above, the analogous sign of chi and we put at its two upper extremes both the antecedent and the consequent of the ratio left out (I mean the antecedent on the left and the consequent on the right, namely, 117;31,15 and 17;37,45) and 120 at its lower left extreme. Next, we

72 The extremes of 'the sign of the letter chi' are meant.

73 Here, as elsewhere, Bryennios inverts the terms of the ratio. The mistake will not be corrected until the end of this first taking of a fourth proportional. The numerical operations are correct, their correlatives in terms of designators are not.

74 The verb 'to apply' (παραβάλλειν) denotes the geometric construction corresponding to a numerical division: a plane region is 'applied' to a straight line when it is transformed into a rectangle having the straight line as a side. The 'result' of the application is the other side of the rectangle.

75 Again, the extremes of 'the sign of the letter chi' are meant.

76 Note that the second and third identification of the numerical values with the compounding ratios are false, as it must be since ΚΛ ≠ ΕΓ.

77 If we consider the compounded ratio '17;37,45 to 24;15,57', the procedure of removal simply amounts to eliminating its consequent.

multiply 120 itself by the said consequent of the ratio left out (namely, 17;37,45) and we divide the number resulting from the multiplication by one another of both such <numbers> by the antecedent of the ratio left out (namely, 117;31,15), and taking the number resulting from such a division (as here 18;0,15), which is the fourth proportional (and in fact, as 17;37,45 is to 117;31,15, so 18;0,15 is to 120) we also exactly have ME, of 18;0,15 parts. Such a transmogrification and manipulation is called transformation of similar ratio starting from a proportion.

Document 12: Some propositions of Barlaam's *Treatise of Logistic*[78]

1. To find the value of a given ratio.

Let there be a given ratio A to B. Then one has to find the value of the ratio which A has to B.

Let A be divided by B and let Γ result.

Now then, Γ multiplying B will make A; but indeed, we call value of ratio a number that, multiplied by the consequent, makes the antecedent; and the antecedent of the ratio which A has to B is A, the consequent is B; therefore Γ is the value of the ratio which A has to B.

8. If a ratio be compounded of two ratios and their values be taken, the same ratio is also compounded of the two ratios which its [*scil.* of the compounded ratio] value has to each of the remaining values.

In fact, the ratio of A, B be compounded both of the ratio of Γ, Δ and of that of E, Z, and let the value of the ratio of A to B be H and that of the <ratio> of Γ to Δ be Θ and that of the <ratio> of E to Z be K. I say that the ratio of A to B is also compounded of two ratios, both that which H has to Θ and that which H has to K.

In fact, since the ratio of A, B is compounded both of that of Γ, Δ and of that of E, Z, and the value of the ratio of A, B is H and that of the <ratio> of Γ, Δ is Θ and that of the <ratio> of E, Z is K, therefore K multiplying Θ makes H. Similarly, also, Θ multiplying K makes H; therefore K is the value of the ratio which H has to Θ and Θ the value of the ratio which H has to K; therefore the value of the ratio which H has to Θ, multiplying the value of the ratio which H has to K, makes the value of the ratio which A has to B; therefore the ratio of A to B is compounded both of that of H to Θ and of that of H to K.

9. The values of identical ratios are equal, and the ratios whose values are equal are identical.

In fact, let the ratio of A to B be identical to that of Γ to Δ, and let the value of the ratio of A, B be E, that of the <ratio> of Γ, Δ be Z. I say that E is equal to Z.

---

78 Translated from Carelos 1996, 69.15–24, 75.4–76.15, 80.10–18 and 81.7–20, 82.22–83.7.

In fact, since E multiplying B makes A, therefore, as E is to the unit, so A is to B. Exactly for the same reasons, as Z is to the unit, so Γ is also to Δ; but indeed, the ratio of A, B has also been supposed identical to that of Γ, Δ; therefore, as E is to the unit, so Z is to the unit; and those having identical ratios to the same are equal to one another; therefore E is equal to Z.

Again, let E be supposed equal to Z. I say that, as A is to B, so Γ is to Δ.

In fact, since, as A is to B, so E is to the unit, and, as E is to the unit, so Z is to the unit—because E has been supposed equal to Z—and, as Z is to the unit, so Γ is to Δ, therefore, as A is to B, so Γ is to Δ.

14. If the values of two ratios be taken, any of them is compounded both of the other and of the ratio which its value has to the value of the former.

In fact, let the values of two ratios—both that which A has to B and that which Γ has to Δ—be taken, and let them be E, Z. I say that the ratio of A to B is compounded both of that which Γ has to Δ and of that which E has to Z.

[*I translate the alternative proof*]

In fact, let the value of the ratio of E, Z be H.

Therefore H multiplying Z turns out to make E; and H is the value of the ratio which E has to Z and Z the value of the ratio of Γ, Δ and E that of A, B; therefore the value of the ratio of E, Z, multiplying the value of the ratio of Γ, Δ, made the value of the ratio of A, B, so that the ratio of A to B is compounded both of the ratio of Γ, Δ and of that of E, Z. Again, let the value of the ratio of Z, E be Θ. Therefore Θ multiplying E turns out to make Z, so that the value of the ratio of Z, E, multiplying the value of the ratio of A, B, turns out to make the value of the ratio of Γ, Δ; therefore the ratio of Γ to Δ is compounded both of the ratio of A, B and of that of E, Z.

17. A given <ratio> being removed from a given ratio, to find which is the ratio left out.

Let the given ratio, from which one has to remove, be that of A, B, the <ratio> that one has to remove be that of Γ, Δ. Then, once the ratio of Γ, Δ is removed from that of A, B, one has to investigate which ratio is left out.

Let the value of the ratio of A, B, be E and that of the <ratio> of Γ, Δ be Z.

And since the ratio of A to B is compounded both of that of Γ to Δ and of that of E to Z, therefore, if from the ratio of A to B we subtract the ratio of Γ to Δ, the ratio of E to Z will be left out for us.

Document 13: Demetrios Cydones' General Rules[79]

Since Ptolemy—even if, in the mathematical theory of what moves spherically, he sets out lemmas for the removals of ratios—does not quite seem to refer there to all removals he appears to employ in the subsequent books, it is necessary to preliminarily set

79 Translated from my own critical text of Cydones' scholia.

out them, too, so that nothing can stand against those who approach such a theory with research aims. Accordingly, let the diagram of the first rectilinear lemma be set out.

Then it is manifest from this that, if a ratio be compounded of 2 ratios, the inverse is also compounded of the inverse ratios; [Ptolemy] turns out to employ this removal in the second theorem of the second book, and elsewhere.

Then it is manifest from this that, if a ratio be compounded of 2 ratios, one of the compounding <ratios> is also compounded both of the compounded <ratio> and of the remaining among the compounding <ratios>, taken in the inverse.

{*in the margin*: this is the inverse of the first porism, for the inverse is compounded of the inverses} Then it is manifest from this that, if a ratio be compounded of 2 ratios, one of the compounding <ratios> taken in the inverse is also compounded both of the remaining among the compounding <ratios> and of the compounded <ratio>, taken in the inverse. [Ptolemy] employs this removal in the theorem in which he seeks the setting angle <contained> by the beginning of Taurus and the horizon.

Document 14: The Definitions of George Gemistus Plethon[80]

A composition of ratios occurs in three terms, once a middle term is taken (either less than one of the extremes and greater than the other, or even greater than both, or even less than both) and this is removed away [ὑπεξαιρουμένου] in the composition of ratios.

A removal of a ratio from a ratio (once three terms have been set out, one of which is shared by the removed ratio and by that <ratio> from which this removed ratio has to be removed, and next a fourth proportional has been found in addition) leaves out the remaining term between the shared one among those previously set out and this fourth <proportional> found in addition, taken as a middle of the terms that contain the ratio from which the removed <ratio> has to be removed (and next either of the extremes has been removed away [ὑπεξῃρημένου]).

The fourth proportional term is found in addition once two terms have been multiplied by one another and what turns out to result from the multiplication has been divided by the remaining <term>; the <number> that turns out to result from such a division is in fact the fourth proportional term, which—whenever the extremes among the original terms (that is, the greatest and the least) multiply one another and the division occur by the middle—will be taken as a middle of either of the extremes and of the middle among the original terms—but whenever the middle among the original terms multiply either of the extremes and the division occur by the remaining <extreme>—the <terms> multiplying one another will be the middle <terms>, the one

80 Translated from Acerbi *et al.* 2016, 443.

by which the division occurs and this one that turns out to result from the division, will be the extremes.

Document 15: George of Trebizond's 'Chiasmus Rule'[81]

I say chiasmus when we get the ratio left out by multiplying the first <term> by the fourth and the second by the third. For instance, in the case of the continuous <compounded ratio>[82] 2 4 8, writing down the compound <ratio> first, I write underneath one of the compounding <ratios>, thus [*diagram*], and multiplying by chiasmus I get 8 16; therefore the <ratio> left out is double, and once either of the extremes is removed, say 8, a double <ratio> is left out. We must always use the chiasmus in this way, whenever the proportion is from three terms, namely, by doubling one of the extremes and by writing underneath one of the compounding <ratios> after writing down the compound <ratio>.

The procedure is as follows. Four numbers being given, if we divide by the first the result of the multiplication of the second by the third, we put it after the fourth and the ratio of the fourth to the result of this division is the <ratio> left out; if, instead, we divide by the fourth, we put it after the first and the sought <ratio> is that of the first to it. And if the multiplication occur to us the first being made by the fourth, if we divide by the second the result of such a multiplication, the result of the division is put before the third; if, instead, the division occur by the third, the result of the division is put before the second.

If one likes to operate by chiasmus, one must know that the compound <ratio>, once the terms are written down in a row, is the first to the third. Let four numbers be given in a row 4 6 8 20; by chiasmus [*diagram*]. And the left out <ratio> is two-thirds-more [ἐπιμερίζων δύο τρίτα], 80 to 48, and this itself is found in four ways according to Ptolemy's practice. In fact, multiplying the second (namely, 6) by 8 (namely, the third) I get 48, and if I divide it by the first (namely, 4), I get 12, and, as 80 is to 48, the fourth (namely, 20) is to 12. If, instead, I divide by the fourth (namely, 20), I get 2;24, and, once I have put it after the first, I get 4 to 2;24, identical to 20 to 12. Again, I multiply the first by the fourth (namely, 4 by 20), and it results 80; if I divide this by the second (namely, 6), it results 13;20, which is to 8 (for one must put it before the third) as 20 is to 12; if, instead, <I divide> by the third (namely, 8), I get 10; once I have put this before the second (namely, 6), the ratio of 10 to 6 is as that of 20 to 12.

---

81 Translated from my own critical text of the *Isagoge* to the *Almagest*.

82 Namely, a compounded ratio written in standard form.

Bibliography

Texts quoted

Apollonius, *AGE*: Heiberg, J. L. 1891–1893. *Apollonii Pergaei quae graece exstant cum commentariis antiquis*. 2 vols. Leipzig.

Archimedes, *AOO*: Heiberg, J. L. 1910–1915. *Archimedis opera omnia cum commentariis Eutocii*. 3 vols. Leipzig.

Barlaam of Seminara, *Treatise of Logistic*: Carelos, P. 1996. *Βαρλαὰμ τοῦ Καλαβροῦ, Λογιστική. Barlaam von Seminara, Logistiké*. Athens.

Euclid, *EOO.*: Heiberg, J. L. & H. Menge 1883–1916. *Euclidis opera omnia*. 8 vols. Leipzig.

Hero, *HOO*: J. L. Heiberg, L. Nix, W. Schmidt & H. Schöne 1899–1914. *Heronis Alexandrini opera quae supersunt omnia*. 5 vols. Leipzig.

Meliteniotes, Theodoros: Leurquin, R. 1990. *Théodore Méliténiote. Tribiblos astronomique. Livre I*. Amsterdam.

Metochites, Theodoros: Derycke, M. 1985. *Études sur 'l'Élément astronomique' de Théodore Métochite*. 2 vols. Louvain-la-Neuve.

Nicomachus, *Introductio Arithmetica*: Hoche, R. 1866. *Nicomachi Geraseni pythagorei Introductionis Arithmeticae libri II*, Leipzig.

Pachymeres, George: Tannery, P. 1940. *Quadrivium de Georges Pachymères*. Città del Vaticano.

Pappus, *Collectio*: Jones, A. 1986. *Pappus of Alexandria: Book 7 of the Collection*. 2 vols. New York.

Pappus, *iA*: Rome, A. 1931. *Commentaires de Pappus et de Théon d'Alexandrie sur l'Almageste*. Città del Vaticano.

Ptolemy, *POO*: Heiberg, J. L. 1898–1907. *Claudii Ptolemaei opera quae exstant omnia*. 2 vols. (in 3 tomes). Leipzig.

Theon, *iA*: Rome, A. 1936–1943. *Commentaires de Pappus et de Théon d'Alexandrie sur l'Almageste*. 2 vols. Città del Vaticano.

Works quoted

Acerbi, F. 2005. "Archimedes and the Angel: Phantom Paths from Problems to Equations." In: *Aestimatio* 2, 169–226.

Acerbi, F. 2017. "The Mathematical *Scholia Vetera* to *Almagest* I.10–15. With a Critical Edition of the Diagrams and an Interpretation of Their Symmetry Properties." In: *SCIAMVS* 18, 133–259.

Acerbi, F., S. Martinelli Tempesta & B. Vitrac 2016. "Gli interventi autografi di Giorgio Gemisto Pletone nel codice matematico Marc. gr. Z. 301." In: *Segno e Testo* 14, 409–454 (with 2 plates).

Acerbi, F. & I. Pérez Martín. 2015. "Gli scolii autografi di Manuele Briennio nel Par. gr. 2390." In: L. Del Corso *et al.* (eds.), *Nel segno del testo. Edizioni, materiali e studi per Oronzo Pecere*. Firenze, 103–143 (with 10 plates).

Acerbi, F. & P. Riedlberger 2014. "Un scolio tardo-antico sulla rimozione di rapporti, fonte dello Pseudo-Domnino." In: *Koinonia* 38, 395–426.

Acerbi, F., N. Vinel & B. Vitrac 2010. "Les *Prolégomènes à l'Almageste*. Une édition à partir des manuscrits les plus anciens: Introduction générale – Parties I–III." In: *SCIAMVS* 11, 53–210.

Allard, A. 1977. "Le premier traité byzantin de calcul indien: classement des manuscrits et édition critique du texte." In: *Revue d'Histoire des Textes* 7, 57–107.

Allard, A. 1981. *Maxime Planude. Le grand calcul selon les Indiens*. Louvain-La-Neuve.

Bulmer-Thomas, I. 1974. "Domninus of Larissa." In: Ch. C. Gillispie (ed.), *Dictionary of Scientific Biography*. Vol. 4. New York, 159–160.

Cufalo, D. 2007. *Scholia Graeca in Platonem. I. Scholia ad dialogos tetralogiarum I–VII continens*, Roma.

Heath, T. L. 1921. *A History of Greek Mathematics*. 2 vols. Oxford.

Knorr, W. R. 1989. *Textual Studies in Ancient and Medieval Geometry*. Boston.

Lorch, R. 2001. *Thābit ibn Qurra on the Sector Figure and Related Texts*. Frankfurt am Main.

Mogenet, J. 1956. *L'introduction à l'Almageste*. Académie Royale de Belgique, Mémoires, 2nd ser. 51.2. Louvain.

Moll, J. 1965. *Étude sur un traité anonyme d'initiation à l'Almageste*. 2 vol. Mémoire de licence dactylographié. Louvain.

Monfasani, J. 1976. *George of Trebizond. A Biography and a Study of His Rhetoric and Logic*. Leiden.

Monfasani, J. 1984. Collectanea Trapezuntiana. *Texts, Documents, and Bibliographies of George of Trebizond*. New York.

Mueller, I. 1981. *Philosophy of Mathematics and Deductive Structure in Euclid's* Elements. Cambridge, MA.

Netz, R. 2004. *The Transformation of Mathematics in the Early Mediterranean World: From Problems to Equations*. Cambridge.

Neugebauer, O. 1975. *A History of Ancient Mathematical Astronomy*. 3 vols. Berlin.

Riedlberger, P. 2013. *Domninus of Larissa, Encheiridion and Spurious Works*. Pisa-Roma.

Ruelle, Ch.-É. 1883. "Texte inédit de Domninus de Larisse sur l'arithmétique avec traduction et commentaire." In: *Revue de philologie, de littérature et d'histoire anciennes* 7, 82–94.

Saito, K. 1986. "Compounded Ratio in Euclid and Apollonius." In: *Historia Scientiarum* 31, 25–59.

Sylla, E. 1984. "Compounding ratios. Bradwardine, Oresme, and the first edition of Newton's *Principia*." In: E. Mendelsohn (ed.), *Transformation and Tradition in the Sciences. Essays in Honor of I. Bernard Cohen*. Cambridge, 11–43.

Tannery, P. 1885. "Notes critiques sur Domninos." In: *Revue de philologie, de littérature et d'histoire anciennes* 9, 129–137 (repr. In J. L. Heiberg & H. G. Zeuthen (eds.) 1912. *Mémoires Scientifiques II*, Toulouse-Paris, 211–222).

Vogel, K. 1936. *Beiträge zur griechischen Logistik. Erster Teil*. Sitzungsberichte der Bayerischen Akademie der Wissenschaften, Mathematisch-naturwissenschaftliche Abteilung. München, 357–472.

Vogel, K. 1960. "Buchstabenrechnung und indische Ziffern in Byzanz." In: *Akten des XI. internationalen Byzantinisten-Kongresses*, München 1958, München, 660–664 (repr. In M. Folkerts (ed.) 1988, *Kleinere Schriften zur Geschichte der Mathematik*. Stuttgart, 452–456).

Henry Mendell

Why Did the Greeks Develop Proportion Theory? A Conjecture

Abstract: Greek mathematicians in the late fifth and fourth centuries BC developed number theory based in whole numbers and proportion theory. Their neighbors, Egyptians and Babylonians, did not. We cannot know why, but one significant difference is that the Greeks had no representation system for expressing and calculating with fractions, although they had no difficulty speaking of them as multiple parts. Greek mathematicians would have worked with the acrophonic system, which lacked a system of fractions, and even if they had used Ionic (alphabetic) numerals, there is no evidence of the Ionic fractional system before the end of the fourth century BC. On the other hand, ratios are conceptually a part of Greek thinking, while employment of sub-units and super-units was foundational to arithmetic calculations and puzzles. My suggestion is Greeks worked with what they had, so that these form the foundation for the theory of numbers that we find in *Elements* VII–IX.20. The evidence for this picture is, admittedly, weak. I consider evidence from Plato's *Republic* VIII for a playing with number puzzles, the Archimedes' *Cattle Problem* for a class of false position puzzles, and the *Anthologia Palatina* for Euclid's *Elements* as providing a foundation for such puzzles, and Euclid's *Sectio Canonis* and Ptolemy's *Harmonica* for working from fractions to sub-units.

Keywords: Number Theory; Ratio; Proportion; Sub-Unit; Fraction; Acrophonic Numerals; Ionic Numerals; Alphabetic Numerals; Harmonics; Sexagesimal Numerals in Greece; Nuptial Number; Euclid; Plato; Ptolemy; Anthologia Palatina; Archimedes Cattle Problem.

I Introduction[1]

The number theory that we find in Euclid's *Elements* represents the culmination of one of several revolutions in mathematics that took place in the fifth and fourth centuries BC and was certainly fundamental to the development of a general theory of proportion and irrationals. It has two foundations and conceptual parts, one built on a theory of proportion (VII.1–IX.20, 34–5) and the other a brief account of odd and even (IX.21–33). The most basic question for the modern reader is why did Greek mathematicians

1 In what follows, all translations, unless stated otherwise, are mine. Inscriptions cited may be found at the Packard Humanities Institute: http://inscriptions.packhum.org.

https://doi.org/10.1515/9783110565959-008

develop theories of either, without there being any precedent in older, neighboring cultures? Of course, we do not really know, and it is probably because we cannot know and so desperately want to know that we fashion stories that we believe in because we find them plausible. In this paper, I shall construct a story about why ratios became important to Greek mathematicians. At the beginning of the last century, there was a popular myth, which no one believes today, about a foundation crisis with the discovery of incommensurables, so that ratios became the method of distinguishing numbers from magnitudes.[2] The foundation crisis, as Wilbur Knorr argued,[3] was more the crisis of the late nineteenth century than of the fifth century BC. Perhaps, my story will be just as anachronistic, and based on as little evidence. More recently, David Fowler argued that Greek number theory was built on the Egyptian conception of fractions.[4] On the contrary, I shall argue that the mark of fifth to fourth century treatment of numbers was the lack of any general way of representing fractions and calculating with them. Thinking about units and sub-units and ratios between numbers and how to calculate with them was natural because they were the best way of thinking generally about measures. From there, they could develop harmonic theory, treatments of irrationals, even odd views about the numbers of the cosmos. On this view, the revolution of fifth and fourth century BC arithmetic emerged out of conceptions and strategies that were part of Greek conceptions and techniques.

Even if we think that the theory of ratios was developed, for example, to build a rich harmonic theory,[5] we can still ask, 'Why ratios?' For example, the fourth, fifth, and octave, can easily be conceived as fractions from a low string, 1: 1 1/3, 1 1/2, and 2; or as a sequence of units from 6: 8, 9, 12; or as a ratio; 4 to 3, 3 to 2, 2 to 1, or even verbally as a third-again (ἐπίτριτος), half-again (ἡμιόλιος), double. Why ratios? In fact, all but the first appear in Classical Greek fragments of harmonic theory.

I shall first argue that the principal way of representing numbers would have been what I shall call the 'Greco-Italian' system (of which the so-called 'Acrophonic' system is the most common). However, even if people had used the 'Ionic' system (sometimes known as the 'Alphabetic' system), it is unlikely, even implausible, that they had the system of fractions that later was associated with Ionic numerals. So, here is a major difference between Babylonian and Egyptian mathematicians and Greek mathematicians. Whereas their neighbors had sophisticated ways of speaking about and calculating with fractions, Greek mathematicians had only a very limited vocabulary of fractions and so were restricted to two related techniques: the one involves introducing and working with subunits, while the other involves ratios. It is important to keep in mind that the issue here is not conceptual. If they needed a fraction, they could

2 Cf. Scholz 1928.
3 Cf. Knorr 2001.
4 Cf. Fowler 1999.
5 Cf. Szabó 1978, 99–184.

write it out, just as they could also write out ordinal numbers. It is also implausible to think that no Greek had ever met sexagesimal numerals or Egyptian parts. However, the technical limitation on a Greek treatment of fractions left them with two possibilities, to invent a new system or adapt one from their neighbors, or to work within the methods available. Early Greek mathematicians chose to work with what they had; later ones chose to adapt, eventually both the Egyptian and Babylonian systems. There may even be a trace of Egyptian fractions in Euclid's *Elements*. Anyway, that is my suggestion. It also has several negative virtues: it does not involve a foundation crisis, religious views of ratios, philosophical numerology, nor philosophical views about numbers and magnitudes. However, it is a conjecture and could be all wrong. But that's the way it is.

II Many Ways of Representing a Number

Greeks in the fifth and fourth century BC had many ways of representing numbers. Numbers could be represented with objects: marks or pebbles representing ones, which are simplest and get reflected in the lore of Pythagorean arithmetic;[6] pebbles arranged on a counting board or abacus;[7] fingers used in counting and calculating.[8] It is not uncommon to find running counts of strokes on vases, |||||, sometimes ending in a numeral for the last element of the tally.[9] That said, Greeks in the fifth century BC had two basic systems for representing numbers, the Greco-Italic and the Ionian systems.[10]

In addition, people could lexically order things according to the alphabet, as was occasionally used from the fourth century on.[11] This is not a number system, if merely because it is not used for calculation, and will be merely a secondary concern here. Some people were certainly aware of other counting systems in the Persian Empire, the common Assyro-Babylonian system, the Phoenician system, and the Demotic

6 Cf. Knorr 1975; Netz 2002.

7 Cf. Turner 1951; Lang 1957; Lang 1968; Keyser 1986; Schärlig 2001a and 2001c.

8 Cf. Turner 1951, 65–69; Williams and Williams 1995. Despite the many texts they furnish to show that Greeks and Romans used fingers for calculating, neither can provide any clear example to show how this was done. For any complicated calculation, either some numbers are presented in some other format, or a good memory is required, in which case ancient calculators are in the same position as modern ones. Those with good memories and skills can do it; those without cannot. I also wonder what happened to those students who could not independently bend the pinky and ring finger.

9 Cf. Lang 1956 and 1976; Lawall 2000; Johnston 1979, 30–31.

10 See Tod 1911/12 and 1913 for seminal papers, but also Chrisomalis 2010, Threatte 1980, Johnston 1979 and 2006.

11 Cf. Threatte 1980, 117–119; Tod 1954.

Egyptian system.[12] I shall argue that there would have been some awareness additionally of the Babylonian sexagesimal system. Besides all these, one could simply write out the words for the number. In this regard, there are three sorts of numbers that one might be concerned with representing: cardinal numbers, fractions, and ordinals. It is a principle of my discussion that the modes of representing numbers has a profound effect on how people calculate and conceive of what they are doing when they calculate.

By the 'Greco-Italic' system,[13] I shall mean any decimal system with iterated units of the form 10^n, and employed in the ancient Greek and Italian world. They probably emerged, sometime in the early sixth century BC, as a Greek version of standard numeral systems prevalent in Mediterranean trade (Aramaic and Phoenician), where a sub-base would be introduced, e.g., to avoid writing, e.g., |||||| or ⦙⦙ for 6. Where a sub-base is employed in the Greco-Italic version, the system is basically the familiar Roman number system with the subtractive option (such as XIIX instead of XVIII in Republican Rome or Etruscan XIX=19 or XIIXXX=38)[14] as a dialectical variation. In Chrisomalis's classification,[15] the most common version, then, is a cumulative-additive decimal system with a sub-base of 5. That is, there are separate symbols for 10^n and

12 Cf. Chrisomalis 2010, 54–56, 70–76, 247–249.

13 Chrisomalis (2010, 93–104, esp. 103) calls the system 'Italic' on the assumption that it first appears in Italy and particularly among the Etruscans, and in following the suggestion of Keyser 1988 that the Etruscan version is based on a tally system where the symbols for 5×10^n are the bottom half of the symbols for 10^{n+1}, and finally an a priori argument that the tally system should precede any acrophonic system. However, given that these and other variants all appear about the same time in the middle to end of the sixth century in many places, the argument is at best one of plausibility. I do not see why someone could not adapt an acrophonic system to a tally system, just as the other way around. Someone might have taken a practice of tallying ||||||||\ (9 strokes with the last crossed), which perhaps could evolve more reliably (or randomly) into |||\|||\, with the merchant then selecting Λ and X to correspond to the acrophonic numerals or as a invention of their own. Or they might have taken a tens tally and made the decision to adapt the acrophonic system by taking the lower half of X for 5. It is even still possible, if less likely, that, with Graham 1969, the values for 5, 10, 100, 1000 were originally taken from the end of the Etruscan alphabet, as Λ, X, Ү, 8, with the last three symbols occurring in fourth-century real estate sales records in Olynthus, which then evolved into Λ, X, ⚹, ⊕, but with the modification, with Keyser and Chrisomalis, that the 5's are regarded as part of the next higher decimal digit, but here on the basis of an analogy with Λ and X (numerals for 5×10^n are not found at Olynthus, but that is just an accident of the sales records—it is just unknown whether Olynthian numerals in the fourth century included a 5 sub-base). Cf. also Johnston 1975 for a related system. My point is that we just do not have enough evidence to do more than guess and too many plausible stories available. All that is essential to my argument is that the Greco-Italic systems were common in Italy, Sicily, and Greece by about 500 BC, before Greek number theory developed. It takes but one person to invent a system of numerals, and it is just not possible, without evidence, to know either his psychology or that of those accepting the system.

14 Cf. examples in Ifrah 2000, 189–190. The Etruscan is to be read right to left.

15 Chrisomalis 2010, 9–14 and Chrisomalis 2004.

$5{\times}10^n$, written in descending order of the values of the signs (whether from left to right or right to left).[16] In all cases, there is a maximum number that can be directly represented, as determined by the largest 10^n and $5{\times}10^n$ in the system. Numbers are the sum of the values of the symbols, while all numbers no greater than the maximum representable are achieved by iterating symbols for 10^n.[17] For emphasis or clarity, the numerals are often separated from the rest of the text by punctuation ':'.

Acrophonic[18] dialects of the Greco-Italic system are the most common in the Greek world, namely taking as a symbol for a digit the first letter of its name, a natural practice that is somewhat older than Greek culture.[19] The Greek acrophonic systems involve taking the first letter of the name of the word for 5 and each $10^n{>}1$ and a ligature of the symbols for 5 and 10^n to form a symbol for $5{\times}10^n$. That is, with 𐅃 as ΠΕΝΤΕ (5) and Δ a ΔΕΚΑ (10), 50 will become 𐅄, as Priscian suggests,[20] πεντάκι δέκα (five-times ten).

For most Greeks, the basic symbols in the most common Greco-Italian acrophonic system are 1 (I), 5 (𐅃), 10 (Δ), 50 (𐅄), 100 (H), 500 (𐅅), 1000 (X), 5000 (𐅆), 10000 (M), 50000 (𐅇), with 1, 10, 1000, 10000 repeated to get to the next level. So 68369 would be written 𐅇M𐅆XXXHHH𐅄Δ𐅃IIII. An important feature of the system, however, is that symbols may differ, depending on what is being counted. For example, ⊢⊢ is two drachmas, where I will then be one obol. Two talents will be TT, two minae MM, and two choae XX (with the possibility of ambiguities arising), etc.

The acrophonic system is never used to represent ordinals.[21] More important for our purposes, there is no general system for representing fractions other than special cases, e.g., C (1/2), which is similar to the Ionic form (see below), and Ɔ or T (1/4).

In addition to representing numbers in this way, ordinary Greeks used abaci or counting tables. The tables involved using pebbles to represent symbols from the Greco-Italic sub-base 5 system, that is, 1's, 5, 10's, etc., placed at different places on a board appropriately divided up. So, it would be easy to read off of a numeral to pebbles on the abacus board, and back again from pebbles to write the resulting

16 Even here there are exceptions. Tod (1911–12, 127) alludes to two Boeotian systems where there are signs for 30 and 300.

17 With exceptions, most prominently *IG* IV 523–5 (Argive Heraeum), cf. Tod (1911–12, 103) and Mitsos (1947, 85 and Plate XIV n. 8); also, cf. Tod 1936–1937, 238–242, 255–257.

18 Keil (1894, 253, n. 1) introduced the term. For early essays, cf. Tod 1911–1912; 1913; 1926/27.

19 So long as we do not limit the term to alphabets, the Assyro-Babylonian common system is acrophonic given the representations of 100, 1000, and occasionally 60. Cf. Ifrah 2000, 137–139.

20 After stating that Attic numerals are derived from the first letter of the word for the number, with I implausibly derived from the occasional feminine form ἴα (it is not just that I is the standard symbol for 1 almost everywhere, but the form itself), Priscian, *De figuris numerorum...*, 1.4-6 (p. 406) goes on to create a story that Latin numerals work somewhat the same (easy for C, M), in refutation of which, cf. Keyser 1988.

21 Tod (1936–7, 254–255) discusses a late possible exception in a day of the month from Aphroditopolis, 57 or 56 BC, which could well be a scribal error, 𐅃 for Γ (third) or I (tenth) in the Ionic system.

numeral. In this way, the number system and the abacus together form a calculating system.[22] We need not here be concerned with the details of this.[23]

It is but one step between the abacus and Pythagorean pebbles, where each number *n* is simply marked by *n* pebbles. Indeed, the most primitive form of representing *n* is by so many *n* strokes. However, in the hands of ancient arithmeticians, at least so it has been supposed, it becomes a powerful too for working out relations by rearranging pebbles into different shapes.[24]

The other principal system for representing numbers is the more familiar Ionic system (more often called the 'alphabetic system' because it is based on an extended form of the Ionian alphabet), where 1, .., 9 are: Α, Β, Γ, Δ, Ε, Ϲ, Ζ, Η, Θ, respectively; 10, ..., 90 are: Ι, Κ, Λ, Μ, Ν, Ξ, Ο, Π, Ϙ, respectively; and 100, ..., 900 are: Ρ, Σ, Τ, Υ, Φ, Χ, Ψ, Ω, ͳ, respectively. It is characteristic of the system that numerals are non-iterative. That is, numerals are never repeated. In the system of Chrisomalis, this is ciphered additive decimal (that is each numeral within the base 10 has its own symbol).

So far, the highest number representable is 999. However, one could add a vertical line under digits for 1, ..., 9 to get 1000 to 9000, that is Α̩ or ͵Α is 1000. This gets us to 9999. Still higher numbers could be written by writing the number over an M for myriad (10000), e.g., $\overset{\text{͵}\alpha\sigma}{\mathrm{M}}$ will be 1200,0000, which will get us up to 9999,9999. For the myriad is seen as a sort of super-unit. Later, Archimedes and Apollonius developed multiplicative systems based on these for writing unlimitedly large numbers, for Archimedes the base 1,0000,0000 system and for Apollonius the simpler 1,0000 base. But this is all outside our concern.[25]

---

22 I do not understand the discussion of Chrisomalis (2010, 116–118) for the claim that Greco-Italic sub-base 5 numerals are not calculative, that is that "people used abaci or finger computation to actually manipulate numbers with the numerals being used only to express the result." First, the move from the numerals to the calculating table is direct and requires no translation (as, for example, is required when one moves from binary calculation to the decimal input/output of a computer). So the numerals are actually represented on the board. Secondly, one could, of course dispense with the table and use the numerals directly: ||| + ΗΓ||| = ΗΓ|||||| = ΗΓ(|||||)| = ΗΓΓ| = ΗΔ|, each |, Γ, Η, replacing a stone on a table in its special place, that is doing exactly what one does on the table, even more clearly if they are appropriately aligned (cf. Turner 1951, 64–65). Even our position-decimal system is more complex, since one needs to memorize an addition table. As to multiplication and division, one can readily construct rules that mimic practice on an abacus. If one did not do this in one's head, one would do it on an ephemeral medium, such as a wax tablet, and it would be strange indeed if no one in the ancient world ever learned to calculate reasonably messy numbers in their head. It would be a great limitation on ancient calculators and businessmen if they always needed to use an abacus or fingers. For a contrast with an Ionic abacus, cf. Schärlig 2001b, but of the Hellenistic Age; cf. also West 1992. Here, the symbols base 10, sub-base 5, need to be translated back into the Ionic system.

23 There are many accounts of how Greek abaci worked. Cf., for example, fn. 7.

24 Cf., for example, Knorr 1975, 135–169.

25 Archimedes gives his system in the *Arenarius*; the suggestion that Apollonius is responsible for the other method is based on the relics of Pappus *Collectio* II, a series of theorems of Apollonius that

Later on, numbers are typically, but not always, written with a line over them. For 1000 to 9000, one writes a dot or line under the numeral for 1, ..., 9. For example, 8369 could be written $\overline{\text{Ͱ}\text{ΤΞΘ}}$. As with letters in geometry, the line above will avoid confusion between, say, $\overline{\text{MH}}$ (47) and MH ('not'). Ionian numerals could be used for ordinals, e.g., in dates, and also come to be used to represent Egyptian style fractions and even fractions in a style akin to how we do it.

It is uncontroversial that the Ionic system of fractions derives from the Egyptian system,[26] where the basic idea behind Egyptian fractions is that every division of two numbers that is less than a whole number is the sum of a non-iterative sequence, $1/n_1+...+1/n_m$, where $n_i < n_{i+1}$. Here, each representation of a (fractional) part will have vertical or slanted lines above each digit. Let us contrast this with an iterative system of conceiving of fractions, as a sum of 1/n+...+1/n, e.g. m n'th parts, which I shall introduce later.[27] In addition, following Egyptian practice, there are two special symbols, Β́ for 2/3 and ∠ for 1/2. Following custom, we can write these nth's as n′, with 2′ for 2/3 and keeping ∠.[28] For example, in P. Hibeh I 27 (line 55), we find ΙΓ Ί Β́ Μ́ Έ. One knows that this is 13 12˜ 45˜, and not, say, 13 10′ 2′ 40′ 5′, because the numbers increase towards the right, or at least that is the rule, so that we would represent this as 13 12′ 45′, i.e., 13 19/180. Since numerals will not iterate, 2/5 will not normally be written Έ Έ. This is not to say that errors and violations will not occur.[29] There is also a ligature for 3/4, as well as other variations, which need not concern us here.[30] I shall call such combined parts, non-iterative. Later, we also find instances of a non-standard form, iterative, common fractions (n/m) but with the denominator on top,[31] e.g. $\frac{\acute{\text{E}}}{\text{B}}$ as 2/5.[32]

---

employ system. They are also semi-symbolic number systems, for which see how the components get combined in Pappus, *Collectio* II 18.23.-28.28. At least, when Proclus, *In Rem Publicam* II 23, reports a number from Sosigenes in Apollonian form, he writes out which myriad each set of Ionic numerals represents, starting from the fourth, i.e., a multiple of 1,0000,0000,0000,0000, to the first myriad, i.e. a multiple of 1,0000, to units.

26 As Fowler observes (1999, 258–262), P. Cair. 89127–30, 37–43 from the third century BC also uses a symbol for 5/6, which he describes as a ligatured 2′ 6′. The question is how one conceives a ligature. Cf. Chrisomalis 2003. Chrisomalis also derives the Ionic system from Demotic; however, assuming that he is right, there is still no necessity that both systems, the whole number system and the fractional system, derived from Demotic at the same time, as I shall argue.

27 Note that I not suggesting that our conception of fractions is non-iterative, e.g., where m/n is the number that results when m is divided by n. The Greeks never conceived of fractions in this way.

28 Cf. Fowler 1999, 224–229. Fowler uses 2′ for 1/2 and 3″ for 2/3, where I use ∠ and 2′, which are closer to the original, if less clear to the modern reader.

29 P. Hibeh i.27 has at least two clear exceptions to this, where 45′ is written 5′40′ (lns. 89, 123), cf. also ln. 153 for 3′10′40′5′5′, clearly an error that was poorly corrected—it should be 3′10′5′40′5′, i.e. 3′ 15′ 45′).

30 Cf. Fowler 1999, 224–229.

31 Here, Fowler (1999, 227) calls fractions of the form *p*/*q*, which can be manipulated, 'common' and argues against these being genuine common fractions. This question is outside my present concern. For a different view, cf. Knorr 1982

32 Cf. Fowler 1999, 255–257 and Vitrac 1992, 162–164. Certainly, by the third century AD, such fractions are common, as in Diophantus. Metrodorus (ca. 300 AD) writes out in his poems even unreduced

In addition to using the supplemented Ionian alphabet for numerals, in limited circumstances one could identify objects sequentially by the alphabet. There are several examples of this in Athens and in the Hellenistic period. Aristotle tells us that law courts in Athens were ordered by the alphabet,[33] while there are lists of groups of items so indexed in warehouses, theater seats (much as we do),[34] and groups of lines in a text being counted to ensure completeness (where the totals are all normally acrophonic).[35] We do not know when most of Aristotle's longer books were ordered in this way, but it is likely that Sappho and Homer were arranged into alphabetically ordered books in the Hellenistic Age.[36] I suspect that all of these alphabetical indices would, if required, be matched with ordinals. For example, people in a later period can refer to 'book H' of an Aristotelian treatise, but also to 'the seventh book', which can also be written out with an Ionic numeral as τὸ ζ̅ βιβλίον, a practice that readily causes confusion. It is reasonable to suppose, however, that this manuscript practice follows an ancient one. Obviously, if one is looking at an inscription, it is impossible to tell whether it is a numeral or a letter unless it either has one of the three supplementary letters, ꞓ, ϙ, or ͳ, or has an otherwise meaningless sequence of symbols, e.g., KB. For example, if one sees on a vase a Δ, and one suspects that it is a count of something, there is no more reason to think that it is part of an alphabet sequence than a numerical one. In fact, only the writer probably knew what they were up to. This difficulty is important for determining the frequency of Ionic numerals in the fourth century.

Obviously, the easiest thing a Greek could do in representing numbers would be to write out their names.[37] We find this in manuscripts, just as we find it in early inscriptions. Writing out a number seems to us cumbersome, but we often do it anyway, for whatever reasons. However, written out numbers do not play well into calculations.

---

fractions, cf. *Anthologia Graeca* XIV 117 (ἕκτα ... δοιά or 2/6), 119 (δύο τέτρατα or 2/4), 140 (δύ' ἕκτα or 2/6), 143 (δύ' ὄγδοα or 2/8). Although he normally will use written out common fractions, I note that Metrodorus once (xiv 121) writes out a unit fraction for 2/15 as "eighth and twelfth of one tenth (ὄγδοον ἠδὲ μιῆς δωδέκατον δεκάτης)", but this way of writing 120′ may be due to metrical constraints.

33 Cf. Aristotle, *Ath. Const*. 63–64. In this regard, it is significant that the ranking comes from the order of the alphabet, not necessarily starting from A, given that the law courts are numbered from Λ, while the jury pool from each tribe is divided into 10 groups lettered A, ..., K, so that the boxes involved in a jury selection will still be A, ..., K, Λ, ..., depending on how many juries are needed. Obviously, this system presupposes the Ionic alphabet, officially adopted in 403/2.

34 Cf. Threatte 1980, 117–119, as well as Tod 1954.

35 Most notably for Homer, Herodotus and Demosthenes. Cf. Graux 1878; Burger 1892; and Wendel 1949 chapter 3 §3.

36 For Sappho, cf. Yatromanolakis 1999. The issue of the number of Sappho's books, 8 or 9, can readily be explained by a confusion of reading Θ as a numeral instead of as a letter. The earliest suggested arrangement of Homer's epics into 24 alphabetically ordered books would seem to be that of Nagy 1996, the period of rule in Athens by Demetrius of Phalerum (317–307 BC). Cf. Haslam 1997, 58 n. 6. Aristotle is more controversial.

37 On the prevalence of written out numbers in Attic inscriptions, cf. Threatte 1996, 412–449.

Nonetheless, it is important to remember even in Archimedes' *Arenarius*, a treatise on a system for writing out large numbers, after he has done several calculations in chapter 1, the only Ionic numerals that appear (cf. 2-4) are ι (10), ρ (100), and ͵α (1000).

In addition to these systems, some people have wondered when Greeks came to know the Babylonian sexagesimal place system, where digits 1, ..., 59 are written decimally, iterations of 𒁹 (1) and 𒌋 (10), but then are grouped in a base 60 arrangement, where the relative value of the numeral is based on position relative to other sexagesimal units. The more minimalist will then argue that we can only say that it was known in the first century BC, as it appears in Hypsicles, *Anaphoricus*, or in the second, according to our reports of Hipparchus. The question is misstated. We need to distinguish between what people are aware of and what they use. To draw a modern analogy, most Americans know of the metric system and many can convert from American (avoirdupois) measures to metric measures. However, few can think with the metric system without translating back or have a feel for it. From the traces of sexagesimal numbers in Greek literature and practice, it is very likely that there was the sort of awareness that one would expect in a culture abutting, trading with, and often ruled by the sexagesimal culture, even if the sexagesimal numerals were in a modified form. These include plethora of measure units dividing into sixty or thirty subunits: for distance, one parasang (Persian, anyway) into 30 stadia and one stadium into 600 feet, or one parasang into 18000 feet = 5,0,0 feet in sexagesimal; for weight, 1 talent = 60 mina (a Semitic word, anyway), and if we include 100 drachmas per mina and 6 obols per drachma, we wind up with 36000 obols = 10,0,0 obols in sexagesimal. Perhaps, this is all accident and can easily be accounted for by the prevalence of the Assyro-Babylonian system, which is cumulative additive with base 100 with sub-bases of 60 and 10. However, it is not at all likely that every merchant in the Persian Empire working with these units failed to understand the basic idea behind them.

The tribute list for the Persian Empire at Herodotus 3.89–95 mostly is not rounded sexagesimal numbers of talents, as one might expect, but a significant number are multiples of 60, namely 7 out of 20, i.e. where the expectation would presumably be 3 or 4, given that all are multiples of 10. Of course, this means little since Herodotus is reporting what may well have been the case. But, in his travels through the Persian Empire, he would have come across, at the very least, someone translating from numbers represented in sexagesimal. However, this would reflect the fact that the dominant numeral systems were base 100 with sub-bases of 60 and 10.

The strongest evidence comes from *IG* I^3 1387, two separated fragments of a single stele from the Acropolis in Athens, ca. 450–440 BC.[38] This is a list of pairs of letters, each column of pairs separated by a vertical line. The only reasonable suggestion, to my mind, for the meaning of the text is that it consists of sexagesimal numbers with

38 So Keil 1901.

digits written with a variant of the Ionic system.[39] On the reasonable assumption that every occurrence of O is a scribal error for Θ, which does occur,[40] all the first letters are A,..., I, including ⊏, while all the second letters are A, I, ... N. Next, we need to deal with the single occurrence of ΔA and the occurrences of IN and IM. Koehler suggested that in the case of, e.g., IΛ, IN, IM, we understand I as an indicator of 10's only, so that these would be 30, 40, 50, respectively.[41] Although Keil thought ΔA a scribal error for ΔΛ (34),[42] given that we otherwise have no indication on the stele of a number less than 10, it is more probable that A indicates, as with Δ and the case of 10's, that there are only units, so that ΔA is 4. So I is a sort of Ionic tens marker to indicate absence of ones and A a sort of Ionic ones marker to indicate absence of tens. In any case, with these assumptions, every pair represents a number less than 60, but written, ones-tens, in reverse order from later practice, something not unknown in sixth-century practice.[43] So, it is very reasonable to that we have lists of sexagesimal numbers, whose purpose remains mysterious. Another is that the list is of minae, but why have a long list of minae less than a talent?

Perhaps, the most charming evidence of all comes from what I hazard is a joke in Plato. The Nuptial number in *Republic* VIII 546BC is $(3\cdot4\cdot5)^4=60^4=$1296,0000, a number well beyond either Greek system before the notation of Archimedes, athough Plato could write: MXXꟼΔΔΔΔΓIIII myriads, that is changing the unit. However, he might well have in mind how a Babylonian, lacking place markers, might write this, as 𒁹, i.e. 1. If so, his choice of number is very amusing, in a discussion that might, in fact, be something of an amusement.[44]

39 As Koehler (1896, 153–154), though we need not infer, with him, that the text is astronomical.

40 As $IG^3$1387 col. W3 and X11. The occurrences of O are only in the left fragment. This may be worrisome—the two fragments might be two related stele.

41 As Koehler 1896, 153–154. Note that he only considers the part of the inscription where ΔA and O do not occur.

42 Keil (1901, 154 n. 1) rightly notes that all the other right digits are 10's, I, ..., N, with K absent, and so infers wrongly, as I suspect, that ΔA must be a mason's error for ΔΛ. If one takes the numerals as Ionic, A cannot have its standard meaning in an Ionic numeral, so that the emendation is accepted in the notes of *IG* I³.

43 Cf. Johnston (2006, 17, 126–7, and fig. 80E), item 11E18a (Princeton 170 and Beazley Database 351077), an Attic Black-Figure hydria in the manner of the Madrid Painter (525–475 BC, from Nola, Italy), with ZΛ (37). Two of the three cases of multiple lettered numerals from the fifth and fourth century BC in Lang 1976 are also in reverse. Cf. fn. 47.

44 There is another odd piece of evidence in Plato for sexagesimal numbers that one would only put forward with the greatest hesitation. In the description of the council at *Laws* VI 756BE, the Stranger divides the council into twelve sections (the tribes corresponding to a division of the year) of thirty each, as 360 is appropriate for divisions (πρέποντες ταῖς διανομαῖς). He then divides these into four sections of 90 and two of 180. Of course, this could all be a felicitous accident arising for the virtues of 360 for division and a longish 30 day month, but it could easily be an homage to the Babylonian division of the zodiac.

Given that sexagesimal numerals are the language of Babylonian mathematics even in the Persian period, but not the language of commerce, it is unclear how widespread knowledge of them would be. My point is, however, that it is extremely likely that some Greeks dealing with the Persian Empire understood the origin and structure of the numeral system. The ancient world is not that large. However, awareness does not entail common usage. *IG* I³ 1387 aside, we do not know of Greeks systematically using sexagesimal numbers before Hipparchus, and there, of course, only for fractions.

As to other numeral systems, Egyptian, Phoenician, and other representation systems that traders, soldiers, ambassadors, refugees, and the occasional tourist might come across in their dealings with locals of the Persian Empire, from Ionia to Babylon, to Susa, to Naucratis, to Heliopolis, to Tyre, we can presume that some Greeks were aware of them. Would they know about Egyptian fractions? We have to presume that some would. Would they have used them? That is a different question.

In sum, Greeks had many ways of representing cardinal numbers, pebbles and scratches, acrophonic numerals, Ionic numerals, as well as foreign systems. It remains to be seen which they would or could have used.

III The Number System of the Fifth and Fourth Century

My strategy will be first to argue that any mathematician involved in the early stages of number theory would have employed the Greco-Italic system which lacked a system of fractions. However, even though a system of fractions was adapted to the Ionic system, any such adaptation did not exist until the late fourth century BC and would have been unavailable to early developers of number theory.

It is ironic that students studying Classical Greek today learn the Ionic system, as it is the dominant system for most of antiquity but not for the classical age. I shall argue that neither the Ionic system and, especially, its system of fractions has anything to do with the development of early Greek arithmetic. The dominant system for Classical mathematicians was the acrophonic system, so that, most significantly, any system of representation of fractions would have to have been based on its poor treatment of fractions.

Although the Ionic system first appears in the early sixth century BC Ionia,[45] it seems to have fallen into disuse around 475 BC, albeit with two striking exceptions, IG I³ 1387 and *SGDI* 5727 (=Halikarnassos 31). The first, I discussed above; the other

45 Cf., inter alia, Johnston 1973.

is from Halicarnassus (last quarter fifth to first quarter fourth century BC).[46] This is a stele on land purchases that is inscribed on four sides, one with acrophonic numerals, and the other three with Ionic numerals indicating drachmas, with iterated |'s for obols and −, =, ≡ as fractions of obols. Other than these, Ionic numerals appear but rarely and trivially, that is as tabulations of measures on pots.[47] One might suspect that, as they seem to occur in mostly tallying (|||||Ϲ), they were actually conceived as ordinals, where, despite their rarity, a tally ending in an Ionic numeral might be clearer than an acrophonic.[48] In any case, their use is not significant except as witnesses to the survival of the notation.

The next known significant occurrence is in markings on five silver vessels found in Macedonian royal Tomb II at Vergina, where a sixth uses acrophonic numerals.[49] Given that the vessels are likely to be of Macedonian manufacture, so that the

46 Blümel 1993, 3; cf. Keil 1894, 259.

47 Cf. Tod 1950 and Johnston 1979, 27 with the items in the catalogue referred to; and Johnston 2006, 17. Besides those here discussed, Johnston mentions one possible Cretan use. For Athens, Lang (1956, 5) lists 5 instances of Ionic numerals (10–14: P2371, P2364, P11386, P11387, and P9340) in a list of 103 instances of numerals. These are all summations of tallies, iterations of |. Only two of these could not be letters instead of numerals (P2364 with Ϲ||||| and P2371 with |||||||[H]), although it is more likely that the others are numerals as well. Lang 1976 expands this list. With the numbering in her catalogue in parentheses, these are: P25983 (E15, fourth century BC in third century BC context, so that the graffito, ΔΙΟ ΟΛΚ Μ (40), could be later), L3042 (F163, the letters are ΜΕΤΩΝΟΣ ΑΙ, but what is 11, if it is a numeral?), P7670 (F170, the graffito is ΣΥΡΟΣ Α, for which the letter could be anything), P26070 (Ha7, with]||||||| Ζ Ε ΚΚ, the Z having a vertical line), P24760 (Ha10, possibly from Mende, with letters ΠΘ on the neck and ΧΧΗΚ[on the opposite shoulder), P11021 (Hd1, the dipinto is ΟΣΧΟΣ Α, where A may indicate the quality of the vinegar and not be a numeral), P20373 (K9, the letters are Δ Κ Υ or Δ Κ ⊢, but the dating is IV–III BC). Lang's list might be taken as an upper bound of Attic graffiti, since, as Johnston (1978, 218 and 1979, 27) points out, it is possible that a single letter or pair might be a trademark or just a letter, and, indeed, Lang treats some isolated letters as abbreviations, e.g., P24698 (F87), P12510 (F98). So, the total of certain cases in the second list is merely 1, i.e., P26070. In any case, all of these, given the total is between 3 and 11, i.e., probably 7 or 8, of which 6 are tallies and some of which could also represent metic or Ionic trading practice as well, it would be odd to claim anything more than that a few people used Ionic numerals in trade and personal accounts for marking measures. Threatte (1980, 23) suggests that it is also possible that some of five known abcdaria with digamma are actually lists of numerals, for which cf. Lang (1976, 7), A1, A2, A3 (all before the mid-fifth century BC) and Young (1940, 6–8) nr. 9, and especially Pervanoglu (1867, 75) for a lead plaque with qoppa and sampi from the Acropolis, which Pervanoglu places after Olymp. 94, though probably, as Threatte suggests, merely on the basis of its having Ionic letters. The sampi rightly suggested to Pervanoglu that these are numerals. However, it would be odd if all of these abcdaria were lists of numerals. Lang (1976, 22) E15 (ΟΛΚ Μ, 40 drachmas in weight), a grafitto on a sherd from a vase of the late fourth century BC, but found in a third century BC context, is likely to be outside our period. The Ionic abacus from Laurion that Schärlig 2001b identifies is dated by West 1992 as being from late fourth to first century BC merely on the basis of its affinity to Greek papyri from Egypt, starting with P. Eleph 1 (311/10), in which case it has no relevance to this discussion.

48 Cf. the tally of 9 drachmas (and possibly an obol) in Lang 1976, P19389 (K8): |||||||||[Γ⊢⊢⊢⊢].

49 Andronikos 1977, 72 image 31; see Andronikos 1977, 39–40, 70, for a brief description.

graphiti would be local,[50] they point to a Macedonian revival of the Ionic system. Obviously, if the tomb is that of Philip II, we have a terminus ante quem of 338 BC,[51] but even if Philip III Arrhidaeus, the murdered brother of Alexander the Great, occupies Tomb II, as may be more likely, we still get a terminus ante quem of 316/5,[52] earlier than the first non-trivial re-appearance of Ionic numerals anywhere else. So, in either case, the only evidence for ordinary fourth-century use of Ionic numerals outside Halicarnassus and Egypt lies in the royal tombs of Macedon. The fact that one of the vessels uses acrophonic numerals is an indication that the use was common but not universal, or rather that we may see here a trace of an impending shift from acrophonic to Ionic numerals.

Ionic numerals next appear ca. 311/10 BC on a marriage contract, P. Eleph. 1, ca. 311/10 BC, as ⊢ Ą (drachmas 1000). It is significant that this contract also contains the first occurrence of the mark of 1000's, just as we do not have a method for writing myriads until later. We have no reason to believe that prior to this time there was any way of writing Ionic numerals with values greater than 999. Afterwards, with the exception of stichometric counts, that is the line counts used to check that a text has been correctly copied and which are written with acrophonic numerals, Ionic numerals come to dominate in Egyptian papyri and contemporary literary mathematical texts. In early Ptolemaic inscriptions, they become the standard form for representing the day and regnal year, with acrophonic numerals never appearing. Towards the end of the Hellenistic age, they become the standard form for representing numbers in all media everywhere.

Because Ionic numerals are common in papyri, one might suppose that, prior to the Macedonian revival, Ionic numerals were standard in papyri and acrophonic numerals in inscriptions.[53] Since papyri are very rare outside Egypt before 300 BC, with numerals non-existent, this hypothesis has not a shred of foundation. However, there are, in fact, three papyri (two unpublished) that have acrophonic numerals. Partly on the basis of their being acrophonic numerals, Turner identified them as

---

50 Gill 2008, 341–344.

51 Andronikos 1977 initially put Philip II (d. 338 BC) as the kingly inhabitant.

52 Philip III, the half-brother of Alexander the Great, was murdered in December 317 BC and could only get a proper burial with the defeat of Alexander's mother, Olympias, by Cassander. For the forensic evidence, cf. Bartsiokas 2000, for a reply ruling out Philip III, cf. Musgrave et al. 2010. Gill (2008, 341–344) argues for a terminus post quem after 338 BC on the basis of the weights, which range 4.05 (a hydria in Tomb III (probably that of Alexander IV, 310/9 BC) to 4.26 grams to the drachma, in conformity with the Athenian standard. Prior to a reform of Alexander III, the expected drachma weight should be about 3.4 grams to the drachma. On the assumption that Alexander would not have changed the standard until after the funeral of his father, Tomb II could not be that of Philip II. Of course, this argument makes several assumptions, that no one ever entered the tomb after the funeral to add to its pelf and that no silver smith would ever use an Athenian measure until after the reform. I leave such questions to my betters.

53 Fowler 1999, 222–223.

late fourth century, a bad criterion if the periods when acrophonic and Ionic numerals were being used is in question. Nonetheless, Turner used other criteria for dating them, and precision here is not necessary, so long as they are from the fourth or very early third century, in that they testify to a use of acrophonic numerals in Egypt in ordinary contexts, a spending list, a scrap of notes.[54]

Acrophonic numerals also appear, as already indicated, in the line counts of literary papyri to ensure all lines are included.[55] Even if they are an archaic survival, much as we use Roman numerals,[56] this would still indicate that at one time they were a standard that survives into Hellenistic texts. Again, the only thing that the Hellenistic practice could indicate is an earlier use.

There is not a shred of evidence that, in the fifth and fourth century BC, Ionic numerals were numerals of literary and formal texts, and that acrophonic were restricted to inscriptions. Of course, there is also barely an ounce of evidence against it. However, it does not seem plausible that there should be separate notation for inscriptions and for documents and literary texts. It is much more plausible that Ionic numerals were revived in Macedon and in Egypt under the Macedonian administration. For example, in Athens, they do not appear at all in inscriptions until the Augustan age,[57] while in Egypt the numerals are Ionic. So, the use of Ionic numerals in inscriptions parallels their use in literary texts. If one system of numerals, with minor exceptions, were standard for all media in Egypt, the safer, if uncertain, bet would be that Greeks elsewhere did not normally distinguish numeral systems by media.

In this regard, it is fair to ask what numeral system would be common in Greek literary texts. As with many inscriptions, it is plausible that they were just written out. The papyrus of Aristotle's *Athenian Constitution* does use Ionic numerals, but it is easy to see that at least some are not original to the text, at least in the way that they appear. For example, at 43.2, the hybrid ϛ̄ καὶ λ̄ ἡμέρας (6 and 30 days) and πέντε καὶ λ̄ ἡμέρας (five and 30 days), for the terms of tribes on the council in Athens,

54 The published papyrus is from Saqqara, labeled Sak GP.9 No. 5676, and was found during in an expedition of 1971/2. Cf. Turner 1975. Turner also implies that other papyri found in the expedition had acrophonic numerals. However, the bulk of the collection has never been published and may never be published since the negatives are lost and some of the papyri stolen (communications of John Tait and Harry Smith). Prof. Smith and the Egypt Exploration Society kindly supplied me with photographs of S.H5-GP.1 n. 1909 and S.H5-GP.2 n. 1910, found in 1966/7. GP.1 on the recto has writing in a language other than Greek. On the verso, it has ƐƐIII, probably 4+4+3 = 11 drachmas (Cyrene). GP.2 has, inter alia, acrophonic representations of 13 drachmas and 2 drachmas, and probably 3 drachmas (ΚΕΓ: I ⊢ I) or 2 drachmas 1 obol (ΚΕΓ: ⊢⊢ I). It should be noted that there is a great difference between the informality of these texts and the formality of the marriage contract, P. Eleph. 1, where each of three occurrences of the numeral, ⊢ Ạ (1000 drachmas) on two copies (one was sealed for prevent forgery), is separated off and written differently from the other alphas.

55 Graux 1878; Burger 1892; Wendel 1949, chapter 3; or McNamee 1981, xv, 122.

56 Fowler 1999, 223 n. 5.

57 Threatte 1980, 117.

would certainly have originally been written out ἓξ καὶ τριάκοντα ἡμέρας and πέντε καὶ τριάκοντα ἡμέρας. Similarly, the mixture of written out numbers and Ionic numerals in the *Problemata* XV 3.910b23-6 is likely to have been the result of some scribal choice, not necessarily original to the text.[58] Manuscripts readily give written out ordinals or Ionic numerals for the headers of books, sometimes even for alphabetically titled books of Aristotle.[59] We have no reason to think that this is anything more than a scribal choice as well.

Finally, the one numerical practice that most probably goes back well into the fourth century is the counting of line numbers to check whether a line was dropped in copying. Since these numerals remain acrophonic, even in the age of codices, if Ionic numerals were the numerals of papyri, we might expect something different.

Even if the employment of Ionic numerals were common in the fifth and fourth century, we have even less reason to suppose that the Egyptian system of fractions was employed in the Greek world much before 300 BC. It is uncontroversial that Ionic fractions derive from Egyptian fractions,[60] while it is almost certain that the Ionic system of fractions derives from the Egyptian Demotic system.[61] But when did they derive, with the invention of the system or much later?

Ionic fractions emerge in extant texts like Athena, in full raiment from the head of Zeus, namely in the table of additions and subtractions of 45′ hour per day for the length of daytime and nighttime in the parapegma, P, Hibeh I 27. There must have been a gestation period, but we have no reason to think that it was a long one. The author begins by saying that he has spent five years in the region of Sais, where he was intimate with a wise man who explained about sun dials and the lengths of day and night (19–33). He goes on to say (34–40), "thus, so that I would collect them as precisely and as few as possible so that the complexity of the parts (i.e., Ionic

58 Διὰ τί πάντες ἄνθρωποι, καὶ βάρβαροι καὶ Ἕλληνες, εἰς̅ τὰ δέκα καταριθμοῦσι, καὶ οὐκ εἰς̅ ἄλλον ἀριθμόν, οἷον β̅, γ̅, δ̅, ε̅, εἶτα πάλιν ἐπαναδιπλοῦσιν, ἓν πέντε, δύο πέντε, ὥσπερ ἕνδεκα, δώδεκα; (Why do all people, barbarians and Greeks, count up to ten, and not to another number, as 2, 3, 4, 5, then again reduplicating them one five, two five, just as eleven, twelve). It also possible, however, that the *Problemata* reflects late fourth/early third century BC practice.

59 For example, the eleventh-century Coislin 330 (242v) has *Topics* Z as ϛ (6), where the late ninth century Vat. Urb.gr. 35 (352r) has Z. The thirteenth century Vat. gr. 1027 (*Naturalia*) identifies the names of books in four different ways: as written out ordinals, 'second (δεύτερον)' (137v); as Ionic ordinal, '8th (H^{ov})' (97r); as written out letter, 'Zed (ζῆτα)' (75v), and once ambiguously as a letter 'ε' (f. 67r, with a second εε written to the right of it).

60 As Fowler observes (1999, 258–262), P. Cair. 89127–30, 37–43 from the third century BC uses a symbol for 5/6, which he describes as a ligatured 2′ 6′. The question is how one conceives a ligature.

61 Cf. Chrisomalis 2003. Chrisomalis also derives the Ionic system from Demotic; however, even if he is right, there is still no necessity that both systems, the whole number system and the fractional system, derived from Demotic at the same time.

fractions) not seem[62] to you something lengthy and strange to understand, we will apportion (only) the required days."[63] While the emphasis is on the solar theory, he does suggest that the fractions themselves will prove difficult, so that putting them all in will be difficult. We may question whether omitting some days, which need to be part of the calculation anyway, does in fact make things easier; however, it also suggests that the reader might not be an adept at Ionic fractions. That would certainly be the case if they are relatively new. Secondly, if they are new, they are probably the product of the world of that wise man in Sais, the former capital of Egypt.

I admit that this argument from plausibility is not strong, yet it fits with the fact that there are no occurrences of Ionic fractions in earlier extant ancient texts,[64] nor are there any clear occurrences of even written out non-iterative fractions.[65] We can see this when we look at earlier putative occurrences of fractions associated with Ionic numerals employed in *SGDI* 5727, mentioned above, namely ‑, =, ≡, which should be 1, 2, and 3 fourths,[66] as in the amphora handle P26070 (Ha7 in Lang 1976), if Lang is right that Z E KK expresses 7 1/2 choes, 2 kotyls (E for εμιχοον), that is, 7 choes 8 kotyls.[67] In effect, both these texts treat fractions no differently from their treatment in purely acrophonic texts; indeed, the numerals, as far as they represent fractions, are acrophonic.[68]

There are two pretty stories that one could concoct. One is that Eudoxus, during his trip to Egypt,[69] both began to use Ionic numerals as closer to Demotic numerals and as familiar from his home town, Cnidus, and himself also adapted the Demotic fraction system to Ionic numerals. These appear in mathematical papyri because of the importance of the school of Eudoxus. The other is that in the time of Alexander (or his father Philip), Macedonian administrators began, for whatever reason, to use Ionic

62 Reading δοξῆι (as Grenfell & Hunt suggest in P. Hibeh). The impossible first person δόξω would seem to arise because the previous result clause is first person (ἠδυνάμην) and the subject occurs 6 words after the verb. Note that this is an error a copyist might make, but not an author.
63 That is the days the reader needs to know because they are important in the calendar.
64 Communication of Prof. Alan Johnston.
65 One might misread a preliminary report by Threatte (1996, 448) as indicating a non-iterative fraction on a then unpublished inscription on grain taxation from 374/3 BC, but the text is, in fact, a description of two separate taxes, a twelfth tax and a fiftieth tax. Cf. Stroud 1998, 4, 6, 9, 26–39, lines 6–8 of the inscription.
66 Cf. Keil 1894.
67 For a difficulty with this interpretation, cf. Johnston 1978, 218.
68 Johnston (1979, 30–31) identifies dots as fractions in two vases by the Swing painter (Attic, late sixth century) of his style 8D, London (BM) B186 and Paris (Louvre) F50 (8D68 in Johnston's list). See the transcription for Figure 7, 8D 68. This latter is a X followed by 6 vertical strokes and 3 dots arranged vertically, much as in Keil's Halicarnassus example, *Halikarnassos* 31, although it is unlikely that there is any relation between the two. Perhaps, to make a wild guess, this too is 3/4.
69 Cf. Diogenes Laertius, *Vitae* VIII 8.87 and Strabo, *Geog*. XVIII 1.30.

numerals and in Egypt began to use also the fraction system as a way of coordinating Egyptian and Greek accounts, or at least some Greek scientists saw the virtues of the local systems of fractions. The two stories are not completely incompatible, but the only evidence we have for either is that by 311 BC the system of Ionic numerals was expanded to handle thousands and soon to handle 1,0000's, and that by 300 BC the adaptation of Egyptian fractions was common enough in a quasi-technical group in Sais and Hibeh to appear in P. Hibeh I 27.

If Eudoxus introduced Egyptian fractions, it is conceivable that they influenced his own thinking on proportion theory. However, they clearly could have had no influence on the development of number theory in Italy and Sicily, especially among the Pythagoreans, since there is barely a trace of Ionic numerals in their world. Nor is it likely to have influenced Theodorus of Cyrene or Theaetetus of Athens, each of whom would have worked with acrophonic numerals.

The numerical system that any Greek mathematician before Eudoxus working on ratios would have involved acrophonic and not Ionic numerals. Of course, a mathematician could represent fractions meta-diagrammatically, as an interpretation of a divided line or a collection of pebbles with some marked off. It is not important then to this argument whether any Greek mathematician in the fifth or fourth century BC represented numbers with an acrophonic system or worked with written out numbers. It is enough that the only way mathematicians could speak of fractions would be by writing them out. Our next question will be how such written out fractions would have been used in calculation and reasoning. This will involve the related notion of subunits.

IV Subunits, Large Numbers, and Fractions

Ordinary Greeks dealt with fractions and probably large numbers just the way we often do in measuring: they switched units. If I have to pay the royal debt of 24,0000 drachmas, I do not even try to think of the amount in this way, especially given that I do not have a symbol for 24 myriads. Instead, I will think of it as the more malleable 2400 mina, still a large number, or a mere 40 talents. Alas, even Midas might not need to record 24,0000 talents, but there are ways of dealing with this too, e.g., by constructing an arbitrary unit of 10000 talents. In other words, one can also create superunits on the fly.[70] In the other direction, I may need to divide 2 drachmas equally

70 Cf. Lang 1957 for a suggestion that Herodotus calculates with large numbers by treating myriads (10000) as units. Although it employs Ionic numerals and might well be later than 300 BC, the Euclid, *Sectio Canonis* 9, could very well be the surface of a calculation that originally was done with acrophonic numerals. In any case, it exhibits the feature of creating a larger unit. To get six eighth-again intervals, the author starts with 8^6 or 26 myriads 2144, and ends with 9^6 or 53 myriads 1441. Myriads (μύριοι) function as units in the text.

among 5 people. Well, I multiply 2 drachmas by 6 obols per drachma to get 12 obols, which, in division, gives me 2 obols per payee with 2 remainder. But I multiply these by 8 coppers per obol to get 16 coppers, so that I can disperse 2 obols 3 coppers. The remaining copper is not worth much, so I will just give it to anyone. Subunits allow for dealing with numbers large and small.[71]

One shifts quickly from one sort of unit to another as one sets up different parts of the calculation, a task easily handled on abaci whose columns could shift values on the basis of need, now a talent, now a drachma, now ten drachmas, etc. So the fact that we find 'half quarter obols', for example, is not more surprising than the other sorts of divisions, e.g., the sixther (ἕκτευς) or a unit that is 1/6 medminos (dry volume). The difference will be: just as one does not write out more than one five-sign (Γ), one does not normally write out more than one half-sign (often Ϲ, a symbol also of the Ionic system) or one quarter sign (often T), while one can write up to five sixthers, or other multiples of halves, i.e., three halves or five halves.[72] Most fractions in extant texts are really subunits.

The lack of a rich system of fractions along with the ubiquitous use of subunits has a deep effect on Greek thought that we can immediately observe. With fractions one can readily represent an amount of any small size no matter how small. Yet, we see Aristotle frequently looking to establish a smallest unit, something least with respect to sensation.[73] He regards such a smallest unit as the most precise, as indeed there would never be a need to break it up into a smaller unit. Even though he works in a world of non-iterative fractions, Archimedes' choice of a grain of sand as a natural least unit with which to measure the world reflects the same idea.

---

71 In Mendell 2009, I argue that this is the basis of Plato's conception of number in *Rep.* VII 525DE.

72 I suspect that the primary reason we do not find ϹϹ is a matter of standard form and not conceptual. Lang (1956, 8), discusses an amphora with IIHH, which she interprets as 2 halves half and argues that the owner was measuring out choes and marking them down on the way. One does find unreduced fractions in late antiquity. However, the Hippocratic corpus can speak of "cooked flower of copper half-portion, of myrrh *two half-portions*, of saffron three portions, a little honey cooked with wine (ἄνθος χαλκοῦ ὀπτὸν ἡμιμοίριον, σμύρνης δύο ἡμιμοίρια, κρόκου τρεῖς μοῖραι, μέλι ὀλίγον, σὺν οἴνῳ ὀπτώμενα)" at *De ulceribus* 12.27-30 for swellings and inflamations around an ulcer. The use of 'two half-portions' is appropriate because one is doubling a half-portion. Cf. notes 32 and 81 for other examples of unreduced fractions. Netz (2002, 11) regards the use of half, third, drachmas, etc. as evidence for unit-fractions. However, it is not uncommon to find three-halves and five-halves in late fifth-third century BC inscriptions. Cf. Threatte 1996, 447; as well as inscriptions Didyma 84, 85, 86, 87 (all 217/6 BC), Delos ID 500 (297 BC), and IMT Kyz Kapu Dag 1486 (Mysia, fourth century BC). Since I am distinguishing iterative and non-iterative parts, one can say that Greeks used parts as their basic fractional units.

73 This is most notable in Met. I 1 and N 1, but we see it elsewhere, e.g. *De sensu* 6.445a33-6a6, and *De caelo* B 4.287a23-30, *Physics* Δ 14.223b18-31.

Units and subunits form the basis of commercial and industrial treatments of fractions in the Greek world. One will always start with a part. Yet, because the acrophonic system cannot handle more arcane fractions, such fractions would have been written out. If Egyptian fractions had been normal in the fifth or fourth centuries, we would have expected to find complicated non-iterative fractions written, 2/5 as "a third part and a fifteenth." We do not find them. We do find, however, the equivalent fraction treated as iterative parts: "two portions of the five (τῶν πέντε τὰς δύο μοίρας)"[74] and "the two of five parts (πέντε μερῶν τὰ δύο)".[75] In each case, the land is divided into five parts, and then two parts are taken. Sometimes one does this implicitly, as the Stranger at Plato's *Laws* VI 758D, taking a twelfth part of the year (τὸ δωδέκατον ... μέρος), and then the remaining eleven parts of the year (τὰ ἕνδεκα ἀναπαυόμενον τοῦ ἐνιαυτοῦ μέρη). Although rare, it is possible to find other cases of iterated parts.[76]

This way of conceiving fractions always deals with a determinate amount. It says nothing about how we are to conceive of the fraction of an indeterminate amount. An Attic inscription, *IG* II^2 1675 lns. 17–20, from 337/6, providing specifications for the dowels and dowel receptors of column drums for a portico in Eleusis, shows how this conception of iterative parts can be conceived abstractly and connected to a general conception of subunits:[77]

---

74 Thucydides, *Hist.* I 10.2.5, on the portion of the Peloponnesus that Sparta rules. The argument here is in response to Fowler 1999, 227ff. His argument against iterative fractions is basically an argument from silence for before 300 BC and all the evidence for after 300 BC. But if 300 BC represents a shift, this should not be surprising. Vitrac (1992, 1) mentions this first example to make a similar point.
75 Aristotle, *Politics* II 9.1270a24, on the portion of Spartan land owned by Spartan women.
76 Two more examples occur in the grain law of 374/3 BC (Stroud 1998) 54: lns. 21–23 have "the buyer (of a tax) weigh out wheat drawing (a balance) five sixth of a talent of wheat (τοὺς πυροὺς ἀ[πο]-/στήσει ὁ πριάμενος ἕλκοντας πέντε ἐ̣[κ]-/τέ<α>ς τὸ τάλαντον)," and 58, 60 lns. 57–60 have "two-tenths (ἐκ τοῖν δυοῖν δεκάτ[]-/ιν and τὼ δύο δεκάτ[])." Here, the language is without a word for 'parts'. Of course, we should distinguish such examples from cases like Aristoxenus, *Harmonica* 32.10-11, "a fourth is composed of three twelfh-parts τὸ δὲ τεταρτημόριον ἐκ τριῶν δωδεκατημορίων σύγκειται", where the three twelfth parts may merely be the operation 12′ + 12′ + 12′.
77 The inscription uses χαλκός to mean both bronze and copper (as clearly in the second occurrence here). The first occurrence here specifies that the χαλκός should be from Marion (a city in Cyprus), whose copper was valued for its purity. Given that the contractor would be expected to smelt the bronze and that the auditor would have little means of determining whether imported bronze was 1/11 tin or 1/14 tin (say that less than 2% change in the mixture is not observable), it would be odd to say that the bronze should be from Marion. However, it is possible that "Marion bronze" was a name for bronze made with copper from Marion. So the first occurrence is probably, but not necessarily, 'copper'. For these and other details on the chemical and economic issues, cf. Varoufakis 1975. 'The twelfth', an internal accusative, is feminine without a modified noun. It is difficult to know what noun should be supplied, whether μοῖραν (cf. Thucydides above) or μερίς (e.g. Stroud 1998, 4–5 lns. 8, 28, 30, 32; *IG* II^2 334), perhaps the word understood in the feminine of a tax amount, ἡ δωδεκάτη (the twelfth). Clearly, the author assumes we know what is to be supplied, which suggests that the phrase is formulaic.

(Χ)αλκοῦ δὲ ἐργάσεται Μαριέως κεκραμένου τὴν δωδεκάτην, τὰ ἕνδεκα μέρη χαλκοῦ, τὸ δὲ δωδέκατον καττιτέρου.

He will work on copper of Marion mixed with regard to the twelfth [portion], the eleven parts of copper, the twelfth of tin.

A twelfth is taken as an indeterminate portion of a unit mixture, with the mixture made up of 11 portions of copper and 1 portion of tin.

Empedocles (*DK* 33 B96), likewise, uses the language of parts to describe a mixture with an unreduced fraction, where bone is constituted by eight parts, two for Nestis (air and water?), four for Hephaestus (fire?) and the remaining two of earth.[78] Here, the parts are not reduced, i.e., one, two, and one parts of four, perhaps because the two for Nestis are one part each for air and water. For our purposes, it is enough that the fraction is given in the same language of parts and, interestingly, in unreduced parts.

Whether or not these examples illustrate the common language of mixture, it should be obvious that it is much easier to work with than a system of Ionic fractions, which might ask the user to take two thirds and a fourth of the total, instead of just

78 The text in *DK* is: ἡ δὲ χθὼν ἐπίηρος ἐν εὐστέρνοις χοάνοισι / τῶ δύο τῶν ὀκτὼ μερέων λάχε Νήστιδος αἴγλης, / τέσσαρα δ' Ἡφαίστοιο· τὰ δ' ὀστέα λευκὰ γένοντο / Ἁρμονίης κόλλῃσιν ἀρηρότα θεσπεσίηθεν (Grateful earth in her broad-chested melting-pots gets possession of the two of the eight parts from the gleam of Nestis and four from Hephaestus. They became white bones, divinely fit together by the gluings of Harmony.). In *De anima* I 5 410^{a}4-6, Aristotle quotes the first three lines, while Simplicius, *In Physica* 300.21-24 adds the fourth line. We can presume that the other seven commentators on Aristotle (except perhaps Alexander) who cite the second line are using *De anima*, namely Themistius (*Paraph. De an.* 33.13), Philoponus (*In De an.* 176.26-27 (lemma), 178.1), Sophonias (*Paraph. De an.* 32.16), Alexander (*In Met*. 135.16), Asclepius (*In Met*. 112.2), Syrianus (*In Met.* N 188.17), and ps.-Alexander (*In Met.* 828.8). There are two issues to focus on, the gender of the article before 'two', where τὼ, a very plausible emendation of Steinhart from τῶν, and whether the text has the neuter 'parts (μερέων)' or the feminine 'portions (μοιράων)' (there are three other manuscript variations that are likely errors, μοιρῶν ἔλαχε (two mss. of Alexander), μοιρέων (most mss. of ps.-Alexander), and μῆρα (thighs) (Vat. 1339 (P) for Aristotle), though μοιρέων (an emendation of μοιράων?) at least scans). Although it probably echoes *Iliad* X 253 (τῶν δύο μοιράων), μοιράων does not scan when τῶν ὀκτὼ are inserted (one long syllable too many). Similarly, of the three variations of the article before δύο that appear in manuscripts, τὰ (Laur. 87.20 (F^b), Simplicius (one ms. F and the Aldine), Themistius (one ms. Z)) won't scan, so that only τὰς (Vat. 1026 (W) and P, Simplicius (two mss. DE), Alexander, Syrianus, ps.-Alex.) and τῶν (the other mss. of Aristotle used by Ross and Jannone in their respective editions of *De anima* and the remaining mss. of the other commentators) are possible. The variations especially in Aristotelian commentators between feminines (τὰς, μοῖρα) and neuters (τὰ, μέρος), even when not possible metrically, show that there is no basic conception among the earliest copies of what the supplied words are to be for 'the two of the eight parts', even though only μερέων scans and we expect 'two (δύο)' to be neuter (μέρη), given that the parts in the next line are also neuter (whence Steinhart's emendation). For the meaning of the text, cf., for example, Wright 1981, 208–210. For the manuscripts in the commentators, see the relevant preface in the edition.

counting out eleven of twelfths. This holds true for any sort of mixture of ingredients, whether in metallurgy, pharmacology,[79] or cooking,[80] and even taxation.[81] We can also see the system of iterative parts as closely akin to the system of subunits. In the example, the twelfth is, in effect, a subunit of the whole. The difficulty is to find the appropriate part that constitutes the division one wants to make.

The system of iterative parts also ties in neatly into the language of ratio.[82] It is easy to forget, in the hunt for the earliest occurrence of the word 'ratio' (λόγος) (Heraclitus, fr. 31), that the language of ratios goes back to the beginning of Greek literature (Homer, *Odyssey* IX 209–10).[83] In fact, were it not for specifying the base portion as a twelfth, we could not tell whether *IG* II^2 1675 were specifying the ingredients of the alloy in the language of ratios or the language of multiple parts. In this regard, there is little difference conceptually between the two prior to the discovery of incommensurability and hence in the language of numbers. Yet, it is easy to see why Greeks would keep the two notions separate. If a fraction is essentially a portion of a whole, the fraction would only be functionally the same as a ratio if the second term of the ratio is the whole. So when the metallurgist prepares the alloy, the ratio of 11:1 would be conceived as relating parts of single unit, which is the sum of the terms of the ratio.

We can see this connection in yet one more occurance of two-fifths, Herodotus (VIII 129), perhaps describing a small tsunami that engulfed a Persian army attacking Potidaea through a shallow lagoon during a 'flowing back' of the sea:

---

79 Many recipes in the Hippocratic corpus do not specify amounts at all, and some others specific measures; however, there are recipes that are given purely in terms of portions. Cf. *De Ulceribus* 12.27-8, 28–33, 33–37, 15.4-6, 9–10, 21.18-21, as one would expect where dosage is irrelevant. I have not made a careful survey of Egyptian medical texts, but a perusal of the Papyrus Ebers, ca. 1500 BC, yields the following: either measures are equal and listed as 1 for each or the measures are in parts, where the parts do not need to add up to one, the only parts not n'ths being 2/3 (2 times) and 5/6 (3 times); otherwise, all measures are 3′, ∠, 4′, 8′, 16′, 32′, or 64′ (recall that ∠ teaspoon = 96′ cup, and that a pinch is less than 8′ teaspoon). Note that it would also be lunatic to take 64′ as the common part and then to measure out 16 units. Contrast this Egyptian measure with the Hippocratic *De morbis* ii 55.21-2: a drachma-ish (handful) of oregano, about what you can hold with three fingers (ὀριγάνου δραχμίδα ὅσην τρισὶ δακτύλοισι περιλαβεῖν).

80 It is tempting to think that Aristophanes, *Equites* 1187 (with the explanation of Suda, *Lexicon* κ 1262), "you have to drink it [wine] mixed three and two ("Εχε καὶ κεκραμένον τρία καὶ δύο)," in other words three parts water and two wine, reflects the same language of parts. Of course, we can read it as a ratio, 3 : 2, but that is not the idiom, or rather we may ask how the language of ratios and of parts are distinguished.

81 Cf. a tax collection of two tenths, Agora I 7557, lns. 58–9 (Athens, 374/3 BC), also Stroud 1998, 80–81, cf. 27–28, for the difficulties in interpreting 'two tithes'.

82 That a late writer, such as Iamblichus (cf. *In Nic. Arith. Intro.* 64.1-14), might call a 'triple ratio' the triple of a number may be dismissed as coming from a period when the distinction between multiple and ratio was already eroding. So, my arguments will not depend on these sorts of texts.

83 ἓν δέπας ἐμπλήσας ὕδατος ἀνὰ εἴκοσι μέτρα χεῦ' (Maron would pour filling one goblet [of his private reserve wine] for twenty measures of water).

Ὡς δὲ τὰς δύο μὲν μοίρας διοδοιπορήκεσαν, ἔτι δὲ τρεῖς ὑπόλοιποι ἦσαν τὰς διελθόντας χρῆν εἶναι ἔσω ἐν τῇ Παλλήνῃ, ἐπῆλθε πλημμυρὶς τῆς θαλάσσης μεγάλη, ὅση οὐδαμά κω, ὡς οἱ ἐπιχώριοι λέγουσι, πολλάκις γινομένη.

As they marched through two portions, while there further remained three needed for those crossing to be there on the Pallene (the peninsula of Potidaea), a great rise of the sea came up on them, such as had never before, although it is frequent, as the inhabitants say.

One could read this merely as providing a ratio of 2:3 for the amount crossed to the amount not crossed, but it is not expressed in that way. The fraction is provided by giving portions of the amount and the portions of the remainder, which we can then add up to five if we want.

Given this evidence, the common idiom of taking two parts (δύο μέρη) with an implicit or explicit third part (τρίτον μέρος) should not be seen as the exceptional 2/3 in a non-iterative Ionic-Egyptian treatment of fractions. It is merely the common way of speaking of a common fraction, two of the three parts.[84]

In measuring and calculating, Greeks worked primarily with a system of units and subunits. The system allowed them to overcome the occasional problem that their standard representation of numbers had a relatively low upper bound, 9999. More commonly, it allowed them to deal with fractional portions of quanties less than their unit, drachmas, choes, etc. For more abstract considerations, they could use the same conception. Treating the n'th part as a unit, they count up the parts. Our examples are very simple, and it may well be there were never employments of fractional portions much more complicated. This is marked by the lack of a flexible and general notation and the resulting need to write portions out. We have also seen that portions can be readily treated as portions, n, m, of a whole divided into n+m parts. And so it is but a small step to tie this together with the notion of a ratio. Mathematically, this is exactly what we see in the earliest, extant Greek mathematical treatises involving numbers.

V Ratios and Fractions

The treatment of ratios in Euclid's *Elements* VII–IX, the books on number, supplemented perhaps by some theorems from *Elements* V, the general account of ratios,

84 For example, Thucydides, I 104.2.4-5, "taking control also of two parts of Memphis they campaigned against the third, which is called White Wall (κρατοῦντες καὶ τῆς Μέμφιδος τῶν δύο μερῶν πρὸς τὸ τρίτον μέρος ὃ καλεῖται Λευκὸν τεῖχος ἐπολέμουν)," but also Thucydides, ii 10.2.2, 47.2.2 (implicit), iii 15.1.5 (implicit); Isocrates, *In Euthynum* 16.3-4; ps.-Demosthenes, *Contra Boeotum 2* 2.5, 48.5. But one should not always so read the expression, e.g. Thucydides, ii 62.2.3 (there are only two parts in consideration: the land and the sea), vi 62.1.2 (the Athenians make two parts of their army). Fowler (1999, 228) treats all instances as written out Egyptian style.

provides a basis for handling fractions, as well as completing a formal treatment of subunits. Even if we all grant that the material well antedates Euclid's composition, it is very difficult to argue for this thesis because of the paucity of fifth and fourth-century BC evidence, in particular, the types of problems that would illustrate it. So, my primary evidence will be Euclid. However, keep in mind that the story is merely a conjecture. On my view, *Elements* VII consists of three major projects. The first is establishing part and parts, then ratios, and finally fractions. Of course, within this is the wealth of theorems about relative primes, but according to the suggestion, these are primarily aimed at establishing the series of problems at the end, finding the least ratio, the least common multiple, and least number containing given parts (anachronistically, the least denominator). Such problems along with finding the greatest common measure are basic for solving numerical puzzles of many sorts. In this sense, fractions are not avoided in the *Elements*, but they are, to use Plato's expression, multiplied to avoid splitting units.[85] On the assumption that *Elements* VII reflects early work on number theory, say, from the late fifth and fourth centuries, the techniques presented should reflect calculations with Acrophonic numerals, with their limitations in treating fractions, namely the interplay of sub-units, super-units, and ratios. After presenting this interpretation, I shall look at six illustrative examples. I acknowledge that one is from late antiquity, two are Hellenistic, and only three are from the fourth century BC (one, perhaps, the least interesting).

Elements VII opens with four propositions on finding the greatest common measure of two numbers. *Elem.* VII.1 is a theorem that if the anthyphairesis (alternate subtraction) of two numbers ends at a unit, the numbers are relatively prime. Then, for non-relatively prime numbers, it uses anthyphairesis to the number to find the greatest common measure (GCMeas) for two (*Elem.* VII.2) or three (*Elem.* VII.3) numbers. The fourth proposition, the corollary to *Elem.* VII.2, is that if a number measures the GCMeas then it measures the numbers for which it is the GCMeas. Hence, Euclid sets up two cases, where the numbers are relatively prime and only measured by a unit, and where they are not, where the notion of 'a measuring b' is a primitive notion.[86]

This fact ameliorates the great peculiarity of *Elem.* VII.4, that every smaller number is a part or parts of any larger number, where, from *Elem.* VII.df.3, 5, a is a part of b iff a measures b iff b is a multiple of a, and, from *Elem.* VII.df.4, a is parts iff a does not measure b, that is, iff a is not a part of b. By these definitions, it should follow that every larger number is parts of any smaller number, but that is not the oddity. The oddity is that Euclid breaks the proof up into two cases, one where the numbers are relatively prime and one where they are not. Let $a<b$. If a, b are relatively prime the units of a will be parts of the units of b. But, by the definition, any unit of a will measure b. I mention this because it shows that Euclid has no problem in speaking of the unit as a part and as measuring a number. The second case also has two cases,

85 Plato, *Rep.* VII 525DE. Cf. Mendell 2009.

86 So Mueller 1981, 58, 59.

where *a* measures *b*, and where it does not. For the second case, he finds the GCMeas *c* of *a* and *b*, and shows that *a* breaks up into numbers equal to *c*, such that each measures *b*. If the point is merely that there is a division of *a* into a_1, ..., a_n such that each a_i measures *b*, then the proof is prolix. The first case would have done it. Even if we say that no part is also parts, it still will be the case that either *a* measures *b* or the units of *a* measure *b*; so, this is clearly not Euclid's goal in the theorem. Unstated, but clearly implicit in the proof, is the notion that where *a* is parts of *b* (or even a part of *b*), those parts are the GCMeas of *a* and *b*. Of course, one can argue that this point is internal to the arguments of VII.6, 8, 10, where one needs to use the GCMeas, so that where *a* is the same parts of *b* as *c* of *d*, one takes the parts of *b* and *d* to get the parts of *a* and *b* to be equinumerous. If one takes the unit for non-relatively prime numbers, this will obviously not be the case. I shall return to this issue, but given that one needs to be able to infer all this without the help of a definition, one may well ask why Euclid takes it as obvious that one chooses the GCMeas as the requisite part. One reasonably suspects that it is part of how one works with numbers, and the reasonable place to go is in setting up subunits for working with all sorts of mixtures and measures that we have glimpsed at through the dearth of evidence.

The definition of 'proportional' in *Elements* VII.df.21 goes: Numbers are in proportion when the first is equal-times a multiple of the second as the third is of the fourth or the same part or the same parts'. The definition is slightly inadequate without much explication. Although one can build a definition of 'equal-times a multiple' from definition 17 of 'a number multiplies another', yet 'equal-times a multiple' plays no role until multiplication gets introduced at *Elem.* VII.16, after the fundamental theorems on ratio (*Elem.* VII.11-14), while there is no definition of 'same part' nor 'same parts' in the *Elements*. The meaning only becomes clear through the first theorems employing the notions, *Elem.* VII.5-6,[87] where, in any case, they turn out also to be undefined, primitive notions. In particular, *a* is the same parts of *b* as *c* of *d*, if there is a part *m* of *a* and *b* and a part *n* of *c* and *d* (and there always will be since the unit is a part of every

87 My analysis differs from Mueller 1981, 62–63, in that I think that there are four distinct and primitive notions, which are not merely notationally different: *a* measures *b*, *a* is the same part of *b* as *c* of *d*; *a* measures *b* as many times as *c* does *d*; and there is a number *m* such that *a* measures *b* according to the units in *m* (or *m*-times). Of course, it turns out that *a* measures *b* iff ∃*m* *a* measures *b* *m*-times, and *a* cannot measure *b* as many times as *c* does *d*, or be the same part, unless there is some *m* such that *a* and *c* measure *b* and *d* *m*-times. However, Euclid is very careful in introducing measuring in VII.1, same part or parts' in VII.5-14, 'measures as many times' in VII.15, and the last notion with the commutivity of multiplication in VII.16. So, a full account of the logical structure of *Elem.* VII would have to trace where these notions appear and, in particular, as I shall point out, the difference between equinumerosity and there being a number of two equal collections. They are obviously equivalent, but that is irrelevant. On this view, it is very clear why Euclid uses at each stage the minimal possible concept for his argument, as the four notions constitute a progression. It goes without saying that this interpretation is yet more distant from views such as Zeuthen (1910, 410–411), which treats the numerical values, *m*, *n*, such that $a = m/n\, b$ as the conception underlying *Elem.* VII.4.

number), such that *m* is the same part of *a* as as *n* is of *c* and *m* is the same part of *b* as *n* is of *d*. As we saw above, the part in each case needs to be the GCMeas.

Euclid also introduces the notion of 'measuring equal-times' at *Elem.* VII.15 (alternando for a unit measuring a number equal-times as another number measures a third number), a theorem notoriously a special case of *Elem.* VII.9 (alternando for two numbers that are the same part of two other numbers) and foundational to the commutivity theorem on multiplication (*Elem.* VII.16) and the theorems on fractions (*Elem.* VII.37-38). But we see from the proofs and dependencies that this too is a primitive notion formally equivalent to 'same part', i.e., *a* is the same part of *b* as *c* of *d* iff *a* measures *b* equal-times as *c* does *d*. For *Elem.* VII.15 infers

> *a* measures *b* equal times as *c* does *d*
> Hence, as many times as *a* measures *b*, so many times does *c* measure *d*
> Hence, $a{:}b{=}c{:}d$

Now, according to the definiton, this must mean either that *b* is equal times a multiple of *a* as *d* of *c* or *a* is the same part of *b* as *c* of *d*. It hardly matters which.[88] Nonetheless, Euclid infers from '*a* measures *b* equal times as *c* does *d*' to '*a* is the same part of *b* as *c* is of *d*' in *Elem.* VII.37 and conversely in *Elem.* VII.38. So, these are equivalent but not thereby identical notions (otherwise, the inferences would be superfluous).

It is important to distinguish arithmoi (numbers as multitudes of units) and plêthê (numbers as multitudes of anything), as I shall do for the rest of my discussion of *Elem.* VII. So, in proving the basic theorems, *Elem.* VII.4-15, two arithmoi, for example, will be divided into arithmoi that are equal in plêthos, where the arithmoi of one plêthos of arithmoi correspond (dare we say 1-1) to the arithmoi in the other plêthos of arithmoi. When Euclid starts to talk about multiplication at *Elem.* VII.16, he will introduce the notion that there is a correspondance between arithmoi in the plêthos of arithmoi and the units in some arithmos. Only then can he speak of an arithmos measuring an arithmos according to some arithmos (or more strictly, according to the units in some arithmos). In other words, 3 arithmoi are not thereby instances of the arithmos 3.

Euclid builds the account of ratio in a minimalist way, where the core of the theory is in the account of same-part and same parts. It is hardly surprising then that the proofs of the three basic theorems on ratios, VII.11-13, are trivial because Euclid prepares them with theorems on part and parts: VII.5-6 for VII.12, the sum of first terms to the sum of last terms: VII.7-8 for VII.11, separando, VII.9-10 for *Elem.* VII.13 alternando, with *Elem.* VII.14, ex aequali, based on *Elem.* VII.13. In each case, the first theorem is a determinate case, where one arithmos is a part of another, which is

---

88 If we follow the definition of 'multiple', then '*a* measures *b* equal-times' should mean '*b* is equal-times a multiple of *a*'. However, this must be equivalent to '*a* is the same part of *b*', or *Elem.* VII.11-13 will be invalid.

then used to prove the second, indeterminate theorem where one arithmos is parts of another. The proportion theorem in each case is then just an application of the two cases to the definition (ignoring 'multiples'). These conceptions are central to what goes on in *Elem.* VII.

The proof of *Elem.* VII.15, that if a unit *u* measures *b* as many times as *c* measures *d*, employs the definition of proportion and the proportion theorem *Elem.* VII.12 (the sum of first terms to the sum of last terms). This is unnecessary. The theorem follows directly from *Elem.* VII.9, that if *a* is the same part of *b* as *c* is of *d*, then *a* is the same part or parts of *c* as *b* is of *d*. Let *a* be a unit. Since the unit is a part of every arithmos (cf. the proof of *Elem.* VII.4), *a* is a part of *c*. Therefore, *b* is a part of *d*. Hence, it follows that *a* measures *c* equal-times as *b* measures *d*. I do not think that we can know why Euclid proves the theorem as he does.[89] It is enough to show how the edifice is built on the notion of parts, say, in contrast to the treatment of proportions in *Elements* V as based in equi-multiplication.

Yet a part is also a fractional part, whence parts are iterative-fractional parts. To see this, we turn to the end of the book, VII.37-39, where Euclid introduces 'homonymous' arithmoi, that is, *n* and *n*'th. If *n* is a part of *m*, *n*'th is a fraction of the arithmos, for us $1/n$ of *m*.[90] Given that, Euclid is certainly writing in a period and location where Ionic/Egyptian fractions are common, so that these propositions about homonymous parts might be tailored for that system. In fact, all three are general to either system of fractions.[91] More important to our purposes is the nature of the two fundamental theorems about fractions (*Elem.* VII.36-37): '*n* measures *m* if (38) and only if (37) there is an *n*'th part of *m*'. The argument for these propositions uses the alteration of parts where one term is a unit (VII.15). Hence, for *Elem.* VII.37, there is an *n*'th, and for

89 See Mueller (1981, 74) for speculation about why Euclid proves the theorem in this way. It would be very odd, as Mueller suggests uncomfortably, for Euclid to want to avoid alternando with one term as a unit, since that is the theorem anyway, while there is nothing special about a unit as part in any of theorems *Elem.* VII.4-10. However, this is not to say that Mueller's speculation is wrong. Nonetheless, the theorem is set up as a lemma for *Elem.* VII.16 and is positioned there for that reason, after four basic theorems on ratios (*Elem.* VII.11-14). So, it is plausible that Euclid merely wants to illustrate that you can use the theorems on proportions now that they are available.

90 I do not think Mueller (1981, 74, 80) correct in thinking of *Elem.* VII.37-8 as 'verbal lemmas' for *Elem.* VII.39, nor of *Elem.* VII.39 as 'a terminological variant of' *Elem.* VII.34-36. It is correct that if *a* is a part of *b*, then *a* must measure *b* so-number of times, i.e. according to the units in some *c*. However, this is never proved. They introduce a different concept, the fraction, and so address, as I shall argue, a different set of arithmetic puzzles.

91 Given that *Elem.* VII.37-9 play no further role in the *Elements* and that this use of 'homonym' does not appear anywhere before Euclid (the word first appears in Homer, *Iliad* 17.720, for 'someone with the same name', here the two Ajaxes), one might well suspect that Euclid added these to the earlier material for his treatise or even that they are later additions to the text. If so, then we can expect even more so that the language of part and parts, with the language of ratios based on part and parts, constitutes the notion of fraction presented in the *Elements*.

Elem. VII.38, some q will be a part of m, whence there is an n such that q is a part of m according to the units in n. The theorems concern the relations between parts and fractions. Nor are the theorems a claim that if n measures m there is an arithmos p according to which n measures m according to the units in p, i.e. m p-times. This does follow for sure and is involved in the proof, but it is not the present concern.

The problem that follows, *Elem.* VII.39, 'To find a number which is least of those having the given parts', is tailored for a general class of arithmetical puzzles, where one wants a least n'th for however many n_1'th, ..., n_m'th. It is true that this theorem is almost a trivial variant on *Elem.* VII.36, finding the least common measured arithmos, in that its solution is the homonym of the solution for *Elem.* VII.39. However, it transforms the greatest common measured theorems into a theorem about fractional parts. As the scholion says (*Scholia ad Eucl. El.*, 237–8), "for example if it happens that ∠ 3′ 4′ 5′ are given, to find the smallest arithmos of all that have the same parts as these". Homonymous parts are not exactly parts as defined at the beginning of *Elem.* VII. This theorem can then be used for finding the minimal whole numerical solution to a broad class of puzzles involving distributions of fractional parts (I give several examples in the next section). In particular, it gives a theoretical foundation to solutions of problems employing false position.[92]

One way of looking at *Elem.* VII–IX.20 is as a treatise on primes and relative primes. No doubt, *Elements* VII–IX conceptually serves many functions besides providing theorems for *Elements* X. *Elem.* IX.21-34, on odd and even, is nigh logically contained,[93] while IX.20, on ever larger primes, and IX.35-6, on perfect numbers, serve, respectively, as capstones to the treatise on primes and relative primes and the entire treatise.[94] In fact, one good way of thinking of the treatise on primes and relative primes is to provide a theoretical and practical foundation for solving numerical puzzles. The basic problems and their allied theorems involve finding the greatest common measure (*Elem.* VII.1-3), arithmoi in least ratio (*Elem.* VII.33), least common measured arithmos (*Elem.* VII.34-36), least arithmos having given parts (*Elem.* VII.39),

92 Although Vitrac (1994, esp. 283) does not regard the *Elem.* VII–IX as a treatise of logistic, he sees the relevance of VII.37-39 for it. I have avoided in this essay discussing what logistic is the fifth-fourth century BC, but cf. Mendell 2009, as well as Vitrac 1992.

93 Mueller (1981, 105–106) points out that only IX.32 depends on a proposition outside, namely IX.13, but cites Becker (presumably, 1936, 535–6) for the point that it could have been derived from IX.31 or from IX.30.

94 Mueller (1981, 106) is inclined, with qualms, towards Becker's view (1936) that IX.21-36 constitute a self-contained treatise on Odd/Even, where originally IX.35-36 depended only on propositions from within IX.21-34. It is a good question why Euclid would have rewritten IX.35-36 to be independent of IX.21-34. For my purposes, it is only important to focus on VII-IX.20, and in fact to IX.19, since IX.20, the unlimited prime numbers theorem is also a capstone theorem. I mean by a 'capstone theorem' one that comes at the end of a treatise and rewards the reader for their effort. If the arrangement has the perfect number theorem at the end, IX.20 serves as a capstone to the treatise on primes and relative primes.

least numbers in continuous but not equal ratios (*Elem.* VIII.4), and whether there is a third or fourth continuous proportional (*Elem.* IX.17-19). Besides the limited logical role they play in VII–IX.20,[95] these problems, several being culminations of the their discussions, feed into techniques for solving arithmetic puzzles.[96]

It is all too easy to see *Elements* VII–IX.20 as a purely theoretical discussion about number, whose main applications lie within the theories of irrationals and harmonics, to see the opening gambit on part and parts as merely feeding into the accounts of ratio, continuous ratio, and continuous ratios involving squares and cubes, with fractions playing little or no role at all. However, if part and parts are the Classical way of thinking of fractions, then a different notion of fraction lies at the foundation of the theory of ratio. The simplest class of true fractions, i.e., *n*'ths appear briefly at the end of *Elem.* VII, and the fundamental theorems, *Elem.* VII.37-38, depend only incidentally on the theory of ratios, namely in the fact that Euclid's proof of *Elem.* VII.15 employs *Elem.* VII.12 instead of *Elem.* VII.9. On the other hand, the basic problem, *Elem.* VII.39 for working with fractions, is of a piece with the other problems of the treatise and is based in reduced ratios.

While the concept of ratio in *Elem.* VII–IX.20 is that of part and parts, and these are a conception of ratio built on a notion of fraction, it is perhaps more surprising that Euclid goes on to build another, more standard notion of fraction. In this sense, the general treatment of number, in so far as it represents an old Greek treatment of ratio, serves as evidence that ratios were developed to handle fractions, so that when he introduces fractions, n'ths, Euclid is showing how to handle fractional parts through ratios.

VI Evidence from Computation (Meagre Stuff)

It is not difficult to show that in late antiquity, some people interpreted the *Elements* as providing practical propositions for handling arithmetic puzzles. Despite the paucity of evidence, it is reasonable to suppose that such puzzles are also a part of Greek mathematical play in the Classical age. I shall briefly point to two examples. It is very difficult to see how fractions are worked out without the employment of non-iterative fractions. For all our evidence of playing with fractions comes no earlier than the Hellenistic age. Nonetheless, it is possible that something like this appears

95 Cf. Mueller 1981, chapter 2.

96 Again, I am not here denying other roles for the techniques involved. For example, there are many ways looking at *Elem.* VII.1-3, which involve anthyphairesis, particularly in their variant in determining when magnitudes are commensurable (*Elem.* X.3-4) or in their role in twentieth-century interpretations of fourth-century BC proportion theory based on this and an illustration of Aristotle, *Topics* Θ 3.158b24-159b2. The literature is enormous, but cf. inter alia Knorr 1975 and Fowler 1999.

in Euclid's *Sectio Canonis*. Finally, the tight connection between subunits and ratios becomes most clear in the harmonic tradition. *Sectio Canonis* 9 and Archytas' division of the tetrachord, as reported by Ptolemy, *Harmonica* I 13, set up subunits instead of ratios. This alternative way of thinking about scales shows that subunits and ratios are two ways of thinking about the same thing from the late fifth century BC on.

The *Anthologia Palatina*, which survives in one manuscript (tenth century, Paris suppl. Gr. 384),[97] contains an anthology of puzzles and oracles in verse (*Anthologia Graeca* xiv), including arithmetical puzzles, 1–13, 48–51, 116–147, many followed by solutions in scholia, though they are not separate and are incorprated into the text of the manuscript, so that we may presume that they were integral parts of the edition used by Cephalas, the anthologist, where it is an ongoing error of all modern editions not to present them as they appear in the manuscript.[98] The manuscript ascribes the group of poems from 1 on (somewhere up to 62) to Socrates[99] and 116–146 or 147 to Metrodorus, whoever they are.[100] The only indication of their dates is that they do not

97 The manuscript was divided at the end of the Napoleonic Wars, with the initial part in Heidelberg, Codex Palatinus 23: http://digi.ub.uni-heidelberg.de/diglit/cpgraec23. The second part, which begins with Anth. Gr. XIV, is Paris Supplément grec 384 and is available on-line (see bibliography).

98 Most recently, Buffière 2002, but cf. page 37. Although Buffière cites in the apparatus non-mathematical aspects of the scholia, he ignores them in the body of his text. Writing 'of the same [poet] (τοῦ αὐτοῦ)', when this does not occur in the manuscript is the most common example.

99 It is normal in the *Anthologia Graeca* to name authors to each poem, including where they are anonymous (ἄδηλον or ἀδεσπότου), and where there is repetition of a known author to say, 'Of the same (τοῦ αὐτοῦ)' or 'another (ἄλλο)' but 'another' where the next author is also unknown. *Anthologia Graeca* xiv seems to violate this, where there are long stretches of unattributed poems or where the scholia state what the poem is about but omit the author. This leaves us with many questions. In any case, since Achilles Tatius quotes XIV 34 (cf. Patton 1916–1918, V 42 n. 1), either Socrates is interspersed (as is probable, see Tannery 1894) or gives out earlier, in which case we do not know who wrote XIV 48–51, or his dates are before the third century, even if he should turn out not to be the epigrammatist referred to by Diogenes Laertius, *Vit.* II 47.9-10.

100 It is not clear when their collection of poems end. Only four poets are mentioned (Mesomedes, starting at 63, and Cleombrotus, starting at 101, are the others), although we don't expect oracles to have a named poet. But the last arithmetic poem is 147, with the last three poems, 148–150 being oracles to named recipients. Since the collection of Metrodorus is 'Metrodorus' Arithmetical Epigrams (Μητροδώρου/ἐπιγράμματα ἀριθμητικά)' (fol. 9r), it would be odd if it also included oracles. So, Metrodorus probably wrote 116–146 or 147. 147 may be too trivial and the scholion introducing it (fol. 14v) resembles those to 148–150, in its giving background for the poem. Most, but not all poems in the group after the first are introduced as 'another (ἄλλο)' and a number in the poem (and not 148). But some numbers are missing, which suggests missing poems. If, as Tannery 1894 conjectures (see also Buffière 2002, 35–36), the missing poems appear elsewhere in the book, i.e. those ascribed to Socrates, the book of Metrodorus contained poems of two authors, with those of Socrates separated out. However, if the only poems of Socrates are the arithmetical ones, then Book XIV is lackadaisical about attributing authorship (see previous note). An aid might be the common use of ἄλλο for another poem or ἄλλος for another oracle of the same poet, but the manuscript is hardly finicky about this as well. As for who Metrodorus is, it is reasonable that he lived somewhere between the third and the

indulge in Christian imagery, while Metrodorus provides an encomium of Diophantus (xiv 23) and refers to *Arithmetica* I.2. The sole edition of the poems with scholia is Tannery, *Diophantus* II, 43–72.[101] He plausibly ascribes the scholia to Metrodorus.[102]

The puzzles and solutions of the *Anthologia Palatina* are distinctive of later Greek, popular mathematics. They typically employ false position and then adjust the answer. *A.G.* xiv 116, the first of the group ascribed to Metrodorus, divides an unknown whole of nuts, here *S*, into 8 portions, 2/7 *S* (ἕβδομα δοιά in the poem and written ζ′ ζ′, i.e., 7′7′ in the scholion to it—iterative fractions are acceptable here), 1/12 *S*, 1/6 *S*, 1/3 *S*, 20 nuts, 12 nuts, 11 nuts, and 1 nut.[103] The scholiast takes the least number that divides into the given parts (*Elem.* VII.39, an application of the numbers coordinate with the parts, of 7, 12, 6, 3), a false sum of 84 nuts. Since this divides up into 2/7·84 = 24, 1/12·84=7, 1/6·84=14, 1/3·84=28, the sum of these parts will be 73, so that the remainder is 84–73=11. Now the actual remainder needs to be 20+12+11+1 = 44 nuts. If we multiply the total and each part by 4, we will get the corrected amount, 96 nuts + 28 nuts + 56 nuts + 112 nuts + (20+12+11+1) nuts = 336 nuts. We can anachronistically describe the problem as $(\mathrm{x}\, n_1/m_1 + \ldots + \mathrm{x}\, n_p/m_p) + c = \mathrm{x}$. By finding the least number which has these parts, in effect, the LCM of $m_1, \ldots, m_p = m = \mathrm{x}_{\text{false}}$, we find the sum of $(n_1/m_1 + \ldots + n_p/m_p) \cdot m = n$, a whole number, so that $c_{\text{false}} = m - n$. Hence, $\mathrm{x} = m \cdot (c/(m-n))$, which can, if one wishes, be trivially verified algebraically. $(n_1/m_1 + \ldots + n_p/m_p) \cdot m \cdot c/(m-n) + c = m \cdot c/(m-n)$.

The scholia refer to five propositions, four from Euclid and one from Diophantus. The Diophantus proposition, *Arith.* I.2, (ad xiv 127, 129, 139), 'It is required to divide the displayed arithmos into two arithmoi in a ratio that's given.' The four from Euclid are: *Elements* VII.39 (ad xiv 2); V.15 (ad ix 3), $a{:}b = na{:}nb$; *Elements* VII.19 (ad xiv 130), $a{:}b = c{:}d \leftrightarrow a \cdot c = b \cdot d$; *Elements* VII.11 (ad xiv 138), $a{:}b = c{:}d \rightarrow a{:}b = a\text{-}c{:}b\text{-}d$. The scholion to xiv 116 (53.17-19), discussed above, also refers to problems in the *Data* (ad xiv 116, discussed above) which Euclid calls 'for a given arithmos or in ratio (δοθέντι ἀριθμῷ ἢ ἐν λόγῳ)'. However, 'arithmos' makes no appearance in the *Data*, while 'in ratio' is a general classification. The scholiast probably has nothing specific in mind. Nonetheless, it is clear that the scholiast looks at Euclid VII, especially, as a practical book that

mid-sixth century. Metrodorus, in 121, mentions a trip from Cadiz to Rome. In 123, he requests 30 talents for his tomb and a sacrifice to Infernal Zeus (Οὐδαίῳ Ζανί), difficult in more Christian or Islamic times. Tannery is likely right that Metrodorus is at least the poet, and very probably author of all the explanatory scholia, but even that is open to question.

101 Tannery's edition of the scholia is inadequate (Anth. Pal. 1885). He only prints mathematical scholia and poems so that one does not get a sense of the context, does not accurately place marginal scholia (e.g., 50.26-51.6 is in a margin a page earlier in the text), expands mathematical ligatures (e.g., 53.16 λείψας), and does not always distinguish between what is in the margin and what is part of the main text, although several of the marginal comments are also integral, e.g., as when the main text leaves a space for the marginal comment.

102 Cf. Tannery 1894 and *An. Pal.* 1895, II 44 n. 1.

103 For this scholion, cf. *An. Pal.* 1895, II 53-4.

provides theorems for working out puzzles. *Elem*. VII.39, in particular, is not just cited at the beginning of the collection, it pervades it.[104]

Of course, the poems with their scholia are from late antiquity, after Diophantus.[105] Yet they are witnesses to the importance of *Elem*. VII (and V) as providing techniques for the solutions of mathematical puzzles. It may be uncontroversial that a tradition of such problems were not merely post-Diophantine number play but are part of a long Greek tradition.

Yet, we can readily push this into Hellenistic age with 'The Cattle Problem' of Archimedes, an entertainment of counting up cattle of eight sorts (four colors: white, black, dappled, and yellow; each for bulls and cows) belonging to the Sun. If we divide the problem into two parts, one difficult to solve, the numbers of the eight different groups given certain rules, and one nigh impossible without computers, that the sum of two sorts of bulls should be square and that the sum of the other two sorts of bulls should be triangular, the solvable part also sticks to these methods.[106] As we would expect in the third century BC, the fractions are non-iterative. Using the initial letters of the colors, upper case for females and lower for males, the 9 equalities of the easier part are:

Bulls	Cows
w = ∠ 3′ b + y (lns. 9–11)	W = 3′ 4′ (B + b) (lns. 18–19)
b = 4′ 5′ d + y (lns. 12–3)	B = 4′ 5′ (D + d) (lns. 20–22)
d = 6′ 7′ w + y (lns. 14–6)	D = 5′ 6′ (Y + y) (lns. 23–24)
	Y = ∠of 3′ 7′ (W + w) = 6′ 7′ (W+w) (lns. 25–6)[107]

The first step to solving the problem is to homogenize the non-iterative fractions. After we have done this, the first difficulty is to find the least solution to the three equalities involving only bulls. Simplest is to use false position to take the number of yellow bulls as 1. One uses *Elem*. VII.39 to find the least arithmos that there can be of the bulls, in this case 891.

w = 5/6 b + y	w = 2 148/297	2226
b = 9/20 d + y	b = 1 79/99	1602
d = 13/42 w + y	d = 1 689/891	1580
	y = 1	891

104 *Elem*. VII.39 is also used in *Anth. Graeca* XIV 1, except that the false position turns out to be the same as the final value, i.e. x/2+x/4+x/7+3=x. Here, the false position will be 28, so that 14+7+4=25, while 28–25=3, so that x=28, and there is no need to adjust the value.

105 There are simple Greek, pre-Diophantine problems, that constitute forms of logistic in *P.Mich*. 144 (three algebraic problems), and *P.Mich*. 145 (a distribution problem among conversion problems), but these do not employ any of the problems of *Elem*. VII–IX, so far as I can tell. Also, cf. Vitrac 1994, 260 n. 73.

106 Cf. Dijksterhuis 1987, 398–401.

107 The odd way of writing 6′ as 'half of third' is probably just metrical.

One can also verify that these four numbers are relatively prime.

Now we turn to the cows. Use the values you have for the bulls thus far, although you will get this nasty result of the second column; however, it is now trivial to apply *Elem*. VII.39 to avoid chopped beef, so that you now get the number of whole cows.[108]

W = 7/12 (B + 1602)	W = 1547 1981/4657	W = 720,6360
B = 9/20 (D + 1580)	B = 1050 3396/4657	B = 489,3246
D = 11/30 (Y + 891)	D = 754 4442/4657	D = 351,5820
Y = 13/42 (W + 2226)	Y = 1167 4494/4657	Y = 543,9213

We also multiply the number of bulls by 4657 to get their number as well.

w = 2226 · 4657 = 1036,6482
b = 1602 · 4657 = 746,0514
d = 1580 · 4657 = 735,8060
y = 891 · 4657 = 414,9387

One can readily, but not easily, verify that the groups of the herd are relatively prime, so that this is a least solution. One can even go further to try to find a multiple such that w + b is a square arithmos and d + y is triangular (equal to a sum of arithmoi from 1), although you will not succeed without tools from the future. There are several things lacking in this presentation. I have not explained how to solve the two sets of problems, although conceptually each is a simple matter of substituting equal terms on the right until one gets a single answer. It is easy to see how errors could arise in trying to clean up the result. Nonetheless, the technique I have used is so close to the sorts of techniques we find in late antiquity and so evident in the context of a written out presentation of the problem, i.e., without the sorts of symbols that we use, that it seems that this is how Archimedes would have expected his Alexandrian audience to proceed before being stumped.

In this regard, it is notable that the one known ancient attempt at the easier part of the problem, the scholion printed in Heiberg's text, comes up with values 80 times the actual minimal solution.[109] The Cattle Problem is so similar to problems from

108 If you think that you need to ensure that the added amounts, 7/12 · 1602, 9/20 · 1580, 11/30 · 891, 13/42 · 2226, need to be multiplied to be whole numbers, you will wind up with an amount 10 times the minimum.

109 Or, 8 times if we take all the parts separately. For example, 1/3 + 1/4 the black bulls and cows gives a whole number, but taken separately requires half a bull and half a cow. However, the poem seems to require taking a herd of black cattle together for this part of the calculation and is using normal non-iterative fractions, and the scholion's description of the problem does not differ from the poem on this point. It is unclear where the error in the scholion comes from, that is, if it is an error, given that the scholiast does not tell us that it is a minimal solution.

late antiquity and so extraordinary in its playfulness, that it is hard to believe that it is not an extravagence based on a well established tradition of numerical puzzles. This may be partly confirmed by the reliqua we have of a treatise of Apollonius on an arithemetic puzzle on the supreme power of Artemis, partly preserved as Pappus, *Collectio* ii.

The first Classical example I have is playful but not so challenging, the nuptial number of Plato (*Republic* VIII 546BC), which I alluded to earlier, is of this ilk yet involves no hint of fraction avoided.[110] It involves a calculation of years and mentions a square of an irrational and so will not be separate from perceptibles and hence cannot be an example from the expert's manual of calculation problems. Nonetheless, as worded, albeit obscurely, the conditions could be seen as a puzzle, to find the first, i.e. least, number for certain conditions set down and a second number built on it. Now, it is very likely that the resulting number is 1296 myriads (1296,0000 or in a Babylonian system, 1, as already noted) and that it is the only number that satisfies the conditions. Socrates only expresses these through calculation puzzles: the larger is a square of a multiple of 100 (36), the product of 100 times the square of the square root of 49 minus 1 or the square of 100 times the square of the square root of 50 minus 2 (4800), by 100 times the cube of three (2700), perhaps $3 \cdot 4 \cdot 5$ multiplied by itself three times, perhaps the last term of the continuous ratio 60:3600:21,6000:1296,0000, and perhaps other properties one might cull from this text. It is even possible that finding puzzles is part of the exercise. One of these seems to be squares of numbers whose double differs by 1 from a square number. The sequence is familiar ({1, 1}, {3, 2}, {7, 5}, ..., or (...,{m, n}, {2 n+m, m+n}), where the actual problem is not quite the same as approximating the square root of two.[111] My point is that Plato presents the puzzle in a way that is consistent with posing a calculation problem, even if it is not of the form that requires multiplying to avoid dividing.

There is another place where we can see manipulations avoiding fractions that is early enough for us to suppose that it may reflect a Classical calculation with fractions evaded, namely *Sectio Canonis* 6–9. Although it is not crucial to my argument, I shall assume that Euclid is the author;[112] moreover, its relation to the *Elements* is obscure, since it seems to refer to the *Elements*, albeit to a text different from the one we have.[113] Just as *Elements* VII–IX reflects earlier treatises on

110 Cf. Adam 1902, 263–86, with reservations, in particular with regard to Adam's derivation of a first number 216 as $3^3 + 4^3 + 5^3 = 216$. I also omit various standard points such as the relation of the passage to the general rule for approximating the square root of two: $p^2 \pm 1 = 2q^2 \Rightarrow (p + 2q)^2 \mp 1 = 2(p + q)^2$.

111 Cf. Proclus, *In rem pub.*, ad loc.

112 So, Heiberg 1882, 50; Barker 1989, 190–191. For a survey of the evidence and litererature with very sceptical conclusions, see Barbera 1991, 3–36.

113 In particular, as Heiberg observes (1882, 53 n. 2) (cf. also Menge 1916, 162), *Sectio Canonis* 2 refers to a proposition: if there are however many numbers in proportion, and the first measures the last, then it will measure all those between (καὶ τοὺς μεταξὺ μετρήσει). *Elem.* VIII.7 is weaker: it will measure number, so the *Sectio Canonis* seems to reflect earlier fifth to fourth century BC work on harmonics, and that is all I need. The text has three basic types of propositions: propositions 1–5, general propositions about doubly composed intervals (in effect, continuous proportions), that depend mostly on the definition of 'same ratio', transitivity of measuring,[114] and *Elem.* VIII.7; propositions 6–9, theorems on particular intervals; propositions 10–20, applications to a system of tones. I shall be concerned with the second group, in particular, propositions 6–8, where we can distinguish two techniques, where the first proof of two proofs of 6 uses fractions and where the second proof of 6 and the proofs of 7, 8 avoid fractions.

In each case, the three theorems concern combining intervals:[115]

6. The double interval is constructed out of the half-again interval and the third-again interval.
7. From the double interval and the half-again a triple interval comes about.
8. If the third-again interval is taken from the half-again interval, the remainder is the eighth-again interval.

Euclid does not define interval, but treats the interval AB as a ratio of numbers A:B in propositions 1–5, and avoids the word 'ratio'.[116] So, the interval where A is an eighth-again B will have to be equivalent to a ratio A:B=9:8. Nonetheless, in propositions 6–8, one infers from 'A is an eighth-again B' (interval) to 'A has B and an eighth of B' (quantity) to '8 A's are equal to 9 B's' (quantity) back to 'A is an eighth-again B'. These cannot all be proportions, while the third is explicitly an equation (sic). What we do not see anywhere, though, is the fractional statement that we would expect if fractional discourse were standard: 'A is equal to 1 8′ B'.

---

the second (καὶ τὸν δεύτερον μετρήσει). However, the proposition cited in *Sectio Canonis* trivially follows from the *Elements*' proposition. It is less significant that the word order of the two propositions is different. Would a treatise of late antiquity have hewn closer to the received text of Euclid's *Elements* (in some version)? On the other hand, the expression, 'we learned (ἐμάθομεν)', is very rare in Euclid outside the *Sectio Canonis*, namely *El.* X 10, where it is suspect (see Vitrac 1998, 131). Indeed, it is rare in Greek mathematics before Heron. Does its occurrence in the *Sectio Canonis* suggest that at least this passage is of later vintage?

114 This is a basic assumption of *Elem.* VII–IX, but is not proved. Cf. Mueller 1981, 64 and *Elem.* VII.2 Cor., 33, 34.

115 The differences between the three versions, the manuscripts of the *Sectio Canonis*, the version of Porphyry, *In Ptol. Harm.* I 5, and the version of Boethius, De *Institutione Musica* 4.1-2 (see Barbera 1991) is not crucial to my argument. Boethius completes the proofs with numerical summaries, Porphyry is slightly more concise but inserts an extra theorem after Thm. 5, and both use only the alternative proof of Thm. 6. I shall mostly use the *Sectio Canonis* based on its mss.

116 Porphyry, *In Ptol. Harm.* 92.29-93.4, observes that Euclid uses 'interval' instead of 'ratio'. Cf. also Barker 1989, 197 n. 19, who uses this fact to argue that in Porphyry's version (*In Ptol. Harm.* 100.26-101.8), the theorem that comes after proposition 6 is not original to the text. All doubles, epimorics, etc. are intervals.

Let us then look at the two techniques through proposition 6 in the first version and proposition 8, where I use proposition 8 instead of proposition 6 in the second version because it is less trivial:

8. If a third-again interval is taken away from a half-again interval, the remainder is left over as eighth-again.

> For let A be half-again of B, and G be a third-again of B. I say that A is an eighth-again of G. For since A is half-again of B, therefore A has B and the half of it. Therefore, eight A's are equal to twelve B's. Again since G is a third-again of B, therefore G has B and the third of it. Therefore, nine G's are equal to twelve B's, but twelve B's are equal to eight A's. Therefore, eight A's are equal to nine G's. Therefore, A is also equal to G and the eighth of it. Therefore, A is an eighth-again of G.

A being half-again of B and G being a third again of B are intervals. Euclid translates the interval into magnitudes, where A has B and half B. We then get the equation.

$8A=12B$

Again, from G has B and a third of it, we get:

$9G=12B$

Euclid sets up two equations where one term is the same in each of the three proofs of this sort.[117] The obvious question is how Euclid gets the common value 12B from $A=B+1/2\,B$, and $G=B+1/3B$. We might well suppose that there is an unmentioned intermediate equation: $2A=3B$ and $3G=4B$. From this it is obvious that the common value, the least common measured, is 12. In other words, *Elem.* VII 34.

It is crucial in the argument that however one understands the fraction, it is immediately replaced by the equation. In other words, one starts with a proportion, turns it into a claim that one value contains wholes-and a -fraction, and then converts this into an equation with whole numbers. Having done this twice, one works with the two equations to get a new equation, that is then converted directly into a proportion. This is very much what we would expect where ratios and fractions are conceptually tied together, but where one avoids working with fractions. Indeed, all fractions are written out.[118]

---

117 6 (alternative). 3 B's equal 4 G's and 3 B's equal 2 A's ⇒ A is 2 G. 7. A equals 2 B's and 2B's = 3 G's ⇒ A equals 3 G's.

118 It would have been awkward not to have written them out, as there would have been ambiguities. On the other hand, there is nothing to prevent a scribe from using notation instead of writing out numbers. So the apparatus of Barbera (1991, 134–135, Porphyry's version 202–212) has several instances of numerals. The first proposition to use numerals is proposition 9, where the numbers would be very messy to write out.

As mentioned, there are two versions of Proposition 6. Here is the first version, which is different from the other proofs.

6. The double interval is constructed from two largest epimoric (intervals), from the half-again, and the third-again.

For let BG be half-again of DZ, and DZ third-again of Q. I say that BG is double Q. For I took away an equal to Q, ZK, and to DZ, GL. Accordingly, since BG is half-again DZ, therefore BL is a third part of BG, and half of DZ. Again, since DZ is a third-again of Q, DK is a fourth-part of DZ, and a third part of Q. Accordingly, since DK is a fourth-part of DZ, but BL is a third part of BG, therefore, DK is a sixth part of BG. But DK is a third part of Q. Therefore, BG is double Q.

The argument here is surprisingly complex and unnecessarily so. Here, the basic idea is, given x is an n'th again y, to subtract the whole of y from x and y+n'th y. We start with BG is half-again DZ and DZ is third-again Q, where we need to show that BG is double Q. For ease of reading, I shall be slightly anachronistic. We subtract from both sides to get these two equations. For clarity, a=BG, b=DZ, c=Q, which implies that BL=a-b and DK=b-c.

1. $a = b + 1/2\, b \Rightarrow (a - b) = 1/2\, b = 1/3\, a$
2. $b = c + 1/3\, c \Rightarrow (b - c) = 1/3\, c = 1/4\, b$
3. $b - c = 1/4\, b$ and $(a - b)\ (= 1/2\, b) = 1/3\, a \Rightarrow b - c = 1/6\, a$
4. $b - c = 1/3\, c$ and $b - c = 1/6\, a \Rightarrow a = 2\, c$

The idea in the proof is to get the two quantities $1/6\, a$ and $1/3\, c$ equal to $b - c$ (DK).[119] Steps (3) and (4) use inferences: from $x = 1/4\, y$ and $1/2\, y = 1/3\, z$ to $x = 1/6\, z$; and from $1/3\, x = 1/6\, y$ to $y = 2\, x$. These steps in the proof are trivial if one treats 1/4, 1/3, and 1/2 as ratios, and one can think of the fractions in this very way. But one distinctive feature of the proof is its initial move, the two subtractions so that one is only working with *n*'ths. That the method is somewhat awkward and is not repeated does not imply that this is an earlier clumsy proof technique replaced by something slicker. It could well be an elementary exercise in working with fractions that gets dropped because once is enough. On the contrary, I think it reasonable to suppose that the proofs where fractions are evaded is more representative of an earlier technique (I take both versions of proposition 6 to be authentic).

119 We could imagine a variant, where the middle term is $a - b$: $2(b - c) = 2/3\, c = 1/2\, b = a - b = 1/3a$.

My fifth example, which comes from Euclid's proof of *Sectio Canonis* 9, that six eighth-again intervals are more than a double interval, is very different and is a lemma for Proposition 14, that the octave is less than 6 notes. Behind the proof is taking a subunit, if you like, of $8^6 = 26{,}2144$, which will be the first arithmos in a series of seven. Then one constructs a series of the next six eighth-again arithmoi, in modern notation $8^5\ 9^1, \ldots\ 8^0\ 9^6$, to see that $9^6 = 53{,}1441$ is larger than $2 \cdot 8^6$, than twice the subunit. After the statement of the theorem, Euclid does not use the word 'interval (διάστημα)' again, so that it is implicit that the seven members of the series are terms of the six eighth-again intervals. This is completely consistent with what we would expect if the initial unit, a whole number, is calculated precisely so as to avoid fractions, especially if the language of fractions is inadequate.[120] Of course, there might be other reasons why Euclid constructs the proof in this way, but this is the most obvious one.

In fact, such subunits are foundational to the harmonic tradition, although the evidence is mostly late. Ptolemy (*Harmonics* I 13 30.3-31.18) discusses Archytas account of the division of the tetrachord[121] and presents Archytas as giving the components in terms of base points, which taken together are least numbers. This is the table that Ptolemy presents, but with the means of calculation that he indicates in *Harm*. I 13.[122]

Base values: the high note is 1512; the tetrachord is 3rd-again; the ratio of the second highest note of the chromatic to the second highest note of the diatonic (CD) is 256:243.

---

120 Euclid states, "Since we learned how to find seven numbers successibly eighth-again from one another, ... (ἐπεὶ ἐμάθομεν εὑρεῖν ἑπτὰ ἀριθμοὺς ἐπογδόους ἀλλήλων, ...)." Obviously, *Elements* VIII 2 will do the job, and I take it that Boethius's version (Barbera 1991, 254–255 and note 26) "in arithemtica modum," is just a way of referring to the arithmetical books of the *Elements*. Odder, we expect what follows to be specifically what we have learned (as props. 4 ln. 7, 13 ln. 3, as well as concerns above (note 113) about prop. 2), and not a particular case of it. It is possible, that the problem alluded to was more particular than *El.* VIII 2.

121 Cf. also Barker 1989, 43–45.

122 Without having examined the manuscripts (except for Vat. 192 (V), which incorrectly fills in the empty slots for the enharmonic), I find the modern transmission of the table for I 13 in the edition of Düring (1930), the two translations of Barker (1989, 45 and 303), and the transmission of Solomon (2000, 45), not a little perplexing in their disharmony. None of the three translations quite translates Düring's text, although the first version of Barker comes closest. This might be important to my argument, since I am assuming that Ptolemy deliberately does not provide the non-epimoric intervals.

Enharmonic			Chromatic			Diatonic		
points	ratio if given	calculation	points	ratio if given	calculation	points	ratio if given	calculation
1512		initial	1512	initial		1512		initial
	4th-again	given		none			8th-again	given
1890		ratio and highest	1792		from 1701 (diatonic) and CD	1701		ratio and highest
	35th again	not used		none			7th-again	not used
1944		lowest and ratio 27th-again	1944		lowest and ratio 27th-again	1944		lowest and ratio 27th-again
	27th-again	given and same for all		27th-again	given and same for all		27th-again	given and same for all
2016		highest and tetra-chord	2016		highest and tetra-chord	2016		highest and tetrachord

It is easy to see that the three progressions form a relatively prime set and are least numbers, just as we saw in the *Sectio Canonis*, in other words, so that the extremes of the tetrachord accommodate all three scales with whole numbers. It seems that the whole numbers are original to Archytas.

One can easily object to this suggestion by observing that Ptolemy standardly selects units for the ratios. Ptolemy does this in two ways. In the theoretical discussion of harmonics, *Harmonica* I, Ptolemy uses ratios of whole numbers. However, in the empirical discussion, *Harmonica* II (cf. esp. the tables of 14 and 15), he presents the kinds mostly in the range 60 to 120 (perhaps not incidentally his values for the standard radius and diameter of a circle), and gives the fractional parts in sexagesimal, also his practice in astronomy, but here restricted to one place. So just as Ptolemy provides approximations of Archytas' genera in *Harmonica* II 14, so in I 13 he calculates a set of numbers that satisfy the given ratios.

There is a reply to this objection. Ptolemy actually calculates the second highest note from the number for the second highest diatonic, which itself is calculated from the interval between the highest and second highest (8th-again) and the points for the highest. The two highest intervals for the chromatic are never needed. Then, before going on to other objections, at the beginning of the next chapter (*Harm.* I 14.2), Ptolemy points out that the ratio of the second highest to the highest and of the third to the second fail to be epimoric, though even here he does not actually state the reduced ratios of 32:27 and 243:224. In other words, these reduced ratios are not part of his account. Of course, it is perfectly possible that Archytas said that the second highest chromatic is the interval composed from 8th again and 256:243, which one can determine to be (via arguments such as we saw in Euclid) the interval 32:27, and, for some reason, Ptolemy chose to ignore it. It is just that there is nothing in Ptolemy to suggest this. Indeed, one probably has to compound the ratios to get the minimum value for the high note 1512. Yet, from the way Ptolemy develops his argument, it is reasonable to suppose that Ptolemy's practice in *Harm.* I is like *Sectio Canonis* 9 and is at least as old.

We can find very limited affirmation in ps.-Plato, *Epinomis* 991AB (mid 4th cent. BC), where half-again and third-again are said to be the two means of the six to twelve, that is the arithmetic and the harmonic (these terms are not used). Clearly, six and twelve provide the basis for the simplified scale.[123] If we find this in the trivial case, it is reasonable that the same method was used in the more complicated case.

My last example lies in the mixing of the concepts of ratios and fractional parts in Archytas' definition of the harmonic mean (*DK* 47 B2):[124]

> And the other (mean) is subalternate, which we call harmonic, those where by the part (μέρει) of itself that the first term exceeds the second, by this part (μέρει) of the third the middle exceeds the third. It comes about that in this proportion (ἐν ταύτᾳ τᾷ ἀναλογίᾳ) the interval of the larger terms is larger and that of the smaller is smaller.

We can write this out as: A-B is the same part of A as B-C is of C. This echoes part of the definition of 'same ratio' for numbers, *Elem.* VII.df.21, and gets described in the next sentence as a proportion. It would also be surprising, however, if built into this conception of the subalternate/harmonic mean, every A-B had to measure A, e.g. that 8 should not be the harmonic mean between 20 and 5 because 12 does not measure 20. 'Part' is used more generally than in the *Elements* and hints at the notion of a fraction.

Let us review what the six examples seem to show. The Palatine manuscript shows that at least some mathematicians in late antiquity regard *Elements* V and VII as providing techniques for the solutions of ordinary arithmetical puzzles, particularly in the use of VII.39 for finding the least arithmos having given fractional parts. Archimedes' Cattle Problem clearly mimics a tradition in the Greek world of such ordinary puzzles. Greek mathematicians in the third century BC would have adopted and even refined Egyptian methods of working with n'ths, so that perhaps there would no need to appeal here to VII.39, but it seems to lie in the background of the base solution to the basic Cattle Problem if we take part of the method of a solution as false position. In any case, Plato also provides in the nuptial number, a tradition of play with ratios of numbers. All this does not get us to a way of solving arithmetical puzzles without fractions. I suggest that three proofs of Euclid's *Sectio Canonis* do exactly this, in that

---

123 Cf. Szabó 1978, 118–119, who however, also uses the passage to make a stronger point about the history of the interval, about which I have very little to say. My view is that twelve as the canon for the trivial division falls naturally out of the method of taking units and subunits.

124 We find language of part in the nearly identical descriptions of the harmonic mean in Plato, *Timaeus* 36A and ps.-Plato, *Epinomis* 991A: a mean that "exceeds and is exceeded by the same part (μέρει) of the extemes themselves."

they convert an interval/ratio problem to a fraction problem to a problem without fractions. The primary proof of proposition 6 draws a contrast for us, especially if it represents a newer technique. The proof of proposition 9 just is an example where long multiplications use the techniques of *Elements* VII to avoid fractions, something especially poignant in a world of bad representations of fractions. However, along with Ptolemy's presentation of Archytas, along with the *Epinomis*, it shows the close relationship between creating subunits and using ratios. The last example, also from Archytas, suggests the close relationship between *Elem.* VII.df.21 and a Greek notion of fraction.

VII Mildly Sceptical Conclusions

There are three components to my speculations about why Greeks developed a theory of ratio rather than a practice for dealing with fractions. The Greeks lacked a symbolic system for calculating with fractions, but they could think of fractions as part or parts of a number. At the same time, they had a language of ratios and abaci on which they could calculate with subunits. Thus, in Euclid's books on number, we see an account of part and parts developed into a theory of ratio. On this view, fractions are at the center of the Euclid's discussion, just treated in two different ways, as part or parts and as ratios. This allows for the introduction of fractions as n'ths, homonyms of a number. I have intentionally omitted the grand issues of fifth and fourth century BC mathematics, the development of incommensurability and general theories of proportion, though I touched lightly on harmonic theory. I looked to a small issue, but one which, I suspect, deserves more prominence. It is curious that Greeks developed a rich number theory, while Babylonians and Egyptians, with rich systems of fractions and subunits, do not seem to have done so, well, at least in the sense of a theory of whole numbers. If Greeks could avail themselves of a system of fractions that was primarily a system of subunits with associated systems of iterative parts and ratios of numbers, then the development of number theory is much more comprehensible, as is the focus on ratio, as is the choice in Euclid of building a theory of numerical ratios on an account of part and parts. On this view, the revolution in number theory might not seem remarkable. Yet, the creation of a demonstrative theory of ratios for the purpose of giving a foundation to working arithmetical problems was no mean feat, even if it works with available methods . However, this is merely a suggestion.[125]

125 This paper is an expansion of some material in Mendell 2007. I thank Michalis Sialaros for giving me the opportunity to expand on these thoughts and for his helpful comments. I thank Leslie Threatte, Alan Johnston, John Tait, and Harry Smith for their notes, and Harry Smith and the Egypt Exploration Society for kindly making available photographs of Saqqara papyri. Much of my ideas developed out of discussions with many people, particularly David Fowler, Reviel Netz and Bernard Vitrac, when I presented aspects of this work at various conferences. They are not responsible for my errors.

Bibliography

Ancient texts cited

Inscriptions: IG (Inscriptiones Graecae), Didyma (McCabe), Halikarnassos (McCabe), *SGDI* (*Sammlung der griechischen Dialekt-Inschriften*), *ID* (*Inscriptions de Délos*), *IMT* (*Inschriften Mysia und Troas*): Packard Humanities Institute: http://epigraphy.packhum.org/

Alexander, *In Met.*: Hayduck, M. 1891. *Alexandri Aphrodisiensis in Aristotelis metaphysica commentaria*. Commentaria in Aristotelem Graeca 1. Berlin.

Anthologia Graeca: Patton, W.R. 1916–1918. *Greek Anthology*. 5 vols. Cambridge (USA) and London. (reprint: 1969, 1970, 1968, 1979, 1970).

Anthologiae Palatini pars altera Codicis. Paris Supplément grec 384: http://gallica.bnf.fr/ark:/12148/btv1b8470199g.r

Anthologiae Palatini (Ad epigrammata arithmetica scholia): Tannery, Paul. 1895. *Diophanti Opera Omnia*. Vol. 2. Leipzig.

Anthologiae Palatini: Buffière, F. 2002. *Anthologie Grecque: Antholigie Palatine Livres XIII–XV*. Paris.

Archimedes. *Arenarius*. Heiberg, I.L. & E. Stamatis, E. 1913. *Opera Omnia*. Vol. 2. Stuttgart (rev., 1972).

Archimedes. *Problema Bovinum*. Heiberg, I.L. and E. Stamatis. 1972. *Opera Omnia*. Vol. 2. Stuttgart (rev., 1972).

Archytas: Diels, H. & W. Kranz. 1951. *Die Fragmente der Vorsokratiker*. Vol 2. Zürich (6th. edition, reprint: 1985)

Aristophanes, *Equites*: Coulon, V. and van Daele, M. 1923. Aristophane. Vol. 1. Paris (reprint with corr, 1967).

Aristotelian Corpus, *Problemata*: Bekker, I. 1831. *Aristoteles Graece*. Vol. 2. Berlin.

Aristotle, *Athenian Constitution*: Kenyon, F.G. 1903. *Res Publica Atheniensium*. Commentaria in Aristotelem Graeca. Supplementum Aristotelicum 3.2. Berlin.

Aristotle, *Athenian Constitiution*: Oppermann, H. 1928. Aristotle, *Ἀθηναίων πολιτεία*. Leipzig (reprint, Stuttgart, 1968).

Aristotle, *De anima*: Jannone, A. 1995. *Aristote: De l'âme*. Paris (2nd ed.).

Aristotle, *De anima*: Ross, W.D. 1961. *Aristotle's On the Soul*. Oxford (reprint, 1967).

Aristotle, *De caelo*: Moraux, P. 1965. *Aristote: Du ciel*. Paris.

Aristotle, De sensu: Ross, W.D. 1955. *Aristotle: Parva Naturalia*. Oxford (reprint, 1970).

Aristotle, *Metaphysics*: Ross, W.D. 1924. *Aristotle's Metaphysics*. Oxford (reprint, 1970).

Aristotle, *Physics*: Ross, W.D. 1950. *Aristotle's Physics*. Oxford (corrected, 1956).

Aristotle, *Politica*: Ross, W.D. 1957. *Aristoteles: Politica*. Oxford.

Aristotle, *Topics*: Ross, W.D. 1958. *Aristoteles: Topica et Sophistica Elenchi*. Oxford.

Aristoxenus, *Harmonica*: da Rios, R. 1954. *Aristoxeni elementa harmonica*. Rome.

Asclepius, *In Met.*: Hayduck, M. 1887. *Asclepii in Aristotelis metaphysicorum libros A–Z commentaria*. Commentaria in Aristotelem Graeca. Berlin.

Demosthenes, *Contra Boeotum 2*: Rennie, W. 1921. *Demosthenis orationes*. Vol. 2. Ed. W. Rennie. Oxford (reprint, 1966).

Diogenes Laertius, *Vitae*: Ed. Long, H.S. 1964. *Diogenis Laertii vitae philosophorum*. 2 vols. Oxford (reprint, 1966).

Diophantus, *Arithmetica*: Tannery, P. 1893. *Diophanti Alexandrini opera omnia*. Vol. 1. Leipzig (reprint, 1974).

Empedocles: Diels, H. & W. Kranz. 1951. *Die Fragmente der Vorsokratiker*. Vol 1. Zürich (6th. edition, reprint: 1985).

Euclid, *Elementa* v–ix: Heiberg, I. & E. Stamatis. 1884. *Euclidis Opera Omnia* vol. 3. Leipzig (rev., 1970).

Euclid, *Elementa* x: Heiberg, I. & E. Stamatis. 1886. *Euclidis Opera Omnia* vol. 3. Leipzig (rev., 1972).

Euclid, *Data*: Menge, H. 1896. *Euclidis opera omnia*. Vol. 6. Leipzig.
Euclid, *Sectio Canonis*: Menge, H. 1916. *Phaenomena et Scripta Musica*; *Frag*: Teubner.
Euclid, *Sectio Canonis*: Barbera, André. 1991. *The Euclidean Division of the Canon*. Lincoln.
Herodotus, *Historiae*: Legrand, Ph.–E. 1932–68. *Hérodote: Histoires*. 9 vols. Paris.
Hippocratic corpus, *De morbis* i–iii: Littré, É. 1849–1851. *Oeuvres complètes d'Hippocrate*. Vols. 6–7. Paris.
Hippocratic corpus, *De ulceribus*: Littré, É. 1849. *Oeuvres complètes d'Hippocrate*. Vols. 6. Paris.
Homer, *Iliad*: Allen, T.W. 1931. *Homeri Ilias*. 2 vols. Oxford.
Homer, *Odyssey*: von der Muhll, P. 1962. *Homeri Odyssea*. Basel.
Iamblichus, *In Nic. Arith.*: Pistelli, H.; rev. Klein, U. 1894. *Iamblichus: In Nicomachi Arithmeticam Introductionem Liber*. Ed. Hermenegildus Pistelli. Leipzig (revised, Stuttgart, 1975).
Isocrates, *In Euthynum*: Mathieu, G. and Brémond, É. 1929. *Isocrate: Discours*. Vol. 1. Paris (reprint, 1963).
Joannes Philoponus, *In de Anima*: Hayduck, M. 1897. *Ioannis Philoponi in Aristotelis de anima libros commentaria*. Commentaria in Aristotelem Graeca 15. Berlin.
P. Cair.: Parker, R.A. 1972 *Demotic Mathematical Papyri*. Brown Egyptological Studies vii. Providence, Brown University Press, 1972.
P. Eleph. 1: http://papyri.info/ddbdp/chr.mitt;;283 and http://ww2.smb.museum/berlpap/index.php/03734/
P. Hibeh I 27: Grenfell, B.P. and Hunt, A.S. 1906. *The Hibeh Papyri*. Part 1. London.
P. Mich.: Winter, J.G. 1936. *Papyri in the University of Michigan Collection: Miscellaneous Papyri*. Michigan Papyri 3. Ann Arbor.
Pappus, *Coll.*: Hultsch, F. 1876–1878. *Pappi Alexandrini Collectionis quae supersunt*. Berlin (reprint: Amsterdam: Hakkert, 1965).
Papyrus Ebers: Smith, G. Elliot (intro.) and Bryan, Cyril P. (trans.). 1930. *Ancient Egyptian Medicine: The Papyrus Ebers*. Translation of H. Joachim, *Papyros Ebers: das älteste Buch über Heilkunde* (Berlin, 1890). New York (reprint, Chicago, 1974).
Plato, *Republic*: Burnet, J. 1902. *Platonis Opera*. Vol. 4. Oxford (repr. 1968).
Plato, *Timaeus*: Burnet, J. 1902. *Platonis Opera*. Vol. 4. Oxford (repr. 1968).
Plato, *Laws*: Burnet, J. 1907. *Platonis Opera*. Vol. 5. Oxford (repr. 1967).
Plato (ps.-), *Epinomis*: Burnet, J. 1907. *Platonis Opera*. Vol. 5. Oxford (repr. 1967).
Porphyry, *Comm. In Ptolemaei Harmonica*: I. Düring. *Porphyrios Kommentar zur Harmonielehre des Ptolemaios*. Göteborgs Högskolas Årsskrift 38, 1932: 2. Göteborg: Erlanders Boktryckeri Aktiebolag, 1932 (reprint: New York: Garland, 1980).
Priscian, *De figuris numerorum*: Keil, H. 1860. *Opuscula*, vol. 1. Leipzig.
Proclus, *In rem pub.*: Kroll, W. 1901. *In Platonis rem publicam commentarii*. Vol. 2. Leipzig.
Ptolemy, *Harmonica*: Düring. I. 1930. *Die Harmonielehre des Klaudios Ptolemaios*. Göteborgs Högskolas Årsskrift 36. Göteborg (reprint.: New York, 1980).
Scholia ad Euclidis Elementa. Heiberg, I.L, and Stamatis, E. 1888. *Euclidis Opera Omnia*. Vol 5.2. Leipzig (rev. 1977).
Simplicius, *In Physica*. Diels, H. 1882, 1895. *Simplicii in Aristotelis physicorum libros octo commentaria*. Commentaria in Aristotelem Graeca 9–10. Berlin.
Sophonias, *In de Anima*: Hayduck, M. 1883. *Sophoniae in libros Aristotelis de anima paraphrasis*. Commentaria in Aristotelem Graeca 23.1. Berlin.
Strabo, *Geographica*: Meineke, A. 1877. *Strabonis geographica*. Leipzig (reprint, 1969).
Suda. *Lexicon*. Adler , A. 1928–1971. *Suidae lexicon*. 4 vols. Leipzig.
Syrianus, *In Met.*: Kroll, G. 1902. *Syriani in Aristotelis Metaphysica (Β, Γ, Μ, Ν) commentaria*. Commentaria in Aristotelem Graeca 6.1. Berlin.
Themistius, *In de Anim.*: Heinze, R. 1899. *Themistii in libros Aristotelis de anima paraphrasis*. Commentaria in Aristotelem Graeca 5.3. Berlin.
Thucydides, *Historiae*: Jones, H.S. & J.E. Powell. 1942. *Thucydidis historiae*. 2 vols. Oxford (reprint, 1970, 1967).

Modern works cited

Adam, J. 1902. *The Republic of Plato*. 2 vols. Cambridge: Cambridge University Press.

Andronikos, M. 1977. "Βεργίνα. Οἱ βασιλικοὶ τάφοι τῆς Μεγάλης Τούμπας," 1–39, and "Vergina The royal graves in the Great Tumulus," 40–72 (translation of previous, but the images are distributed through both). In *Ἀρχαιλογικὰ Ἀνάλεκτα ἐξ Ἀθηνῶν* (*Athens Annals of Archaeology*) 10.

Barker, A. 1989. *Greek Musical Writings II: Harmonic and Acoustic Theory*. Cambridge.

Bartsiokas, A. 2000. "The Eye Injury of Philip II and the Skeletal Evidence from the Royal Tomb II at Vergina." In *Science* n.s. 288: 511–514.

Beazley Database (Classical Art Research Database) http://www.cvaonline.org/xdb/ASP/default.asp

Becker, O. 1936. "Die Lehre vom Geraden und Ungeraden im Neunten Buch der Euklidischen Elemente." In *Quellen und Studien zur Geschichte der mathematik, astronomie und physik*, abt. B (3): 533–553.

Blümel, W. 1993. "*SGDI* 5727 (Halikarnassos): eine Revision." In *Kadmos* 32: 1–18.

Burger, F. 1892. *Stichometrische Untersuchungen auf Demosthenes und Herodot*. Inaugural-Dissertation K. Friedrich-Alexanders-Universität, München.

Chrisomalis, S. 2003. "The Egyptian Origin of the Greek Alphabetic Numerals." In *Antiquity* 77: 485–96.

Chrisomalis, S. 2004. "A cognitive typology for numerical notation." In *Cambridge Archaeological Journal* 14: 37–52.

Chrisomalis, S. 2010. *Numerical Notation: A Comparative History*. Cambridge.

Dijksterhuis, E. J. 1987. *Archimedes*. Trans. by C. Dikshoorn. Forword (this edition) by Wilbur Knorr. Princeton.

Fowler, D. H. 1999. *The Mathematics of Plato's Academy*. 2nd ed. Oxford (1st ed., 1987).

Gill, David W. J. 2008. "Inscribed Silver Plate from Tomb II at Vergina: Chronological Implications." In *Hesperia* 77: 335–358.

Graham, J. W. 1969. "X = 10." In *Phoenix* 23: 347–358.

Graux, C. 1878. "Nouvelles rescherches sur la stichométrie." In *Revue de Philologie de Littérature et d'Histoire Anciennes*: 97–143.

Haslam, M. 1997. "Homeric Papyri and the Transmission of the Text." In Morris, I. and Powell, B., *A New Companion to Homer*. Leiden, 55–100.

Heiberg, I. L. 1882. *Litterarische Studien über Euklid*. Leipzig: Teubner.

Ifrah, G.. 2000. *The Universal History of Numbers from Prehistory to the Invention of the Computer*. Trans. from French by D. Bellos, E. F. Harding, S. WWood, and I. Monk. New York.

Johnston, A. 1973. "Two-and-a-half Corinthian Dipinti." In *Annual of the British School at Athens* 68: 181–9

Johnston, A. 1975. "A South Italian Numeral System." In *La Parola del Passato* 30: 360–366

Johnston, A. 1978. Review of Lang [1976]. In *Journal of Hellenic Studies* 98: 218–219.

Johnston, A. 1979. *Trademarks on Greek Vases*. Warminster.

Johnston, A. 2006. *Trademarks on Greek Vases, Addenda*. Warminster.

Keil, B. 1894. "Eine Halikarnassische Inschrift." In *Hermes* 29: 249–280.

Keil, B. 1901. "Eine Zahlentafel von der athenischen Akropolis." In *Strassburger Festschrift zur Versammlung Deutsher Philologen und Schulmänner*. 117–142.

Keyser, P. 1986. "Errors of Calculation in Herodotus." In *Classical Journal* 81: 230–242.

Keyser, P. 1988. "The Origin of Latin Numerals 1 to 1000." In *American Journal of Archeology* 92: 529–546.

Knorr, W. R. 1975. *The Evolution of the Euclidean Elements: a Study of the Theory of Incommensurable Magnitudes and its Significance for Early Greek Geometry. Synthese Historical Library*, vol. 15. Dordrecht.

Knorr, W. R. 1982. "Techniques of Fractions in Ancient Egypt and Greece," *Historia Mathematica* 9: 133–171

Knorr, W. R. 2001. "The Impact of Modern Mathematics on Ancient Mathematics." Edited by H. Mendell and K. Saito. In *Revue d'histoire des mathématiques* 7: 121–35.

Koehler, U. 1896. "Attische Inschriften des Fuenften Jahrhunderts." In *Hermes* 31: 137–154.

Lang, M. and Crosby, M. 1964. *Weights, Measures and Tokens. The Athenian Agora*, vol. 10. Princeton.

Lang, M. 1956. "Numerical Notation on Greek Vases." In *Hesperia* 25: 1–24

Lang, M. 1957. "Herodotus and the Abacus." In *Hesperia* 26, 271–288.

Lang, M. 1968. "Abaci from the Athenian Agora." In *Hesperia* 37, 241–243.

Lang, M. 1976. *Graffiti and Dipinti. The Athenian Agora*, vol. 21. Princeton.

Lawall, M. L. 2000. "Graffiti, Wine Selling, and the Reuse of Amphoras in the Athenian Agora, CA. 430 to 400 B.C." In *Hesperia* 69: 3–90.

McNamee, K. 1981. *Abbreviations in Greek Lieterary Papyri and Ostraca. Bulletin of the American Society of Papyrologists* Suppl. 3.

Mendell, H. 2007. "Two Traces of Two-Step Eudoxan Proportion Theory in Aristotle: A Tale of Definitions in Aristotle, With a Moral." In *Archive for History of Exact Sciences* 61: 3–37

Mendell, H. 2009. "Plato by the Numbers." In Føllesdal, D. and Woods, J. (eds.), *Logos and Language: Essays in Honour of Julius Moravacsik*. London, 141–76.

Mitsos, Th. M. 1947. "Greek Inscriptions." In *Hesperia* 16: 82–88.

Mueller, I. 1981. *Philosophy of Mathematics and Deductive Structure in Euclid's Elements*. Cambridge.

Musgrave, J.; Prag, A. J. N.; Neave, R.; Fox, R. L.; White, H. 2010. The Ocupants of Tomb II at Vergina. Why Arrhidaios and Eurydice must be excluded." In *International Journal of Medical Science* 7: 1–15.

Nagy, G. 1996. *Poetry as Performance: Homer and Beyond*. Cambridge.

Netz, R. 2002. "Counter Culture: Towards a History of Greek Numeracy." In *History of Science* 40: 1–32.

Pervanoglu, P. 1867. "Scavi sull' Acropoli d' Atene nel 1866." In *Bullettino dell' Instituto di Corresponenza Archeologica per l'anno 1867*: 72–82

Schärlig, A. 2001a. Compter avec des cailloux, le calcul élémentaire sur l'abaque chez les anciens Grecs. Lausanne.

Schärlig, A. 2001b. "Les deux types d'abaques des anciens grecs et leurs jetons quinaires." In *Archives des sciences* 54: 69–75.

Schärlig, A. 2001c. "Une piece trouvée à Laurion: C'est un abaque!" In *Zeitschrift für Papyrologie und Epigraphik* 134: 137–138.

Scholz, H. 1928. "Warum haben die Griechen die Irrationalzahlen nicht aufgebaut." In Hasse, H. and Scholz, H., *Die Grundlagenkrisis der Griechischen Mathematik*. Berlin, 35–72.

Solomon, J. 2000. *Ptolemy's Harmonics: Translation and Commentary*. Leiden.

Stroud, R., 1998. "The Athenian Grain-Tax Law of 374/3 B.C." *Hesperia Suppements* 29.

Suppes, P.; Moravcsik, J.; and Mendell, H. (eds.). 2000. *Ancient & Medieval Traditions in the Exact Sciences: Essays in Memory of Wilbur Knorr*. Stanford.

Szabó, Á. 1978. *The Beginnings of Greek Mathematics*. Dordrecht. Trans. by P. Unger of *Anfänge griechischen Mathematik*, Budapest.

Tannery, P. 1894. "Sur les épigrammes arithmétiques de l'Anthologie palatine." Reprinted in *Mémoires scientifiques*, vol. 2. Toulouses (1912), 442–6.

Threatte, L. 1980. *The Grammar of Attic Inscriptions*, vol. 1: *Phonology*. Berlin.

Threatte, L. 1996. *The Grammar of Attic Inscriptions*, vol. 2: *Morphology*. Berlin.

Tod, M. N. 1911/12. "The Greek Numeral Notation." *British School of Archeology*. 18: 98–132 (reprinted in Tod 1969).

Tod, M. N. 1913. "Three Numerical Systems." In *Journal of Hellenic Studies* 33: 27–34 (reprinted in Tod 1969).

Tod, M. N. 1926/27. "Further Notes on the Greek Acrophonic Numerals." In *British School of Archeology* 28: 141–157 (reprinted in Tod 1969).

Tod, M. N. 1936/37. "The Greek Acrophonic Numerals." In *British School of Archeology* 37: 236–258 (reprinted in Tod 1969).

Tod, M. N. 1950. "The Alphabetic Numeral System in Attica" In *British School of Archeology* 45: 126–132 (reprinted in Tod 1969).

Tod, M. N. 1954. "Letter-Labels in Greek Inscriptions" In *British School of Archeology* 49: 1–8 (reprinted in Tod 1969).

Tod, M. N. 1979. *Ancient Greek Numerical Systems: Six Studies*. Preface by J. Breslin. Chicago.

Turner, E. G. 1975. "Four Obols a Day Men at Saqqara." In Binen, J., Cambier, G., and Nachtergael, G. (eds.), *Le Monde Grec: Hommage à Claire Préaux*. Brussels, 573–77, and Pl. X.

Turner, J. H. 1951. "Roman Elementary Mathematics: the Operations." In *The Classical Journal* 47: 63–74.

Varoufakis, G. 1975. "Materials Testing in Classical Greece: Technical Specifications of the 4th Century B.C." In *Historical Metallurgy* 9: 57–63.

Vitrac, B. 1992. "Logistique et fractions dans le monde hellénistique." In Benoit, et al., 1992: 149–72.

Wendel, C. 1949. *Die Griechisch-Römische Buchbeschreibung Verglichen mit der des Vorderen Orients*. Halle.

Williams, B. P. and Williams, R. S. 1995. "Finger Numbers in the Roman World and the Early Middle Ages," In *Isis* 86: 587–608.

Wright., M. R. 1981. *Empedocles, the extant fragments*. New Haven.

Yatromanolakis, D. 1999. "Alexandrian Sappho Revisited." In *Harvard Studies in Classical Philolology* 99: 179–195.

Young, R. 1940. "Excavation on Mount Hymettos, 1939." In *American Journal of Archeology* 44: 1–9.

West, M. L. 1992. "An Alleged Musical Inscription." In *Zeitschrift für Papyrologie und Epigraphik* 93: 27–28.

Vitrac, B. 1992. "Logistique et fractions dans le monde hellénistique." In Benoit, P., Karine Ch., and Jim R. 1992. *Histoire de fractions, fractions d'histoire*. Basel, 149–72.

Vitrac, B. 1994. *Les Éléments d'Euclide. Vol. 2. Livres V à IX*. Paris.

Vitrac, B. 1998. *Les Éléments d'Euclide. Vol. 3. Livre X*. Paris.

Zeuthen, H. G. 1910. "Sur la constitution des livres arithmétique des Éléments d'Euclide et leur rapport à la question de l'irrationalité." In *Oversigt over det Kongelige Danske Videnskabernes Selskabs Forhandlinger*, 395–435.

Geoff Lehman and Michael Weinman

Recursive Knowledge Procedures Informing the Design of the Parthenon: One Instance of Continuity between Greek and Near Eastern Mathematical Practices

Abstract: This paper argues for reconsideration of two generally held views in the historiography of Greek mathematics, a narrow one about fourth and fifth-century Greek mathematics and a more general one about the continuity between Greek and earlier near Eastern mathematics. The narrower view called into question, owed to Knorr 1975, holds that two mathematical notions crucial for the development of Greek mathematics—(1) continuous proportion (and its non-reducible-to-a-unit-or-multiples-of-a-unit correlate, *anthyphairesis*) and (2) the reflective awareness of incommensurability owing to a misfit between magnitudes and multitudes—were understood for the first time in the mathematics of the first decades of the fourth century. Against this view, we propose to show that the Parthenon's construction displays at least a *foundational* comprehensive understanding of both of these mathematical ideas. Building on this narrower reconsideration, we further claim that the specific manner, which can be called 'algorithmic-problematic', of the Parthenon's engagement with these themes gives us reason to believe that the relationship of fifth-century Greek mathematics to its near Eastern precedents shows more continuity than is generally held.

Keywords: Greek Miracle, Incommensurability, Knorr, Szabó, Anthyphairesis.

I Introduction

In what follows, we shall re-examine two views in the historiography of Greek mathematics that have enjoyed relative stability for about a generation. First, that the development of a specifically theoretical understanding about two mathematical notions of great importance not only for Greek mathematics but for the history of mathematics as such—(1) continuous proportion (and its non-reducible-to-a-unit-or-multiples-of-a-unit correlate, *anthyphairesis*), and (2) the relationship of the geometric and harmonic mean (crucial for the theorization of incommensurability)—can be traced to the first decades of the fourth century. We hope to show that this dating might unnecessarily rule out genuinely theoretical knowledge at least one or two generations older in provenance, arguing that the Parthenon's construction (447-432 BC) entails a properly theoretical (not 'merely' practical) understanding of both notions. On this basis, we shall also call into question a more general historiographical standpoint which holds

https://doi.org/10.1515/9783110565959-009

that one can find much more discontinuity than continuity between Greek mathematics and its near Eastern predecessors. For, we shall see, the specific manner in which the mathematical principles of the Parthenon are put into practice (we shall call it 'algorithmic-problematic', as opposed to 'proof-theoretical') gives us reason to believe that the relationship of fifth century Greek mathematics to its near Eastern precedents shows more continuity than is generally held.

Already 50 years ago, Arpad Szabó and Wilbur Knorr debated on when Greek mathematics transitioned from a 'merely' practical to a properly theoretical understanding of the key mathematical notions later enshrined in Euclid's *Elements* as the basis of mathematical knowledge. If the position of Knorr in that debate is right—and there is a relative consensus that he was—then the kind of theoretical understanding we believe rests beneath the Parthenon's construction would not have been possible at that time. Yet, notwithstanding the scholarly consensus adduced, we will suggest that they were known in this reflective way at that time.

It is crucial to speak tentatively here given the fact that, as Waschkies (2004, 16) notes, any "idea concerning the origins of Greek mathematics" is effectively a guess—hopefully an educated one—given the lack of original sources prior to the time around the turn of the fifth to the fourth century. Yet, as Netz (2014, 167–71) recounts, even if definitive statements about the pre-Euclidian period are impossible, we can be relatively sure that the path toward the consolidation offered by *Elements* relies on knowledge networks that can be read out of Proclus' summary of the early mathematicians. The reconstruction of the earlier historiographical debate in section II hopes on this basis to suggest that it is at least not impossible that one such network was active in Athens in the mid-fifth century though largely unrecorded, Proclus' mention of Oenopides of Chios (fl. in Athens around 450 BC) notwithstanding.[1]

Building on this first part, the balance of the present work will suggest reading the Parthenon itself as a source that demonstrates the work of this network and how it fits into what Netz (2014, 181) calls the "gears" by which the knowledge networks generate and transmit the practical mathematical knowledge that came to be canonized only a century or more later. First, in section III, we identify one potential source of the insufficiency of choosing between the views of Szabó and Knorr: namely, that both authors understand the relationship between practical mathematical knowledge and theoretical reflection on it in an overly narrow way. We examine two potential correctives to this failing: (1) following Fowler 1999 in paying more attention to the role *logistike*, *anthyphairesis* in particular, plays as part of the theoretical mathematics of the early Academy; (2) re-thinking the possible 'legacies' of near Eastern mathematics in the reflective comprehension (*theory*) as well as the *practice* of mathematical procedures and their results.

1 For an exhaustive discussion on Oenopides, about whom quite little is known definitively, see Bodnar 2007.

Finally, the concluding third part (section IV) offers a new reading of the mathematical principles of the Parthenon's design that emerge from, but cannot be answered by, the *practical* application of procedural mathematical knowledge. If we begin with the view that such features could not have been consciously implemented in the Parthenon's construction absent a rich reflective awareness of these principles, then we will find that the designers involved could not but have had a *theoretical* understanding of 'cutting edge' mathematics of the kind being developed in the networks that Netz 2014 describes, to one of which those responsible for the design of the Parthenon would have belonged. Crucially, we will discover, such mathematical knowledge was sufficiently advanced to have accomplished something that Knorr believes to have only been possible only 30–50 years after the period of the Parthenon's construction, after the work of Theaetetus, Archytas and Eudoxus.

II The Szabó/Knorr Debate

A return to the Szabó/Knorr debate is crucial for telling a story about the historical development of the theory of incommensurability, which itself seems central to pre-Euclidian mathematical knowledge. There are two central components of Szabó's analysis that are relevant for us. First, is his thesis that there was much more of a theoretical development in Greek mathematics during the fifth century than is acknowledged by the conventional view, which holds that theoretically advanced mathematics began only in the early fourth century—that is, during Plato's intellectual career. It is important to note, though, that views do differ as to how central Plato and the Academy were in this development, and also about just how far theoretical advances went at what time in the relevant period, i.e., 410-370 BC.[2] Second, is his view of the priority of music theory, both in terms of time of circulation and in terms of intellectual formation. This priority finds primary expression in Szabó's chief dissent from the consensus that developed after his debate with Knorr. Namely: if he dates significant advances further back chronologically in the development of Greek mathematics than is commonly done in the literature, this is not because of a disagreement about when the integration of work on number and work on geometry occurred. Indeed, all parties agree that this is the work of the early decades of the fourth century, and it is here that formal proof and all the features of "Greek formal-deductive mathematics"

2 For more on the debate on Plato and the Academy, and also on the claim about the stability of the dating of the rise of theoretical mathematics, see Zhmud 2006 and his bibliography. Particularly relevant are Dillon (2003, 16–25), who addresses the role of mathematics in cementing an intellectual legacy in the early Academy, and Horky (2013, 3–35, 201–260), who offers a discussion on how appropriate the moniker 'mathematical Pythagorean(ism)' is with reference to Plato, colleagues of Plato and members of the early Academy after Plato.

truly take hold.[3] Rather, as we will see in some detail just below, Szabó 1978 insists that this theoretical integration and advance was built on the basis of a prior theoretical achievement within music theory that developed out of musical practice and was already mature in the first half of the fifth century. We do not hope to decisively determine anything in particular about Szabó's case for the priority of music theory, but we do hope to show it is plausible that there was a theoretical understanding of interdisciplinary issues in mathematics at work in the Parthenon's design.

How does Szabó hope to substantiate the claim that theoretical advance occurred by 430 BC, some decades before the conventional account allows? He hopes to show that even if nothing like the formal (i.e., 'axiomatic-and-deductive') proof, and surely nothing like 'proof-theoretical truth' had developed in the mid-fifth century, certainly it is the case that by the time of Philolaus and Hippocrates there was already a rich tradition of truly theoretical mathematics. Such a truly theoretical mathematics is one whose practitioners possessed a kind of familiarity with the interdisciplinary relevance of incommensurability that transcends a mere 'working' or practical knowledge within the mathematical disciplines of harmonics, geometry, or number and thus merits appreciation as a deep theoretical engagement with this issue.[4] But why does Szabó's account focus on the development of the three central mathematical disciplines named here? His whole historiographical framework, elaborated just below, is based on articulating a timeline that makes sense of how these disciplines came to be identified, together with astronomy, as the core mathematical disciplines in the education of philosopher-kings in *Republic* 7, and canonized soon thereafter as the center of mathematical, and thereby liberal, learning. For Szabó's overarching picture to hold, much must have been known in the middle of the fifth century. Corroboration for such a dating, earlier than allowed by Knorr, is provided by Waschkies 2004, who however sceptical he might be of Szabo's account, does argue that theorem sets are likely to have been "composed from about 475 BC onward," with "an order induced by stating some [theorems] to be principles and deducing all others logically."[5]

Szabo's central claim hinges on three subsidiary claims. First, owing in part to the fact that Propositions II.14 and VI.13 in *Elements* are more or less the same, and in part to the fact that II.14 is demonstrated so as to avoid any mention of propositionality, Szabó is confident that the construction of the mean proportional on a straight line

3 See Netz 1999, and the discussion of same below.

4 Hahn 2017 argues that already Thales was presenting such a theoretical understanding of incommensurability, complete with proofs that transcend the ostensive character of a diagram. If that's right, then the chronology debate at issue here becomes moot with respect to establishing plausibility for the notion that such a theoretical awareness can plausibly be attributed to mathematicians working on the Parthenon in 447-432 BC.

5 Waschkies 2004, 16. This article provides an in-depth analysis of the Szabo-Knorr debate. The juxtaposition of both the latter authors' view with Mittelstrass 2004 is highly relevant for the consideration of the continuity or discontinuity between Greek and earlier near Eastern mathematical practice.

(Proposition VI.13) was already known by the time of Hippocrates, whom he dates to having been active in Athens around 430 BC.[6] Second, that the oldest demonstrative reasoning in proving mathematical truth is Epicharmus' theory of odd and even, dated fairly accurately to ca. 500 BC, which is crucial because "this theory clearly culminated in the proof of the incommensurability of the diagonal and sides of a square".[7] Third, that provided sufficient attention to the integration of the three mathematical disciplines, a chronology can be established that proceeds in four stages: First, the musical theory of proportion develops, from which the terms of Eudoxean proportionality were borrowed; this occurs in two phases, with experiments with the monochord, giving rise to terminology for 2:1, 3:2, 4:3, followed by the development, by this means, of the technique of 'reciprocal subtraction' (*anthyphairesis*; Szabó: "successive subtraction").[8] Second, this musical theory of proportions is applied to arithmetic, in a manner later discernable in the presentation of numerical ratios in Book VII and especially the discussion of continuous proportionality in Book VIII.[9] Only then, Szabó holds, was this proportion theory then applied to geometry, at "the time of the early Pythagoreans", working with the construction of the mean proportional.[10] Finally, "mathematics within a deductive framework" developed; here speaking of *Elements* and its concomitant theory of proof.[11]

Szabó stresses two points about this chronology:[12] first, if things did proceed this way, then quadratic incommensurability "must have been known well before the time of Archytas"; second, "the discovery of incommensurability is due to a problem which arose originally in the theory of music". Of course 'knowing well' can mean a number of things, as critics of Szabó have pointed out.[13] Even if Szabó himself was not explicit, we think we can take his meaning here to coincide with the gloss provided for what is meant by a truly theoretical mathematics above; by 'knowing well' we mean being theoretically aware: the practitioners of mathematical knowledge, such as those at work in the design of the Parthenon, possessed a kind of familiarity with the interdisciplinary relevance of incommensurability that transcends a mere practical knowledge within the mathematical disciplines of harmonics, geometry, or arithmetic.

---

6 Ibid., 14, 48–9; in the latter argument, Szabo points out (in conversation with Heiberg) that II.14 seems to be a later elaboration that expressly excludes the mean proportional from consideration, in the way done in VI.13: "Indeed the proof of Proposition II.14 gives the impression that knowledge of the mean proportional is perhaps unnecessary for carrying out *tetragonismos*".

7 Ibid., 25.

8 Ibid., 26–27.

9 Ibid., 27–28.

10 Ibid., 28–29.

11 Ibid., 29.

12 Ibid., 29.

13 We have Berggren 1984 and Bowen 1984 in mind.

Szabó's chronology, concerning the development of a proof of incommensurability that it entails, involves him in an extended argument to the effect that it is a mistake to attribute to Theodorus and to Theaetetus the decisive role in the proof of incommensurability in the period 410-370 BC.[14] Szabó's claim rests primarily on two grounds: (1) a philological argument focused on his understanding of the use of the term *dynamis* (as both 'square' and 'power'), in mathematical texts, from the earliest surviving fragments down to the fourth century; (2) an interpretive argument concerning Plato's complicated intentions in associating the discovery of a *proof* of incommensurability with these two men in the *Theaetetus*. Here, we briefly state the decisive moment in each of these arguments for Szabó, and then discuss the basis of criticism thereof, and why we think we ought to consider this an open question.

(1) Szabó's philological argument against dating development of a theory of incommensurability to after 410 BC. Szabó summarizes the philological argument as follows: [15]

> "Thus my previous conjecture to the effect that *dynamis* and *tetragonismos* originated at the same time inevitably leads to the conclusion that the creation of the concept of *dynamis* must have coincided with the discovery of how to construct a mean proportional between any two line segments".

This is the crucial link for Szabó, insofar as he holds that Hippocrates (and not Archytas) originated the proofs about mean proportionality. Szabó argues that the term 'dynamis' began signifying extent-in-square at the same time that it was learned that one could *practically* construct a square while using the mean proportional between two line segments.[16] If this is true, and if it is also true that Hippocrates first circulated proofs of some kind concerning the mean proportional prior to 420 BC, then we can be confident that a crucial step in *proving* incommensurability dates to around 430 BC and not 410-370 BC.[17]

(2) Szabó's hermeneutical arguments against dating development of a theory of incommensurability to after 410 BC were not especially well-received by Knorr himself, nor by Bowen 1984, for instance. This results from differing attitudes to interpreting the dialogues, however, and not at all to matters directly historical or historiographical. If, for the purposes of historical discussion, we can keep an open mind on how to interpret Plato—Griswold 2002, 1–18 and Nikulin 2010, 1–22 can help us to see why this is necessary—, then we find that this is an open matter. Szabó cannot be said to have proven that Theodorus and Theaetetus are not crucial for the development of

---

14 Szabo 1978, 33–84.

15 Ibid., 48.

16 Ibid., 50–54.

17 Bowen 1984 is particularly unconvinced by the *dynamis* argument and, hence, by the view that incommensurability was understood theoretically by the period of 450-430 BC. To the best of our knowledge, however, this matter has not received much scholarly attention.

a proper *theory* of incommensurability, that is, but he provides good reasons to doubt their centrality in this story. First, Szabó deploys a reading of the pun on '*dunamei dipous*' in the *Statesman* 266a5-b7 to confirm that what Theaetetus describes working through with young Socrates was common knowledge.[18] Second, Szabó (1978, 76) argues that the mistaken attribution of significance mostly derives from two sources, which he spends five pages trying to debunk: first, an ancient Scholium on *Elements* X.9; second, "a report which probably stems from Pappus' commentary on Book X and survives only in an Arabic translation".[19] Since both of these sources themselves rely mostly on a wrong minded reading of the mathematics section (147c-148b) of *Theaetetus*, there is no reason to see this attribution as anything other than a 'false tradition'.[20] Finally, Szabó presents a hermeneutical and philological analysis of the dialogue in the service of this answer: Plato presents Theaetetus as a very talented, but also young and naive, and over-eager, researcher. The relevant section about the discovery of a theory of irrationals must be read in this light.[21]

Holding these arguments to be clearly interpretive, but possibly right and in any case not disproven, we move to the second of the relevant points in Szabó's analysis: the central importance and chronological priority of musical theory, which inspired an analytical program in number theory, which in turn inspired a reflection on the (im) possibility of fully integrating number theory and geometry—the latter of which had developed separately, without prior integration with the other two disciplines. Szabó (1978, 108–78) argues that those working in music theory arrived at the three means and realized there is no geometrical mean for the fourth, fifth and octave, and on this point his view is corroborated in the literature on Greek music.[22] From this tradition, mathematicians approached the double square and discovered mean proportionality between two lengths, numbers or not. Thus: music then arithmetic, then geometry.

Against Szabó, Knorr argues that "by the time of Hippocrates both [the theory of congruence and the theory of similarity (based on proportion)] were already well developed", but that the combination of these two traditions began "only about fifty years later", with the contributions of Eudoxus.[23] Knorr insists that "evidence in the pre-Socratic literature discourages dating [the] discovery [of the theory of incommensurability] before ca. 430".[24] One may recognize only after that time signs of a dialectical interest in the problem of incommensurability. The center of the debate

18 Szabo 1978, 68–71.
19 See Szabo 1978, 76, nn. 57–59, for full bibliographical detail on this matter. For the Greek of the Scholium in question, see ibid., 76n59.
20 Ibid., 79.
21 Ibid., 79–84.
22 See, especially, Barbera 1991 and Barker 2007.
23 Knorr 1975, 7–9.
24 Knorr 1975, 49.

is in the interpretation of Plato's *Theaetetus*.[25] Knorr here provides three arguments against Szabó. First, that even if something like the statement to be proven in Euclid VIII.18 is at work here, this only shows that "there is no *integer* which is the mean proportional between two terms that are not similar numbers", which is not enough to establish incommensurability.[26] Second, Knorr claims that Szabó's "view of the antiquity of the mathematics of the dialogue is an assumption not well supportable by the available documentary evidence, against which he has offered an argument.[27] And, finally, given Knorr's own view of the relative novelty of any theoretical understanding (and certainly proof) of incommensurability, it is "at least a reasonable counter-assumption that the number-theoretic foundation, upon which Theaetetus and his successors built their theory of incommensurability, had not yet achieved an advanced form at Theodorus' time".[28] Knorr here underscores what Szabó himself acknowledges: the relative novelty or antiquity of the discovery within the range of 430-360 BC is a matter of speculation that existing sources will never definitively settle. Thus, we are dealing with the relative merit of a later or earlier date within this half century timeframe.

Given that the participants in the debate both acknowledge that definitive knowledge is impossible here, and that the matter is really about who knew what when, within a fairly narrow research program in a relatively short time period, this question might seem pointless, especially as the alternatives do not strongly differ. Nevertheless, determining which, if either, is correct would be very telling for the argument we make about the Parthenon. For, if Knorr is right that the kind of integration Szabó sees in Hippocrates was only done in the time of Eudoxus, then it becomes difficult-to-impossible to believe that this integration, which we argue is integral to the design of the temple, would have been thinkable—let alone achievable—for its designers in the period 447-432 BC.

In order to advance this deadlock, we turn now to the role of *anthyphairesis*. Fowler (1999, 1–24), asking different questions for different reasons but responding to the issues at stake in the Szabó-Knorr debate, focuses on this very knowledge procedure. Fowler notes that Knorr agrees with Szabó that "the theory of incommensurability will be perceived as contributing to important aspects of every part of the *Elements*, save for the oldest geometrical materials contained in Books I and III", because they are both committed to the standard story about the history of Greek

25 It is notable here that even though this debate proceeded and was received as a debate in the history of mathematics, a careful consideration of Knorr's argument—see Knorr 1975, nn. 71, 74—shows that, again, the decision comes down to how one interprets Plato's work, and not historical knowledge of independent sources. Thus, if one finds Szabo's reading of Plato more convincing than Knorr's, then there is no reason yet given *not* to adopt Szabo's historical hypothesis.

26 Knorr 1975, 116–117.

27 Ibid., 82–86.

28 Ibid., 117.

mathematics. This traditional account, Fowler (1999, 2) continues, holds that the "early Pythagoreans based their mathematics on commensurable magnitudes (or on rational numbers, or on common fractions *m/n*), but their discovery of [...] incommensurability (or the irrationality of √2) showed that this was inadequate". This, as Fowler concludes his retelling of this view, "provoked problems in the foundation of mathematics that were not resolved before the discovery of proportion theory that we find in Book V of Euclid's *Elements*".[29] Fowler, for his part, "disagrees with everything in this line of interpretation".[30] Fowler, thought indebted to Knorr, provides grounds for arguing that Szabó is right to identify the sophistication belonging to Greek mathematics already before 430 BC. This is so, even though Fowler does not agree with Szabó that theoretical advances occurred earlier than Knorr allows, rather insisting that mathematical *practice* was well "ahead" of its theorization. Especially relevant for us as we approach the mathematics of the Parthenon is Fowler's central proposal that *anthyphairesis* and its insights were familiar to and fascinating for mathematicians of the mid-fifth century. It may be that Fowler is entirely correct that a lot of what we—following the ancient and neo-Platonic commentaries—call 'Pythagorean' had nothing to do with Pythagoras and his followers.[31] It may also be that nothing like (i) the formal theory of proportion in *Elements*, V[32] or (ii) the formal system of deductive proof presented in *Elements* as a whole had been seriously developed during the fourth century, let alone the fifth.[33] All the same, by underscoring the centrality of "reciprocal subtraction" (*anthyphairesis*) as a *practice*—especially in addressing problems in geometry (Theaetetus), in music theory (Archytas), and in astronomy (Eudoxus)—Fowler provides the welcome corroboration of our suggestion that the practicing mathematicians and mathematically informed artisans of the generations working at the time of the Parthenon's construction used reciprocal subtraction as a means by which to test, geometrically, solutions to problems in harmonics.

The chronology debate we have reviewed here points to a more fundamental debate about the very nature of Greek mathematics and its logical and methodological (not to say ontological) foundations. In the next section, we will argue on this basis that Greek mathematics shares much more with its near Eastern predecessors than is often believed.

---

29 Fowler 1999, 4.
30 Ibid., 4.
31 See especially Fowler 1999, 289–302. Fowler 1999, 356–401 is also telling for how he hopes his work to serve as a corrective to the standard story. More recently, Netz 2014 has reviewed the serious difficulties involved in the historiography of 'Pythagorean' mathematics.
32 Fowler 1999, 15–20, but also Chapter 10.1.
33 Ibid., 204–217, but also Chapter 10.4.

III Beyond the 'Greek Miracle'

This section adduces one contextual argument against the idea of a Greek Miracle in which Greek thinkers introduced *sui generis* and from their own genius, as it were, patterns of thought utterly unknown prior to the "awakening" of (Greek) science, as Bartel L. van der Waerden (1961, 5) famously referred to the manner in which Greek mathematicians "took their start from Babylonian mathematics but gave it a very different, a specifically Greek character". The work of Walter Burkert is of central importance in the academic interrogation of the self-sufficiency of Greek culture, while also just as clear as those who found his arguments and evidence so unwelcome regarding the magnificence of the Greeks. In a late synoptic work that reflects his whole body of work over the previous 40 years, Burkert (2004, 12) paradigmatically concludes that "although Greeks had been on the receiving side for a long time, there is no doubt that the result is Greek. It is Greek art and architecture that have become classical, and Greek literature that has become world literature". More recently, Josiah Ober has offered a comprehensive account of how the 'Greek Miracle' arose as a way of describing the rise of Greek-speaking city-states from peripheral backwaters to centers of thought and of economic and political power in the period between 800 and 400 BC, which he calls "the efflorescence of Ancient Greece". Ober (2015, 15) argues that, rather than a miracle, the primary cause is "the powerful role of specialization, innovation, and creative destruction in the decentralized world of ancient Greece".

With this critique of the Greek Miracle and the established need for the contextualization of Greek culture that has occurred with respect to literature and religion in mind, we pose the wider of the two revisionist claims outlined in the introduction. Namely, what can the practice of *anthyphairesis* as developed from the earliest (scantily) recorded sources of the sixth century through to its presentation in Euclid's *Elements* tell us about the supposedly revolutionary character of classical Greek mathematics? In answering this question, it is worth noting that while much is controversial about the degree of novelty in Greek mathematics, no one denies that there are some indications of contacts between Greek mathematicians (especially astronomers) and near Eastern counterparts, certainly by the fifth century (Meton), maybe even earlier (Hesiod).[34] Waschkies (2004, 15) shows that rather than some logical or philosophical principles being at work—as Szabo had argued—it is perfectly possible that a problematic approach to an arithmetical issue such as the *psephoi*-arithmeticians used would lead to the recognition that 'the side and diagonal of a single square are

34 As pointed out to me by Jeanette Fricke and Lis Brack-Bernsen, one possible point of direct influence that has not yet been thoroughly checked would be an analysis of the 'Works and Days' section of *Works and Days* with Column 6 of the 'Astronomical Diary Text' known as BM47762/BM49107. In general, direct influence by at least the sixth and fifth centuries, perhaps as early as the eighth century, is attested by Høyrup 2010 and Rowe 2013.

a pair of incommensurable magnitudes, and therefore the method of proof by *reductio ad absurdum* might be an invention by early Greek mathematicians, if it had to be invented at all'. In other words, the algorithmic-problematic style of mathematics might very well have generated the practice of indirect proof quite independently of any philosophical interest, and this might just as well have been done by the early Greek mathematicians operating in this style as by their near Eastern predecessors, for all we know.

To our best knowledge, the most firmly established early point of contact between near Eastern and Greek mathematics is in the discipline of astronomy, particularly the dates of appearance of stars and constellation during the year, comparable to material in MUL.APIN, and already present in Greek sources of the 8th century. As Evans (1998, 15–16) points out, "the data gathered during the reign of Nabonassar (747-733 BC) is especially important from the viewpoint of later Greek astronomy".[35] Given the evidence of overlap of 'research problems' and particularly of calculative techniques/knowledge procedures, we suggest it seems more likely than not that early theoretical advances in Greek mathematics developed with no small degree of continuity with the earlier work in near Eastern mathematics.

All the same, we quite agree with Netz (1999, 272–5) that the formal-deductive framework of Greek mathematics does in fact give every appearance of having been *sui generis* and becomes canonical only with and after Aristotle, with a handful of proleptic attempts (now lost) such as that of Oenopides, discussed above, preceding. Based on this overview, Netz concludes that since "mathematics appears suddenly, in full force" in Aristotle's works, one should believe that "as a recognisable scientific activity", it "started somewhere after the middle of the fifth century", and "since dates are a useful tool", he proposed 440 BC as the point of emergence. In keeping with Netz's conclusion, it seems to us that there are two significant respects in which slightly earlier, specifically mid-fifth century, Greek mathematics needs to be *read backward* in conversation with the non-formal and non-deductive procedures of mathematical knowledge from other traditions. First, we must see that both theoretical objects of mathematical knowledge (such as mean proportionality and periodicity) and intricate mathematical procedures (such as reciprocal subtraction) of great importance for the Parthenon were studied to a high degree of comprehension already by the mid-fifth century. In this respect, it is valuable to note that Proclus's account of Oenopides as the first to distinguish between theorems and problems, to insist on geometry done only with a compass and a straight edge, and to draw a perpendicular straight line from a given point to a given straight line is generally accepted. If this is true, it certainly signals a mature context of theoretical mathematical knowledge production *in Athens*—i.e., knowledge pursued solely for its own interest and

35 We thank Alexander Jones for the reference.

without relation to physical and practical production—by the time of the design of the Parthenon.[36]

Second, we must notice that (1) *if* this first claim is true, *then* there must have been some basis for such a (relatively) advanced state of development, and (2) it seems *prima facie* unlikely that such development would occur prior to the development of Greek-style mathematics as a largely fourth-century phenomenon. Thus, Waschkies 2004 seems right when, returning to an argument made at greater length in Waschkies (1989, 312–26), he argues that these objects and methods might well have been introduced not because of a burgeoning influence of Greek-style systematic, proof-theoretical knowledge procedures, but rather through the reception of near Eastern antecedents executed with an eye toward generalization. As he argues, both the near Eastern antecedents and their generalization were interesting for Greek mathematicians in their own non-deductive and not logically rigorous practice.

This second suggestion is, we know, controversial. The possibility of near Eastern influences on Greek mathematics of the Euclid type is both very difficult to establish and controversial to explore in the first place. Establishing connections is difficult for two main reasons, first chronology (most of the evidence for Old Babylonian mathematics is early second millennium BC), and secondly the manifest difference between the deductive-demonstrative style of mathematics of classical Greek mathematics (from the fourth century BC onward) and the algorithmic problem-oriented style of mathematicians in ancient Iraq, whether Old or Late. Indeed, if Fabio Acerbi (2010, 153) is right that writings in the Greek mathematical tradition of 'a foundational character' show us that 'the Greek way' of doing mathematics is very much through the establishment of a 'literary genre', which he attempts to delineate and interpret through the works of Apollonius and Geminus, then it would be hard to find a tradition more unlike the tradition of court-based astral science in ancient Iraq of the same period, as the latter is characterized by Eleanor Robson (2008, 214–62).

But more than just difficult to establish, possible links are suspect in advance as they seem both to belie the differences noted and to be driven by an *a priori* teleology where classical Greek mathematics must have been the direction that mathematical disciplines were headed, wherever practiced for whatever reasons. Indeed, in discussing one of the central foci on which the debate concerning the influence of near Eastern mathematics on Greek mathematics has focused—the right triangle problem at the heart of what came to be called the Pythagorean Theorem—Robson (2008, 110) concludes that the famous 'showings' of the relationship between rectangles and their diagonals with specific examples that express quantities later readers focus on as 'irrational' in Plimpton 322 and other early second millennium sources are "a student exercise in calculating lines and areas, not an anachronistic anticipation of the supposed

36 Thanks to Michalis Sialaros and Dmitri Nikulin for independently calling our attention to this piece of 'possible corroboration'.

classical Greek obsession with irrationality and incommensurability as is sometimes implied".

Robson is surely right that there is much suspect in twentieth-century attempts to place Old Babylonian sources in a specific kind of relation with classical Greek mathematics, with incommensurability at the heart of the story. Yet, as Høyrup (2009, 67) notes in his overwhelmingly positive review of the book, when Robson discusses

> "possible links to Greek mathematics, Robson restricts herself, on one hand, to sweeping arguments [and] on the other, to rather unspecific references to renowned publications that take the pertinence of these for granted in a way that unwittingly supports the myth of the Greek genius that invented everything on its own without interaction with other cultures".

Thus, given the current state of knowledge and the norms about historiographical methodology and the challenges of changing orthodoxy, it remains the case that what is largely a nineteenth-century prejudice holds firm in place of a better and firmer hypothesis. This prejudice has it that while Egyptian and Mesopotamian cultures developed practical mathematical understanding much earlier than the Greeks, it was the Greeks alone who sought to develop a proper theoretical understanding of these mathematical objects. This prejudice, as Høyrup has extensively argued, is quite hard to overcome; Steele 2006 has also discussed the ways in which cultural expectations can inform historical analysis, especially when original sources are scarce and questions of influence are involved. And, as Robson (2008, 274) notes, even as "(some) Old Babylonian mathematics was written into the Western tradition", anything from the 3000 year history of mathematics in ancient Iraq that "was irrelevant to the standard teleological narrative by virtue of being contemporaneous with, or even later than, Classical Greek mathematics" went ignored.

If we heed Robson's advice, as well as the work of Waschkies noted above, however, there are other reasons to be less than certain that the manifest difference in style between Greek mathematics, once it comes to be well-attested in the fourth and third centuries BC, and its near Eastern predecessors preponderates against the possibility of meaningful influence of the latter on the former. Robson herself finds that work, like that of Høyrup and Netz referred to elsewhere herein, which tries to uncover the workings of ancient mathematics in either the Greek or the near Eastern traditions in their own contexts and with less 'conforming' translations and explications "reveals three very different mathematical cultures", meaning Old Babylonian, Late Babylonian and Greek.[37] Surely there are profound and meaningful differences, and surely too, as Robson (2008, 272–4) argues, there has been much 'orientalizing' in the historiography of ancient mathematics. Yet, a granular and context-aware cognitive history of late first

---

37 Robson 2008, 275, 281.

millennium near Eastern and archaic (just pre-Classical) Greek mathematics is neither impossible nor without promise for those two reasons.

Philolaus (ca. 475-385 BC) is a central figure for the questions raised in this section, and thus makes for an excellent case study in attempting to address them.[38] First, his intellectual career closely coincides with, if largely slightly postdates, the design of the Parthenon, and shows a concern for precisely the kind of integration of the mathematical disciplines organized thematically around 'harmony' that we find there. Still more important, whenever exactly he published his one treatise on the cosmos, as Graham (2014, 48) notes, he did so by entering into a "larger conversation" that was already ongoing; a strong indication that the specifically mathematical elaboration of principles about nature already was directed at an audience sufficiently trained and large enough to receive such an account. While we will focus on another discovery/invention attributed to Philolaus, the Pythagorean tuning preserved in what Huffman 1993 refers to as Fragments 6/6A is integral to the mathematics of the Parthenon. Philolaus' work carries a distinct trace of what near Eastern mathematics did best, astral science, specifically the theory of the moon and the sun, and the position of the ecliptic.

Take the case of Philolaus' great year. Huffman provides a thorough account of what is known about this intellectual achievement, which purports to name the period within which solar years coincide with lunar months: namely, 59 solar years (where each solar year has a value of 364 1/2 days), or 729 lunar months (where each lunar month has a value of 29 1/2 days).[39] As Huffman pointedly asks: "The crucial question is: how did Philolaus arrive at this set of numbers?"[40] Huffman considers two main alternatives, one in which Philolaus is adopting and adapting the value that Oenopides (mentioned above, and a slightly older contemporary of Philolaus) had arrived at: 730; the other in which he derives it from recorded observations directly. If he revised an earlier value, the supposition goes, this could be either because of some preference in working with the observations—like the value it gives for the solar year (of 364 1/2 days, rather than 365 22/55 days—or because of the inherent appeal of 729 as a cube of the cube of 3.[41] The 'numerological' reading seems to gain credence from the tradition, going back to Aristotle (*Metaph.* 986a3-12, *Cael.*293a23-7), that holds Philolaus introduced the counter-earth solely to get a perfect number of ten heavenly bodies. As Graham (2014, 57–60) shows, however, Philolaus has good observational

38 For what little information there is on Philolaus' life, including his dates, see Huffman 1993, 5–6. See also Graham 2014, who builds on and updates this account.

39 Huffman 1993, 276–279.

40 Huffman 1993, 277.

41 This appeal was clear to Plato, who chooses to make the number crucial for the central problem of *Republic*—showing that the life of the just is better than the life of the unjust—by having Socrates claim that the life of the just man is 729 times more pleasant than that of an unjust man, and expressly citing Philolaus' great year in the process (*Resp.* IX, 587b-588a).

reasons to introduce the counter-earth—this would explain the greater occurrence of lunar than solar eclipses within his "central hearth" structured cosmos.

Building on Graham's work with the counter-earth, we would extend this way of reading the interplay of a priori reasoning and observational awareness to the enterprise that would lead one to posit a great year, a topic which Graham 2014 does not address. Huffman identifies this as "an attempt to harmonize two important ways of measuring time, the lunar month and the solar year"—and as not *sui generis* within Philolaus' cosmology, or within Pythagoreanism.[42] Both the observational data and the values with which Philolaus and his Greek contemporaries were working came to them from Egypt and Mesopotamia. Given this fact, Huffman's analysis—where the interpreter is left to decide between the possibility that Philolaus, as a mid-fifth century Greek mathematician/cosmologist, would base a certain key finding on either (a) fit with an overarching theoretical commitment or (b) fit with observational data—leaves aside a possibility worth serious consideration, even if it is difficult to substantiate fully. Namely, is it not possible that solar year and lunar month periodicity presses itself on these fifth-century Greek sources through the determination of the same issue in the recent Babylonian astronomical work of the seventh and sixth centuries BC?

While it is difficult to establish connections for the reasons stated above, here is an instance where we have Babylonian source materials—specifically 'Goal-year tablets', which make predictions of where certain heavenly bodies will be at certain times in a given year, and 'Lunar prediction tablets', which focus on the moon over long periods—from the time in question. Since the 1990s, a great deal of work has been done on exemplars of this tradition determinately dated to the period (e.g., tablets from ca. 642-640 BC, 593 BC, 523 BC) and others whose dating is not precisely known but date from some time not earlier than the fifth century BC and not later than the third century BC.[43] These texts display two features that relate directly to the question Huffman raises concerning Philolaus' process in arriving at his value for the great year.

The first point of close contact between near Eastern astral science of the mid-first millennium and Philolaus' interest in the great year rests in the fact that, as Huffman (1993, 277) notes, the latter "is an attempt to harmonize two important ways of measuring time, the lunar month and the solar year". For our comparative purposes, it is crucial to note with Robson (2008, 214) that a "significant quantity of mathematics survives from the Old Babylonian period" through the Persian (Achaemenid) period, during the fifth and fourth centuries BC, and down to Alexander's conquest and its

42 Huffman 1993, 276.

43 Britton, Brack-Bernsen, Huber, and Steele have published on these sources. Especially relevant for the consideration of a possible near Eastern influence on Philolaus' great year are: Brack-Bernsen 2014, Britton 2002, Huber & Britton 2007, and Huber & Steele 2007. For a more complete bibliography of works in this field, see Brack-Bernsen 2014.

aftermath. This transmission history includes a long tradition of interest in precisely this sort of solar year/lunar month harmonization at issue with Philolaus, among other topics of interest for reasons both mathematical and practical.[44] It is true that the underlying motivation in Late Babylonian mathematical practice displays a significant difference from what we can glean as being Philolaus' motivation, namely the political and religious importance of the eclipse in the near Eastern case as opposed to a more ontological and cosmological set of considerations for Philolaus.[45]

All the same, the observational data with which Philolaus and his contemporaries could seek out significant number patterns is owed to this tradition. But Greek astronomers and mathematicians likely received a good deal more than just the data. For instance, bearing in mind the work of Huber and Steele 2007 on the so-called 'Saros function'—an 18-year solar cycle—that is transmitted in lunar prediction tables dated to 642-640 BC, and also the work of Britton 2002 on a lunar prediction table (dated to ca. 620 BC) that uses a 27-year solar cycle, one detects a strong consonance between Philolaus' interest in powers of three and the repeated thematic and methodological use of powers of three in these near Eastern antecedents of which Philolaus and/or those with whom he worked might well have been aware. Unfortunately one cannot corroborate this claim more closely with comparison to Philolaus' work, for (as Huffman 2014, 276–7 notes) we have only a few scant later testimonies with which to work. But it can help to recall two facts concerning Plato's reference to Philolaus' great year at *Resp.* 587d-588a. First, we are told there that the number 729 is both 'true' and 'appropriate to lives' both because of days and nights (729 days and nights in one 364 ½ day long year) *and* months and years (729 lunar months coincides with 59 years having 21 intercalary months). Second, this passage precisely proceeds by means of taking the number three in square and then in cube. We can see, then, that both in the near Eastern astral science of the relevant time period and in the work of Philolaus, the harmonization of the solar year and the lunar month is of primary interest and best understood through the power series of the number three.

The second close point of contact between the great year as a paradigmatic instance of Philolaus' interest in observational astronomy and near Eastern practices of the first millennium is the way in which he directly interprets data, in this case charts of the positions of the sun and moon, as themselves objects of observation. This appears to have been the hallmark of Babylonian astral science in particular: already by the seventh century, we see the development of what is called the 'linear zigzag function', which describes the pattern that emerges on a table inscribed on these tablets, given the values for mean speed (for moon and/or sun) through the

44 Our account here would not have been possible without the assistance of Lis Brack-Bernsen. We are solely responsible for what errors remain despite her patient guidance and assistance.

45 For a discussion of the ontological and epistemological focus of Philolaus' work in cosmology and observational astronomy, see Graham 2014, 64–68.

phases of the moon over a period of solar years.[46] From especially the fifth century on, this sort of function is used in order to give sequences of numbers tabulated for equidistant intervals of time, allowing for the determination of the relevant period and thus the prediction (for instance) of eclipses. This back-and-forth procedure, from tabular data to worldly phenomena, became integral to Greek mathematics, especially astronomy. One striking, albeit later, example of this is the work of Hypsicles on the mean speed of the sun and the moon. Hypsicles used Greek mathematical rhetoric of a Euclidian kind to present precisely Babylonian values of the 'linear zigzag function' for the mean speed of the sun and the moon. As Evans (1998, 124) notes, Hypsicles' "whole scheme of using an arithmetic progression to represent the rising times of the signs is of Babylonian origin". We do not have a record of when precisely Greek mathematicians began working with the periodic functions of near Eastern astronomy. Given the contacts, known to have been established between the seventh and fifth centuries, however, there is no reason to believe that Hypsicles or his contemporaries were receiving the zigzag function for the first time; in any case, the interest in periodicity, and the habit of noticing periodicity in data directly, and then bringing it to objects of observation or construction is shared by the near Eastern astral science of the seventh century and Philolaus.

This brief investigation of Philolaus' great year can at best provide a plausibility test for our claim that fifth-century Greek mathematical practices were continuous with and expressly borrowed from near Eastern predecessors. This comparison of the great year with goal-year tablets does not *prove* anything. We would insist, though, that it shows questions concerning the development of Greek mathematics at the time of the design of the Parthenon remain open and scholarship could benefit from a more extended comparative analysis of the kind we have initiated here. What's more, it should at least be clear—however much we might doubt a direct reception of near Eastern astral science by Philolaus in developing his great year—that the kind of knowledge procedures employed in the 'algorithmic problem-oriented' style of near Eastern mathematics in this time period is quite similar to the ever-more-advanced, but still algorithmic and problem-based knowledge procedures in Greek mathematics prior to formalization and the introduction of deductive proof in the fourth century.

This brings us back to the Parthenon and the nature of mathematical theory and practice in Athens in the mid-fifth century. We see in the temple's design precisely the kind of algorithmic problem-oriented style we know to be the hallmark of near Eastern approaches. Analyzing the mathematically-informed features of the Parthenon's design, it seems quite likely that the *process* of coming to these design features is through the *reflective* deployment of a variety of instruments derived from just this (putatively near Eastern) style. If that was happening in Athens at the

46 Brack-Bernsen 2014, 37.

time of the temple's design, it seems natural to ask: how did they know how to do this? Why did they choose to do so? This article concludes with a first approximation of an answer to this question, which serves as the basis for the sustained analysis in the forthcoming publication of the findings of the research we have conducted on the Parthenon. In that further and deeper elaboration, we hope to show in detail how the continuity with, but also innovation when measured against, the sixth and earlier fifth-century tradition of Doric temple architecture indicates the underlying continuity of a relatively advanced fifth-century Greek mathematics with its near Eastern precedents.

IV The Parthenon: Continuity with Near Eastern Mathematical Knowledge Procedures

This closing section focuses on three related mathematical features manifestly present in the design of the Parthenon: periodicity, the application of the unit, and the constructed harmony of the whole. Each of these features displays significant connections with emerging themes known to be crucial for the agenda of fourth-century Greek mathematics. At the same time, each expresses the algorithmic-problematic style long associated with near Eastern mathematical traditions in contrast to the formal-deductive style of classical Greek mathematics. Noticing these two facts about these three features, we conclude by suggesting that their integral presence in the Parthenon's design shows how advanced the interdisciplinary practice of mathematics, theoretical as well as practical, was by the mid-fifth century. We further show that this advance neither depends on nor correlates with the development of formalization or deductive apparatus.

In order to see why the presence of these features could be telling in this way for our conceptualization of the relationship between theoretical and practical mathematics on the one hand and the two styles of mathematical practice discussed here on the other, let us begin by recalling the Plato and the mathematicians discussion, in which the use of the difference in methodology between one kind of mathematics and another is key for understanding Plato's critique of the (Greek) mathematicians (of his time) in *Republic* 6 and 7 (510c-d, 528b-d, 529b-c, 530a-b, 531b-c).[47] Plato here seems to be critical of mathematicians who simply apply their procedures, which they treat as granted setting-stones (hypotheses), and do not question their principles. In acting this way, Plato argues, they deprive mathematical practice of its

47 The literature here is very extensive. A good beginning might be Barbera 1981, Benson 2012, and Burnyeat 1987, 2000. For a more detailed discussion, see Cornford 1932, Cherniss 1951, Lloyd 1978, 1992, Mueller 1992, Taylor 1967, Zhmud 1998.

greatest potential: they make it impossible for mathematics to lead us to the forms. If we bear in mind the central suggestion of the previous section of this article concerning possible near Eastern influence on the Greek mathematical practices before ca. 375 BC, this critique takes on a new dimension. Might it not be that the nature of Plato's concern here is precisely the prevalence of the algorithmic-problematic style of mathematics argued here to be integral to the design of the Parthenon? Further interpretive attention is surely needed here. Yet should it prove true that there are substantial continuities between the (pre-Euclidian Greek) mathematicians Plato is criticizing and the near Eastern mathematicians that the second section has tried to show at least might have had an influence on them, this might help us to thicken our understanding of a critique that remains opaque to many of Plato's readers right down through controversies in the contemporary secondary literature.

By teasing out the background of Plato's critique, we can see how his worry about certain algorithmic-problematic practices points toward the development of formal-deductive practices that are more consistent with anti-materialistic ontology and with the forms in general. Nikulin 2012, discussing 'indivisible lines', and Negrepontis (in this volume), discussing the periodic *anthyphairesis*, make clear how constructions produce the tension that arises when trying to define and use a unit both arithmetically and geometrically. This tension, as we will see in a moment, correlates strongly with the central problems the designers of the Parthenon were attempting to resolve related to inevitable remainders that arise when trying to maintain strictly numerical ratios in constructing a three-dimensional figure (a temple in fact) according to the canon of proportions used in the Doric order.

This vindicates Negrepontis's finding—itself an echo of Szabó 1978 and Fowler 1999—about Aristotle's claim from *Topics* (158b-159a) that there was a pre-Eudoxan approach to proportions in Greek mathematics defined with reference to finite and infinite *anthyphairesis*. Such a pre-Eudoxan approach to proportions by an algorithmic procedure rather than a definition and demonstration suggests that there already existed by the time of the Parthenon's design a robust, *theoretical* reflection on (for instance) why some recursive expansions in square yield continuous proportions of whole numbers, while others arrive at the 'infinite *anthyphairesis*' Plato is worried about in dialogues like the *Theaetetus*, *Philebus*, and *Statesman*. A synoptic discussion of three crucial elements of the Parthenon's design mentioned above—periodicity, the application of the unit, and the constructed harmony of the whole—will show this quite clearly.

The first is periodicity. The Parthenon's construction seems to have begun with a literal, physical given: the column drums of 1.905m in diameter. However, another equally important given, or at least a starting point, for the Parthenon's exterior design was the 3:2 ratio of metope width to triglyph width towards which the aesthetic modifications and adjustments made to the Doric order over the sixth and early fifth centuries had been converging: indeed, by the time of the Parthenon's immediate predecessors—the Temple of Zeus at Olympia, the Temple of Hera Lacinia at Akragas, and a few others—the 3:2 ratio seems to have become canonical.[48] With these two starting points in mind, one can imagine a design process for the flanks and façades (figure 1) in which an initial continuous proportion is constructed to yield the building's governing ratio of 9:4. Beginning with the lower column diameter of 1.905m, the average or 'ideal' metope width is then established at 2/3 the column diameter (1.27m). Then applying the recently canonized 3:2 ratio of metope to triglyph, the triglyph width becomes 0.8467m (corresponding extremely closely to the actual standard triglyph width of 0.845m), creating a continuous proportion of column diameter to metope width to triglyph width of 9:6:4, with the 9:4 ratio—so central to the Parthenon in all of its measurements—holding between the column widths and their visually analogous form at the level of the frieze, the triglyph. Moving in the other direction, and applying this newly constructed 9:4 ratio to the next larger principal visual element of the exterior (whether façade or flank), the intercolumniation, in its relation to the columns (the 1.905m column drums), gives a standard intercolumniation of 4.29m, i.e., with a 9:4 ratio of intercolumniation to column diameter. At this point, the further extension of the façade's continuous proportionality appears, the one that, as we shall see, will ultimately define the cubic dimensions of the building as a whole: 81:36:16, here the ratio of intercolumniation to column diameter to triglyph. Thus, the design of the Parthenon's façades and flanks suggests the way these proportions are unfolded out of each other, and are in fact constructed, both in the literal sense—in terms of the measuring and cutting of individual stones, the making of the building itself, i.e., in terms of *technē* as craft—and in a musical or mathematical sense, viz., in a manner similar to the construction of a musical scale out of series of related small number ratios.

Because all of these elements are so prominently visible—indeed, they provide the principal visible horizontal articulation of the rectangular field (that is, the main divisions of its left-to-right extension)—they give the whole ensemble the character of a manifest demonstration of a simple mathematical law, the law of continuous proportionality. In addition, on a more profound if less explicit level, they collectively contribute to the aesthetic quality of harmony and of unity that so strongly characterizes the building as a whole. In other words, the pre-existing and recently canonized forms that defined the proportions of the Doric order have been pressed into service to make visible, in geometric form, some of the same harmonic principles, specifically the relationships among small integer ratios, by which the fifth-century Greek musical scale was constructed. Likewise, those ratios together produce an ineffable aesthetic effect parallel to the one produced by the same combinations of small integer ratios in music. This 'aesthetic' experience, it is worth stressing once again, could be more fully described as an embodied one—again, parallel to music and to its effects on the body. Thus the predominance of continuous proportions can be understood to

48 See Waddell 2002, 2–3 and Mertens 2006, 270–271, 389.

engage the viewer in a bodily as well as an intellectual fashion. Approaching one of the façades, passing between columns, and circumambulating just outside or within the outer peristyle, one may experience, even if pre-consciously, one's body as a '4' in relation to the '9' of the intercolumniations (since the width of the column drums, 1.905m, corresponds to the scale of the human body), just as one is encouraged to feel one's body scaled as a 9 in relation to the 4 of the triglyphs, or a 3 in relation to the 2 of the metopes, when visually following the line of the columns to the entablature above.

If this construction of *symmetria* (the proportionality of the whole) in the Parthenon can indeed be understood as analogous to the constructive methods, the aesthetics, and the ontology (the mathematical being) of the fifth-century Greek musical scale, one would expect the *symmetria* of the building's interior as a whole to relate to that of the exterior (the stylobate, flanks, and façades) in a significant way. And indeed it does: the overall ratio of principal exterior dimensions to principal cella dimensions is precisely 13:9 (figure 2). This is a small integer ratio, though not an obviously musical one, and as such it raises interesting new questions relevant for appreciating the crucial importance of *anthyphairesis* in the building's design. Just as subtraction of the *diapason* (octave) from 9:4 yields the *tonos* (whole tone); just as, in the musical scale, subtraction of the *diapente* (fifth) from the *diapason* produces the *diatesseron* (fourth) (and the *lemma* is produced by subtracting two *tonoi* from the *diatessaron*, etc.); so in the case of the 13:9 ratio, a single application of *anthyphairesis*, subtracting the smaller measure from the larger, produces a remainder of 4 and a new ratio of 4:9, the ratio of this remainder to the smaller of the original elements. One can imagine, and this is of course purely speculative, that the designers could have arrived at 13:9 as a desirable ratio quite easily, given the predominance of *anthyphairesis* as a working method in the Parthenon's construction, as well as the tendency to work from "both sides" of a problem, i.e., not just to start from a given, but to start from dimensions that, through a known procedure, will arrive at a given.

In addition, however, *anthyphairesis* not only has an arguably central importance, just as it does in the case of the musical scale, as a constructive method for the Parthenon, whose design employs the same algorithms and engages the same problems as musical scale construction; its use also has a self-reflexive quality. Indeed, the problems raised by *anthyphairesis*, *qua* problems, are made evident in various parts of the building, most notably in the treatment of the well-known 'corner problem' (discussed below), in ways that give the building a pedagogical character. This appears, arguably, in its simplest form in the superimposed Doric orders of the cella interior (figure 3), where the lower order relates to the upper one in a ratio of 13:9. Certainly there may have been aesthetic reasons for this particular choice of proportions, as well as structural ones (for instance, if it provided the appropriate reduction of mass of the upper order to avoid its being top-heavy but still strong enough to hold up the roof). However, one cannot help but notice that precisely here in the building, one finds two visible elements that are proportionally and formally identical but at two different scales, inviting comparison: two unequal but analogous magnitudes, emphasizing a single dimension of extension, inviting the viewer to size one up against the other, i.e., to perform a kind of intuitive *anthyphairesis* in thought. Walking around the Parthenon interior, as one can still experience it today in the twentieth-century reproduction of the building—begun in the 1920s, based on Dinsmoor's specifications— in Nashville, Tennessee, one feels (pre-consciously or otherwise) the simultaneous presence of the 9:4 proportions of the central space's ground plan and the 13:9 ratio of the two orders of Doric columns, along with the invitation to compare, and perhaps to calculate (figure 4).

To return to the corner problem, i.e., to the question of how to negotiate the necessary adjustments to proportions in the Doric entablature in order to ensure that a triglyph and not a partial metope (that is, a remainder) is placed at the building's corner (figure 5): the remarkable solution to this problem in the Parthenon seems to involve a particularly sophisticated use of *anthyphairesis* as a construction method. In the Parthenon, in contrast to earlier buildings, there is an astonishing variation in the widths of metopes on all four sides of the building, an apparent irregularity that is unprecedented in Doric design, and that suggests a complex process of reciprocal small adjustments. Furthermore, the adjustments both to the intercolumniations and to the frieze were coordinated in a way that implies orientation towards a larger goal, that of *harmonia* (harmony) in the corner articulation, and in the building as a whole. With this in mind, the sense of repeated reciprocal adjustments producing what seems like a slight but unaccountable irregularity in the dimensions and positions of elements in the Parthenon's entablature becomes intelligible. And those adjustments have the character of a procedure by which progressively smaller intervals are produced using *anthyphairesis*, as in the construction of the musical scale, paired with the recognition that such a process can only be an incomplete solution, given the existence of irrational magnitudes, but that, nevertheless, such incompleteness is preferable to ignoring the problem altogether. Thus, the complex process of interrelated adjustments to the Parthenon's 'ideal' proportions that seems to characterize the approach of the Parthenon's designers and builders to the corner problem—evidently an ongoing working out of an unsolved(/unsolvable) problem at the time the building was being constructed—[49]could be understood as parallel in method to the building of a musical scale, where the component numerical ratios do not *perfectly* fit together, in the sense that they cannot measure (i.e., subdivide) each other, requiring small fractional adjustments (resulting, among other things, in the *lemma*, the *apotome*, and the *comma*) generated by a method of reciprocal subtraction that inevitably leaves a remainder. This 'imperfection'—the impossibility of subdividing the octave into any number of equal units (whether *diapente*, *tonos*, or any other interval, however small)—involves a conscious acknowledgement of the existence of the irrational. And the use of *anthyphairesis*, both in the musical scale and in

49 Cf. Yeroulanou 1998, 414–416, on the evidence for modifications to the building's design during the course of construction, based on the variations of metope widths on the west and east façades.

the Parthenon's entablature, constitutes an algorithmic-problematic approach to the negotiation of this problem.

To give just one example: early analyses of the building's refinements, specifically by Guido Hauck 1879 and William Goodyear 1912, revealed minute adjustments not only to metope widths but to the guttae underneath the triglyphs and to the centering of the abaci over the columns in counterbalancing ways.[50] For instance, considering the south side of the entablature on the Parthenon's east façade (figure 6), if a triglyph is decentered to the left with respect to the intercolumniation beneath it (as is the case with the second triglyph from the south), the guttae underneath the triglyph are decentered a still smaller amount to the right to compensate, i.e., to produce a slightly more centered overall ensemble. Or, since the triglyph over the corner intercolumniation appears noticeably closer to one column than another, the triglyph over the adjacent intercolumniation is shifted off-center in the same direction, but by a smaller amount (and, correspondingly, the abaci of the columns below are shifted the opposite way to further, slightly, reinforce this decentering), in order to reduce the difference between the degrees of decentering in the two adjacent sections of the entablature. Just as the use of a 13:9 proportion on the largest scale invites one, in comparing them, to subtract the smaller from the larger and arrive at the normative 9:4 ratio, so too on this microcosmic scale the most subtle of the adjustments to the entablature proportions imply a constructive method that has the specific, procedural characteristics of *anthyphairesis*.

The approach taken on the west façade in fact differs from that of the east façade, and that of the northern half of the east front differs from that of its southern half, indicating an ongoing working-out process in which different algorithms were tried, but also (given the likely design chronology of east to west)[51] demonstrating an increasing tendency towards a harmonization of the corner intercolumniation with the corner of the frieze, with both being visibly contracted—an approach most fully realized on the west façade.[52] This progression towards greater harmony at the corners, culminating in the west façade (see figure 1)—with peristyle and entablature refinements coordinated, as well as the visible strengthening of the corners as a whole through contraction—seems consonant with the Parthenon's overall prioritization of *harmonia* as a value (cf. the 'third principle' discussed above). It is worth emphasizing the importance of corner definition for the impression of unity in the building, and both the harmonizing of elements and the greater density and strength at the corners contribute to this definition of the corner *qua* corner. However, a similar kind of experimental, working-out process could certainly be reconstructed with an alternative chronology, going from

50 Hauck 1879, 130–133 and Goodyear 1912, 200–202.

51 See Korres 1994, 115, Yeroulanou 1998, 423, and Goodyear 1912, 191–192. For an alternative chronology, see Wesenberg 1983, 60–62.

52 On corner column contraction in the Parthenon, see Coulton 1974, 66, 76, Bundgaard 1976, 63–64, and Wesenberg 1982.

west to east. What is important is not the specific reconstruction, but the sense of an experimentation through reciprocal adjustments to create the most satisfying solution to the corner problem, based on the interdependence of *all* elements in the design, and not only on the modification of an isolated few.

The second key feature is how modular construction relied on the practice of 'the application of the unit'. The achievement of continuous proportion (part of a larger goal to achieve *symmetria*, i.e., commensurability) both macrocosmically and microcosmically on the exterior of the Parthenon seems, indeed, an overriding concern of the building's designers, and makes a compelling case for the idea that the unusual octastyle façade was chosen to make these proportions possible and not, as has often been argued, for other reasons, such as to accommodate a larger cult statue in the cella.[53] One consequence of the unique method used to determine the Parthenon's stylobate dimensions, combined with the similarly atypical choice of octastyle fronts—and indeed as a likely motivation for such—is that the building's overall proportions of 81:36:16 (length:width:height) has as its modular unit the actual, visible triglyph width (figure 7). The mainland rule of design—five triglyphs per intercolumniation plus one additional triglyph defining the measure of each side of the stylobate—produces, for an 8 x 17 temple, a stylobate dimension of 36 x 81 triglyph widths. And with the height of the order (4/9 of the stylobate width) thus measuring 16 triglyph widths, the unit for the 81:36:16 cubic volume of the building becomes the basic formal unit of the entablature: the triglyph, so prominently visible at each of the building's corners, that articulates the horizontal extension of the building, in 2:3 rhythm with the metopes, at the level of the frieze. In effect, the Parthenon's design not only produces a rigorously commensurable building; it also foregrounds the unit of measure that defines it, seeming to invite the viewer to consider those very measurements reflectively, as well as the whole problem of *symmetria* and its significance.

However, the way in which the Parthenon uses the inward inclination of the columns of the peristyle to mediate between the proportions of the stylobate, based on an 'ideal' triglyph module, and those of the entablature, based on the actual triglyph module, indicates that the unity of the building as a whole, the joining together of its different parts (its *harmonia*), takes precedence even over *symmetria*. In other words, the designers created small deviations from the pure *symmetria* of stylobate and entablature for the sake of balancing and adjusting multiple elements and of harmonizing slightly different proportional systems, as well as to bring the corner into harmony with the rest of the building. This refinement—producing a slight discrepancy between the actual triglyph width and that of an 'ideal' triglyph module—allows the Parthenon's designers to bring the continuous proportionality of the facades and flanks (81:36:16, discussed above) into harmony with the traditional method for modular construction of the stylobate dimensions.

53 Cf. Bundgaard 1976, 69–70.

As Coulton 1974 has discussed at some length, in the middle of the fifth century there were effectively two rules for determining the dimensions of the stylobate, one based on the intercolumniation module, prevalent in mainland Greece and another based on the number of columns in the peristyle, prevalent in the western colonies, particularly Sicily. With its unusual design of an 8 column by 17 column peristyle—unprecedented in mainland Greece, where Doric temples invariably had 6 columns on the façades—the Parthenon, though based primarily on the 'mainland' rule, effectively harmonizes the two approaches by also conforming to the 'Sicilian' rule (where an 8:17 peristyle yields a 9:18, or 4:9, stylobate); at the same time this approach to the design achieves a precise 9:4 ratio for the stylobate (measuring 69.50m by 30.88m).[54] As Büsing 1988 has shown, the Parthenon's unusual 8 x 17 peristyle yields a stylobate, when following the mainland rule, of precisely a 9:4 proportion ([17+1/5]:[8+1/5] = 81:36 = 9:4). This rule assumes a triglyph module 1/5 the width of the intercolumniation, increasingly standardized as a modular basis for calculating the dimensions of the Doric stylobate, a problem discussed in illuminating fashion by Gene Waddell 2002 and Mark Wilson Jones 2001, 2006.[55] In the Parthenon, however, as we have seen, intercolumniation and triglyph are instead in an 81:16 ratio. What the designers seem to have done to reconcile these two givens is to construct the stylobate as if the triglyph width were 1/5 that of the intercolumniation (0.858m, a kind of 'ideal' triglyph module), but then account for the difference from the actual average triglyph width in the entablature, 16/81 of an intercolumniation (0.845m), through the inclination of the columns, leaning inward from the stylobate to the slightly smaller entablature. If this is calculated based on the flanks, which are 81 ideal triglyph modules in length, with the column inclination equal to half a triglyph width on each side, then the entablature, when 'compressed' by one triglyph module, would be 80 ideal triglyph modules in length, and at the same time of course, 81 actual triglyphs in length. And this in fact corresponds exactly to the 81:80 ratio of the ideal triglyph module—the one that determines the stylobate dimensions and the large scale *symmetria* of the building —to the actual average triglyph width of 0.845 (0.858:0.845 ≃ 81:80), while this difference between the ideal module of the stylobate and the real triglyph width in the entablature precisely corresponds to,

---

54 Coulton 1974, 76.

55 The reconstruction of Doric design methods in Waddell 2002 suggests that first the krepis and stylobate dimensions were determined in this fashion, and from there the dimensions of the module (the triglyph) were derived, while Wilson Jones 2001 proposes a somewhat different, but related, procedure for the determination of the proportions in Doric temples, emphasizing a modular design that begins not with the overall krepis dimensions but with the physical module, i.e., the triglyph. For his more recent work on mathematical problems relating to Doric design, see Wilson Jones 2006. In the broader context of Doric temple design, however, 1/5 is only one of a number of different small fractions that could be added to the number of intercolumniations to determine stylobate dimensions, as Coulton 1974, 76, 83–84 points out.

and counteracts, the difference between the 5:1 (=80:16) ratio of intercolumniation to triglyph module and the 81:16 ratio between actual intercolumniation and actual triglyph width with which the Parthenon was built, with its overriding concern for 3:2 and 9:4 continuous proportionality. Furthermore, it suggests that the holding together, indeed the very being, of the building, as *one thing*—what the refinements seem to address—takes precedence over proportion, per se: as we see here with the refinement of column inclination, the refinements are not just added onto pre-existing proportions, but instead engender adjustments to the building's overall *symmetria*—in fact, a reciprocal back and forth of adjustments—[56] for the sake of the unity of the whole.

One could look to the brief but challenging discussion of 'the one' in Resp. 7 (probably a later theoretical reflection on a mid-fifth-century mathematical problem), as an indication of the ontological depths of this problem of unity with respect to harmonics, and of the particular character of what is at stake in the question of the unit, or module, in the Parthenon. Just as Socrates and Glaucon, in *Resp.* 7, 524d-526c, discuss 'the one' as holding together as a unit despite the apparent divisibility of any specific 'one' into innumerable parts, so the Parthenon is a building made up of carefully coordinated parts involving progressively smaller adjustments (suggesting the possibility of an *anthyphairesis* without limit) that also has a powerful presence as a single, harmonious whole. Creating unity out of disparate parts is, of course, a fundamental aesthetic value in many contexts and is even, one might argue, the defining value of aesthetics, from the point of view of the object (as opposed to that of the subject, of subjective judgment). However, the construction of a unified whole in the Parthenon engages a much more specific set of concerns, namely, the consideration of the relationship of *harmonia* (the wholeness, or oneness, of the object) to a constituent module, or unit, made visible in the triglyph that measures the building—and, in effect, defines and anchors its corners at the level of the frieze.

In additional to being a practical issue of constructive design, this is a theoretical issue of a mathematical sort, and, in the case of the Parthenon, a question of direct experience as well. A viewer or worshipper in the presence of the Parthenon experiences the building as an ontological and aesthetic whole, and at the same time is engaged, directly and intuitively, by an interplay of related parts (see figure 7) whose proportional relationships function as a kind of visual analogue for the effects of harmony in music, i.e., the relationships among pitches, built from some of the very same numerical ratios present in the Parthenon, with respect to a whole. The tension between wholeness and multiplicity remains a tension, as it does in the experience of music, precisely because the multiplicity of proportional parts is ultimately irreducible to the aesthetic/ontological whole, since that whole necessarily incorporates

56 Consider, in this context, the analysis that Korres 1999 and Haselberger 2005, 127–133, 142–146 provide of the concept of 'refinements of refinements'.

the irrational. It is this tension that one experiences as harmony, that underlies what various scholars have identified as the building's uniquely organic quality,[57] and that shapes the intuitive experience of the building.

The third and last principle, which is mathematical but not merely mathematical, already alluded to above, is the harmony of the whole. As in the construction of the ancient Greek musical scale some decades before the Parthenon was built, and as Nikulin 2012 details in his discussion of the divided line and its importance to Plato some decades after it was built, so too in the temple's construction, the consideration of a geometric object in arithmetic terms as a means of forging an aesthetic (and ontological) whole gave rise to an irreducible tension between magnitude and multitude. The analogy between the architecture of the Parthenon and musical harmonics is based not only, or even primarily, on the simple presence of musically significant ratios throughout the building; rather, it is the coordination and, even more so, the construction of interdependent ratios to form the *symmetria* of the whole that gives the building its musical quality. We can properly speak of the harmonics of the building here. In terms of harmonics, however, *symmetria* is really only a starting point for a much more complex engagement of the problems raised by musical harmonics, including the necessary acknowledgement of the irrational, of magnitudes that cannot be reduced to multitudes. It is the notion of *harmonia*, in the fullest and most rigorous sense, that encompasses the irrational—as well as joining ratios and *symmetriai*—in order to make a whole that is central to understanding the Parthenon and its invitation to dialectical thought ('dialectical' in the sense discussed in *Resp.* 7, and which we will discuss further briefly below). However, just as with the construction of the musical scale, the deepest engagement with these problems in harmonics begins with the attempt to construct *symmetria*—itself already more than just a series of proportions, or even continuous proportionality, but rather the resonance of each part with all the others and with the whole. Indeed, as Pollitt 1964 has observed, *symmetria* for the Greeks was not only a fundamental aesthetic concept, but arguably the predominant term of value in all of classical Greek and Roman aesthetics.[58] *Symmetria* not only forms the basis of the Parthenon's aesthetics, but, as we have begun to outline here, was the starting point of the inquiry into the nature of the building's being itself.

However, as we have argued above, *symmetria* alone could not fully account for the building's being, as its designers understood it: the principle of *harmonia* was even more crucial, and ultimately given priority within an ontological/aesthetic understanding that explicitly and self-reflexively incorporated the irrational. In the inclination of the peristyle columns (which by the time of the Parthenon were a standard

---

57 See Haselberger 2005, 105, Gruben 2001, 187–188, Pollitt 1972, 74–78, Goodyear 1912, 81–103, and Hoffer 1838, esp. 370.

58 Pollitt 1964, 14–23, 256–258.

refinement in Attic Doric temples)[59] and the slight discrepancy that they introduce between the dimensions of the stylobate and the dimensions of the entablature, the Parthenon in fact strove to bring into harmony two different and not entirely compatible principles of *symmetria*. These two *symmetriai* were, once again, the 5:1 intercolumniation to triglyph ratio used in the modular design of the Doric stylobate in the mid-fifth century and the 81:16 ratio between intercolumniation and triglyph (intercolumniation: column diameter: triglyph :: 81:36:16) on which the continuous proportions specific to the Parthenon are based. Thus, the refinement of column inclination was thoroughly integrated into the design of the building: as a mediating factor, it effectively 'solved' the incompatibility of these two *symmetriai*, both so crucial to the temple's design. Or rather, since this 'solution' also introduced approximations and adjustments that departed from perfect measure (for instance, the visible triglyph was now very close to, but not exactly, the width of the module used for the stylobate), it was clearly part of the larger working-out process in response to the irreducible tension between magnitude and multitude, the process of joining the parts of the building as much as possible into a geometric whole defined by numerical ratios—that is, into a harmonic unity. As should be clear, this raises issues quite similar to those at stake in the Parthenon's approach to the corner problem, and in ways that make evident how the treatment of column inclination—with the adjustments it necessarily introduces into the entablature—is integrally bound up with the negotiation of the corner problem itself. This harmony shows, for instance, in the way the contraction of metopes at the corners, virtually unique to the Parthenon, relates both to the excessive column contractions (another unusual feature of the building) and to the now standard refinement of thickening of the corner columns: all three now work together to create the impression of a strengthened corner (see figure 5). And it is crucial to note that the emphasis on corners has the effect of reinforcing the unity and definition of the building as a three-dimensional object, i.e., as a harmonious whole, since the corners are what articulate the most general contours of the building's three-dimensional geometric form. Indeed, the method, discussed in this section, of joining slightly incompatible *symmetriai* and of integrating different but related problems (column inclinations, the corner problem) suggests an overriding concern with *harmonia*, even at the expense of perfect *symmetria*.

It is a defining characteristic of art objects, as opposed to ordinary objects of manufacture, not only to relate formal—and even material—qualities to the transcendent realm of thought and meaning, but also to reflect upon themselves, upon their own conditions of being (or, more precisely, to explicitly provoke such reflection in viewers).

---

59 On the integration of column inclinations with the other refinements, as an emerging canonical part of Doric design, in one of the Parthenon's most important predecessors, the Temple of Aphaia at Aegina, see Haselberger in Neils, ed. 2005, 119, 122, 124 and Haselberger in Haselberger, ed. 1999, 16, 32.

This self-reflection, embedded within the work through a kind of dialogue between the artist and the artwork, becomes, in the afterlife of the work, a dialogue between artwork and viewer. The Parthenon creates such a dialogue, simultaneously in the way it engages the senses—and, indeed, the whole body—and in the way it engages thought. Consider, for instance, the refinement of curvature: the curvature of all the apparently straight lines in the Parthenon, a feature without any evident constructive necessity, seems designed to introduce a discrepancy between the (straight) mathematical idea of the temple and its (curved) physical reality, and thus raises ontological questions in an entirely self-reflexive manner. Indeed, the introduction of this discrepancy, the product of extensive and precise technical labor, makes questions of (mathematical) ontology and of the relationship of mathematical to physical being explicitly, and self-reflexively, present in the building. Furthermore, there is a sense in which the beautiful, in its representation of a perfection that cannot be perfectly described—i.e., that contains a mysterious, irreducible remainder—both engages the viewer directly through the presentness of the object and leads her or him beyond it. In this sense, there is a resonance between the Parthenon's pedagogical function—its invitation to (what is described in *Resp.* as) dialectical thought and, to put it this way, to an education in the liberal arts—and its aesthetic function as a work of art: both involve an open-ended dialogue with the viewer or interlocutor, inviting him or her to engage with what is present in order also to go beyond it—affectively, interpretively, imaginatively, intellectually. And naturally, as we have indicated above, both the pedagogical and the aesthetic modes resonate meaningfully with the Parthenon's principal, religious function, its call to worship and to piety. Furthermore, both the building's beauty—and by extension, its power as a work of art—and its pedagogical character arise, in some fundamental way, from the dialogue between arithmetic and geometry, the kind of dialogue between mathematical arts discussed in *Resp.* as the basis for dialectic. In the Parthenon, these two mathematical arts are brought together with a third art, musical harmonics, and all are pressed into the service of trans-disciplinary questions and aesthetic expression alike—that is, the three arts together work to engage a viewer in philosophical and value-oriented questions (the latter with respect to beauty)—and in that sense could more properly be described in this context as liberal arts.

V Conclusion

The sophistication with which the Parthenon manifests an algorithmic approach to constructive problems makes clear that there is a robust reflective (theoretical) awareness of the mathematical principles involved, and no mere rules of thumb being arbitrarily instituted. Yet, this theoretical treatment is neither formal nor systematic, and it surely is not based on concerns related to proofs or the desire to offer them. If it is true both that (a) the Parthenon's design displays a theoretical sophistication in its mathematical principles and (b) there is clearly not the concern with formal deductive proof distinctive of fourth-century and later Greek mathematics, then we submit that our reading of the Parthenon substantiates both of the two revisionist claims in the historiography of Greek mathematics advanced above.

The first relates to dating the development of an understanding of key mathematical notions, such as incommensurability and continuous proportionality, within Greek mathematical thought. Here, given the dates of the Parthenon's construction (confidently set at 447-432 BC), we can conclude that if the Parthenon's design displays a reflective awareness of the knowledge procedures related to continuous proportion and the relationship of magnitudes and multitudes, then there must have been advances of this kind made before 430 BC, as Szabó argued. Second, with respect to the wider claim about how far Greek mathematics is continuous with earlier thought and practice, we must take note that the reflective awareness at work in the Parthenon is *not* formal or deductive, but rather algorithmic and problematic. As such we surely notice (at a minimum) the close analogy with Mesopotamian astral science of the seventh to fifth centuries. This, we conclude, is good reason to believe that, at least up to the time of the Parthenon, Greek mathematics was borrowing from and surely extending earlier mathematics. Much will remain undecidable in what we have presented, but interpreting the mathematical features of the Parthenon gives us reason to believe that the development of Greek mathematics in the first decades of the second half of the fifth century is, at least in part, a story of continuity with earlier mathematical traditions. If one is clear about the relatively advanced state of *practical* mathematics in mid-fifth century Athens, and agrees that these advances are a function of the algorithm, a development of recursive practice and not of theoretical reflection, then we have good reason to think again about the possibility that there was real continuity between these practicing mathematicians and their near Eastern predecessors. We must, that is, develop a deeper appreciation of the role that the algorithmic problem-oriented style was playing for pre-Euclidian mathematicians. Our monograph on the Parthenon tries to demonstrate that there is a conscious attempt to introduce this mathematics into the building in a thematized and programmatic way, and the intellectual background of that attempt is precisely the problem-based approach to ontological questions about the misfit between magnitude and multitude that arises from the reception of near Eastern mathematics in the Greek world by theorists and practitioners understood to be Pythagoreans.

Bibliography

Texts quoted

Aristoteles, *Metaphysica*. Jaeger, ed. 1957. Oxford.
Aristoteles, *De Caelo*. Ross, ed., 1953. Oxford.
Aristoteles, *Topica*. Ross, ed., 1958. Oxford.

Euclides, *Elementa*. Densmore, ed.; Heath, tr. 2007. Sante Fe. (repr. 1999)
Plato, *Politicus*. Fowler, ed., 1925. Cambridge, MA.
Plato, *Respublica*. Emlyn-Jones, ed., 2013. Cambridge, MA.
Plato, Theaetetus. Fowler, ed., 1921. Cambridge, MA.

Works quoted

Acerbi, F. 2010. "Two Approaches to Foundations in Greek Mathematics: Apollonius and Geminus." In: *Science in Context* 23, 151–186.
Barbera, A. 1981. "Republic 530c–531c: Another Look at Plato and the Pythagoreans." In: *The American Journal of Philology* 102, 395–410.
Barbera, A. 1991. *The Euclidian Division of the Canon: Greek and Latin Sources*. Lincoln.
Barker, A. (ed.) 1984–1989. *Greek Musical Writings*. 2 vols. Cambridge.
Barker, A. 2007. *The Science of Harmonics in Classical Greece*. Cambridge.
Benson, H. H. 2012. "The Problem is not Mathematics, but Mathematicians: Plato and the Mathematicians Again." In: *Philosophia Mathematica* 3, 170–199.
Berggren, L. J. 1984. "History of Greek Mathematics: A Survey of Recent Research." In: *Historia Mathematica* 11, 394–410.
Bodnar, we. 2007. "Oenopides of Chius: A survey of the modern literature with a collection of the ancient testimonia." Berlin: *Max Planck Institute Preprint Series*, No. 327.
Bowen, A. C. 1984. Review of Szabó 1978. In: *Historia Mathematica* 11, 335–345.
Brack-Bernsen, L. 1999. "Goal-Year Tablets: Lunar Data and Predictions." In: N. M. Swerdlow (ed.), *Ancient Astronomy and Celestial Divination*. Cambridge, MA, 149–177.
Brack-Bernsen, L. 2014. "The Lunar Four in Connection with Concepts and Methods in Early Babylonian Astronomy." Unpublished manuscript.
Britton, J. P. 2002. "Treatments of Annual Phenomena in Cuneiform Sources." In: Steele, J. M. & A. Imhausen (eds.), *Under One Sky: Astronomy and Mathematics in the Ancient Near East*. Münster, 21–78.
Bruchmüller, U. (ed.) 2012. *Platons Hermeneutik und Prinzipiendenken im Licht der Dialoge und der antiken Tradition*. Hildesheim.
Bundgaard, J. A. 1976. *The Parthenon and the Mycenaean City on the Heights*. Copenhagen: The National Museum of Denmark.
Büsing, H. 1988. "Einheitjoch und Triglyphon am Parthenon." In: M. Schmidt (ed.), *Kanon: Festschrift Ernst Berger*. Basel, 2–3.
Burkert, W. 2004. *Babylon Memphis Persepolis: Eastern Contexts of Greek Culture*. Cambridge, MA.
Burnyeat, M. 1987. "Platonism and Mathematics: A Prelude to Discussion." In: A. Graeser (ed.), *Mathematics and Metaphysics in Aristotle*. Bern, 213–240.
Burnyeat, M. F. 2000. "Plato on Why Mathematics is Good for the Soul." In: T. Smiley (ed.), *Mathematics and Necessity*. Oxford, 1–82.
Cherniss, H. 1951. "Plato as Mathematician." In: *Review of Metaphysics* 4, 395–425.
Christianidis, J. (ed.) 2004. *Classics in the History of Greek Mathematics*. Dordrecht.
Cornford, F. 1932. "Mathematics and Dialectic in the Republic VI-VII." In: *Mind* 41:37–52, 173–190.
Coulton, J. J. 1974. "Towards Understanding Doric Design: The Stylobate and the Intercolumniations." *Annual of the British School in Athens* 69, 61–86.
Cuomo, S. 2001. *Ancient Mathematics*. London.
Dillon, J. M. 2003. *The Heirs of Plato: A Study of the Old Academy (347–274 BC)*. Oxford.
Evans, J. 1998. *The History and Practice of Ancient Astronomy*. Oxford.
Friburg, J. 2007. *Amazing Traces of a Babylonian Origin in Greek Mathematics*. Hackensack, NJ.
Fowler, D. H. 1999. *The mathematics of Plato's Academy: A new reconstruction* (orig. 1987). Oxford.

Goodyear, W.H. 1912. Greek *Refinements: Studies in Temperamental Architecture*. New Haven: Yale University Press.
Graham, D. 2014. "Philolaus." In: C. A. Huffman (ed.), *A History of Pythagoreanism*. Cambridge, 46–67.
Griswold, C. (ed.) 2002. *Platonic Writings/Platonic Readings* (orig. New York 1988). Pennsylvania.
Gruben, G. 2001. *Griechische Tempel und Heiligtümer*. Munich: Hirmer Verlag.
Hahn, R. 2017. *The Metaphysics of the Pythagorean Theorem*. Albany, NY.
Haselberger, L. 2005. "Bending the Truth: Curvature and Other Refinements of the Parthenon". In: Neils, J., ed. *The Parthenon: From Antiquity to the Present*. Cambridge: Cambridge University Press, 101–58.
Hauck, G. 1879. *Die subjektive Perspektive und die horizontalen Curvaturen des dorischen Styls: eine perspektivisch-äesthetische Studie*. Stuttgart: Konrad Wittwer.
Hoffer, J. 1838. "Das Parthenon zu Athen, in seinen Hauptteilen neu gemessen." *Allgemeine Bauzeitung* 3: 249–51, 371–5, 379–83, and 387–91.
Horky, P. S. 2013. *Plato and Pythagoreanism*. Oxford.
Høyrup, J. 2009. "Review of Robson, Mathematics in Ancient Iraq: A Social History." In: *Math Intelligencer* 32, 65–67.
Høyrup, J. 2010. "Old Babylonian 'Algebra', and What It Teaches Us about Possible Kinds of Mathematics." In: *Ganita Bharati* 32, 87–110.
Huber, P. J. & J. Britton 2007. "A lunar six text from 591 BC." *Wiener Zeitschrift für die Kunde des Morgenslandes* 97, 213–18.
Huber, P. J. & J. M. Steele 2007. "Babylonian lunar six tablets." *SCIAMVS* 8, 8–36.
Huffman, C. 1993. *Philolaus of Croton: Pythagorean and Presocratic*. Cambridge.
Huffman, C. A. (ed.) 2014. *A History of Pythagoreanism*. Cambridge.
Knorr, W. R. 1975. *The Evolution of the Euclidean Elements: A Study of the Theory of Incommensurable Magnitudes and Its Significance for Early Greek Geometry*. Dordrecht.
Korres, M. 1994. "Der Plan des Parthenon." *Mitteilungen des Deutschen Archäologischen Instituts, Athenische Abteilung* 109, 53–120.
Korres, M. 1999. "Refinement of Refinements". In: Haselberger, L., ed. *Appearance and Essence. Refinements of Classical Architecture: Curvature. Proceedings of the Second Williams Symposium on Classical Architecture, held at the University of Pennsylvania, Philadelphia, April 2–4, 1993*. Philadelphia: The University of Pennsylvania Museum, 79–104.
Lloyd, G. E. R. 1978. "Saving the Appearances." In: *The Classical Quarterly* 28, 202–22.
Lloyd, G. E. R. 1992. "The Meno and the mysteries of mathematics." In: *Phronesis* 37, 166–183.
Mahoney, M. S. 1968. "Another Look at Greek Geometrical Analysis." In: *Archive for the History of the Exact Sciences* 5, 318–348.
Mertens, D. 2006. *Städte und Bauten der Westgriechen*. Munich: Hirmer Verlag.
Mittelstrass, J. 2004 [1962-66]. "Die Entdeckung der Möglichkeit von Wissenschaft." In: J. Christianidis, *Classics in the History of Greek Mathematics*. Dordrecht, 19–44.
Mueller, Ian. 1992. "Mathematical Method and Philosophical Truth." In: R. Kraut, ed., *Cambridge Companion to Plato*. New York, 170–199.
Negrepontis, S. 2012. "Plato's theory of knowledge of Forms by Division and Collection in the Sophistes is a philosophic analogue of periodic anthyphairesis." Unpublished manuscript.
Neils, J., ed. 2005. *The Parthenon: From Antiquity to the Present*. Cambridge: Cambridge University Press.
Netz, R. 1999. *The Shaping of Deduction in Greek Mathematics: A Study in Cognitive History*. Cambridge: Cambridge University Press.
Netz, R. 2002. "Counter Culture: Towards a History of Greek Numeracy." In: *History of Science* 40, 321–352.
Netz, R. 2014. "The Problem of Pythagorean Mathematics." In: C. A. Huffman (ed.), *A History of Pythagoreanism*. Cambridge, 167–184.
Neugebauer, O. E. 1969. *The Exact Sciences in Antiquity* (orig. Providence 1957). New York.
Nikulin, D. 2010. *Dialectic and Dialogue*. Stanford.

Nikulin, D. 2012. "Indivisible Lines and Plato's Mathematical Ontology." In: U. Bruchmüller (ed.), *Platons Hermeneutik und Prinzipiendenken im Licht der Dialoge und der antiken Tradition*. Hildesheim, 292–326.
Ober, J. 2015. *The Rise and Fall of Classical Greece*. Princeton.
Pollitt, J. J. 1964. *Ancient View of Greek Art: Criticism, History, and Terminology*. New Haven, CT: Yale University Press.
Pollitt, J. J. 1972. *Art and Experience in Classical Greece*. Cambridge: Cambridge University Press.
Robson, E. 2008. *Mathematics in Ancient Iraq: A Social History*. Princeton.
Rowe, D. 2013. "Otto Neugebauer's Vision for Rewriting the History of Ancient Mathematics." Unpublished manuscript.
Steele, J. M. 2006. "Greek Influence on Babylonian Astronomy?" In: *Mediterranean Archaeology and Archaeometry* 6, 153–60.
Szabó, Á. 1978. *The Beginnings of Greek Mathematics* (orig. Munich 1969). Transl. A. M. Ungar. Dordrecht.
Taylor, C. W. 1967. "Plato and the Mathematicians: An Examination of Professor Hare's Views." In: *The Philosophical Quarterly* 17, 193–203.
Waddell, G. 2002. "The Principal Design Methods for Greek Doric Temples and Their Modification for the Parthenon." In: *Architectural History* 45, 1–31.
Waschkies, H-J. 1989. *Anfänge der Arithmetik im Alten Orient und bei den Greichen*. Amsterdam.
Waschkies, H-J. 2004. "Introduction." In: J. Christianidis, *Classics in the History of Greek Mathematics*. Dordrecht, 3–18.
Wesenberg, B. 1982. "Wer erbaute den Parthenon?" In: *Mitteilungen des Deutschen Archäologischen Instituts, Athenische Abteilung* 97, 99–125.
Wesenberg, B. 1983. "Parthenongebälk und Südmetopenproblem." *Jahrbuch des Deutschen Archäologischen Instituts* 98: 57–86.
Wilson Jones, M. 2001. "Doric Measure and Architectural Design 2: A Modular Reading of the Classical Temples." *American Journal of Archaeology* 105, 675–713.
Wilson Jones, M. 2006. "Ancient Architecture and Mathematics: Methodology and the Doric Temple." In: S. Duvernoy and O. Pedemonte, eds., *Nexus VI: Architecture and Mathematics*. Turin, 1–20.
Yeroulanou, M. 1998. "Metopes and Architecture: The Hepaisteion and the Parthenon." In: *Annual of the British School at Athens* 93, 401–25.
Zhmud, L. 1998. "Plato as "Architect of Science"." In: *Phronesis* 43, 211–244.
Zhmud, L. 2006. *The Origin of the History of Science in Classical Antiquity*. Berlin.

Illustrations

Fig. 1: The Parthenon, Athens, 447-432 BC: west façade.

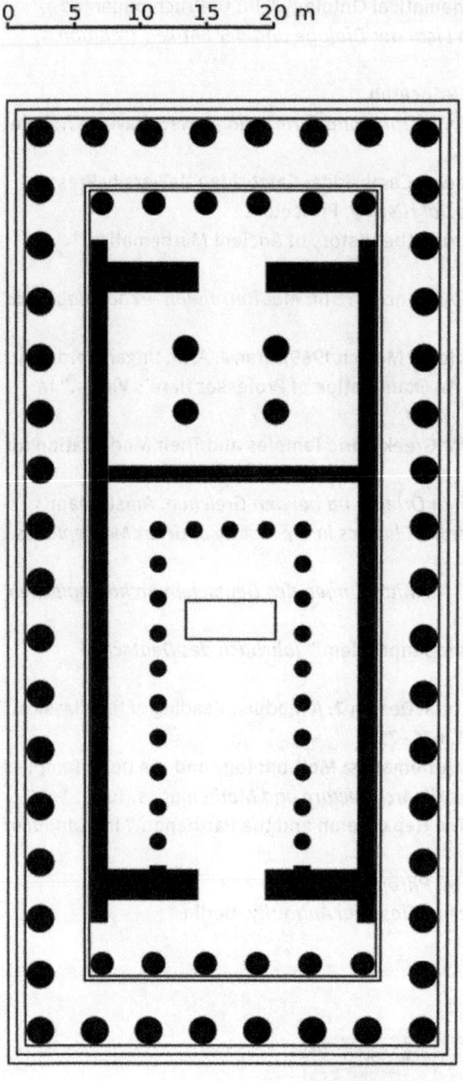

Fig. 2: The Parthenon: ground plan.

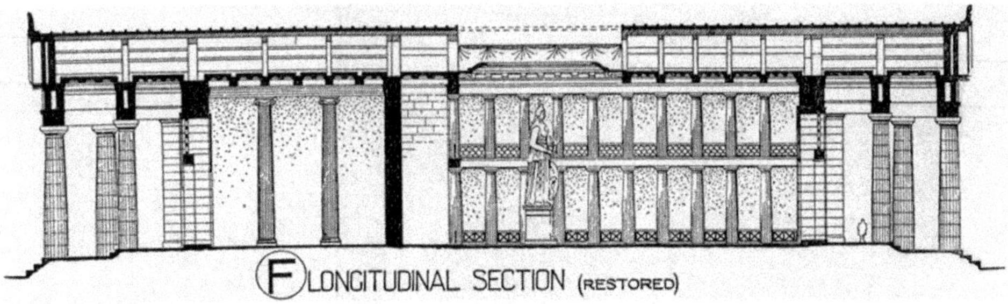

Fig. 3: The Parthenon: section, showing the Doric and Ionic orders of the interior.

Fig. 4: Parthenon replica, Nashville (begun 1920s): interior of the naos.

Fig. 5: Joseph-Philibert Girault de Prangey, *Façade and North Colonnade of the Parthenon on the Acropolis, Athens*, daguerreotype, 1842.

Fig. 6: The Parthenon: east façade, south side of the entablature.

Fig. 7: The Parthenon: view from the northwest.

Jeffrey A. Oaks[1]

Diophantus, al-Karajī, and Quadratic Equations

Abstract: In the beginning of his *Arithmetica*, Diophantus promises to show how to solve three-term quadratic equations. The rules are not in the extant part of his work, but al-Karajī (early eleventh century) provides clues as to how Diophantus may have derived and presented them. In his algebra book *al-Fakhrī* al-Karajī gives both the standard Arabic rule for solving three-term equations and the variation practiced by Diophantus. After proofs of the rules he gives a derivation that he calls "the method of Diophantus". By examining proofs and problems in several of al-Karajī's works we propose that Diophantus derived his rules for solving three term equations by this "method of Diophantus", and that al-Karajī borrowed both the derivations and the rules from the *Arithmetica*.

That we are able to speak of Diophantus and al-Karajī both solving quadratic equations is due to the common conceptual foundations and procedures shared by Greek and Arabic algebra. Continuity from one to the other derives not from any chain of texts, but rather from oral tradition that apparently flourished across the old world, and which has left traces in our surviving texts. It is within this conceptual setting that a variation in rules for solving equations enables us to sort out al-Karajī's rules and postulate Diophantus's lost exposition.

Keywords: Diophantus, al-Karajī, Premodern Algebra, Quadratic Equations.

I Introduction

In a well-known passage in the introduction to his *Arithmetica*, Diophantus promises to explain the solutions to composite quadratic equations: "Later I will also show you how to solve the case in which two terms are left equal to one". The general rule is not found in any extant part of his book, in Greek or in Arabic. But he does solve composite equations in the Greek books 'IV' to 'VI', and in two occasions the rule by which it is calculated is spelled out.

The rules Diophantus follows are not the standard rules for solving composite equations in Arabic algebra. The difference lies with the fact that in Arabic the equation is first normalized, that is, the 'coefficient' of the square term is set to 1, while Diophantus's rules do not call for normalization. This difference becomes important when we turn to al-Karajī's early eleventh century algebra book *al-Fakhrī*. Al-Karajī

1 I thank Jean Christianidis for his comments on a draft of this paper, and especially for adjusting my translations from Diophantus.

https://doi.org/10.1515/9783110565959-010

borrowed heavily from Diophantus, and he presents both the standard Arabic rules and the rules followed by Diophantus, though without naming him. Then, after his geometric proofs, al-Karajī gives another procedure for two equation types based on completing the square that he calls the 'method of Diophantus'. But these methods are not applied anywhere in the extant books of the *Arithmetica*. There is a problem, then, to reconcile these two 'Diophantine' procedures.

Through a reading of the solutions of problems in three of his books, I argue that al-Karajī learned the rules for non-normalized equations from the *Arithmetica*. And in a little-studied work of his, the *Causes of Calculation in Algebra*, the 'method of Diophantus' is transformed into a proof to the standard Arabic rule for each type of composite equation. This leads me to propose that somewhere between Book VII and the Greek Book 'IV' Diophantus derived his rules for solving normalized equations via this 'method of Diophantus'.

Suggesting that both Diophantus and al-Karajī solved 'composite equations', and that the latter borrowed from the former in his own work, implicitly situates the two mathematicians in the same conceptual sphere. Often times cross-cultural comparisons rest on anachronistic readings in which radically different conceptual foundations go entirely unnoticed.[2] This is not the case here. A series of recent articles has shown that while the structure and underlying concepts of Arabic algebra are indeed fundamentally different from those of modern elementary algebra, they match very well those at work in Diophantus's *Arithmetica*. For this reason, we refer to the method common to Diophantus and Arabic authors as 'premodern algebra', to distinguish it from modern algebra.[3] What we should be careful about, then, is to distinguish between premodern equations in Diophantus and al-Karajī on the one hand, and modern equations on the other.

The near isomorphism between Diophantus's vocabulary, rules, and concepts with those of Arabic algebra strongly suggest a historical connection between them. Unfortunately we lack textual evidence linking Alexandria with Baghdad, but this is no reason to dismiss the suggestion. Premodern algebra had circulated orally among professional groups before al-Khwārizmī and Ibn Turk wrote the first Arabic books on the topic, so it is entirely plausible that the method was just not written down in (enough) books to have left the traces we would like.[4] If there was in fact a continuous tradition of algebraic problem solving from which both Diophantus and al-Karajī drew, it is the fortuitous *discontinuity* somewhere along the line in the particular ways that composite equations were solved that allows us to refine our understanding of the influences of Diophantus on al-Karajī, and to propose the way Diophantus derived his rules.

I present the basic features of premodern algebra in section 2. In section 3 I describe Diophantus's *Arithmetica* and Abū Kāmil's *Book on Algebra*, the two main

---

2 Such is the case with Roshdi Rashed's accounts of Arabic algebra. See Oaks 2014b.

3 See Christianidis, this volume, for more on premodern algebra and for references to relevant studies.

4 See Oaks 2014; King 1988, for evidence of this oral tradition.

influences on al-Karajī, as well as *al-Fakhrī* itself. Composite equations in Diophantus and in Arabic algebra up to al-Karajī's *al-Fakhrī* take up sections 4 through 6. Relevant sections of al-Karajī's other works are reviewed in sections 7 and 8, and it is all brought together in the final section 9.

II Premodern Algebra

Arabic algebra was a numerical method of problem-solving that differs in significant ways from modern algebra in both its structure and its concepts. The brief overview in this section covers the critical elements that are described in more detail in Oaks 2009 and Christianidis & Oaks 2013, and are summarized in Christianidis, this volume.

In both the *Arithmetica* and in Arabic algebra names were given to the powers of the unknown. The names of the first few powers, together with the corresponding (though not conceptually identical) modern notation, are listed in this table:

Diophantus	Arabic	Modern
monas (unit)	*dirham* (a silver coin), *min al-ʿadad* (in number), or *aḥad* (unit)	1
arithmos (number)	*jidhr* (root) or *shayʾ* (thing)	x
dynamis (power)	*māl*[5] (sum of money)	x^2
kybos (cube)	*kaʿb* (cube)	x^3
dynamodynamis	*māl māl*	x^4

Equations are stated in terms of these names, like Diophantus's "1 *dynamis*, 16 units lacking 8 *arithmoi* are thus equal to 1 *dynamis*, 12 units".[6] This corresponds to our modern $x^2 + 16 - 8x = x^2 + 12$. An Arabic example is this equation from Abū Kāmil's late ninth century *Book on Algebra*: "sixty-two things and half a thing less four *māls* and a fourth of a *māl* equal a hundred dirhams and two *māls*",[7] corresponding to $62\frac{1}{2}x - 4\frac{1}{4}x^2 = 100 + 2x^2$.

Premodern polynomials, and thus also equations, were not conceived the same way as their modern counterparts. Instead of a linear combination assembled from the operations of addition, subtraction, scalar multiplication, and exponentiation, polynomials in Greek and Arabic were treated as aggregations of the names of the powers, with no operations present. Diophantus's "1 *dynamis*, 12 units", for example, is a collection of thirteen objects of two kinds, like saying '1 table, 12 chairs'. For

5 The plural of *māl* is *amwāl*, but I write it with the English suffix: *māls*

6 Diophantus, II.12.

7 From problem (2). Abū Kāmil, *Alg.*(a) 50.16; Abū Kāmil *Alg.*(b) 339.9. Rashed numbers this as problem <8>. I will translate into English all powers but *māl*.

convenience I often represent premodern algebraic expressions and equations with modern notation, but the premodern concepts should be kept in mind.

Diophantus instructs the reader to simplify equations with two steps: "to add the lacking species on both sides" and "to subtract like from like on both sides".[8] These same steps are also performed in Arabic algebra, where they are called *al-jabr* (restoration) and *al-muqābala* (confrontation).

The rules for solving simplified equations take the numbers of the terms (i.e. the 'coefficients') as parameters, and because only positive numbers were acknowledged, the Arabic books classify six types of simplified equation. They are given here in the order in al-Karajī's *al-Fakhrī*:

<u>Simple equations</u>
1) things equal number ($bx = c$)
2) *māls* equal things ($ax^2 = bx$)
3) *māls* equal number ($ax^2 = c$)

<u>Composite (*murakkabāt*) or connected (*muqtarināt*) equations</u>
1) *māls* and things equal number ($ax^2 + bx = c$)
2) *māls* and number equal things ($ax^2 + c = bx$)
3) things and number equal *māls* ($bx + c = ax^2$)

Each type of equation is given its own solution, and these solutions are accompanied in many books by geometric proofs.

To solve a problem by algebra one, or sometimes more than one, unknown quantity is named in terms of the designated names. The conditions of the enunciation are worked out to set up a determinate equation in terms of these names, which is then simplified and solved. Here is a simple indeterminate problem from ʿAlī al-Sulamī's tenth-century *Sufficient Introduction on Calculation by Algebra and What One Can Learn from Its Examples*:[9]

> [*Enunciation*]
> So if [someone] said, a quantity[10]: if you added to it a dirham the result has a root, and if you subtracted from it a dirham the result has a root.
> [*Setting up the equation*]
> So make your quantity a *māl* and a dirham [$x^2 + 1$], so that if you subtracted from it a dirham, the remainder has a root. So it is the original quantity. Then add to it a dirham to get a *māl* and

---

8 Diophantus, 14.11-20.

9 ʿAlī al-Sulamī f. 50b.3. I numbered this as problem I.31 in Oaks 2015. The nine indeterminate problems in this book, and also the 38 in Abū Kāmil's *Book on Algebra*, do not derive from Diophantus. Cf. Sesiano 1977b; Oaks 2015. Diophantus's problem II.11 is nearly the same. Instead of adding and subtracting 1, he adds 2 and 3. His approach in his second solution matches al-Sulamī's solution.

10 The word behind 'quantity' is *māl*, used here in a non-algebraic sense as a common noun. Throughout the solution I translate this arithmetical *māl* as 'quantity' and I leave the algebraic *māl* (x^2) untranslated. See Oaks & Alkhateeb 2005, for the different meanings of *māl* in Arabic arithmetic and algebra.

two dirhams [$x^2 + 2$]. So look for a root of this. So if you made it a thing and a dirham [$x + 1$], you multiplied that by itself to get a *māl* and two things and a dirham. So this equals a *māl* and two dirhams [$x^2 + 2x + 1 = x^2 + 2$].
[*Simplifying and solving the equation*]
So cast away a *māl* by a *māl* and a dirham by a dirham, leaving: a dirham equals two things [$1 = 2x$]. So the thing is a half and the *māl* is a fourth. And thus the *māl* and a dirham is a dirham and a fourth. So it is the sought-after [quantity].
[*Another solution*]
And if you wanted, the root[11] is a thing and half a dirham [$x + \frac{1}{2}$] so the root [x] is one and a half and a fourth [$1\frac{3}{4}$] and the *māl* is three and half of an eighth [$3\frac{1}{16}$], and adding to it a dirham gives four and half of an eighth, which is the sought-after [quantity].

In the first solution, the quantity is named "a *māl* and a dirham" ($x^2 + 1$), and the root of the larger square is named "a thing and a dirham" ($x + 1$), resulting in the determinate equation $x^2 + 2x + 1 = x^2 + 2$. This is then simplified and solved. The second solution proceeds in a similar manner.

The same structure holds for the problems in Diophantus's *Arithmetica*. Problem I.1 is a simple determinate example. The enunciation is "to divide a given number into two having a given difference. Let the given number be 100, and the the difference be 40; to find the numbers."[12] The smaller number is named "1 *arithmos*" [x], and the equation is set up as "100 units are equal to 2 *arithmoi*, 40 units" [$100 = 2x + 40$]. This is then simplified and solved.

As Arabic authors inform us, the purpose of algebra is to find unknown numbers, that is, to solve problems.[13] Diophantus's 'way' is likewise a method of solving problems. In both cases this method consists of naming an unknown or unknowns in terms of pre-assigned names of the powers, then working the conditions of the enunciation to set up a determinate equation in terms of these names. This equation is then simplified and solved. This core method, with the same aggregations interpretation of polynomials, is common to both Diophantus's work and Arabic algebra.[14] It is the method I intend by the term 'algebra' in this article.

III Diophantus, Abū Kāmil, and al-Karajī

The two main influences on al-Karajī's *al-Fakhrī* are Diophantus's *Arithmetica* and Abū Kāmil's *Book on Algebra*.[15] The three books are described here in chronological order.

---

11 I.e., the square root of the square $x^2 + 2$.
12 Diophantus, I.1. For a schematic rendering of this problem, see Christianidis, this volume.
13 See Oaks forthcoming, for translations of such passages from early Arabic books.
14 Christianidis & Oaks 2013.
15 Al-Karajī only mentions Diophantus by name in the two places where he explains the 'method of Diophantus'. He never cites Abū Kāmil by name.

Diophantus of Alexandria wrote his *Arithmetica* in Greek some time in late antiquity, perhaps in the fourth century CE. Six of the original thirteen books are extant in Greek: Books I through III and three other books that come after Book VII. I follow Sesiano 1982 in calling these last three Books 'IV' to 'VI', with quotes, preserving the numbering they were assigned in the Renaissance. The *Arithmetica* was translated into Arabic in the mid- to late-ninth century CE by Qusṭā ibn Lūqā. Because Diophantus's method is structurally and conceptually compatible with Arabic algebra, Qusṭā translated Diophantus's book using the algebraic terminology of his time. Only Books IV to VII survive of this translation.[16]

In his introduction, Diophantus reviews the terminology and rules in the order in which they are applied in the solutions to problems.[17] He begins with the terms appearing in the enunciations of problems, then he gives the names of the powers up to the sixth degree and their reciprocals, and he explains how to multiply them. This is followed by a brief explanation of operations on polynomials and how to set up and simplify equations. It is at this point that he says he will later show how to solve three-term equations. The rest of the extant treatise consists of worked-out problems.

Abū Kāmil worked in Egypt in the late ninth century. His *Book on Algebra* (*Kitāb fī al-jabr wa'l-muqābala*) builds on the *Book of Algebra* of al-Khwārizmī, which was written in Baghdad sometime 813–833 CE.[18] The two treatises begin with what I call the 'algebra proper', which is divided into two parts. The first part gives the definitions and rules necessary to solve the collection of worked-out problems in the second part. In both books the first part covers in order the names of the first two powers, the classification, solutions, and proofs to the six simplified equations, and operations on polynomials and roots. Abū Kāmil includes more rules, examples, and proofs than al-Khwārizmī. And where al-Khwārizmī gives 39 worked-out problems in the second part of the algebra proper, Abū Kāmil gives 74.[19]

In both books, the algebra proper is followed by related material. al-Khwārizmī adds chapters on the rule of three and mensuration and an extensive book (*kitāb*) of inheritance problems solved mainly by algebra. Abū Kāmil appends to his algebra proper a book of twenty geometry problems, most solved by algebra (his *On the Pentagon and Decagon*), and then a collection of 84 assorted problems, many of which are indeterminate, and again most solved by algebra.

Abū Bakr ibn Muḥammad ibn al-Ḥasan (or al-Ḥusayn) al-Karajī (or al-Karkhī) worked in Baghdad in the late tenth and early eleventh centuries. His main work on algebra is his *al-Fakhrī fī ṣināʿat al-jabr wa'l-muqābala* (*[Book of] al-Fakhrī on the Art*

---

16 Tannery 1893–1895 contains an edition of the Greek text, and the Arabic translation has been edited and translated twice, in Sesiano 1982 and Rashed 1984.

17 Christianidis 2007.

18 Edited in Rashed 2009.

19 See Oaks 2015, for the structure of these treatises and on how this 'algebra proper' is characterized.

of Algebra, or *al-Fakhrī* for short), written in 401H (1010/11 CE).[20] The overall organization of the book is modeled on Abū Kāmil's algebra proper, with the rules of algebra first, then a collection of worked-out problems. He made some changes to the organization of the rules, partly based on Diophantus's introduction. Instead of presenting only the first two powers as al-Khwārizmī and Abū Kāmil had done, al-Karajī gives the names of the powers up to the ninth degree as well as their reciprocals. And where the earlier Arabic books passed next to the equations, he continues with numerous examples illustrating operations on polynomials and roots, a section on summing series, and some theorems in arithmetic.

At this point, al-Karajī explains the steps of solving problems by algebra, incuding the naming of the unknown, the working out of the conditions of the enunciation to form an equation, and the simplification of the equation.[21] This leads directly to his classification, solutions, and proofs to the six equations. He concludes the first part of his book with some remarks on setting up equations for indeterminate problems.

Al-Karajī then gives 255 worked-out problems divided into five 'categories'. Just over 100 of these problems are taken from Diophantus's *Arithmetica*, and some others seem to be variations of other problems of Diophantus.[22] Many problems are taken from Abū Kāmil, too. Twenty-two are direct borrowings and fourteen more appear to be modifications of problems from the *Book on Algebra*. The borrowings are most evident when a sequence of problems is taken in order. Abū Kāmil's problems (18) to (22) appear as al-Karajī's III.12 to III.16 respectively, with only (21) given a minor adjustment. Also, Abū Kāmil's (57) to (59) are the same as al-Karajī's IV.21 to IV.23.[23] Like al-Khwārizmī and Abū Kāmil before him, al-Karajī also took problems from other sources, and he probably invented some of his own.

Three other works of al-Karajī come into play in this article, all of which were written after *al-Fakhrī*:

- *Marvelous [book] of Calculation* (*al-Badīʿ fī al-ḥisāb*). This book covers number theory, calculations with polynomials, and algebraic solutions to both determinate and indeterminate problems. al-Karajī 1964.
- *The Sufficient [book] on Calculation* (*al-Kāfī fī'l ḥisāb*). A book on arithmetic and mensuration with an extensive chapter on algebra at the end. al-Karajī 1986.
- *Causes of Calculation in Algebra and Its Proof* (*ʿIlal ḥisāb al-jabr wa'l-muqābala wa'l-Burhān ʿalayhi*). A short work giving arithmetical proofs for the solutions to equations and to rules for operating on roots. Saidan 1986, 353–369.

20 Edited in Saidan 1986, 95–308.

21 al-Karajī, *Kafī* 145.9. See Oaks forthcoming, for an English translation.

22 Woepcke 1853, 18–21 gives a correspondence of the problems from *Arithmetica* Books I-III. Sesiano notes that almost all of Book IV is reproduced in *al-Fakhrī*: Sesiano 1982, 10.

23 Add 6 to the problem numbers for Abū Kāmil to get Rashed's numbering in Rashed 2012.

IV Composite Equations in Diophantus

Diophantus writes in his introduction that one should set up and simplify equations so that "there remains, whenever possible, one term equal to one term. Later I will show you how to solve the case in which two terms are left equal to one."[24] The promised explanation is not in the surviving books, either in Greek or Arabic. But Diophantus does solve three-term equations in the Greek Books 'IV' to 'VI'.[25] In each case he simply gives the answer without showing the calculations. For example, in problem 'VI'.9 he has "630 *dynameis*, wanting 73 *arithmoi* are equal to 6 units [$630x^2 - 73x = 6$], from which one finds the *arithmos* to be $\frac{6}{35}$."

But there are two problems in which the calculations are worked out for an inequality treated as an equation. The first is in problem 'IV'.39, where Diophantus is faced with the inequality $2x^2 > 6x + 18$:

> When we solve this as an equation we multiply half of the *arithmoi* by itself, which gives 9, and we multiply the quantity 2 of *dynameis* by the 18 units, which gives 36. Add this to 9, which gives 45, whose side is not less than 7 units.[26] Add the half of the *arithmoi*, [it becomes not less than 10 units, and divide by the *dynameis*]; it becomes not less than 5 units.

If we were to contort this rule into a modern formula for solving $ax^2 = bx + c$, it would look something like this:

$$x = \frac{\sqrt{\left(\frac{1}{2}b\right)^2 + ac} + \frac{1}{2}b}{a}.$$

One critical difference between Diophantus's procedure and the modern formula is that the former is a *sequence of operations* that begins with the halving of the *arithmoi*, while the latter expresses it all simultaneously with modern algebraic notation. In other words, one is spoken while the other is displayed.

The second example occurs in problem 'V'.10, this time for the inequality $72x > 17x^2 + 17$:

> Half of the *arithmoi* multiplied by itself gives 1296. Subtracting the product of the *dynameis* and the units, which is 289, leaves 1007. Its root is not greater than 31. Adding half of the *arithmoi* to this root, it is not greater than 67. Dividing by the *dynameis* makes it not greater than $\frac{67}{17}$.

24 Diophantus, 14.21-24.

25 Heath lists them: 'IV'.22, 31, 'VI'.6, 7, 8, 9 in Heath 1910, 60.

26 The 'side' is the square root. Diophantus is looking for an integral solution to the inequality, so making this root at least 7 will work.

The modern version of this solution, for $ax^2 + c = bx$, would be

$$x = \frac{\sqrt{\left(\frac{1}{2}b\right)^2 - ac} + \frac{1}{2}b}{a}.$$

In two other problems, 'VI'.6 and 'VI'.22, Diophantus explains that the root of the discriminant of the solution must be rational, and its calculation is consistent with the two solutions just given. Faced with $6x^2 + 3x = 7$ in problem 'VI'.6 he writes: "it is necessary that half the quantity of *arithmoi* multiplied by itself, and that added to the quantity of *dynameis* [multiplied by the quantity of units] forms a square". In modern notation, $\sqrt{\left(\frac{1}{2}b\right)^2 + ac}$ must be rational. For $172x = 336x^2 + 24$ in problem 'VI'.22: "this is not possible if half the quantity of *arithmoi* multiplied by itself, and subtracting from this the quantity of *dynameis* multiplied by the quantity of units, does not form a square".[27] In modern notation, $\sqrt{\left(\frac{1}{2}b\right)^2 - ac}$ must be rational.

Diophantus must have explained the solutions to composite equations somewhere between Book VII and the extant part of Book 'IV'.[28] His instruction in the introduction "to add the lacking species on both sides" and "to subtract like from like on both sides" suggests that his composite equations were classified into the same three types we see in Arabic. That these would have each received their own solution is supported by the variations in addition/subtraction in the rules he follows in the quotations just translated from problems 'IV'.39, 'V'.10, 'VI'.6, and 'VI'.22.[29]

V Composite Equations in Arabic Algebra Before al-Karajī

Diophantus's rules for solving three-term equations are not the standard rules we find in Arabic algebra. The Arabic rules require one to first normalize the equation, that is, to first set the 'coefficient' of the highest power to 1. For example, in problem (T6)

---

27 In this case there is not even a real solution, since the discriminant is negative.

28 I follow Sesiano (1982, 77) on this point.

29 Sesiano implicitly supports the idea that Diophantus classified the three composite equations in this 'classical' manner: Sesiano 1982, 76–7. This is not the only way to classify three-term quadratic equations. François Viète regarded $x^2 + bx = c$, $x^2 - bc = c$ and $bx - x^2 = c$ to be the three simplified composite quadratic equations. Viète wrote them in his notation as "A quadratum plus B in A, æquetur Z quadrato", "A quadratum minus B in A, æquetur Z quadrato", and "B in A minus A quadrato, æquetur Z quadrato" in Viète 1615, 3–4.

al-Khwārizmī solves the equation "half a sixth of a *māl* equal a thing and twenty-four dirhams" $\left(\frac{1}{12}x^2 = x + 24\right)$:

> Then you multiply half a sixth of a *māl* by twelve so that you complete your *māl*, and multiply the thing by twelve to get twelve things, and multiply the twenty-four by twelve so you get: two hundred eighty-eight dirhams and twelve roots equal a *māl* [$288 + 12x = x^2$].
>
> So halve the roots to get six, and multiply it by itself and add it to two hundred eighty-eight to get a total of three hundred twenty-four. Then take its root, which is eighteen. Add it to half the roots, which is six, to get twenty-four, which is the [sought-after] quantity.[30]

In modern form this solution to the normalized equation $x^2 = bx + c$ can be expressed as

$$x = \sqrt{\left(\frac{1}{2}b\right)^2 + c} + \frac{1}{2}b.$$

Normalization comes in two kinds. If the number of *māls* is less than one *māl*, then one 'completes' (*ikmāl*) the fraction of a *māl*, as al-Khwārizmī does in the passage above. If the *māls* are more than one *māl*, then one 'returns' (*radda*) them to one *māl*. Al-Khwārizmī, Abū Kāmil, al-Karajī, and most other algebraists give examples of both when they explain the solutions to composite equations.

This method by which one normalizes the equation before applying the algorithm was practiced throughout the course of Arabic algebra. It is the rule given in every extant book before al-Karajī,[31] and in every book I have seen written after him. The latest example I know is the *Supercommentary of al-Ḥifnī on al-Yāsamīn* by Muḥammad ibn Sālim Ḥifnī, who died in 1767/8.[32]

Abū Kāmil is the only algebraist before al-Karajī who offers alternative rules for solving equations. For each type of equation he gives the standard rule first, followed by a rule to find the *māl* directly. These rules also requires normalization. Here is his solution for finding the *māl* for "three roots and four in number equal a *māl*" ($3x + 4 = x^2$):[33]

> And the way that you find the *māl* is that you multiply the three roots by itself to get nine, and you multiply this nine by the four dirhams which are with the roots to get thirty-six. Then you halve the nine to get four and a half. Multiply it by itself to get twenty and a fourth, and add it to the thirty-six to get fifty-six and a fourth. Take its root, to get seven and a half. Add it to the four and a half, which is half of the nine, and the four dirhams which are with the roots, and that is eight and a half, and that is sixteen, which is the *māl*.

30 al-Khwārizmī, 155.8. Solving this by Diophantus's method is easier, since the discriminant is just $2\frac{1}{4}$, making its root $1\frac{1}{2}$. And again, I translate the arithmetical *māl* as 'quantity'.
31 That is, in the algebra books of al-Khwārizmī, Abū Kāmil, 'Alī al-Sulamī, and Sinān ibn al-Fatḥ. Other books, like those of al-Khwārizmī's contemporary Ibn Turk and the late ninth century mathematician Thābit ibn Qurra, show proofs to the rules for already normalized equations.
32 Ḥifnī shows both completion and returning, which he calls by their arithmetical terms *al-jabr* and *al-ḥatt*: Ḥifnī, fol. 9b.
33 Abū Kāmil, *Alg.*(a) 20.11; Abū Kāmil, *Alg.*(b) 273.18.

The modern version of this, to solve for x^2 in $x^2 = bx + c$, would be

$$x^2 = \sqrt{b^2c + \left(\frac{1}{2}b^2\right)^2} + \frac{1}{2}b^2 + c.$$

This rule should be preferred in cases where the sought-after unknown is the *māl*. But Abū Kāmil only applies it in the three introductory problems designed to illustrate the solutions to the three composite equations, and there only *after* first finding the 'thing' by the standard rule.[34] He does not apply it in practice, that is, in the remaining 68 problems in the algebra proper or in the problems that he solves later in the book. In particular, in the six problems in which the *māl* is the sought-after unknown and Abū Kāmil works through the solution to the three-term simplified equation, he follows the standard rule to find the 'thing' and calculates the *māl* by squaring it.[35]

Proofs of the solutions to the composite equations are given in five extant texts before al-Karajī, all by geometry. Al-Khwārizmī's and Ibn Turk's proofs show the 'thing' as a line and the *māl* as the square on that line. The arguments unfold by comparing lines and areas with no references to Euclid. Abū Kāmil gives proofs in the style of al-Khwārizmī and also proofs that cite Propositions II.5 and II.6 of Euclid's *Elements*. Thābit ibn Qurra also appeals to *Elements* II.5 and II.6, but his proofs are framed in the language of givens in the style of Euclid's *Data*. Last, Nuʿaim ibn Muḥammad ibn Mūsā proved the solutions to the first and second composite types also with the language of givens, this time appealing to propositions in *Elements* Book VI as well as II.5. There is no indication that al-Karajī knew Thābit's or Nuʿaim's works.[36]

VI Composite Equations in al-Karajī's *al-Fakhrī*

Al-Karajī solves the composite equations in his chapter titled "On the six problems", i.e., the six equations. For each type he first gives the standard rule (labeled '(1)' in the translation below), then the rule for solving the non-normalized case as in Diophantus (labeled '(2)'),[37] and then Abū Kāmil's solution for finding the *māl* directly (labeled '(3)'). Al-Karajī gives examples after each rule. Following this he gives proofs of all three rules based on *Elements* II.5 or II.6. He finishes his treatment of the first two composite equations with what he calls the "manner followed by

---

34 Problems (T4), (T5), and (T6), numbered by Rashed as <4>, <5>, and <6>.

35 Problems (13), (45), and (50) in the algebra proper (Rashed's <19>, <51>, and <56>) and three more problems later in the treatise: <44.8>, <45.5>, and <45.6>. Abū Kāmil, *Alg.*(a) 69.12, 96.10, 107.1, 204.9, 206.17, 207.7; Abū Kāmil, *Alg.*(b) 381, 439, 443, 689, 695.

36 For Nuʿaim, cf. Hogendijk 2003.

37 As I will note below, this part is missing for the third type.

Diophantus" or "method of Diophantus" (labeled '(4)'), which amounts to an algebraic solution based on completing the square. Here is the full translation for the first composite type, which in modern form is $ax^2 + bx = c$:[38]

Seeking the first of the connected [problems].[39]

(1) So for the *māl*s and the things that you equated to the number, the way to find the one root is that you return the *māl*s to one *māl* if there is more than one, or you complete it to a *māl* if it is short of one *māl*, or you leave it alone if there is one *māl*. And whatever you do to the *māl*, whether it is one *māl*, more than one *māl*, or less than one *māl*, you must necessarily do to the things which are with it and to the numbers which they are equal to. So once the *māl* became one *māl* by this calculation, and by working everything that is with it, by returning it or completing it or not, you halved the things, and you multiplied the number of its half it by itself. Then you added the outcome to the number, and you took the root of the result of that, and you subtracted from it half the roots. What remained is the root of the *māl*.

$$\left[\text{once } a = 1,\ x = \sqrt{\left(\frac{1}{2}b\right)^2 + c} - \frac{1}{2}b\right]$$

For example, two *māl*s and twenty things equal a hundred twelve dirhams [$2x^2 + 20x = 112$]. So return the *māl*s to one *māl*. You take its half and half of everything with it, so it becomes: a *māl* and ten things equal fifty-six dirhams [$x^2 + 10x = 56$]. Halve the things to get five, multiply it by itself and add it to fifty-six, to get eighty-one. Take its root to get nine, and cast away from it half the roots, leaving four, which is the root of the *māl*, and the *māl* is sixteen.

Likewise, if [someone] said, a fifth of a *māl* and two roots equals fifteen dirhams [$\frac{1}{5}x^2 + 2x = 15$]. So if you completed the *māl*, it all became after completing: a *māl* and ten roots equals seventy-five dirhams [$x^2 + 10x = 75$]. So if you multiplied half the roots by itself, and you added it to the number, and you took its root, and you subtracted from it half the roots, it left five, which is the root of the *māl*.

(2) If you don't want to complete a portion of the *māl* or return the *māl*s to one *māl*, you multiplied half the roots which are with the *māl* by itself, and you multiply the number by the number of *māl*s, and you join to its outcome what arose from half the number of roots by itself, and you take the root of the outcome of that. You subtract from the result half the roots, and you divide the remainder by the number of *māl*s, and what results from that is the root of the *māl*.

$$\left[x = \frac{\sqrt{\left(\frac{1}{2}b\right)^2 + ac} - \frac{1}{2}b}{a}\right]$$

For example, three *māl*s and ten roots equal thirty-two dirhams [$3x^2 + 10x = 32$]. Halve the roots, and multiply it by itself to get twenty-five. Add it to what arose from multiplying thirty-two by three, which is the number of *māl*s, which gives rise to a hundred twenty-one. Take its root, and cast away from it half the roots, leaving six. Divide it by the three which is the number of *māl*s to get two, which is the root of the *māl*.

38 al-Karajī, *Alg.* 149.4.
39 I.e. composite equations.

This way is easier in all problems with many different fractions, since these parts[40] are difficult. If [someone] said, a third and a fourth of a *māl* and two roots equals thirty-three dirhams $\left[\left(\frac{1}{3}+\frac{1}{4}\right)x^2+2x=33\right]$, halve the roots to get one, multiply it by itself, and add it to what arises from multiplying a third and a fourth by thirty-three, to get twenty and a fourth. Take its root to get four and a half. Cast away from it half the roots, leaving three and a half. Divide that by a third and a fourth, which is, you recall, the quantity in this example. So you find it is six, which is the root of the *māl*.

(3) This problem can be worked out another way that leads to the *māl* before the root. The way to do this, after returning and completing, is that you multiply the number of roots by itself, then by the number, and you remember it. Then you multiply half the square of the number of roots by itself, and you add it to the remembered [number]. And you subtract the root of the result of that from the sum of half the square of the roots joined to the number which is equal to the *māl* and the things. What comes out of that is the *māl* itself, and its root is the root of the *māl*.

$$\left[\text{once } a=1,\ x^2=\left(\frac{1}{2}b^2+c\right)-\sqrt{b^2c+\left(\frac{1}{2}b^2\right)^2}\,\right]$$

For example, a *māl* and five things equals twenty-four dirhams [$x^2+5x=24$]. Multiply five by five, then [by] twenty-four to get six hundred, and remember it. Then multiply half the square of the five, which is twelve and a half, by itself, to get a hundred fifty-six and a fourth. Add to that the remembered [number]. It becomes seven hundred fifty-six and a fourth. Take its root to get twenty-seven and a half. Remember it. Then sum the number, which is twenty-four, and half the square of the roots, which is twelve and a half. It becomes thirty-six and a half. Subtract from it the remembered [number], which is twenty-seven and a half, leaving nine, which is the *māl*, and its root is three, which is the root of the *māl*.

The proofs for the rules are given in the same order. The only innovation over Abū Kāmil in the proof of (1) is that where Abū Kāmil drew the square representing the *māl*, al-Karajī's diagram is a single line.[41]

I made a determination in this book to strip it of proofs, lengthy explanations, and numerous examples.[42] But I cannot avoid giving a brief summary of proofs for the connected problems[43] and the causes of halving the roots and what is associated with it.[44]

(1p) So for this, a *māl* and ten roots equal thirty-nine units [$x^2+10x=39$]. The way to find the root is that you multiply half the roots by itself, and you add it to the number, and you take the root of the outcome, and you subtract from it half the roots.

Proof of this is that you make line BG a thing, and line AB ten in number, and we divide it into two halves at point D. We want to extend it with line BG. You knew that if any line is

40 Fractions were often expressed in 'parts' (*ajzāʾ*), like 'three parts of eleven parts of a unit' for $\frac{3}{11}$.

41 It may be that he gave a one-line diagram because he needed the second dimension for the number of *māls* in (2). The other algebraists before al-Karajī also draw the square: al-Khwārizmī, Ibn Turk, Thābit ibn Qurra, and Nuʿaim ibn Muḥammad ibn Mūsā.

42 This is an implicit criticism of Abū Kāmil, who included fifty proofs in his book, including fifteen proofs for the rules for solving equations. See Oaks 2011.

43 I.e., the composite equations.

44 The "what is associated with it" is the rest of the rule, after taking half the roots.

appended with an extension, then multiplying the line with the extension by the extension, and adding to it the square of half the line, gives the square on all of the half line with the extension. So in this problem line *AB* is extended by line *BG*, so the product of all of line *AG* by line *BG* and line *DB* by itself is equal to line *DG* by itself, and Euclid demonstrated in his work.[45] And we knew that line *AB* is ten, and line *BG* is a root of the *māl*, and if you multiplied all of line *AB* by line *BG* it gave thirty-nine units, which is equal to the *māl* with ten of its roots. So if you added to that line *DB*, which is five, by itself, it gave sixty-four, and a root of that is line *DG*. So line *DG* is known, which is eight, and line *DB* is five, leaving line *BG* three, which is the root of the *māl*, and the *māl* is nine. And this is its figure:

(2p) And if [someone] said, three *māls* and six roots equal twenty-four units [$3x^2 + 6x = 24$], and you wanted to find the thing without returning the *māls* to one *māl*, you made line *BG* three things and line *AB* six in number, and you attached at point *G* line *DG* equal to line *BG*, and you drew line *HṬ* parallel to line *BG* and line *ḤY* likewise parallel to it, so that each one of the parts of line *DG* is a root of the *māl*. Thus the product of line *AG* by *GH* is three *māls* and six roots, since it is six and three roots by one root, which is surface *AH*. But three *māls* and six roots are twenty-four units, so the whole of surface *AD*, the entire surface, is seventy-two, which comes from multiplying line *AG* by line *GD*. But line *GD* is equal to *BG*, so all of line *AG* by line *BG* is seventy-two.

So we divide line *AB*, which is six, into two halves at point *S*, to get line *SB* is three. You multiplied it by itself to get nine, and if you added all of that to line *AG* by *BG*, which is seventy-two, it yielded eighty-one. And the root of that is nine, which is equal to line *SG*.[46] But line *SB* is three, leaving line *BG* six, and because we made line *BG* three roots, each root is two. And this is its figure:

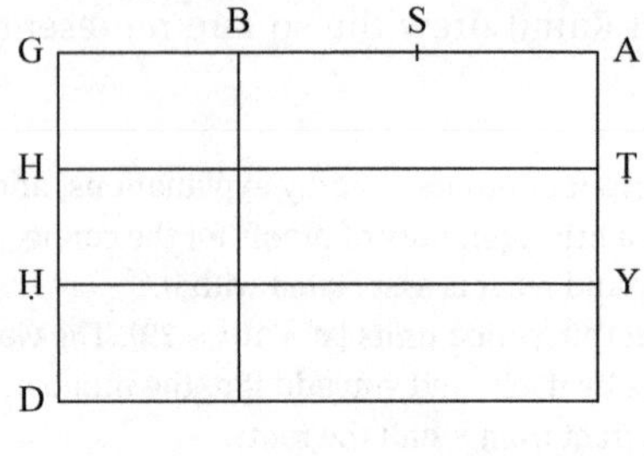

And if there were a part of a *māl* and roots equal a number, like half a *māl* and two roots equal six dirhams [$\frac{1}{2}x^2 + 2x = 6$], and you wanted to find the one root without completing the *māl*, you made line *AB* half a thing and line *BD* two, and line *AY* a root of the *māl*, so the product of line *AY* by line *AD* is half of the *māl* and two things, since it comes from multiplying a thing by half a thing and two [$x \cdot (\frac{1}{2}x + 2)$]. Thus all of surface *AZ* is half a *māl* and two roots, which is six in number, since half a *māl* and two roots are equal to six.

45 *Elements* II.6.
46 *Elements* II.6.

So if you divided line *AY* into two halves at point *S*, and you drew line *SṬ* parallel to line *AD*, then surface *SD* is three in number, since it is half of the larger surface, which comes from multiplying *AD* by *AS*. But *AS* is equal to *AB*, so all of line *DA* by *AB* is three in number.

So we must divide line *BD* into two halves at point *G*. *AB* is an extension of line *BD*, so the product of *AD* by *AB*, which is three, with the product of *BG* by itself, which is one, all of which is four, is equal to line *AG* by itself.[47] So line *AG* is two. But line *BG* is one, so line *AB* is likewise one, and since that is half of the thing, the entire thing is two, which is the root of the *māl*. And this is its figure:

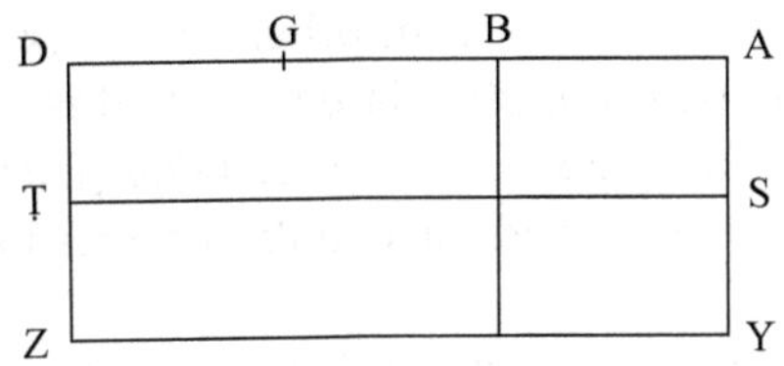

(3p) And if it were a *māl* and ten roots equal thirty-nine units [$x^2 + 10x = 39$], and you wanted to find the *māl*, you made line *GD* a *māl*, and line *DH* ten roots, so all of line *GH* is thirty-nine units. And we make line *DA* equal to line *DH*, and you produce on it square *DB*, and it is broken into a hundred *māls*, because ten roots of something, if multiplied by itself, gave a hundred times that same something. And we make surface *GṬ* equal to surface *DB*, so you break surface *GṬ* into a hundred *māls*. And because line *GD* is a *māl*, line *DṬ* is a hundred, and all of surface *GY* is three thousand nine hundred, since it comes from multiplying line *GH*, which is thirty-nine, by a hundred, which is *GZ*. And likewise all of surface *ṬB* is three thousand nine hundred, since surface *GṬ* is equal to surface *DB*. But surface *ṬB* arose from multiplying line *YB*[48] by line *HD*, which is equal to line *AB*.

So we divide *HY* into two halves at point *S*, and we say it comes to be that all of line *YB* by *BA*, which is three thousand nine hundred, with the product of *HS* by itself, which is two thousand five hundred, all that being six thousand four hundred, is equal to line *BS* by itself, by the condition which you saw earlier.[49] So line *BS* is eighty. But line *DH* is equal to line *HB*, so the sum of lines *DH*, *HS* is eighty. But line *HS* is fifty, so the remaining line *DH* is thirty. And line *GH* was already thirty-nine, so line *GD* is nine, which is the *māl*. And this is its figure:

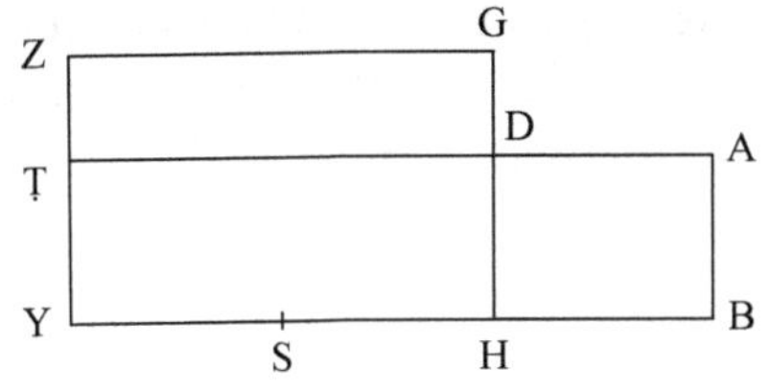

47 *Elements* II.6.
48 Saidan's edition says *DB*.
49 *Elements* II.6.

(4) And if you wanted to find the root of the *māl*[50] in the manner followed by (*madhab*) Diophantus, you searched for a number which, if added to a *māl* and ten things [$x^2 + 10x$], has a root. It is nothing but twenty-five, which added to a *māl* and ten things has a root that is a thing and five dirhams [$x + 5$]. And you knew that a *māl* and ten things are thirty-nine units [$x^2 + 10x = 39$], so if you removed the *māl* and ten things, and you put in its place thirty-nine units, they became sixty-four units. So its root is eight, and that equals a thing and five dirhams. So the thing equals three dirhams, which is the root of the *māl*.[51]

In the "manner followed by Diophantus" al-Karajī observes that if 25 is added to the $x^2 + 10x$ then the sum will have a root, which is $x + 5$. In other words, $(x + 5)^2 \rightarrow x^2 + 10x + 25$.[52] Substituting the $x^2 + 10x$ with the 39 it is equal to, this becomes $(x + 5)^2 \rightarrow 64$. The root of 64 is 8, which equals the $x + 5$. The work takes place within the context of the multiplication of $x + 5$ by itself, and the equation serves to substitute the $x^2 + 10x$ with the 39.

For the second composite type, $ax^2 + c = bx$, al-Karajī follows the same pattern. He gives the rules, with examples, for (1), (2), and (3), and then the proofs in the same order. And again he finishes with the "method of Diophantus", this time for the equation $x^2 + 21 = 10x$:

(4) And if you wanted to solve this problem according to the method (*ṭarīqa*) of Diophantus, you looked for a square which, if you subtracted from it [and a number] ten roots which are equal to the *māl* and twenty-one units, then the remainder is a square. So make the side of the square[53] a thing less five [$x - 5$], or five less a thing [$5 - x$], and each of them leads to a quantity of units, and that quantity is a *māl* and twenty-five units less ten roots [$x^2 + 25 - 10x$]. Replace ten roots with a *māl* and twenty-one units, since it equals it, leaving four in number. And the root of that is two, so [if] five less a thing is two, then the thing is three. And [if] a thing less five is two, then the thing is seven.[54]

Here $(x - 5)^2$ and $(5 - x)^2$ both yield $x^2 + 25 - 10x$. Since $x^2 + 21 = 10x$, we can replace the $10x$ in $x^2 + 25 - 10x$ with $x^2 + 21$ to get $x^2 + 25 - (x^2 + 21) \rightarrow 4$. The two solutions then come from $(5 - x)^2 \rightarrow 4$ and $(x - 5)^2 \rightarrow 4$.

For the third composite type, $ax^2 = bx + c$, the same pattern is followed again, except two parts are missing: the rule and examples for solving the non-normalized equations (2), and the "method of Diophantus" (4). But (2p), the proofs of (2), *are* given, so al-Karajī must have originally included the rule and its examples, but at some point

50 For the same equation $x^2 + 10x = 39$.
51 al-Karajī, *Alg.* 154.13.
52 I use the '$\rightarrow$' to denote the result of an operation, and '=' for equations. The two were distinguished in premodern algebra.
53 The metaphorical use of geometric language to describe numerical operations was common in Arabic arithmetic. There was no underlying geometric conception of numbers.
54 al-Karajī, *Alg.* 159.17.

these were omitted in the transmission of the text.[55] This suggests that the "method of Diophantus" for the third composite type may have suffered the same fate.

For rules (1), (2), and (3) al-Karajī follows the usual practice of giving the statement first and then the proof separately. The "method of Diophantus", though, is more a derivation for rule (1) than a distinct method. It requires no separate proof because the steps are justified along the way. For comparison, Thābit ibn Qurra similarly combined rules and proofs into a single piece of text in his late ninth century *Reasoning on Establishment of Correctness of Algebra Problems by Geometric Proofs*. There he derives the rules (1) for solving the three composite normalized equations through geometric constructions relying on *Elements* II.5 and II.6. In the title of the work he calls them 'proofs' (*barāhīn*), but in the text itself he calls them 'solutions' (*istikhrāj*). Like al-Karajī's "method of Diophantus", the rules are not given first, but emerge in the course of the operations (here the construction) that also justify them. The treatise begins: "The first principle is: a *māl* and roots equal number [$x^2 + bx = c$]. The way to solve this, by proposition six of the second book of the work of Euclid, is as I describe. Let the *māl* be a square *ABCD* and. . ."[56]

VII Solutions to Composite Equations in al-Karajī's Problems

The problems in *al-Fakhrī* that al-Karajī took from Diophantus are all from *Arithmetica* Books I-IV, so none of them show composite equations. But among the ca. 150 'Arabic' problems in that book and the few problems solved in his *al-Kāfī*, al-Karajī has the opportunity to solve composite, non-normalized equations 22 times.[57] In every instance he follows the standard Arabic way (1) of first normalizing the equation by completing or returning the *māls* to one *māl*. Nowhere in these books does he apply the rule for non-normalized equations (2), the rule for finding the *māl* directly (3), or the "method of Diophantus" (4).

But in two instances in his *al-Badīʿ* al-Karajī does apply the rule (2) for non-normalized equations. As part of his instructions on how to solve indeterminate problems he explains how to choose the appropriate square when setting up equations to

55 Saidan consulted six of the nine known manuscripts for his edition of the text. He did not check Hyderabad Osm. 510, Bursa Genel 1169/2, or Tunis Ahmad 5464. Saidan 1986, 92; Rosenfeld & Ihsanoğlu 2003, 119.

56 Thābit ibn Qurra, 161.7. The translation is a mild adjustment of Rashed 2009, 34.

57 al-Karajī, *al-Fakhrī*: problems II.11, 19, 38, 41, 43; III.8,10 (twice), 11, 16, 17, 18, 19, 50; IV.16, 19, 20, 21, 22, 25; al-Karajī, *al-Kāfī* 179.12, 205.10.

guarantee a rational solution. The rule is invoked in cases in which the sum or difference of cubes and things equal a square:[58]

> So if [someone] said, three cubes and five things equals a square. You confronted[59] it with *māls* so that if you halved its number and you multiplied it by itself, and you cast away from the outcome what you get from multiplying the number of cubes by the number of things, the remainder is a square. So if you confronted the three cubes and five things with sixteen *māls* [$3x^3 + 5x = 16x^2$], you halve the sixteen and you square its half, and you subtract from the outcome the fifteen which comes from the multiplication of three by five, leaving forty-nine. Take its root, which is seven, add to it half the roots so it becomes fifteen, divide it by the number of cubes to get five, which is the side of the cube. And if you wanted, work it out by subtracting to get the root of the *māl* is a third of a unit.

In modern notation the two solutions to $ax^2 + c = bx$ are $x = \dfrac{\frac{1}{2}b \pm \sqrt{\left(\frac{1}{2}b\right)^2 - ac}}{a}$.

> So if [someone] said, three cubes less three roots and a third of a root equal a square. You confronted it with nine *māls* $\left[3x^3 - 3\frac{1}{3}x = 9x^2\right]$, so the resulting root is three units and a third of a unit $\left[x \text{ is } 3\frac{1}{3}\right]$. And for this rule, if the cubes were diminished from the things, like three things and a third less three cubes, you confronted it with nine *māls* $\left[3\frac{1}{3}x - 3x^3 = 9x^2\right]$, which is that you restore so it becomes three things and a third equal three cubes and nine *māls* $\left[3\frac{1}{3}x = 3x^3 + 9x^2\right]$. Then halve the number of *māls* and square the half, and add to it the product of the number of cubes by the number of things, namely the ten, to get thirty units and a fourth of a unit, and its root less half the roots is one. Divide it by the number of cubes, resulting in the root being a third of a unit, which is the answer.

In modern notation this solution to $ax^2 + bx = c$ is $\dfrac{\sqrt{\left(\frac{1}{2}b\right)^2 + ac} - \frac{1}{2}b}{a}$.

Searching for the appropriate square in these situations is a characteristically Diophantine problem,[60] and the method of solution, without normalizing, is precisely the method used by Diophantus himself. Because al-Karajī applied the standard Arabic rule in all other instances in his books, it is evident that he took this method from the *Arithmetica*. So potentially both rules (2) *and* (4) come from Diophantus.

---

58 al-Karajī, *al-Badī*ʿ 69.6, 69.13. In these problems it is not possible to set up an equation that simplifies to two terms. Al-Karajī does not bother to divide each term by a 'thing' to make them quadratic equations.

59 Arabic algebraists often set up equations by 'confronting' (from *qabila*) the two sides. The word also sometimes includes the solving of the equation.

60 This kind of problem, with cubes and numbers, is not found among the indeterminate problems in Arabic algebra before al-Karajī.

VIII al-Karajī's *Causes of Calculation in Algebra and Its Proof*

Al-Karajī composed his short work *Causes of Calculation* after *al-Fakhrī*. In it he explains that some people had a difficult time understanding his geometric proofs, so he gives here arithmetical proofs to several propositions. These include proofs for the solutions to equations of type $ax^2 = bx$ and the three composite types, and also rules for adding, subtracting, multiplying, and dividing square roots of numbers.

The arguments for the three composite equations follow the "method of Diophantus" from *al-Fakhrī*. But here, instead of being derivations of solutions to the equations, they are transformed into the 'causes', or proofs, for the standard Arabic rules (1) after normalization. Now the rule is stated first, and the proof follows. This treatise shows us how the 'method' worked for the third type, which was missing in *al-Fakhrī*.[61]

The first problem on halving the roots
[$x^2 + bx = c$. The solution corresponds to our $x = \sqrt{\left(\frac{1}{2}b\right)^2 + c} - \frac{1}{2}b$.]
If you wanted to show the reason for saying, in the first problem, which is the *māl* and roots equal number, that the way [to solve] it is that you halve the roots, and you multiply [it] by itself, and add to the total the number equal to the *māl* and roots, then take the root of the sum, then subtract from it half the roots, then the remainder is the root of the *māl*, we said:

Suppose a *māl* and ten roots equal twenty-four dirhams [$x^2 + 10x = 24$]. So we wanted the agreement between the roots and the number. We added to the root the same as the number of half the roots, which is five, since it is appended in this first problem,[62] as we explained before, to get a root and five in number [$x + 5$]. So we multiplied that by itself to get a *māl* and ten roots and twenty-five dirhams [$(x+5)^2 \rightarrow x^2 + 10x + 25$]. And we already had the *māl* and ten roots in the problem equal to twenty-four dirhams. So if we put the twenty-four in this multiplication in place of a *māl* and ten roots, then the multiplication of a root and five by itself is forty-nine in number [$(x+5)^2 \rightarrow 49$]. So the root of forty-nine, which is seven, is a root and five. So you subtract from it five, leaving the root to be two, which is the root of the *māl*, and the *māl* is four, and ten roots are twenty. If you added them, it gave twenty-four. So this rule is proven true in the realm of number. And if this problem were unchanged except that the *māl* and the roots equal a number other than the twenty-four so that the sum is not a number with a root,[63] we said that the root of that number, subtracting from it five, is the root of the *māl*. For example, if we said: a *māl* and ten roots equal thirty dirhams [$x^2 + 10x = 30$]. So if we added the thirty to the twenty-five, it gave a total of fifty-five. So we said that the root of fifty-five, subtracting from it five [$\sqrt{55} - 5$], is the root of the *māl* that is asked for. And this is the rule to work out this type [of equation].

The second problem on halving the roots.
[$x^2 + c = bx$. The solution corresponds to our $x = \frac{1}{2}b \pm \sqrt{\left(\frac{1}{2}b\right)^2 - c}$.]
[Here is] the reason that we said, in the second problem, that the way [to solve] this is that you halve the roots, and you multiply [it] by itself, and subtract from it the number which is with the

61 al-Karajī, *al-Fakhrī* 359.22.

62 The roots are appended to the *māl* rather than on the opposite side of the equation.

63 I.e., if the discriminant is not a square.

māl, then take the root of the remainder, and subtract [it] from half the roots, or add [it] to it. What comes after the addition or the subtraction is the root of the *māl*.

Suppose we said: a *māl* and sixteen dirhams equal ten roots [$x^2 + 16 = 10x$]. We wanted the agreement between the roots and the number, and we found a root of the *māl* removing[64] the same as the number of half the roots, or one subtracts the root from the same as the number of half the roots, which is five. So the outcome is a root except five [$x - 5$] or five except a root [$5 - x$]. So we multiplied that by itself, to get a *māl* and twenty-five except ten roots [$x^2 + 25 - 10x$]. And we knew that the ten roots are a *māl* and sixteen dirhams. So if one subtracts from the *māl* and twenty-five the deleted ten roots, which are a *māl* and sixteen, it leaves nine [$x^2 + 25 - (x^2 + 16) \to 9$]. So a root of nine, which is three, is the multiplication by itself. So if we had made it a thing except five, then the thing is eight, since the thing except five resulted in three. And if we had made it five except a thing, then the thing is two. So the outcome in either direction you wanted is the asked-for root.

And if you made it eight, then the *māl* is sixty-four. And if you added to it sixteen, it gave eighty, which is equal to ten roots. And if you made it two, then the *māl* is four. So if you added to it sixteen, it gave twenty, which is the same as ten roots.

[*He next gives advice on how to choose between adding and subtracting.*]

And if your multiplication of the root removing from it the number of half the roots $\left[x - \frac{1}{2}b\right]$, or the number of half the roots removing from it a root $\left[\frac{1}{2}b - x\right]$ by itself leads to a *māl* and a number equal to the number which is with the *māl* equated to the roots, except the roots equal to the roots in the equation,[65] then the number of half the roots is the root of the *māl*.

For example, if we said: a *māl* and twenty-five dirhams equal ten roots [$x^2 + 25 = 10x$]. So if we multiplied a root except five or five except a root by itself, it gave a total of a *māl* and twenty-five except ten roots [$x^2 + 25 - 10x$]. And these removed ten roots are a *māl* and twenty-five, so it gives the twenty-five removing twenty-five. So if you continued with the confrontation,[66] then they are not what we get from multiplying something by something, or multiplying less something by less something. And if so, then it is as if we said: a root except five leaves nothing, or five except a root leaves nothing, so we have excluded the whole thing or the whole five. So all of the five is all of the root, and all of the root is the five.

[*He then covers the case in which the square of half the roots is less than the number, which means the equation has no solution.*]

The third problem on halving the roots.

[$x^2 = bx + c$. The solution corresponds to our $x = \sqrt{\left(\frac{1}{2}b\right)^2 + c} + \frac{1}{2}b$.]

[Here is] the reason that we say, in the third problem, that the way [to solve] this is that you halve the roots and you multiply it by itself and add to it the number which is with the roots and take the root of the sum and add to it half the roots, so the sum is the root of the *māl*.

---

64 The text has "not quite" or "not without" (*illā khuluw*). I write "removing" to fit the mathematical meaning.

65 The "number which is with the *māl* equated to the roots" is c, and the "roots equal to the roots in the equation" is bx. The conditions state that $\left(x - \frac{1}{2}b\right)^2$ and $\left(\frac{1}{2}b - x\right)^2 \to x^2 + c - bx$, or to put it more simply, that $\left(\frac{1}{2}b\right)^2 = c$.

66 Sometimes al-Karajī uses the word *qabila* ('confront') in solving problems by algebra to mean the solving of the simplified equation by the designated rule, so the 'confrontation' here likewise means the completion of the assigned procedure.

So it is as if we said: five dirhams and four roots equal a *māl* [$5 + 4x = x^2$]. So we wanted the agreement between the roots and the number, and it is necessary here that from the root one subtracts half the number of roots and nothing more,[67] since the *māl* equals roots and number.

[*He then explains that the root x* must be larger than $\frac{1}{2}b$.]

So if we subtracted from the root the same as half the number of roots, it left a root except two dirhams [$x - 2$]. We multiplied that by itself to get a *māl* and four dirhams except four roots [$(x - 2)^2 \rightarrow x^2 + 4 - 4x$]. And you supposed that the *māl* in this problem is four roots and five dirhams. So if we put four roots and five dirhams in this multiplication in place of the *māl* [$x^2 + 4 - 4x \rightarrow (4x + 5) + 4 - 4x$], then we multiplied a root except two dirhams by itself [to get] nine dirhams [$(x - 2)^2 \rightarrow 9$]. So we take the root of the nine, which is three, which is a root except two dirhams. If we added to it the two dirhams to get the root, it is five. So it is the asked-for root. And if we multiplied it by itself it gave twenty-five, so it is the *māl*, which equals four of its roots, which are twenty, and five dirhams.

Al-Karajī gave another set of arithmetical proofs in his arithmetic book *al-Kāfī*. The proofs for the types 4 and 6 equations are similar but not identical to the ones in *Causes of Calculation*, and the proof for type 5 is entirely different. With these works al-Karajī seems to have initiated a shift among some algebraists away from geometric proof. Four later mathematicians who were influenced directly or indirectly by al-Karajī and who gave arithmetical proofs to these rules are Ibn al-Yāsamīn (d. 1204), Ibn al-Bannā᾽ (late thirteenth century), al-Fārisī (d. ca. 1320), and Ibn al-Hā᾽im (1387). They grounded their proofs either in (a) the rule for squaring a binomial or completing the square, (b) restatements of *Elements* II.5 and II.6 in arithmetical terms, or (c) a particular handy rule for multiplying numbers that comes from finger-reckoning.[68]

IX Piecing the Evidence Together

Few historians have taken note of al-Karajī's "method of Diophantus". Franz Woepcke translated the two passages from *al-Fakhrī* without any comment.[69] Both Thomas Heath and Paul Ver Eecke knew Woepcke's work, but they do not mention the 'method' in their books.[70] Roshdi Rashed brings it up, but he is misled in associating it with the naming of the parts of ten as "five and a thing" and "five less a thing" at the

---

67 I.e. use $x - \frac{1}{2}b$ and not $\frac{1}{2}b - x$.

68 Oaks forthcoming2.

69 Woepcke 1853, 66–68.

70 Heath 1910, 5; Ver Eecke 1959, xviii note 3.

beginning of the solutions to some problems in Abū Kāmil.[71] The only historian who has examined the issue seriously is Jacques Sesiano. He writes:

> It is unlikely that al-Karajī knew of any Diophantine method for solving complete quadratic equations, for not only do these equations occur in the *later* three Greek Books, with which al-Karajī was apparently not acquainted, but also the approach in their resolutions in the *Arithmetica* is not that used by al-Karajī and explicitly attributed by him to Diophantus. One might hypothesize that al-Karajī knew of some other treatise by Diophantus or even a pseudepigraph on this subject, but we have no source which associates the name of Diophantus with any such work.[72]

The argument that al-Karajī was not acquainted with the Greek Books 'IV' to 'VI' is presented by Adel Anbouba in the summary to his edition of al-Karajī's *al-Badī'*.[73] Even if al-Karajī had not read those books, it does not mean that he had not read the now-lost Book in which the solutions to three-term equations are explained. The two Diophantine-style solutions to composite equations in *al-Badī'* translated above in §7 show that al-Karajī *did* know Diophantus's method of working out these equations, and it saves us from turning to Sesiano's unlikely hypotheses. The problem then becomes one of reconciling the "method of Diophantus" with Diophantus's actual practice.

Diophantus most likely classified (or somehow differentiated) and solved the same three composite equations that we find in Arabic algebra somewhere in the *Arithmetica* between the extant parts of Book VII and Book 'IV'. Like the rules in Arabic algebra, the rules Diophantus follows in solving, as equations, the inequalities in problems 'IV'.39 and 'V'.10 give no indication of why they work. It would have been against the pedagogical spirit of his book for him to give the rules without any derivations or explanations justifying them, and indeed he promised such explanations in his introduction. The "method of Diophantus" in al-Karajī fills that role nicely and it comes with Diophantus's name already attached to it.

I conjecture, then, that for each equation type Diophantus derived the rules (2) for non-normalized equations by the 'method' (4). The 'method' indeed also works for non-normalized equations. For example, to solve equations of the type $ax^2 + bx = c$, one can multiply all terms by a to get $a^2x^2 + abx = ac$, and then search for a number which, when added to the left side, makes a square. (The number will be $\left(\frac{1}{2}b\right)^2$, and the square will be $\left(ax + \frac{1}{2}b\right)^2$.) By this method one arrives at Diophantus's rules directly. Of course this is merely speculation, and it may be that Diophantus instead divided by a and worked with normalized equations as al-Karajī shows. Both derivations were in character with Diophantus's approach, and can lead to the rules he followed in the extant books.

---

71 Abū Kāmil, *Alg.* 58. I explain his error in Oaks 2014b, 36–38.

72 Sesiano 1982, 11.

73 Anbouba 1964, 44, by way of Sesiano 1977, 317–318.

For his part, al-Karajī took from Abū Kāmil the rule for solving for the *māl* directly (3) and the idea to prove rules using *Elements* II.5 and II.6, and from Diophantus he took the rule for the non-normalized case (2) and the "method of Diophantus" (4). Because the geometric proofs were not clear to the practitioners who read his *al-Fakhrī*, he recrafted the "method Diophantus" into alternate, arithmetical proofs to the standard rules (1) in his *Causes of Calculation*.

This reorientation away from geometry in favor of arithmetical proofs became a trend that was followed by several later algebraists, including Ibn al-Yāsamīn, Ibn al-Bannā', al-Fārisī, and Ibn al-Hā'im. The only published Arabic books showing geometric proofs after al-Karajī are those of al-Khayyām (ca. 1075), a geometer, and al-Samaw'al (d. ca. 1175), who builds on Abū Kāmil as well as al-Karajī. Even Sharaf al-Dīn al-Ṭūsī (d. 1213), in adjusting al-Khayyām's work, relies on a numerical algorithm for the numerical solutions to equations. Al-Karajī, with inspiration from Diophantus, seems to have initiated this trend.

Bibliography

Texts quoted

Abū Kāmil, *Alg.*(a): Hogendijk, J. P. 1986. *The Book of Algebra: Kitāb fī al-jabr wa'l-muqābala*. Frankfurt am Main.

Abū Kāmil, *Alg.*(b): Rashed, R. 2012. *Algèbre et Analyse Diophantienne*. Berlin.

al-Karajī, *al-Badī'*: Anbouba, A. 1964. *al-Badī' fī al-ḥisāb*. Beirut.

al-Karajī, *al-Kāfī*: Chalhoub, S. 1986. *al-Kāfī fī'l ḥisāb*. Aleppo.

al-Karajī, *al-Fakhrī*: Saidan, A. S. 1986. *Tārīkh 'ilm al-jabr fī l-'ālam al-'Arabī*. (*History of Algebra in Medieval Islam*). 2 vols. Kuwait.

al-Khwārizmī: Rashed, R. 2009. *Al-Khwārizmī: The Beginnings of Algebra*. London.

'Alī al-Sulamī: *Al-muqaddima al-kāfiyya fī ḥisāb al-jabr wa'l-muqābala wa mā yu'rafu bihi qiyāsuhū min al-amthila*. MS Vatican, Sbath 5.

Diophantus: Tannery, P. 1893–1895. *Diophanti Alexandrini Opera Omnia cum Graeciis Commentariis*. Leipzig. (repr. Stuttgart: 1974).

Ḥifnī, M. b. S.: *Hāshīyat al-Ḥifnī' ala al-Yāsmīnīyah*. MS Library of Congress, Mansuri Collection 5–731.

Ibn Turk: Sayili, A. 1962. *Logical Necessities in Mixed Equations by 'Abd al Hamīd ibn Turk and the Algebra of his Time*. Ankara.

Sinān ibn al-Fatḥ: *Kitāb al-ka'b wa'l-māl wa'l-a'dād al-mutanāsiba*. MS Cairo, Riyadāt 260/4 ff. 95r-104v.

Thābit ibn Qurra: Rashed, R. 2009. *Science and Philosophy in Ninth-Century Baghdad*. Berlin.

Viète, F. 1615. *De Æquationum Recognitione et Emendatione Tractatus Duo*. Paris.

Works quoted

Christianidis, J. 2007."The way of Diophantus: Some Clarifications on Diophantus' Method of Solution." **In**: *Historia Mathematica* 34, 289–305.

Christianidis, J., & J. Oaks 2013."Practicing Algebra in Late Antiquity: The Problem-Solving of Diophantus of Alexandria." **In**: *Historia Mathematica* 40, 127–163.

Eeck, P. V. (ed.) 1959. *Les Six Livres Arithmétiques et Le Livre des Nombres Polygones* (orig. Bruges 1926). Paris.

Heath, T. L. (ed.) 1964. *Diophantus of Alexandria: A Study in the History of Greek Algebra* (orig. Cambridge 1910). New York.

Hogendijk, J. P. 2003. "The Geometrical Problems of Nuʿaim ibn Muḥammad ibn Mūsā (Ninth Century)." **In:** *SCIAMVS* 4, 59–136.

King, D. A. 1988. "A Medieval Arabic Report on Algebra Before al-Khwārizmī." **In:** *al-Masāq* 1, 29–32.

Oaks, J. A. 2009. "Polynomials and Equations in Arabic Algebra." **In:** *Archive for History of Exact Sciences* 63, 169–203.

Oaks, J. A. 2011. "Geometry and Proof in Abū Kāmil's Algebra." **In:** Association Tunisienne des Sciences Mathématiques (ed.), *Actes du 10ème Colloque Maghrébin sur l'Histoire des Mathématiques Arabes (Tunis, 29-30-31 mai 2010)*. Tunis, 234–256.

Oaks, J. A. 2014. "The Series of Problems in al-Khwārizmī's *Algebra*" **In:** *Neusis: Journal for the History and Philosophy of Science and Technology* 22, 149–168. (also available in English in: http://problemata.hypotheses.org/157.)

Oaks, J. A. 2014b. Review of Rashed 2012. **In:** Aestimatio 11, 24–49.

Oaks, J. A. 2015. "Series of Problems in Arabic Algebra: The Example of ʿAlī al-Sulamī." **In:** A. Bernard *et al.* (eds.), *Les Séries de Problèmes, un Genre au Carrefour des Cultures*. SHS Web of Conferences 22, 00005. https://www.shs-conferences.org/articles/shsconf/abs/2015/09/contents/contents.html.

Oaks, J. A. forthcoming. "Algebra According to Early Arabic Authors." **In:** K. Chemla and T. Miao (edd.), *A Worldwide Approach to the Early History of Algebra*. Heidelberg.

Oaks, J. A. forthcoming2. "Arithmetical Proofs in Arabic Algebra." In: Actes du 12ème Colloque Maghrébin sur l'Histoire des Mathématiques Arabes (Marrakech, 26-27-28 mai 2016).

Oaks, J. A. & H. M. Alkhateeb 2005. "*Māl*, Enunciations, and the Prehistory of Arabic Algebra." **In:** *Historia Mathematica* 32, 400–425.

Rashed, R. (ed.) 1984. *Les Arithmétiques*. Paris.

Rosenfeld, B. A. and E. Ihsanoğlu 2003. *Mathematicians, Astronomers, and Other Scholars of Islamic Civilisation and Their Works (7th-19th c.)*. Istanbul.

Sesiano, J. 1977. "Le Traitement des Equations Indéterminées dans le Badīʿ fī'l-Ḥisāb d'Abū Bakr Al-Karajī." **In:** *Archive for History of Exact Sciences* 17, 297–379.

Sesiano, J. 1977b. "Les Méthodes d'Analyse Indéterminée Chez Abū Kāmil." **In:** *Centarurus* 21, 89–105.

Sesiano, J. 1982. *Books IV to VII of Diophantus'* Arithmetica *in the Arabic Translation Attributed to Qusṭā ibn Lūqā*. New York.

Woepcke, F. 1853. *Extrait du Fakhrî, Traité d'Algèbre par Aboù Bekr Mohammed ben Alhaçan Alkarkhî (Manuscrit 952, Supplément Arabe de la Bibliothèque Impériale); Précédé d'un Mémorie sur l'Algèbre Indéterminée chez les Arabes*. Paris.

Gabriele Galluzzo

Substantiae sunt sicut numeri: Aristotle on the Structure of Numbers

Abstract: Aristotle's contribution to the metaphysics of numbers is often described in terms of a critical response to the Platonist paradigm. Plato, we are told, conceives of numbers as abstract entities entirely distinct from the physical objects around us, while Aristotle takes the more mundane view that numbers are pluralities of physical objects considered in a particular way, a way relevant to mathematics. Without rejecting altogether this familiar picture, this paper aims to show that Aristotle has another major contribution to offer to the history of philosophy of mathematics. In the *Metaphysics*, he claims that numbers too can be analysed in terms of matter and form (hylomorphism). On the hylomorphic model, a number has both a material component (the units in the number) and a formal one (the structure that keeps the units together). The paper fully explores the motivations behind Aristotle's hylomorphic conception of numbers, as well as its most significant implications.

Keywords: Numbers, Substances, Hylomorphism, Platonism.

I Introduction

Is there an Aristotelian paradigm in the philosophy of arithmetic? Answers to this question are normally phrased in terms of discontinuity with Platonist ideas about the existence and nature of mathematical objects in general. Plato, we are told, conceives of mathematical objects as entities entirely distinct from the physical objects around us, i.e. as abstract entities inhabiting a world accessible through intellectual knowledge only and inaccessible to the senses. Aristotle, by contrast, takes a more mundane view by claiming that mathematics talks about the physical objects around us considered in a particular way, a way relevant to mathematics. Whether or not mathematical objects are for Aristotle quite the physical objects around us will depend on the force one attaches to the idea that physical objects are considered *in a particular way* by mathematicians. While one may well think that emphasis should be put on the claim that mathematical objects are just physical objects, though considered in a particular way, the option is open to interpreters to say that the way in which mathematicians consider physical objects affects the nature of those objects and so mathematical objects end up being a peculiar kind of objects, dependent on but distinct from

https://doi.org/10.1515/9783110565959-011

ordinary physical objects.[1] Be that as it may, Aristotle's major contribution to the history of the ontology of mathematical objects would be on this picture to bring such objects back from the Platonic heaven to the earth of the ordinary physical world. And what applies to mathematical objects in general applies to numbers in particular. Numbers are for Aristotle just pluralities of physical objects considered in a particular way, a way relevant to mathematics.

It is not my intention to deny the truth of this traditional story, or to downplay the importance of Platonism in the development of Aristotle's views on the nature of numbers. In this paper, however, I wish to explore a different line of thought, which has so far been neglected by Aristotelian scholars and historians of mathematics alike. This is simply the idea that Aristotle's thoughts about the nature of numbers do not only stem from his critical attitude towards Platonism, but can also be seen as a development from his most peculiar and controversial metaphysical claims. More particularly, I shall argue that, in the *Metaphysics*, Aristotle is prepared to apply one of his most characteristic metaphysical distinctions, i.e. the distinction between matter and form (hylomorphism), to the case of numbers. The application of hylomorphism to the case of numbers has significant consequences for Aristotle's conception of arithmetic, the most important of which is that numbers are viewed as structured wholes, having both a material and a formal component. On this view, the familiar idea that numbers are, for Aristotle, pluralities of physical objects considered in a certain way is only the beginning and not the end of the story. For on the hylomorphic model, numbers are pluralities of countable physical objects *arranged in a certain numerical structure*. More generally, if I am right, Aristotle's original contribution to the philosophy of arithmetic does not simply consist in a response to Platonism, but also in a sophisticated metaphysical treatment of the nature of numbers. In the rest of this introduction, I wish to sketch out the nature of the problems that will be dealt with in this paper and indicate the general line of my argument.

The central text for my reconstruction is *Met.* H 3. In this chapter, Aristotle comes up with the rather surprising claim that, if substances are in some sense numbers, they are so not by being numbers of units, as some people (the Platonists) tend to describe numbers. Aristotle's wording makes it clear that the antecedent of this claim should be taken to be true and hence substances are indeed like numbers—provided, of course, that the nature of numbers is properly understood. Although the substance-number analogy which Aristotle develops in H 3 contains several points, the core claim is that both substances and numbers are not aggregates, but real unities. Just as substances, in other words, are not aggregates of their parts, so numbers are not aggregates of their parts, i.e. of units. Thus, substances are something over and above their parts. Likewise, numbers are something over and above the plurality of units that constitute them.

1 For two excellent discussions of this point, see Annas 1987 and Mendell 2008.

In spite of its interest, the substance-number analogy has not received much attention from contemporary scholars.[2] Not so in medieval and early modern times, when the analogy was quite popular and got condensed in the slogan 'substantiae sunt sicut numeri'. Curiously enough, medieval and modern philosophers seem to have taken up only one side of Aristotle's analogy and gone, as it were, from numbers to substances. Their interest lay in how the analogy might contribute to our understanding of *substances*: what structural properties of numbers can be carried over to substances? In this paper, I wish to take the substance-number analogy very seriously, and make it the starting point for a reconstruction of Aristotle's mature thoughts about the nature of number. Consequently, I also wish to take a course different from medieval and modern philosophers, and go in the other direction, from substances to numbers: what can the analogy tell us about Aristotle's conception of *numbers*? More particularly, these are the two questions I shall try to answer:

(Q1) What does it mean that numbers are not aggregates of units?
(Q2) What philosophical motivations does Aristotle have to make his claim (i.e. what aspects of the nature of numbers are not captured by describing them as aggregates of units)?

My response to (Q1) will be that Aristotle tries to apply to numbers the matter-form analysis that has proved to be so useful in understanding the metaphysical structure of material substances. On this analysis, a certain number will be analysable into a material aspect (the several units that compose it) and a formal aspect (some sort of structure, arrangement or grouping of such units). Thus, numbers are not aggregates, but rather structured wholes by analogy with ordinary material objects.

Answering (Q2) is more complicated. But the gist of it is that Aristotle primarily conceives of numbers as *groups* of things. The group language invites us to raise problems concerning the unity and identity of numbers: what makes a group of six units *one* group and not six unrelated things? What makes a group of six things a group of six things and not a group of seven? Aristotle's view is that a certain plurality of units makes up a number only if the units in the plurality somehow hold together and are a unity distinguishable from other groups of things. This, in turn, is possible only if numbers do possess a formal aspect that keeps together the units in the different numbers and so accounts for the numbers' unity and identity.

My discussion will fall into four parts. In Section II, I shall present a sketchy reconstruction of the text (*Met.* H 3, 1043b32-1044a11) in which Aristotle presents the substance-number analogy. In Section III, I shall move on to Aristotle's evaluation of the standard intuition that numbers are composed of units. What I intend to show here is that there are several aspects of this intuition that Aristotle is inclined to agree with,

---

2 Partial exceptions are: Gaukroger 1982; Bostock 1994, 267–270.

but also others that he regards as potentially misleading. In particular, Aristotle does not want us to construe the claim that numbers are composed of units as equivalent to the different one that numbers are aggregates of units. Questions Q1 and Q2 above will be taken up explicitly and discussed at length in Section IV. Finally, I shall raise and solve a difficulty for my interpretation of the text in *Met.* H 3. My reading presupposes that there is indeed a strong symmetry between the structure of material objects and the structure of numbers. In *Met.* H 3, however, Aristotle illustrates the substance-number analogy with reference to the *definition* of substances and not to material substances themselves. In Section V, therefore, I shall try to show why this is no obstacle to my contention that Aristotle applies to numbers the hylomorphic analysis.

II The Substance-Number Analogy in *Met.* H 3

Let me start off with an outline of Aristotle's argument in *Met.* H 3, 1043b32-1044a11:

(A) First, Aristotle makes the general claim that, if it is true that substances are in some sense numbers, this is not true on any understanding of numbers. In particular, it is not true if numbers are thought of as "numbers of units", i.e. presumably as aggregates or sums of units (H 3, 1043b32-34).

(B) Then, Aristotle brings out four points of analogy between substances and numbers:

 (1) A definition is some sort of number because, exactly like numbers, it is divisible into indivisibles (1043b34-36).

 (2) If we add something to or take away something from a certain number (no matter how small the addition or the subtraction is), then the number will no longer be the same. Likewise, if we add something to or take away something from a definition and an essence, the definition and the essence in question will no longer be the same (1044a36-1044a2).

 (3) In the case of both numbers and definitions, it is necessary that there be something in virtue of which they are unities. For, either numbers and definitions are mere aggregates (which is not the case), or they are unities, in which case we must give an account of why they are so. But people usually cannot provide such an account either for numbers or for definitions. This happens naturally enough, because a substance has the kind of unity that a number, not a unit, has, i.e. it is an actuality and a definite nature (1044a2-9).

 (4) Just as number does not take the more and the less, neither does substance taken according to form. It is only substance taken together with matter that takes the more and the less (1044a9-11).

In (A), Aristotle makes a general claim about the relationship between substances and numbers: substances are like numbers, provided that we do not construe numbers as numbers of units, i.e. aggregates of units. Then, in (B), he introduces four specific points of analogy which are supposed to confirm the general claim made in (A). One first thing to note is that, although the analogy is introduced as one between numbers and substances, it is then spelt out in (B) with reference to *definitions*. This is problematic, and I shall try to say something more about it in Section V. Suffice it to observe, for now, that the context makes it clear that it is the definition of substances that Aristotle has in mind. Whether or not there is some resemblance between numbers and definitions in general, it is clearly with reference to the definition of substances that the resemblance mostly shows. One reason Aristotle may have had to press this point is that it is only substances that are definable or, at least, it is only they that are definable in the strict sense of the term, as emerges from the discussion in *Met.* Z 4-5.[3] Things other than substances are definable only in a secondary sense—if they are definable at all. Thus, it is natural for Aristotle to assume that he is entitled to use the example of definitions to make a point about substances in general.

It takes just a quick look at Aristotle's text to see that the four points of analogy (B1-B4) between substances and numbers are far from being unproblematic. One difficulty is that B1-B4 are supposed to provide some support for Aristotle's thoughts about the nature of numbers, i.e. that they are not aggregates of units. But it seems that only B3 actually does so. In B3, Aristotle remarks that neither definitions nor numbers are aggregates, but real unities. If this is so, however, it is natural to conclude that there must be something in virtue of which they both are unities. This suggests that both definitions and numbers must possess, in addition to the parts that get somehow unified, a principle in virtue of which the parts are unified. It is not clear, by contrast, how B1, B2, and B4 should help us to go beyond the idea that numbers are aggregates of units. What is worse, B1 and B2 seem to stand even if one *does* construe numbers as mere aggregates of units. B1, for instance, says that a definition is a sort of number because, just like numbers, a definition is divisible into indivisibles. I take it that Aristotle's point here is that a definition must be analysed into undefinables, i.e. terms that we have to take as primitive and not susceptible of definition. Admittedly, many of the definitions Aristotle standardly presents us with seem to contain terms that are further definable: in the definition 'a man is a two-footed animal', for instance, 'animal' is certainly susceptible of further definition. But Aristotle's point must be that a definition must *ultimately* be analysable into undefinables, and so we must stop defining at some point or other. The process of definition cannot go on indefinitely on pain of not grasping the nature of the object defined. Were the process to be carried on indefinitely, we would never reach the point at which we could say we know the object we are trying to define.

Now, Aristotle's remark on definitions carries over perfectly well to numbers even if numbers are aggregates of units. Just like definitions, numbers too are divisible

3 Aristotle, *Met.*, Z 4, 1030a5-7; 1030b4-7; 5, 1031a1-11. See also Z 13, 1039a19-20.

into indivisibles, that is, the units of which a number is composed. Suppose now that numbers are just aggregates of units. Thus any collection of six units, regardless of whether they are grouped together or unified in some particular way, is an instance of number 6. There seems to be no reason why number 6 so conceived should not be thought of as divisible into indivisibles, i.e. the units of which it is composed. Thus, it is not clear how the divisibility issue contributes to promoting the conception of numbers which Aristotle advocates.

Something similar can be said about B2. B2 claims that definitions and numbers are alike because, if we add something to or take away something from a definition or a number, the definition or the number will no longer be the same. The point is straightforward. If one adds a term to or takes away a term from a definition, one will end up defining not the object one intended to define, but something else. Analogously, if we add to a number or we take away from it even one unit, the number will no longer be the same. In this case as well, however, the analogy goes through even if numbers are just aggregates of units. For B2 only requires that numbers be composed of units, and hence it is not clear how thinking of them as structured wholes, i.e. as things endowed with some form or structure, could make the analogy with definitions any tighter.

B4 is problematic in another respect. In B4, Aristotle says that neither numbers nor substances take the more and the less. He hastens to add, however, that the analogy holds only for substance taken according to form, and not for substance taken together with matter. Since the discussion so far has been concerned with definitions, it is surprising that Aristotle should switch here to talk of matter and form. I shall postpone my answer to this difficulty to Section V. There is, however, a more immediate concern. Does Aristotle want us to take the point of analogy generically or specifically?[4] In other words: is he saying that any number is not more of a number than any other? Or is he rather trying to make the point that any instance of number 6 is not more 6 than any other? I am inclined to take the analogy specifically, for the following reason: in *Phys.* IV, 12, in the course of his discussion of time, Aristotle remarks that the number of a hundred horses and that of a hundred men is the same if we consider the number by which we count, and not the same if we consider the number which is counted.[5] In other words, the two numbers are the same number because the number of objects in a collection of a hundred horses is the same as the number of objects in a collection of a hundred men. The two numbers differ, by contrast, because they are numbers of different things, the first being the number of horses, the second the number of men. It is natural to think that, when Aristotle says in B4 that substances and numbers do not take the more and the less, he is thinking of the sense in which the number of two equinumerous collections is the same. Thus,

---

4 See Bostock 1994, 269–270, for an analysis of this difficulty.

5 Aristotle, *Phys.*, IV, 12, 220b8-12.

a collection of six objects is not more an instance of the number 6 than any other. Be that as it may, it is not evident how considerations about the more and the less should cast some light on Aristotle's general idea that numbers are structured wholes and not aggregates of units.

What are we to make of the argument in H 3? The substance-number analogy might seem rather disappointing. Three of the four points of analogy may not make any contribution toward supporting the claim that numbers are not aggregates of units. Moreover, B1 and B2 may be taken to show that there is nothing more to the nature of numbers than their being aggregates of units, after all. I suggest we wait before jumping to conclusions. For one thing, it is clear that Aristotle has a dossier on the analogy between substances and numbers. The most important claim in the dossier is that there is something more to the nature of numbers than their being composed of units. But this does not imply that other points of analogy cannot hold true on a weaker and less demanding conception of the nature of numbers. Second, and most importantly, Aristotle rejects the view that numbers are *aggregates* of units. He does not object to the intuitive idea that numbers have units as their parts. Thus, it is natural to expect that many of the characteristic properties of numbers do, in fact, depend on their being made of units. In the next section, I shall clarify what Aristotle believes is right and what wrong in the intuitive association between numbers and collections of units.

III Aristotle on Numbers, Units and Unities

Aristotle's official definition of a number is that it is "a plurality of indivisibles",[6] i.e. "a plurality of units".[7] Moreover, Aristotle seems to identify the unit with the number 1 and believes, together with the mathematicians of his time, that the unit/1 is not a number, but rather the principle of numbers.[8] 1, in other words, is the principle that generates the number series by being added on over and over again. Thus, a number always involves a *plurality* of units.

There are many aspects of the conception of numbers as pluralities of units that Aristotle is willing to accept. In *Met.* M 6-9, in the course of his violent polemic against Plato's conception of Ideal Numbers or Form Numbers, Aristotle often makes reference to 'the mathematical number'.[9] His aim is to point out that Ideal Numbers, whether they exist or not, cannot possibly be the numbers mathematicians talk

6 Aristotle, *Met.*, M 9, 1085b22.
7 Aristotle, *Met.*, I 1, 1053a30. For other, equivalent definitions see: *Met.*, Δ 13, 1020a13; Z 13, 1039a12; I 6, 1057a3-4; N 1, 1088a5-6.
8 Aristotle, *Met.*, N 1, 1088a6-8; I 1, 1052b23-24. For more detail on these ideas see Mendell 2008, § 10.
9 Aristotle, *Met.*, M 6, 1080a21-22; 1080a30; 1080a36; 1080b13; 1080b14-17; 1080b22-23; 7, 1081a6; 1081a18-19; 8, 1083a28; 1083b2-5; 1083b12-13; 9, 1086a5; 1086a7-8; 1086a10; 1086a30.

about. Thus, for Aristotle, any philosophical account of the nature and structure of numbers should do justice to actual mathematical practice.[10] Such a practice centres upon the nature of units and the role they play in numbers. There are at least four features of mathematical numbers that emerge from Aristotle's discussion in *Met.*, M 6-9:

F1) Numbers are counted by adding units so that after 1 we get 2, after 2 we get 3, and so on (M 6, 1080a30-33).

F2) There are limitless numbers of the same kind, i.e. there are limitlessly many 2's, 3's, and so on (M 7, 1081a10-11).

F3) A number is composed of undifferentiated units, i.e. of units that are perfectly alike both quantitatively and qualitatively (M 6, 1080a20-21; M 7, 1081a19-21).

F4) The units in a number are 'associable' or 'comparable' (συμβληταί) with both the units in the same number and those in any other number, i.e. they enter into arithmetic operations (addition, subtraction etc.) with any other units (M 6, 1080a20-21; M 7, 1081a5-7).

Aristotle seems to think that F1)–F4) represent the way mathematicians normally think of numbers. F1) captures the idea that the number series is generated by adding one unit after another. Aristotle observes that this implies that, for instance, number 4 contains number 3, on the grounds that number 4 is generated by adding one unit to number 3.[11] F2) guarantees that, in normal mathematical practice, we may also work with equinumerous groups of units: '2+3=5' and '3+3=6' are equally legitimate arithmetic operations. But '3+3=6', Aristotle believes, presupposes that there are two groups of three units, two 3's, which we may add to one another.

F3) and F4) are closely connected. F4) establishes that the units in a number can enter into arithmetic operations both with one another and with the units in another number. Take a group of four units, for instance. I can take away two units from that group and so obtain two units. But I can also add to the group of four units a group of three units, thereby obtaining a group of seven units, i.e. a 7. Aristotle seems to think that F4) is true only if F3) is true as well, i.e. only if the units in all numbers are all alike, where 'all alike' means quantitatively and qualitatively undifferentiated. Part of the idea behind F3) must be that only things of the same kind can be counted together—a principle which Aristotle subscribes to.[12] To take an example from concrete pluralities, while it is perfectly normal to add three apples to four apples, because the things that we add belong to the same kind, there would not be much point in adding three apples to four cats, because apples and cats belong to different kinds. Of course, apples and cats can be counted together if we come by a higher kind

---

10 For this important point see Burnyeat 1987, 213–259 and Mendell 2008, §§ 2 and 6.

11 Aristotle, *Met.*, M 7, 1081a18-20.

12 See for instance: Aristotle, *Met.*, I 1, 1053a24-30.

under which both fall, e.g. substance, in which case we count seven substances. On the basis of Aristotle's principle of homogeneity, if mathematical units were different in kind, we could not perform any kind of operation with and on them.

At the beginning of *Met.*, M 8, Aristotle provides us with a sustained argument in favour of F3). He argues that units cannot be different from one another either quantitatively or qualitatively.[13] That one unit does not differ from another in quantity seems to be obvious. But the point is made even more evident by the thought that, if units differed in quantity, two numbers composed of the same number of units would differ in quantity, one being greater than the other – which is paradoxical, because the common intuition is precisely that sameness in the number of units is just sameness in number.[14] Neither can units differ in quality either, considering what Aristotle means by 'quality' in this instance.[15] A quality of a number is a standard mathematical property such as being prime, being square and so on. Thus, Aristotle's reasonable point is that, while the numbers resulting from the units may differ in quality, the units in the numbers do not. Number 3 and number 4 differ in quality (4 is a square number, while 3 is not), and so do 2 and 6 (2 is a prime number, while 6 is not), but the units in all such numbers have no properties that might mark them off from one another.

The conclusion of my brief survey of what Aristotle has to say about mathematical numbers in *Met.* M 6-9 is that there are many features of numbers as they are conceived by mathematicians that should be preserved in a philosophical analysis of the nature of numbers. What is more, many of these features seem to be linked with the intuitive idea that numbers are composed of undifferentiated units, which may enter into arithmetic operations. So, why does Aristotle think that the traditional conception of numbers as pluralities of units is, as it stands, insufficient? The answer should be that, as our passage from H 3 shows, this conception could easily mislead one into thinking that numbers are just aggregates of units and nothing else. Fine, but why should we not think that numbers are aggregates of units after all, given that this view seems to sit rather comfortably with mathematical practice?

In various places in the *Metaphysics*, Aristotle complains that Platonists cannot give an account of the *unity* of numbers. In *Met.* A 9, for instance, he asks Plato the following question: "Why is number taken as a whole one" (i.e. a unity)?[16] A similar complaint is voiced in Λ 10 (1075b34-36): "No one says a word about why number is one, or the body is one, or in general the form and the thing". Thus, it is clear that the conception of numbers as aggregates of units cannot, in Aristotle's eyes, explain why a certain number is a unity, that is, why the units in a certain

13 Aristotle, *Met.*, M 8, 1083a1-17.
14 Aristotle, *Met.*, M 8, 1083a4-8.
15 Aristotle, *Met.*, M 8, 1083a8-17.
16 Aristotle, *Met.*, A 9, 992a1-2.

number hold together and so are one thing, one number. Aristotle's assumption here seems clearly to be that not any collections of units make up a number, but only those collections in which the units are somehow unified and hold together. Aristotle's concern with the issue of the unity of numbers is by no means episodic and is often coupled with a similar concern with the *identity* of numbers: what makes a particular number that number and not another? What makes six units an instance of number 6 as opposed to two instances of number 3? This emerges if we consider a text from *Met.*, Δ 14, the chapter on quality (τὸ ποιόν). In this text, Aristotle remarks that there is a sense of 'quality' that also concerns the objects of mathematics (1020b2-8). We have already come across the idea that numbers have qualities, i.e. such mathematical properties as being prime or square. But I think it is a slightly different sense of 'quality' that Aristotle has in mind in Δ 14, i.e. 'quality' in the sense of the essential differentia of an object. And numbers do have a quality in the sense of an essential differentia. The essential differentia of each number is what belongs to the essence of that number in addition to its quantitative aspect.[17] Aristotle exemplifies his general point with the case of number 6: number 6 is not what it is twice or thrice, but what it is once. For number 6 is once 6. Number 6 as 6, in other words, is not twice 3 or thrice 2, but once 6. Aristotle's general point in the passage seems to be the following. When we sum three groups of two units, by saying for instance '2+2+2=6', we are presupposing that each group of two units is a distinct group of two objects somehow possessing some form of unity or togetherness. When we talk about number 6 as a group of six units, by contrast, we are presupposing that it is the six units of number 6 that somehow hold together to form a unity, with the result that it is incorrect to identify number 6 as twice 3 or thrice 2. For that would somehow break the unity of the number.

The passage from *Met.*, Δ 14 also hints at Aristotle's solution to the difficulty. If numbers have a unity and an identity then it is natural to suppose that there must be something that accounts for their unity and identity. Aristotle's proposal, therefore, is that each number has both a quantitative or material component, i.e. the units that compose it, and a qualitative or formal aspect, which is supposed to explain the unity and identity of the number in question. In conclusion, Aristotle seems to be prepared to apply to numbers some version of the hylomorphic model that has proved so useful in accounting for the structure of material objects. In the following section, I shall try to explain what the application consists in and what motivations Aristotle has for proceeding this way.

17 In *Met.*, M 8, 1083a1-2, Aristotle starts his argument in favour of F3), by remarking that it is important to establish what is the differentia of numbers and that of units. F3) implies that units have no differentia, while the passage on quality from *Met.* Δ 14 suggests that numbers have one.

IV Aristotle on the Structure of Numbers

On introducing the general topic of this paper, I threw on the table two questions about the meaning of the substance-number analogy:

(Q1) What does it mean that numbers are not aggregates of units?
(Q2) What philosophical motivations does Aristotle have to make his claim?

The discussion in Section III has provided us with a solution to Q1. Aristotle wishes to apply the hylomorphic model he has employed in the metaphysical analysis of sensible substances to numbers. On this conception, numbers are structured wholes, which can be analysed into a material aspect (the units composing them) and a formal one (the structure or form that keeps the units together). This discussion has also given us some hints at how Q2 might be answered. Aristotle is worried about problems concerning the unity and identity of numbers: what makes a group of six units one thing? What makes a group of six units that group (i.e. a group of *six* units and so an instance of number 6) and not another?

Aristotle's general intuitions about numbers and their structure raise several difficulties. Take Q1 and the application of the hylomorphic model to numbers. One thing we would like to know is whether the form or structure that keeps together a certain number of units is generic or specific. In other words, is a group of six units kept together by the same unifying principle as a group of seven units, i.e. some sort of generic togetherness? Or is the unifying principle different for different numbers, so that a group of six units is kept together by some sort of 6-structure and a group of seven units by some sort of 7-structure? There are also problems with the answer to Q2. Why is Aristotle led to think that we need a unifying principle for numbers in the first place? What general conception of numbers underlies his inclination to describe numbers as structured wholes?

Let me take up the more general issue first, the conception of numbers underlying Q2 and my solution to it. Much ink has been spilled over Aristotle's philosophy of mathematics.[18] Discussions have mainly revolved around Aristotle's rejection of a Platonist metaphysics of numbers, as well as the extent to which his view can be characterized as 'abstractionist'. I do not wish to go into such debates here. I shall rather confine myself to recalling a few points that may be of some help in solving the problems raised by the substance-number analogy. We may start by observing that for Aristotle numerical predicates apply to and are true of sensible objects.[19] We truly say

18 Besides the papers which have already been mentioned, I take the liberty of making the following selection: Heath 1949; Mueller 1970; Annas, 1975; Lear 1982; Hussey 1983; Barnes 1985; Annas 1987; Mignucci 1987; Hussey 1991. See also Mendell 2008 for an accurate survey of the debate.
19 Aristotle, *Met.*, M 3, 1077b17-22.

that there are three cows in the stable and four people in the room. Sensible objects, therefore, like cows and people, have numerical properties: they do come in groups of two, three, four etc. Of course, arithmetic is not interested in all the features of sensible objects. It is not concerned, for instance, with what they essentially are or with how they come into being and pass away. Arithmetic studies sensible objects only insofar as they have arithmetical properties. The relevant arithmetical property here is countability. Thus, sensible objects are studied by arithmetic only *qua* (i.e. insofar as they are) countable.

Aristotle provides us with a fairly clear description of how we count sensible things. Counting presupposes that we fix a unit by which we count and then proceed to ascertain how many units there are in the group we wish to count. The unit will normally be an instance of a certain kind. If we want to count human beings, we pick one human being as the relevant unit. Thus, the number of human beings in a certain group will be the number of human being-units in the group. This procedure is legitimate, Aristotle thinks, because, even though a human being is divisible into many parts, a human being *qua* human is indivisible.[20] If we divide a human being into parts, none of the parts is a human being and, hence, a human being is, in this sense, indivisible. Thus, human beings, cows etc. can be treated as indivisibles, i.e. as units. Aristotle's application of the *qua*-logic (arithmetic studies sensible objects *qua* countable, a human being *qua* human being is indivisible) explains how he can give an account of the actual practice in mathematics. Although arithmetic is *about* sensible objects, arithmeticians clearly do not talk about human beings, cows etc., but about numbers and units. But this is understandable enough, because human beings and cows are, from the point of view of their arithmetical properties, just groups of units. Mathematicians, therefore, can harmlessly enough disregard what kind of things the units they reason about are. For nothing of what they have to say about units depends on the nature of the objects they consider as indivisible in the first place. The foregoing account provides us with a clear picture of the nature of mathematical numbers. In the concrete, numbers are just groups or pluralities of sensible objects *qua* countable, groups in which each object is taken as a unit. Numbers, of course, can also be considered in the abstract, as when we say that the number of cows in the stable is the same as the number of people in the room.[21] In this case, there is a number that we associate with both the group of cows and the group of human beings. But Aristotle seems to suggest that numbers in the abstract are derivative with respect to concrete numbers.[22] He observes, for instance, that '3' used as a name is derivative from 'three' used as an adjective. Thus, the expression 'the number of cows in the stable is 3' is derivative from 'there are three cows in the stable'. All this suggests that

20 Aristotle, *Met.*, M 3, 1078a23-25.
21 Aristotle, *Phys.*, IV, 12, 220b8-12.
22 See Aristotle, *Phys.*, IV, 7, 207b8-10 and Mignucci 1987 for discussion.

it is the concrete way of speaking that best captures the nature of numbers. Numbers are groups of sensible things *qua* countable, i.e. groups of units.

Now, it is the notion of *group* that is particularly relevant to the problems that concern us here. For, if we push the group language a little farther, we shall see why Aristotle may have been led to think that numbers are unities and so need a principle in virtue of which they are unities. Suppose that there is a group of six people in the dining room. Aristotle's point seems to be that, when we consider the six people as a group of six people, we are taking the six people as a unity. In doing so, we are treating the group as *one* group, distinct from other groups of objects. If there are six more people in another room, they constitute a different group of six objects and so a different unity. But in order for the six people to be distinct from other groups of things, they must somehow hang together and so there must be something that keeps them together. In my example, I made things easy by supposing that the two different groups of people are in different rooms. The spatial dispersal of the two groups renders it intuitively more plausible that the units in each group are somehow together. However, Aristotle's idea should be regarded as a generalized version of my example. If a group of n things is a distinct group of n things, the n things must be grouped together, and so a principle is required that groups them together.

Even more clarity on Aristotle's idea can be gained if we consider that, when we are treating the six people in the dining room as a group of six people, we are thereby excluding other possible groupings within the six people. The group of six people is certainly divisible into two groups of three people, just as it is divisible into six person-units. But when we consider the six people as a group of six things, we are excluding that there actually are two groups of three people or six person-units there. Were we to split the group into two groups of three people, we would immediately destroy the unity of the original group and create two distinct and new unities, i.e. two groups of three people. Another way of making this point is to say that both the two groups of three people and the six person-units exist potentially and not actually in the group of six people. There is some analogy here with the case of the division of the line. In the latter case as well, the parts of the line, i.e. the segments, exist in the line only potentially and not actually. Were we actually to divide a line, we would immediately break its continuity and so create a plurality of distinct segments.[23] The analogy is all the more telling if we bear in mind that in *Met.* Z 13 Aristotle uses the example of the line to explain his claim that a substance cannot be composed of

23 Admittedly, the analogy is only partial because lines are continuous quantities, while numbers are discrete quantities (cf. *Cat.*, 6, 4b20-5a4). The point here, however, is not that all real unities are unities in the same way, but that in the case of all real unities the actual existence of the parts is incompatible with their unity. This leaves room for the idea that what it means to be a unity for a line is something quite different from what it means to be a unity for a number.

substances existing in actuality.[24] The actuality of the parts of a substance is incompatible with its unity, just as the actual division of a line is incompatible with its unity, i.e. with its being continuous. That Aristotle's pattern of reasoning also applies to substances is a further indication that it may be easily carried over to numbers, if indeed substances are in a way like numbers. Two groups of three people, as well as six person-units, exist only potentially in a group of six people.

We can see that the potentiality-actuality model is relevant to understanding Aristotle's thoughts about the structure of numbers if we turn once again to the polemic against Platonism. In *Met.* M 8, 1084b2 ff., Aristotle takes issue with the Platonist understanding of the nature of the one.[25] He protests that Platonists want to have it both ways, by making the one both a material and a formal principle of number. For Platonists maintain both that a number is composed of units, and so the one taken as a unit is a material principle of the different numbers, and that each number is a unity, and so the one is a formal principle that is predicated of each number. For each number is one number, distinct from all the others. Certainly, part of Aristotle's intent is to charge Platonists with equivocation and inconsistency. When they say that the one is the principle of number(s) and so as a principle is prior to number(s), they do not specify whether it is the material or the formal principle that they are talking about. The result is attributing incompatible properties to the one. For being a material principle is not the same thing as being a formal principle. Neither is the formal principle prior according to the same sense of 'prior' as the material principle. The material principle is prior in generation or time, while the formal principle is prior in definition and substance.[26] The one taken as material principle is prior in generation, because numbers are somehow generated by adding units. More difficult to understand is what Aristotle means by saying that as a formal principle the one is prior in definition and substance (to the units). But one reasonable suggestion is that, formally speaking, a certain number is one thing of a certain kind (for instance a 6 or a 7), and so it is its particular nature, its being a 6 or a 7, that determines how many units it is composed of.[27] Thus, the number taken as one thing is prior to the units of which it is composed because it is the nature of a particular number that determines how many units it contains and not the other way round. In any event, the text shows that more than equivocation and inconsistency is at stake here. What Platonists did not see is that, once we realize that numbers have a formal unity, we should reconsider the sense in which units are material parts of a number.

24 Aristotle, Z 13, 1039a3-14.

25 On this passage, see the reconstruction in Annas 1976, 181–184.

26 See *Met.*, M 8, 1084b2-20.

27 See, for this suggestion, Annas 1976, 182 (cf. Aristotle's remarks at 1084b1-6).

Once we realize, in other words, that a certain number is one thing, we should also see that the units in the number exist only potentially and not actually.[28]

Aristotle's remarks square with my reconstruction of the philosophical motivations behind the claim that numbers are unities. More importantly, in the M 8 passage, Aristotle is clearly assimilating numbers to hylomorphic composites, which are analysable into a material aspect (the units that compose them) and a formal one (the principle that keeps the material parts together and so ensures the unity and identity of the number).[29] On this perspective, it is not surprising that he should apply considerations about the structure of hylomorphic composites to numbers. The parts of a hylomorphic composite cannot exist in it in actuality on pain of destroying the unity of the composite. The material parts of a hylomorphic composite exist only potentially. And so do, by analogy, the units of a number as well.

Against this background, we should now be in a position to answer the question as to whether the formal principle that keeps together the different units of a number is specific or generic. To say that the unifying principle is generic means that what makes six units an instance of number 6 and what makes seven units an instance of number 7 is one and the same principle. Presumably, in that case the unifying principle would be some sort of togetherness: a certain plurality of units is a number (no matter what number) when they are somehow together. There are indications, however, that Aristotle wants us to construe the unifying principle as specific to each number. We have seen that in *Met.* Δ 14 Aristotle observes that each number is characterized by some sort of essential differentia. This thought crops up again in *Met.* M 8, where it is suggested that units do not have a differentia, i.e. are qualitatively undifferentiated, while the numbers resulting from units do have one.[30] Aristotle's reference to the differentia chimes very well with the idea that the unifying principle of a number is specific, for a differentia is normally taken to be characteristic of a certain species or kind. Thus, the principle that keeps together a group of six things is specific to number 6 and hence is different from the principle that keeps together a group of seven things, which is specific to number 7. But what could that possibly mean? My answer is the following. Take a group of six units. On Aristotle's view, the six units are an instance of number 6. Now ask the question: are the six units an instance of number 6 because there are six units out there? Or, is it rather the case that there are six units out there because there is an instance of number 6? Aristotle's argument against Platonism teaches us that there are two

28 Aristotle, *Met.*, M 8, 1084b20-23. The claim that units exist in a number only potentially is presented as an unavoidable consequence of the fact that numbers have a formal unity. Thus, the claim is both something Platonists should accept as true and a consequence that Aristotle himself is prepared to endorse.

29 Aristotle, *Met.*, M 8, 1084b12-13.

30 Aristotle, *Met.*, M 8, 1083a1-17.

possible responses to this question, depending on whether it is material or formal explanation that we are after. Materially speaking, the group of six units is there because there are six units. After all, numbers are somehow constructed out of units, as Aristotle is ready to concede. Things are different, however, if it is formal explanation that we are after. Formally speaking, there being six units out there depends on there being an instance of number 6, and not the other way round. For it is built into the nature of number 6 that it is composed of six units, and not of five or seven. This explains in what sense the formal principle that keeps together six units and makes them an instance of number 6 is different from the principle that keeps together seven units and makes them an instance of number 7. For, at the level of formal explanation, it is the nature of a certain number that determines how many units an instance of the number will be composed of, and not the other way round, and hence it is perfectly natural to expect the nature of number 6 to be different from the nature of number 7 with respect to the number of units they respectively contain.

This suggestion may explain some aspects of Aristotle's discussion of numbers that might seem otherwise problematic. According to Aristotle's discussion in H 3, one point of analogy between definition and number (B2 from Section II) is that definition and number are alike, because any addition to or subtraction from either of them will change their nature and identity. As I remarked in Section II, the problem with this point of analogy is that it seems to make no contribution to Aristotle's claim that numbers are not aggregates of units. What is worse, the analogy holds good even if numbers are indeed aggregates of units, or so it might seem. But we can now see that B2 does make a contribution, after all, to the idea that numbers are not aggregates of units. For, if my suggestion is right, the reason why any addition to or any subtraction from a certain number will change the identity of the number is not only or not primarily that a unit is added to or taken away from the number. This is just the material explanation of what is going on. Formally speaking, it is the nature of a certain number that determines how many units it must be made of. And so it is the nature of a certain number that forbids any addition to or subtraction from the number of units.

Finally, my suggestion may also explain Aristotle's obscure remark in *Met.* Δ 14 that number 6 is essentially what it is once and not what it is twice or thrice. It is built into the nature of number 6 not only that it be composed of six units, but also that it be composed of six units taken, as it were, *all together*. Thus, even though it is true that two units taken three times (or three units taken twice) make six units, it is not as the sum of groups of two units or as the sum of groups of three units that number 6 should be identified, but as six units taken together. This thought squares with the previously explored idea that dividing a number breaks its unity and creates a plurality of different numbers. It also gives further intuitive content to the claim that both the sub-groups and the units in a number exist only potentially. Any actual division of a certain number is incompatible with the nature of the number, i.e. with its being that number and not another.

V Numbers and Substances

Most of what I said so far presupposes that the substance-number analogy centres on a comparison between numbers and material substances, i.e. composites of matter and form. The units in a certain number play a role analogous to that played by matter, while the principle of unity or differentia of the number is analogous to form. The trouble with this reading is that the substance-number analogy is spelt out in H 3 with reference to *definitions* and not (at least directly) to hylomorphic composites. The first three of the four points of analogy Aristotle draws our attention to (B1-B3 in Section II) make reference to the definitions of substances and not to hylomorphic composites. One might hope to get some help from the fourth point of comparison (B4), i.e. the claim that neither substances nor numbers take the more and the less. For Aristotle hastens to add that this claim is true only of substance taken as form and not of substance taken together with matter. Thus, one might reason, matter and form are relevant to the substance-number analogy after all. It is not difficult to see, however, that the claim that substances and numbers do not take the more and the less has nothing to do with the issue of the unity of a hylomorphic composite. On the interpretation I favour, the claim should be taken specifically and not generically. Just as one instance of number 6 is not more an instance of 6 than another, thus one human being is not more a human being than another. Aristotle's further remark that this is true only if a human being is taken according to form and not together with matter could and probably should be taken to imply that there is a sense in which the form of two human beings is common.[31] There is no difference between one human being and another insofar as their form is concerned. Differences, including quantitative differences such as the more and the less, enter the picture only when matter is taken into account, i.e. only when it is individual human beings that we are considering. One human being, for instance, can be bigger than another on account of his or her matter. If this is the meaning of Aristotle's observation about the more and the less, hylomorphic composites seem to be marginal to the discussion of the substance-number analogy and the issue of their unity never at stake. Thus, one might easily get the impression that the reading I was offering of the parallelism between substance and number is entirely off track.

One easy way out of this difficulty could be to insist that to talk about the definitions of substances simply is to talk about hylomorphic composites. For one thing, in Book H of the *Metaphysics*, Aristotle seems to be inclined to maintain that the definition of a material substance must mention both the kind of form and the kind of

31 The question as to whether Aristotelian forms are universal or particular is one of the most debated issues in recent Aristotelian scholarship. For a review of the debate I take the liberty of referring to Galluzzo & Mariani 2006, 167–211 and Galluzzo 2013.

matter of which the material substance is made.[32] Moreover, in H 6, Aristotle suggests that the problem of the unity of a material substance and that of the unity of definition are solved by making the same move, that is, by having recourse to the potentiality-actuality machinery.[33] Just as matter is potentially what form is actually, thus the genus in a definition is potentially what the differentia is actually. And what is potential and what is actual are in a sense one thing.[34] Finally, in *Met.* Z 12, Aristotle argues for the conclusion that, in a definition composed of a genus and a plurality of differentiae, it is the last differentia that is the form and substance of the thing defined, while the genus and the intermediate differentiae are material or quasi-material parts of a definition.[35] The last differentia, in other words, absorbs in itself and expresses the whole content of a definition. Putting all these pieces together, one could come up with the following reconstruction of the argument in H 3. The unity of a definition is guaranteed by the fact that one component of the definition, the last differentia, plays the role of form and actuality, while all the other components, i.e. the first genus and the intermediate differentiae, play the role of matter and potentiality. Thus, it is perfectly natural for Aristotle to illustrate the substance-number analogy through the example of definitions, because the unity of a definition perfectly mirrors the unity of a composite of matter and form.

Unfortunately, this solution is difficult. For Aristotle does not seem to say in H 3 that the principle that keeps together the different parts of a definition is one of those parts, just as he does not say that the principle that keeps together a certain plurality of units is just one of them.[36] On the contrary, Aristotle seems to go by the assumption that the genus and the differentia in a definition are the parts that must be unified. In H 3, 1043b10-12, for instance, we read:

> Nor is man animal plus two-footed, but there must be something beside (παρά) these, if these are matter, something that is not an element nor is composed of elements, but is the substance.

The passage clearly suggests that the principle that unifies the parts of a definition is distinct from both the genus and the differentia, just as the principle that unifies

32 Aristotle, *Met.*, H 2, 1043a7-28; 3, 1043b23-32. Actually, Aristotle seems to present in *Met.* Z-H conflicting lines of thought concerning the question of whether matter should figure in the definition of material substances. For two opposed interpretations, see, for instance: Frede 1990 (matter falls outside the definition of substances) and Morrison 1990 (matter cannot be left out of the definition of substances). I side with Frede's reading, even though nothing of what I shall be saying in the following rests upon taking one view or the other.

33 Aristotle, *Met.*, H 6, 1045a23-25; 1045b17-22.

34 See Lewis 1995, Charles 1994 and Loux 1995 for further discussion of this claim.

35 Aristotle, *Met.*, Z 12, 1038a18-20; a25-26.

36 Admittedly, a rather oblique and obscure reference to the genus-as-matter (and differentia-as-form) doctrine appears in *Met.*, H 3, 1043b28-32. But it is hard to see what to make of this passage, as it is part of Aristotle's (partly sympathetic) assessment of Antisthenes' views on definition (1043b23-32) and does not clearly point to genuine Aristotelian doctrine. See Morel 2015, 150–154 for discussion.

a number is distinct from the units, and the principle that keeps together the different material parts of a sensible substance is distinct from each of them. Now, it has been argued that Aristotle is not speaking *in propria persona* here.[37] In other words, the claim that both genus and differentia are material parts of the definition would be the unwelcome consequence of a model of explanation of the unity of definition which Aristotle rejects, i.e. the Platonist model, and not Aristotle's own position. It is only if genus and differentia are two independent entities and Forms, as they are on the Platonist model, that they can appear to be the material parts of a definition, i.e. the parts to be unified. But the general line of argument in *Met.* H 3 suggests that, whether or not he is really committed to it, Aristotle is *exploring* in the chapter a line of thought different from that developed in *Met.* Z 12 and H 6.[38] This line of thought sits more comfortably with the parallelism he wishes to establish between substances and numbers. And it is this line of thought that I want to pursue to show that hylomorphic composites, and not only their definitions, are relevant to the number-substance analogy.

In H 2-3, Aristotle tries to turn the hylomorphic model into a general explanatory scheme, which might explain simultaneously the unity of at least three different kinds of items: (i) material substances, (ii) definitions and (iii) numbers. Aristotle's move is more carefully described as an extension of the hylomorphic model from the case of material substances to those of definitions and numbers. In the second part of *Met.* Z 17, Aristotle argues at some length for the claim that the formal principle that keeps together the material parts of a sensible substance cannot be one of those parts, but must be of a different nature.[39] Aristotle calls the material parts of a substance "elements" and so formulates his claim in its most general form by saying that what keeps together the elements of a substance is not itself an element nor is composed of elements. As can be seen, this general formulation perfectly matches the one Aristotle employs in H 3 to explain his point concerning the unity of definition: what keeps together the parts of a definition (in this case the genus and the differentia) is not an element (i.e. one of the material parts) nor is composed of elements. What guarantees the unity of a definition, in other words, is different in nature from the parts that it unifies. Sameness of wording suggests that Aristotle is trying to apply to definitions the same pattern of reasoning he has employed in the case of sensible substances. This also makes it plausible that he is willing to apply the pattern to the unity of numbers as well.

37 See Morel 2015, 41–59.

38 It is not clear whether Aristotle intends to defend in H 6 the same solution to the problem of the unity of definition as he considers in Z 12. See, for discussion: Frede 1990; Halper 1984 and 1989; Steinfat 1996. What is common, however, to Z 12 and H 6 strategies is that in neither place does Aristotle treat the differentia as a material part of the definition, which he clearly does, instead, in H 3.

39 Aristotle, *Met.*, Z 17, 1041b11-33.

The general context of *Met.* H 2-3 seems to bear out my suggestion. These chapters are dominated by the thought that each thing possesses a formal principle that univocally characterizes and defines its nature. Aristotle describes this principle as a “differentia” (διαφορά). Aristotle’s examples in *Met.* H 2 of things possessing an essential differentia are rather heterogeneous: honey-water, bundles, books, caskets, thresholds, lintels, the day, the night, ice, etc., are all good examples of things that have an essential differentia. The heterogeneity of the examples is justified by the fact that the essential differentia is specific, that is, is different for different kinds of things. Moreover, as Aristotle makes clear, the examples only serve the purpose of providing an analogical basis for the case he is most interested in, the case of material substances.[40] In many, if not all, of the examples, the essential differentia is some sort of accidental property, be it an external relation that keeps together the different parts of an artefact or the characteristic state of a certain kind of body. In the case of material substances, by contrast, the differentia is substantial in nature, i.e. it is the form characteristic of a certain kind of substance. The whole discussion, in other words, is supposed to give further strength to *Met.* Z’s characteristic view that the form of a material object is the substance of the object, i.e. is that which characterizes and defines its nature.[41] It is no coincidence, for instance, that Aristotle observes that the essential differentia is predicated of matter – which is just a reparsing of Z’s characteristic claim that form is predicated of matter.[42] Thus, the essential differentia Aristotle presents us with in H 2 is nothing but the unifying principle he argues for in Z 17, i.e. what keeps together the different material parts of a substance and hence cannot be any of them.

The reference to the Z 17 argument becomes even more explicit in H 3. In the first part of the chapter, Aristotle remarks that a syllable is not the letters *plus* their arrangement, just as a house is not the stones *plus* their arrangement.[43] The example of the syllable is one of Aristotle’s favourite examples in Z 17. The gist of Aristotle’s remark also comes from Z 17: the principle that keeps together and unifies the parts of a thing cannot be one of those parts and must rather be different in nature from each and all of them. What Aristotle wants to show next is that similar considerations apply to a definition and its parts. Aristotle’s use of the word ‘differentia’ to describe the formal principle of a thing may mislead one into thinking that he has in mind the same line of thought as he explores in *Met.* Z 12 (and possibly in H 6): what guarantees the unity of a definition is that the last differentia plays the role

40 Aristotle, *Met.*, H 2, 1043a4-7.
41 Aristotle, *Met.*, Z 7, 1032b1–2; 10, 1035b15–17; 11, 1037a29–30; 17, 1041b8–9.
42 See Aristotle, *Met.*, Z 3, 1029a23-24 (see also Θ 7, 1049a27-36). For the form-matter predication see: Chappell 1973; Dancy 1978; Brunschwig 1979; Page 1985; Lewis 1991; Loux 1991, 147–196. For a recent assessment of the various interpretations of this doctrine, see Galluzzo & Mariani 2006, 89–134.
43 Aristotle, *Met.*, H 3, 1043b4-6.

of substance and form, while the genus and the intermediate differentiae play a material role. But, as I have observed earlier on, this cannot possibly be the case. In the model Aristotle has in mind in H 3, the genus and the differentia are both analogous to the letters of a syllable or to the stones of a house, and none of them is analogous to the characteristic arrangement or combination of either a syllable or a house. Thus, what unifies the different parts of a definition must be distinct from the parts it unifies. But what could possibly keep together the parts of a definition on this conception? Certainly not some sort of external tie or relation, because that would make the resulting unity too weak. In general, things that are kept together by an external tie or relation, be it some binding or gluing or what have you, are not regarded by Aristotle as substances.[44] The reason is that such wholes are not real unities, since the external tie is only accidentally related to the things that it ties up. But definitions are standard examples of real unities. For definitions are only or chiefly definitions of substances, and substances are real unities.[45]

Possibly, what Aristotle has in mind in H 3 is the following idea. Normally, we obtain the definition of a substance by putting together a certain number of predicates: we get the definition of a human being, for instance, by putting together 'animal' and 'two-footed'. This procedure, however, may easily mislead us into thinking that the nature of a definition is merely aggregative or summative. But this is not the case. For two or more predicates make up a definition only if they are the right predicates, that is, those expressing the nature of the object defined. If we put together, for instance, 'black', 'tall', and 'animal' we shall certainly obtain a complex property (being a black tall animal), but not a nature or an essence. Another way of pressing this line of thought is to say that there is something in virtue of which 'animal' and 'two-footed' hang together, something that makes them the right predicates for the definition of a human being. That something cannot but be the nature of human beings which both 'animal' and 'two-footed' contribute to define. Thus, on this line of thought, what guarantees the unity of a definition is not one of its parts, but the nature of the object that the parts jointly contribute to define.

Be that as it may, the outcome of Aristotle's discussion in *Met.* H 2-3 is that neither material substances nor definitions are aggregates of their parts. In the case of both, by contrast, there is something, a formal principle or structure, which keeps together the different material parts and hence is different in nature from each and all of them. Thus, it comes as no surprise that Aristotle wishes to apply the hylomorphic model to the case of numbers as well. As with material objects and definitions, numbers too are not aggregates of their material or quasi-material parts: numbers are not aggregates of units. For numbers, too, we need a formal principle or structure that keeps together and unifies the plurality of units of which a number is composed. This principle will

44 See, in particular, *Met.*, H 6, 1045b7-16.

45 Cf. *Met.*, Z 12, 1037b24-27.

be the ‘differentia’ that characterizes and defines a certain number. Things being so, it becomes less disconcerting that Aristotle should draw a comparison between numbers and the *definitions* of substances instead of making directly the point that numbers are like material substances. For one thing, my reconstruction of the argument in H 2-3 shows that definitions are just as suitable for comparison as material substances. For the hylomorphic model, as well as the anti-reductionist claim it presupposes, apply equally well to definitions and material objects. Neither of them are aggregates, but rather hylomorphic composites of some sort. For another thing, the substance-number analogy contains different points. And there is no doubt that at least some of them are best illustrated by having recourse to the parallel case of definitions. For instance, B1 (both numbers and substances are divisible into indivisibles) and B2 (any subtraction from or addition to numbers and substances will change the identity of the number or substance in question) fit in very well with definitions, while they make little sense, if any, when applied to material substances. What point would there be in saying that individual material objects are divisible into indivisibles? Or that the identity of a material object is changed with any addition or subtraction of parts? Both claims appear to be pointless in the context, if not straightaway false for Aristotle.[46] Thus, it is by no means surprising that Aristotle should choose definitions to make the comparison with number tighter. Some of the points of comparison are better understood if one thinks of the *definition* of substances. The core claim, that both substances and numbers are not aggregates of parts but real unities, holds regardless of whether one chooses definitions or material substances as paradigmatic. For definitions, material substances and numbers all lend themselves to analysis in terms of matter and form.

VI Final Remarks

We can now take up again the question I raised at the very beginning of this paper: is there an Aristotelian paradigm in the philosophy of arithmetic? This question can even be given larger scope: what is Aristotle’s contribution to the philosophy of arithmetic? Now, it is undeniable that, as the traditional story has it, the critique of the Platonist metaphysics of numbers is one of the distinguishing features of Aristotle’s

46 The claim that a material object’s identity changes with any addition or subtraction of parts is plainly false if by “parts” we mean material parts. For Aristotle would certainly allow for a substance to remain one and the same in spite of the replacement of at least some of its material parts. The claim may be true if by “parts” we mean formal parts. But in that case, this claim may be taken to be equivalent to the view that the *definition* of a substance does not admit of any subtraction or addition. For a case can be made for the view that the definition of a material substance is the definition of its form (see footnote 32 above).

account. Aristotle does wish to criticize Plato's claim that numbers are separate and independent entities and does wish to bring numbers back from the Platonic heaven to the earth of physical objects. It also undeniable that Aristotle's view is determined, to a large extent, by the desire to provide a satisfactory account of standard practices in Greek mathematics, as I have tried to show. Platonism can provide no such account, while Aristotle's more down-to-earth conception of numbers can, or so he thinks. If my argument in this paper is correct, however, Aristotle's contribution goes deeper. For, at least in the *Metaphysics*, Aristotle seems to be prepared to apply to the case of numbers his characteristic hylomorphic analysis of sensible objects. The introduction of hylomorphism considerably shifts the emphasis with respect to the traditional story and offers a new paradigm for understanding the nature of numbers. On the traditional picture, numbers are pluralities of physical objects considered in a certain way, for instance as countable or numerable. Although it does not entirely reject this picture, the hylomorphic model invites us to think of number in a different way, that is, as pluralities of countable physical objects (units) *arranged in a certain numerical structure*. The introduction of a structural or formal component to numbers is significant, as Aristotle's view may be taken to imply that equinumerous groups of things share a common structure or form. Of course, this numerical structure or form will not exist separately from and independently of the pluralities of objects that are grouped in the relevant way, but somehow in them: Aristotle has no intention to go back to Platonism and his talk of structure or form should be read by analogy with his standard talk of matter and form, which avoids Platonist consequences. The fact remains, however, that, on Aristotle's mature account, numbers do have a structure and form. What is more, it is the structure or form of a number, more than the units that constitute it, that determines its nature and identity. This is not the place to explore similarities and differences between Aristotle's and structuralist accounts of the nature of numbers. But Aristotle's talk of the structure and form of numbers is an important innovation, which can hardly be underestimated.

Bibliography

Texts quoted

Aristotle, *Met.*: Ross, W.D. 1924. 2 vols. Oxford
Aristotle, *Phys.*: Ross, W.D. 1936. Oxford
Aristotle, *Cat.*: Minio Paluello, L. 1949. Oxford

Works quoted

Annas, J. 1975. "Aristotle, Number and Time." In: *Philosophical Quarterly* 25, 97–113.
Annas, J. (ed.) 1976. *Aristotle's Metaphysics, Books M and N*. Oxford.

Annas, J. 1987. "Die Gegenstände der Mathematik bei Aristoteles." In: A. Graeser (ed.), *Mathematics and Metaphysics in Aristotle*. Bern, 131–147.
Barnes, J. 1985. "Aristotle's Arithmetic." In: *Revue de Philosophie Ancienne* 3, 97–133.
Bostock, D. (ed.) 1994. *Aristotle. Metaphysics, Books Z and H*. Oxford.
Brunschwig, J. 1979. "Form, prédicat de la matière?" In: P. Aubenque (ed.), *Études sur la Métaphysique d'Aristote*. Paris, 131–158.
Burnyeat, M. 1987. "Platonism and Mathematics: A Prelude to Discussion." In: A. Graeser (ed.), *Mathematics and Metaphysics in Aristotle*. Bern, 213–240.
Chappell, V. C. 1973. "Aristotle on Matter." In: *Journal of Philosophy* 70, 679–696.
Charles, D. 1994. "Matter and Form: Unity, Persistence and Identity." In: T. Scaltsas *et al.* (eds.), *Unity, Identity, and Explanation in Aristotle's Metaphysics*. Oxford, 75–105.
Dancy, R. 1978. "On Some of Aristotle's Second Thoughts about Substances." In: *Philosophical Review* 87, 372–413.
Devereux, D. & P. Pellegrin 1990. *Biologie, Logique et Métaphysique chez Aristote*. Paris.
Frede, M. 1990. "The Definition of Sensible Substances in Metaphysics Z." In: Devereux, D. & P. Pellegrin (eds.), *Biologie, Logique et Métaphysique chez Aristote*. Paris, 113–129.
Galluzzo, G. 2013. "Universals in Aristotle's Metaphysics." In: Chiaradonna, R. & G. Galluzzo (eds.), *Universals in Ancient Philosophy*. Pisa, 209–253.
Galluzzo, G. & M. Mariani 2006. *Aristotle's Metaphysics, Book Zeta: The Contemporary Debate*. Pisa.
Gaukroger, S. 1982. "The One and The Many. Aristotle on The Individuation of Number." In: *The Classical Quaterly* 32, 312–322.
Graeser, A. (ed.) 1987. *Mathematics and Metaphysics in Aristotle*. Bern.
Halper, E. 1984. "Metaphysics Z 12 and H 6: The Unity of Form and Composite." In: *Ancient Philosophy* 4, 146–159.
Halper, E. 1989. *One and Many in Aristotle's Metaphysics: The Central Books*. Columbus, OH.
Heath, T. L. (ed.) 1949. *Mathematics in Aristotle*. Oxford.
Hussey, E. (ed.) 1983. *Aristotle Physics: Books III and IV*. Oxford.
Hussey, E. 1991. "Aristotle on Mathematical Objects." In: I. Mueller (ed.), *ΠΕΡΙ ΤΩΝ ΜΑΘΗΜΑΤΩΝ (Peri tōn Mathēmatōn)*. *Apeiron* 24, 105–134.
Lear, J. 1982. "Aristotle's Philosophy of Mathematics." In: *Philosophical Review* 91, 161–92.
Lewis, F. A. 1991. *Substance and Predication in Aristotle*. Cambridge.
Lewis, F. A. 1995. "Aristotle on the Unity of Substance." In: *Pacific Philosophical Quarterly* 76, 222–265.
Loux, M. J. 1991. *Primary Ousia. An Essay on Aristotle's Metaphysics Z and H*. Ithaca.
Loux, M. J. 1995. "Composition and Unity: An Examination of Metaphysics H 6." In: M. Sim (ed.), *The Crossroads of Norm and Nature: Essays on Aristotle's Ethics and Metaphysics*. Lanham, MD, 247–279.
Mendell, H. 2008. "Aristotle and Mathematics " In: N. Zalta (ed.), *The Stanford Encyclopedia of Philosophy* (Winter 2008 edition). URL = <http://plato.stanford.edu/archives/win2008/entries/aristotle-mathematics/>.
Mignucci, M. 1987. "Aristotle's Arithmetic." In: A. Graeser (ed.), *Mathematics and Metaphysics in Aristotle*. Bern, 175–211.
Morel, P.-M. (ed.) 2015. *Aristote. Métaphysique Éta*. Paris.
Morrison, D. 1990. "Some Remarks on Definition in Metaphysics Z." In: D. Devereux & P. Pellegrin (eds.), *Biologie, Logique et Métaphysique chez Aristote*. Paris, 131–144.
Mueller, I. 1970. "Aristotle on Geometrical Objects." In: *Archiv für Geschichte der Philosophie* 52, 156–71.
Page, C. 1985. "Predicating Forms of Matter in Aristotle's Metaphysics." In: *The Review of Metaphysics* 39, 57–82.
Steinfath, H. 1996. "Die Einheit der Definition und die Einheit der Substanz. Zum Verhältnis von Met. Z 12 und H 6." In: C. Rapp (ed.), *Aristoteles, Metaphysik. Die Substanzbücher*. Berlin, 229–252.

Vassilis Karasmanis

The Axiomatization of Mathematics and Plato's Conception of Knowledge in the *Meno* and the *Republic*[1]

Abstract: It is almost a common place of Platonic scholarship that, in the *Republic*, we are presented with an ideal of knowledge rather similar to that of the Euclidean geometry. The main evidence is found in Books VI and VII where we find a hierarchy among the objects of knowledge and their relations. The metaphors of 'up' and 'down' suggest the hierarchical character of this model of knowledge which may be represented as a pyramid with the non-hypothetical first principle at its apex. In the *Republic* we have also the earliest evidence that Greek mathematics were organized axiomatically. In this paper I question the belief that Plato adopts the above model in the *Meno*. I shall argue that it is legitimate to read the *Meno* as presenting a different conception of knowledge that might remind us of modern coherentist theories. In the *Meno* all knowledge seems to be inferential, requiring justification or reasoning into the causes (*aitias logismon* – 98a), but there is no need of fixed and indemonstrable first principles from which such reasoning proceeds. His system of knowledge consists of interrelated elements and several accounts can explain the interrelations among the elements of the system. Such a model is better represented as a *network* rather than as a pyramid. I shall also support the view that Plato's apparent change of his conception of knowledge in the *Republic* was due to the axiomatization of mathematics, a process which took place, for the first time, in the time period, precisely, between the composition of the *Meno* and that of the *Republic*.

Keywords: Knowledge, True Belief, Recollection, Justification, First Principles, Axiomatization.

I

Euclidean geometry is a system of hierarchically ordered propositions. A typical proposition of this system is inferred deductively from others, 'prior' or more elementary ones. But, not all geometrical knowledge is demonstrative. In an axiomatized

---

1 A first, compact version of the views set out in this paper appeared in my unpublished doctoral thesis (cf. Karasmanis 1987, 294–301). Since then, I have presented preliminary versions of this paper on various occasions. I am grateful to many people who asked me then fruitful questions and especially to David Charles and Pantazis Tselemanis for their useful comments.

https://doi.org/10.1515/9783110565959-012

geometry everything starts from some basic principles which, as Plato says, are considered "plain to all" (*Rep.* 510d1). This may mean that these principles either do not need any justification or justify themselves. In Euclid's *Elements*, principles are never demonstrated by appeal to other propositions. Therefore, there is an asymmetry in our knowledge of geometrical propositions between a) principles, which are unprovable and 'self-evident' or "plain to all", and b) propositions, which are derivative from and justified by, the principles or other 'more elementary' theorems by demonstration. We can schematically represent the Euclidean model (hereafter 'model-e') as a pyramid with the principles at its top.

It is almost commonplace of classical Greek philosophy scholarship, that Plato adopts model-e in his theory of knowledge. The main evidence is provided by Books VI and VII of the *Republic*, where we are presented with a hierarchy of the objects of knowledge (Forms) and their relations. The metaphors of 'up' and 'down' suggest the hierarchical character of his model of knowledge which may be represented as a pyramid with the non-hypothetical first principle (the Good) at its apex.

In the *Republic*, Plato says about the mathematical sciences that: a) they are compelled to begin from hypotheses, not proceeding to a starting point, but to an end (510b5-6); b) assuming knowledge of these hypotheses, the mathematicians do not expect to give any account of them either to themselves or to others, as if they were plain to all (παντί φανερῶν, 510c6-d1); c) beginning from these hypotheses, they go through the rest, and end consistently with that which they set out to examine (510d1-3, cf. 533c1-5); d) mathematics (or better mathematical propositions) form a body consisting of first principles, conclusions and intermediate propositions woven together (συμπλέκεσθαι) (533c1-4).

The characteristics that Plato ascribes to mathematics constitute a picture of a deductively organized science. According to Plato's description, mathematics is a science rationally organized in a system where one is able to see what is basic, what is derived, and how each conclusion is derived from previous propositions. Mathematics proceeds downwards from the hypotheses (first principles) to the end. Those first principles are viewed by the mathematicians as self-evident or "plain to all" (510d1) All these features are characteristics of a science already organized in an axiomatic system. Therefore we can conclude that at the time Plato was writing the *Republic*, mathematics had already been organized in an elementary, of course, axiomatic system.[2]

The *Meno* is the first dialogue in which Plato discusses the problem of knowledge in general. In his early dialogues, Plato does not seem to discuss the problem directly, indirectly, or via analogies. Given that Plato introduces the problem of knowledge in the *Meno*, it is reasonable to ask if he sees knowledge as a system, and, if so, how this system is structured. In this paper, I am going to cast doubt on the idea that Plato

2 For this topic, see Karasmanis 1990, 122–123.

adopts, if only implicitly, model-e in the *Meno*. I shall argue that it is preferable to read the *Meno* as presenting a different conception of *knowledge* that might remind us of modern coherentist theories of *justification*. More specifically, I claim that there is no evidence in the *Meno* of an asymmetry (either in knowability or in status) between first principles and derivative propositions. In the *Meno* all knowledge seems to be inferential, requiring justification or reasoning into the causes (αἰτίας λογισμὸς, 98a), but there is no need for any fixed and unprovable first principles from which such reasoning proceeds. His system of knowledge consists of interrelated elements, and several accounts can explain the way in which these elements interact with one another. Such a model can be represented better as a network, rather than a pyramid (hereafter, model-n). In a model like this, circular regress does not need to be vicious. I am not suggesting that Plato consciously proposes such a model of knowledge, but rather that he is exploring in this dialogue some options that had not yet been developed into a specific model of knowledge. I shall also provide support for the view that Plato's apparent change of his conception of knowledge in the *Republic* was due to the axiomatization of mathematics, a process which took place, for the first time, in the time period between the writing of the *Meno* (c. 387-385) and the *Republic* (c. 375-370).[3]

II

In *Meno* 81cff, Plato introduces the theory of recollection in order to answer Meno's paradox (80d-e) of the impossibility of any inquiry.[4] According to this theory, the soul is immortal, has been born many times, has 'seen' all things both here and in the other world, and has learned everything. So, the soul can recall the knowledge of virtue or of anything else it once possessed. In this way, seeking and learning are in fact nothing else but recollection (81c-d). One, who does not know something, has in oneself true beliefs about it (85c6-7). With the help of appropriate questions, the true beliefs are elicited; and, if questions are put to one on many occasions and in different ways, at the end, one will have knowledge (85c9-d1).

Roughly speaking, we can say that the theory of recollection tries to resolve Meno's paradox by introducing true belief as an intermediate stage between knowledge and ignorance.[5] And, since the soul has learned everything and has true beliefs about all things, we have something to serve as a starting-point of our inquiry and are

3 See Karasmanis 1987, 5–7.

4 For some recent discussions of the paradox, see Scott 2006, chapter 7; Fine 2009; McCabe 2009; Franklin 2009. According to Fine, the theory of recollection explains how, relying on true beliefs, we can start and conduct inquiry. According to Scott, the main issue is discovery not inquiry.

5 Many scholars believe that the theory of recollection does not give an explicit answer to the paradox; see, e.g., Weiss 2001, 64; Scott 2006, 79–80.

probably able to recognize its object. From the discussion that Socrates has with the slave-boy (82b-85c) we can see that recollection is a process in which we can distinguish three stages: a) the elimination of false beliefs, b) the eliciting of true beliefs, and c) the acquisition of knowledge.

According to Plato, one's knowledge is recovered by oneself. This knowledge is not empirical, although some empirical contribution or input, like the perception of equals in the *Phaedo* (74a-b) or sensible diagrams may help recollection to begin or proceed further on. However, eliciting knowledge from ourselves does not involve sense-experience.[6] Socrates disregards the definition of colour (76d-e) that is based on empirical observation. Nor do we have to suppose that the soul's knowledge is empirical knowledge from previous lives.[7] Such an explanation of knowledge by recollection is ruled out by Plato's clear statement (86a9) that the soul has been forever in a state of knowledge.

Furthermore, it is emphasized that recollection a) recovers "virtue and other things" (81c8), indeed "all things" (d3), and b) "all seeking and learning" (81d4-5) is recollection. From a) we can infer that these 'things' are similar in kind or 'akin' to virtue. And b) excludes knowledge that demands no inquiry thereby restricting knowledge to the things that the soul once knew.

At 98a, Socrates says that knowledge is true belief "bound by working out the reason" (δήσῃ αἰτίας λογισμῷ). According to Vlastos (1965, 154–5), the above phrase implies rational thought in contrast to sense perception and is used "for knowledge reached and justified by formal inference and analysis in emphatic contrast to sensory cognition". In this way, "a statement becomes known when it is seen to follow logically from premisses sufficient for this purpose; to recollect it, then, would be to see that these premisses entail it". According to this interpretation, recollection in the *Meno* is primarily about our acquisition of necessary truths and not of concepts (like in the *Phaedo*).[8] This view seems to be supported by the text. The theory of recollection is followed by two geometrical problems and the two arguments of the dialogue, after its introduction, deal with proofs of propositions and not with definitions or clarifications of concepts. However, although Plato's examples concern proofs of propositions, it seems that the theory is also posited in order to answer questions like "what is x?" (cf. 81c8, 81e2). In this case, to recollect probably means to gain

6 Vlastos (1965, 143–167) has persuasively argued that even the use of sensible diagrams in the mathematical proof with the slave-boy does not involve empirical knowledge. The evidence from the dialogue shows that the knowledge that Plato had in mind was related to mathematical and ethical notions.

7 See also, Scott 2006, 96.

8 Cf. Ackrill 1973, 177; Franklin 2009, 350. Moravcsic (1971, 60ff) thinks that Plato's theory of recollection in the *Meno* deals primarily with innate concepts in the human mind, rather than with propositions. Hare's (1965) interpretation relies on the distinction between the ability to use a term and the ability to offer a satisfactory definition of it, and restricts recollection to the domain of meaning.

insight into a concept and to find intra-propositional logical relations between terms.[9] Moreover, in Plato, 'knowing x' means 'knowing what x is' (cf. 79c8-9), and the latter is transformed in practice into 'knowing that x is F'. So, the question of knowing x is reduced to the question of finding if one (or some) proposition concerning x is true. Therefore, knowledge of things is description-dependent.[10]

We have just argued that recollecting something is to see that this is entailed by certain premisses. The premisses, too, can be recollected in the same manner. Is there an end to this procedure? If a proposition, in order to be known, must be entailed by some premisses, then we end up either with an infinite regress of proofs and premisses or with circular demonstrations (cf. Aristotle, *An. Post.* 72b15ff). The same happens in defining concepts. Plato, in the *Meno*, says nothing about the need of unproved first principles, or that there is no demonstrative knowledge of some things or some propositions. For him, in order for something to be known, it must be "bound by working out the reason" (98a3-4). How, then, is it possible to acquire knowledge of something?

III

At this point we need to examine the problem more closely. At 81c-d, Plato says that "all nature is akin (τῆς φύσεως ἁπάσης συγγενοῦς οὔσης) and the soul has learned everything, so that when a man has recalled only one thing (ἓν μόνον) - what men call learning - nothing prevents him from discovering all the rest." There are three problems connected with this statement: a) What does "all nature is akin" mean? b) Which is this 'one' from which we may start our process of knowledge? c) How can we acquire knowledge of, or recollect the one thing that we first come to know?

a) The word 'nature' cannot refer to everything that exists,[11] since that would imply recollection of particulars, and kinship between particulars and eternal realities. Moreover, the word 'kinship' is superfluous under this interpretation. There can be kinship among things only if they are different from one another. Klein (1965, 65) has suggested that "by virtue of this assumption [kinship] everything, every bit the soul recollects can be understood as a 'part' of a 'whole' and can be traced back to common origin". In fact the word συγγενής means 'of the same descent or family' and usually συγγένεια (kinship) is due to a common ancestor. However, the relation between part and whole, and the common origin relation are quite different. Moreover, Klein does not explain how an understanding of this part-whole relation or the common origin relation makes recollection possible.

9 See Vlastos 1965, 157.
10 See Fine 1979, 366–367.
11 Bluck (1961, 288) seems to suggest such an interpretation.

"All nature is akin" could mean that either 1) all things are akin, in the sense they are ontologically similar, or 2) that all things that constitute 'nature' are interrelated, or 3) both. Things that constitute 'nature' are things that can be recollected, that is, eternal realities like "virtue and the rest" that the soul has seen (81c6-8) and learned "throughout all time" (τὸν ἀεὶ χρόνον - 86a8). In this way, if we are able to find out one of those things, it is possible to assume that we can similarly find all others.[12] However, the phrase "so that when a man has recalled one thing nothing prevents him from discovering all the rest" suggests something stronger. It implies not only that one can find all the rest in the same way as the first one, but also that the recollection of this first thing will facilitate the recollection of the rest. This may mean that things constituting 'nature' are interrelated and that these relations can be expressed by a set of beliefs coherent with each other. So, the finding of one of them may facilitate the finding of the others.[13]

We could also say that things that constitute 'nature' are 'akin' to the soul (cf. *Phaedo* 77-80), since the soul once learned all these things. In this way, the concepts that the soul forms and the relations found among these concepts reflect the objects of nature and their relations.[14]

b) Plato does not say which or what this first 'one' must be. Reading the phrase "ἓν μόνον ἀναμιμνησθέντα" (if one has recollected just one thing), it appears that, if one recollects this first one (whichever that might be), then all the rest can be known only by processes of thought. The word μόνον (only or just) supports this interpretation.[15] Moreover, we have no evidence (in the *Meno* or in the early dialogues) of the need of first principles from which all knowledge must begin as opposed to relative principles or *homologêmata* peculiar only to particular arguments or discussions. Therefore, it seems that the *Meno* allows us to start from anywhere, i.e. not necessarily from fixed first principles.

c) How can we recollect the 'first thing'? A possible answer could be by some irrational inspiration or intuition. However, it is almost certain that Plato rules out this kind of knowledge in the *Meno*. Knowledge involves αἰτίας λογισμὸν and logical reasoning. At 99e, Plato suggests that intuition might be useful only for obtaining true beliefs, although even that is not certain, because he might be speaking ironically here. As Aristotle says (*An. Post.* 72b15ff), if we accept that all our knowledge is demonstrative and does not start from first principles known independently and in a different way, then there are two possibilities: a) an infinite regress of proofs that

12 See Tinger 1970, 3; Sharples 1985, 149.

13 Cf. Crombie 1963, 140; Cornford 1971, 121; Moravcsic 1971, 60; Gulley 1962, 9; Tinger 1970, 4; Tarrant 2005, 44–46.

14 See Crombie 1963, 140: "The way our minds work corresponds to the way things are, and this is the kinship of the soul to the 'universal' nature."

15 Similarly, cf. Tarrant 2005, 44.

results in the conclusion that demonstrative knowledge is impossible, and b) circular demonstrations. We can plausibly suppose that Plato rejects the first alternative (cf. *Theaetetus* 201d ff.), but we have no evidence regarding the acceptance or rejection of the second.

How is it then possible to start our process of knowing? A plausible answer could be that we could obtain knowledge of the first thing 'A' after reasoning that begins from some true beliefs. In this case, our 'knowledge' relies on true beliefs. But it seems that Plato does not regard such 'knowledge' as genuine knowledge. At 85c, after the proof of the double square, Socrates insists that the slave-boy has not yet acquired knowledge.

Moreover, there is another problem: that of distinguishing true from false beliefs. As we can see from the experiment with the slave-boy, every man has quite a few false beliefs about things. And the first stage of recollection consists in the elenchtic destruction of these false beliefs. Suppose we have some things (or objects of inquiry) A1, A2, A3,.., Ak, ..., An. Then, for some of them (say A1,..., Ak) we have false beliefs. However, according to the theory of recollection, the soul does have true beliefs for all of them (A1,..., Ak, ..., An). Hence, true beliefs are of two kinds: a) actual or conscious true beliefs that we entertain for Ak+1, ..., An, and b) latent or unconscious true beliefs[16] that we have for A1...Ak, for which we entertain false beliefs.[17] Suppose now that we have an object of inquiry for which we have some beliefs (B1,B2,.., Bn). How can we check whether these beliefs are true or false, and so eliminate the false ones?[18] The platonic answer is: through elenchus. But any elenchtic argument involves premises C1, C2, ..., Cn, which in turn require an elenchus, in order to assert their truth or falsity, and so on. In this way, we again end up in an infinite regress.

To escape from this regress, we must accept that a) we always possess some true beliefs in actuality, and b) we are able to recognize, in some way, our true beliefs as such (or at least some of them). If we accept condition a), the problem is how condition b) is fulfilled. There are two possible answers. It is possible to recognize (although this does not amount to knowledge) some of our beliefs as true by some sort of intuition or divine inspiration, but "without any understanding" (99c, cf. 100b3-4). However, as I have already mentioned, I am not convinced that such an answer would satisfy

16 I do not speak of 'latent knowledge' because the knowledge that the soul had was forgotten at the time of birth. This means that the soul forgets both that proposition A is true and also why it is true (αἰτίας λογισμὸς). Therefore, if something remains in the soul in a latent or unconscious condition, this should be true beliefs (85e5-86a7). Scott (2006, 108–112; also Tarrant 2005, 42) thinks that the soul possesses 'latent knowledge' from which we elicit actual or conscious knowledge. Against this view, see Fine 1992, 213 with n.40 and Fine 2009, 348–357.

17 There are also latent beliefs in our souls for which we do not actually entertain either true or false beliefs.

18 Nehamas (1999, 14) has noted that when we begin inquiry our true beliefs are mixed with a number of false beliefs and there is no guarantee that we shall choose at the beginning a true belief.

Plato, and certainly it would not satisfy us. What does 'intuition', 'insight' or 'divine dispensation' mean? It could also be argued that even intuition does not count without further argumentative support. A second, and more plausible, answer would be to see elenchus as a process of obtaining true beliefs in the following way: Elenchus is a process of checking *compatibility* of beliefs. Through that process, we eliminate false beliefs because they are incompatible with others. Beliefs that resist elimination by elenchus can be considered as true.[19] In this way the elenchtic process enables us to obtain a coherent set of beliefs, which are recognized as true. But in order for the elenchtic process to begin in the first place, we need some prerequisites. At 82b, Socrates asks Meno whether his slave-boy is fluent in Greek. For Socrates, this is the only prerequisite in order to start the discussion. But someone who is a competent speaker of a language has the capacity to employ correctly some related notions. So, the boy understands very well what a line, a mid-point, an angle, an area, the relation half-double, etc., is.[20] He also knows how to use the vocabulary of numbers, and the operations of addition and multiplication. Thus, the only things we need in order to start a discussion or an inquiry are a) the ability to use language correctly and b) rational thinking and indeed inferential capacities.[21]

IV

Having discussed the above three questions, let me now return to the problem of the possibility of acquiring knowledge. At 85cff Plato states something that seems to be paradoxical. Although he insists that the slave-boy has recollected properties of the double square by having recovered knowledge from himself (85d3-7), he says that the true beliefs that the boy had in him "have been aroused as in a dream; but if someone asks him these things many times and in many ways, you can see that in the end he will come to know exactly (accurately) of these things, not less than from anyone else" (85c9-d1).[22] As we said, recollection is a long process; and it seems that

19 See Vlastos 1994, 27; Nehamas 1999, 15. Such beliefs are usually very simple and obvious propositions which Socrates uses as agreed propositions (*homologêmata*) in the course of the elenchus. Such an example is the proposition "virtue is good" that Socrates inserts as a premiss in 87d.

20 The slave-boy understands also what a square is (82b). After that, Socrates explains to him two basic properties of the square (equal sides and equal mid-parallels) needed for the proof. Some scholars argue that Socrates does not speak about mid-parallels but about diagonals (cf. Weiss 2001, 84–94; Ebert 1973, 178–81), because otherwise Socrates defines rhombus. But in either case we do not have the proper definition of a square (equal sides and right angles).

21 Franklin (2001, 145 and 2009, 353–355) insists that one must be linguistically competent in order to participate in dialectic.

22 "Οἶσθ' ὅτι τελευτῶν οὐδενὸς ἧττον ἀκριβῶς ἐπιστήσεται περὶ τούτων". The word ἀκριβῶς qualifies ἐπιστήσεται. Cf. Plato *Euthyphro* 4e5-6, 14b1; *Lysis* 205b 4; *Sophist* 243b9.

in this passage, Plato speaks only about the final stage of it. It is probably because of this that he speaks of exact knowledge (ἀκριβῶς ἐπιστήσεται) and not simply of knowledge. Plato does not distinguish here between two different kinds of knowledge, but he probably wants to show that the process of knowing (recollection) does not end with this concrete proof of the double square. In this way, although the slave-boy has succeeded in recollecting and recovering knowledge from himself (85d3-7), he has not yet arrived at 'perfect' or 'exact' knowledge. However, at the end of the above process "he will have no less exact knowledge than anybody's". This means that his knowledge on this matter will be the same as that of an expert geometer who knows very well the whole field of geometry. For any geometrical proposition that may profitably be taken under consideration, both the expert geometer and the novice grasp, at least in some cases, the very same formulation of the geometrical truth under discussion. The difference is that, while the expert possesses a certain highly sophisticated conceptual structure, in which he sees the content of the sentence thus formulated as securely, the novice does not yet see the sentence in this way.

The problem that now arises is the following: If knowledge is obtained by "working out the reasons", and since the boy did provide a correct proof of the theorem, why is Plato not satisfied? Why does he think that the boy needs "more questions and in many ways" (85c10-11) in order to obtain 'exact' knowledge? A suggestion is that "repeated consideration of the problem, and perhaps of others, will enable the slave to grasp more clearly the general principles which underlie it and other problems".[23] According to this interpretation the final stage of recollection consists in an upward movement from our problem (and others) to the first principles, in order to achieve a clear grasp and understanding of them. This is an interpretation of the *Meno* in the light of the *Republic*, which presupposes the adoption of a model-e of knowledge by Plato. But, in the passage 85c-d, Plato says nothing about principles or upward movement towards them. What he says is that the slave, in order to obtain 'exact' knowledge, has to be questioned "on the same things, many times and in many ways" (πολλάκις τὰ αὐτὰ ταῦτα καὶ πολλαχῇ). This seems to mean rather that the slave has to prove his theorem many times and in many different ways in order to explore all the possible relations of this theorem with other geometrical propositions. There are, therefore, many ways of proving or providing a justification of x, and we only have 'exact' knowledge after we have gone through all of them. After the proof, the slave-boy 'recollected' a single property about squares. However, the proof was based on some unexamined assumptions about the nature of the square (82b9-c4) and other geometrical entities. Moreover, geometry is a field of closely interrelated propositions. Therefore, to 'know' a single theorem with one of its possible proofs, does not, probably, amount to knowledge. In order to

23 See Sharples 1985, 155; similarly, Gulley 1962, 14–5; Bedu-Addo 1983, 236; McCabe 1994, 53ff; Nehamas 1999, 15 and perhaps Vlastos 1965, 156–157.

attain 'exact' knowledge, we have to relate this theorem to other geometrical propositions and to continue the same process with the whole of geometry (and after that with other subjects too).[24]

Now we can also explain another problem. At 85c3 Socrates says that the slave-boy already possessed true beliefs in himself. Moreover, at 85c9-10, he says that, after the proof, these beliefs "have been newly aroused in him". Immediately afterwards, at 85c11-d1, he says that the slave will have exact knowledge after many questions and in many ways. Finally, at 85d9, Socrates says that the slave-boy has now (*νῦν*) knowledge (*ἐπιστήμην*). These passages have perplexed scholars. Is it that the boy has true belief after the proof or is it knowledge? Also, what kind of knowledge is that of 85d9? The true beliefs at 85c3 are latent true beliefs that had already been in the soul of the boy. The beliefs that had been aroused in the boy's soul after the proof are actual or conscious true beliefs. But, in another sense, and according to Plato's definition of knowledge at 98a, he should have knowledge and not true belief, because he managed to prove or to justify the true belief about the double square. It is, perhaps, because of this that Socrates says that the boy has knowledge at 85d9. The obtaining of full or exact knowledge is a long process that requires many more questions and in many ways. After the proof, the boy is midway through this process. So, he possesses something less than exact knowledge and something more than simple true beliefs. At 85c9-10 Socrates speaks about true beliefs because he wants to stress the point that the slave does not yet have exact knowledge. But at 85d9, he speaks of knowledge because he wants to stress the point that the boy succeeded in recollecting something.[25]

V

All the remarks so far, may lead us to interpret Plato's theory of knowledge in the *Meno* in a way different from the traditional one. Having found through elenchus a coherent set of beliefs, which may reasonably be considered as true, we can proceed by logical reasoning to 'recollect' a 'first thing', or to infer a first proposition. Using

---

24 See 85e1-3: "For he will do the same with all geometrical knowledge, and all other subjects too."

25 Scholars do not pay attention to the word 'exact' (ἀκριβῶς) and fail to distinguish between a) the 'knowledge' that the soul has forever, b) the 'knowledge' that the boy has after the proof and c) the knowledge that the boy will have in the future after more questioning. So, Fine (1992, 223–224 with n. 40; 2007, 358–361) thinks that the 'knowledge' at 85d9 is the future knowledge that the boy will obtain, although the text says clearly that the boy has this knowledge 'now'. Scott (2006, 111–112) supports the view that this 'knowledge' is the latent knowledge that the soul had forever. But again, from the text it is clear that Socrates speaks here of the actual or conscious knowledge that the boy has after the proof, and not of 'latent knowledge'.

now this 'first thing' we can proceed to infer other propositions and so forth. We arrive at perfect or 'exact' knowledge only when we cover the whole field of knowledge and come back to our initial point of departure. Thus, the system of knowledge is illuminated progressively and the perfect knowledge and illumination comes only after we proceed rationally through all the inner relations of the system, attaining in this way a complete understanding of it. This system of knowledge is not similar to a pyramid with one or more principles at its apex; there is no first principle (*archē*). The system looks rather like a network, in which there exists no particular beginning or end. Anything may serve as an *archē* on a particular occasion. In this system there is no asymmetry in knowability. All knowledge is demonstrative requiring the same kind of justification (*aitias logismos*). Things that are in the centre of the network are more relevant and exhibit more relations than others.[26] The system is coherent, structured, but not hierarchically structured in the model of Euclidean geometry.[27] The view that some propositions (or concepts) are 'higher' than others, in the sense of the *Republic*, seems to be absent in the *Meno*. In this model-n, knowledge of x involves several accounts of x, which explain its interrelations with other things. We cannot have knowledge of isolated things. Knowledge requires a systematical interrelation of the objects of a field via a series of accounts.[28] The circularity resulting from this model is not vicious. Circular accounts contribute to the coherence of the system.

I think that model-n describes well Plato's conception of knowledge in the *Meno*. There is, however, further evidence for the above claim, regarding Plato's acceptance of circular accounts. At 75b, Socrates proposes a definition of a figure (or a shape) as "the only thing which always accompanies colour". The concepts 'figure' and 'colour' seem to be co-extensive.[29] This definition does not have the standard, most familiar from Aristotle of course, but undeniably Platonic in origin, pattern of a definition by genus and differentia. Meno, however, does not attack Socrates' pattern of definition but only the fact that 'colour' is something unknown and thus, in this case, we

26 For a similar modern conception of knowledge, see Quine 1961, 42ff.

27 Many scholars, who, explicitly or implicitly, interpret the *Meno* in the light of the *Republic* (or of axiomatized Euclidean geometry), speak of a hierarchically and deductively organized system of knowledge with first principles at its top. See Vlastos 1965, 155–156; Gulley 1962, 14–15; Cornford 1971, 125; Bedu-Addo 1983, 236; Nehamas 1999, 15. Moravcsic (1971, 60–61) objects to this model claiming that the objects of recollection are concepts that "form a 'field' of some sort, so that if one has brought one element to consciousness, then this will bring with it the ability to bring to consciousness other members of the field as well". However, he does not explain what he means by this 'field', how we 'grasp' the first thing, and how the 'grasping' of it will facilitate the finding of others

28 Plato claims in the *Philebus* (18c) that one cannot know one letter until one knows them all.

29 The word 'always' rules out the possibility of colour without figure, but not the possibility of figure without colour; however, in this last case, the definition of figure would not really be a definition because it would not include all things that fall under the name 'figure'. Others who support the co-extensiveness of the two terms are: Klein 1965, 59; Irwin 1977, 136; Sharples 1985, 131; Thomas 1980, 98–99.

have a definition of 'figure' *ignotum per ignotius*. Nevertheless, Socrates thinks that there is nothing wrong with this pattern of definition (75d, and perhaps 76e3-6).[30] In the above definition, 'figure' is not defined through 'higher' concepts, and the fact that Plato recognizes here such definitions as valid, may shows that, according to his view of knowledge, concepts are not hierarchically ordered as in Euclidean geometry. Moreover, this definition of 'figure' allows us to define 'colour' in the same way; but, then, the one definition would be the conversion of the other. We may, thus, suppose that Plato, in the *Meno*, permits definitions where things are defined reciprocally. If we extend this to proofs, we might suppose that circular demonstrations are not excluded from Plato's conception of knowledge in the *Meno*. However, we have no evidence regarding this last statement.[31]

The problem of circular demonstrations was the subject of considerable debate among philosophers and mathematicians in antiquity. Aristotle refers to some people who hoped to safeguard demonstrative knowledge by maintaining the validity of circular demonstration (*An. Post.* 72b15ff). Cherniss suggested that Aristotle does not refer here to Plato, Speusippus or Xenocrates,[32] but to some followers of Xenocrates. Barnes[33] suggested that Aristotle refers to Plato's pupil Menaechmus (mathematician and philosopher) who worked on convertible propositions (see Proclus, *In Eucl.* 183, 194-4) and used the term 'element' in a very wide sense. According to Proclus (*In Eucl.* 72-73), "in this sense many propositions can be called elements of one another when they can be established reciprocally (κατασκευάζεται ἐξ ἀλλήλων)". Aristotle (*An Post.* 64b39-65a9) also informs us that mathematicians tried to prove the 5th postulate of the parallels by using circular proofs, and Proclus (*In Eucl.* 194-5) says that Apollonius

---

30 In the three definitions offered by Socrates (two of 'figure' and one of 'colour'), we have three different patterns. Socrates dismisses openly only the third definition (that of 'colour', 76d-e). Many scholars claim that Plato finds the type of the first definition unsatisfactory, or at least less satisfactory than the second definition of 'figure' as the "limit of solid", although Plato says nothing about this. For a different view, see Nakhnikian 1971, 127. Hoerber (1960, 96–97) goes further and claims that the first definition is superior to the second, the "deepest philosophically" and the only one that reveals an insight into the cause.

31 At 87d-89a, Socrates 'proves' that virtue is knowledge. The argument, roughly speaking, is the following: Virtue is good; every good is profitable; profitable involves right use; and this last *phronēsis,* which is taken to be synonymous with knowledge - *epistēme*. But, at 73b-c and 74b, he says that *phronēsis* or *sophia* is a part of virtue. The argument, besides its other defects, seems to be circular. However, I am not convinced that Plato deliberately used or even noticed the circularity of the argument.

32 See Cherniss 1944, 67–68 and 1951, 417 n.50. However, it is quite probable that it was Xenocrates himself who attempted circular proofs. Aristotle (*An. Post.*, 91a35-b11) speaks of some people who tried to prove definitions of the 'soul', 'man' etc, by conversion. Their view that the soul is a self -moving number (cf also *On the Soul*, 404b29, 408b32) is ascribed by Plutarch (*De Animae Procr. Tim.* 1012 D) to Xenocrates.

33 Barnes, 1976.

tried to prove axioms in that way. We have no evidence of circular demonstrations in the mathematics of the 5th or early 4th century, although it is very probable that in that early period (before the axiomatization of geometry), more informal proofs that involved some kind of circularity were in use.

Concluding, I think that it is possible to read the *Meno* as presenting a model-n of knowledge, and that this reading solves more interpretative problems of the text than a reading that presupposes a model-e. Commentators who interpret the *Meno* in the light of the *Republic*, or the model of Euclidean geometry, face difficulties in explaining two things: a) Why there is no evidence in the *Meno* about the existence of first principles and of an asymmetry of knowability between two kinds of propositions, and b) why the slave-boy, after he proved the theorem, has not yet exact knowledge but he needs to be asked further about the same thing, many times and in many ways. However, I cannot claim that Plato consciously proposed such a model of knowledge in the *Meno*. Perhaps, we are nearer to the truth if we suppose that he had not a clear conception of this problem when he wrote that dialogue. It is possible that he confronted the problem of the model of knowledge and explored directions similar to the above that are quite different from those he set out in the *Republic*. But, even this claim may be too strong. Perhaps Plato did not confront directly or consciously such problems in the *Meno* nor had he formed any idea about possible models of knowledge.

Nevertheless, I think that we are justified in questioning the validity of the traditional interpretation that sees a model-e of knowledge in the *Meno*. We may also claim that the problem of the model of knowledge in the *Meno* is at least open, and that the dialogue allows us to propose an alternative model-n.

VI

The problem that now arises is that if our reading of the *Meno* is correct, why does Plato change his view of knowledge in the *Republic*? I think that the answer is quite simple. In the time between the writing of the *Meno* and that of the *Republic*, mathematics was axiomatized for the first time. The *Meno* is considered to be a transitional dialogue between Plato's early and middle period, or the first of the middle dialogues, probably written at the time of the opening of the Academy (c. 387 B.C.). We have seen that when Plato writes book VI of the *Republic*, mathematicians were already positing at the beginning of their treatises some first principles. But, although the *Meno* is the most 'mathematical' dialogue, it contains no evidence of mathematical principles. This is the case with all Plato's dialogues preceding the *Republic*. We know that Plato attracted in the Academy almost all the well-known mathematicians of the fourth century. In Proclus' "summary of geometers", which is attributed to Eudemus (a pupil of Aristotle who wrote the first history of geometry),[34] we learn that Leon, one of the early members of the Academy, had written *Elements*. I find it very probable – as I have argued elsewhere[35] – that Plato had Leon's *Elements* in mind in the *Republic*.[36] In this way, Plato, having now the model of a science hierarchically ordered and depending on first principles, adopted this model in his theory of knowledge.

Bibliography

Texts quoted

Aristotle, *An. Post. (Posterior Analytics)*: Ross, W.D. 1964. *Aristotelis Analytica Priora et Posteriora*, Oxford.

Aristotle, *On the Soul*: Ross, W.D. 1963. *Aristotelis De Anima*, Oxford.

Plato, *Euthyphro*: Burnet, J. 1900-1907. *Platonis Opera*. Vol. I. Oxford (repr. 1967).

Plato, *Lysis*: Burnet, J. 1900-1907. *Platonis Opera*. Vol. III. Oxford (repr. 1967).

Plato, *Meno*: Burnet, J. 1900-1907. *Platonis Opera*. Vol. III. Oxford (repr. 1967).

Plato, *Republic*: Burnet, J. 1900-1907. *Platonis Opera*. Vol. IV. Oxford (repr. 1967).

Plato, *Sophist*: Burnet, J. 1900-1907. *Platonis Opera*. Vol. I. Oxford (repr. 1967).

Plutarch, *De Animae Procr. Tim.*: Hubert, C. 1954. *De animae procreatione in Timaeo, Plurarchi Moralia*, Vol. 6.1. Leipzig.

Proclus, *In Eucl.*: Friedlein, G. 1873. *Commentarii in Primum Euclidis Elementorum Librum*. Leipzig.

Simplicius, *In Physics*: Diels, H. 1882. *In Aristotelis physicorum libros quattuor priores Commentaria*, Berolini.

Works quoted

Ackrill, J. L. 1973. "Anamnesis in the Phaedo: Remarks on 73c-75c." In: E. N. Lee *et al.* (eds.), *Exegesis and Argument: Studies in Greek Philosophy Presented to Gregory Vlastos*. Assen, 177–195.

Barnes, J. 1976. "Aristotle, Menaechmus, and Circular Proofs." In: *The Classical Quarterly* 26, 278–292.

34 Proclus *In Eucl.* 66.20.

35 Karasmanis 2009, 153–159.

36 According to Eudemus (*In Eucl.* 65) the first who wrote *Elements* was Hippocrates of Chios, who worked in Athens at the end of the fifth century. I believe that Hippocrates' *Elements* did not involve axiomatization, for two reasons. First, because of Plato's silence in his dialogues before the *Republic* about first principles in geometry, but mainly because of the way in which Hippocrates used the word 'principle' (*archē*). In the text of Eudemus (reported by Simplicius, *In Physics*, ed. Diels, 60-8) where he speaks about Hippoctates' quadratures of lunes, the word *archē* is used to describe nor first principles but theorems useful to solve his problems. Moreover, there is no evidence that Hippocrates worked on definitions or any other kind of principles (see, Karasmanis, 2009, 151–152). Therefore, I think that when Plato wrote the *Meno* geometry had not yet been organized into an axiomatic system. According to Eudemus (*In Eucl.* 67), after Leon, another member of the Academy, Theudius wrote another, better book of *Elements*. It seems that the book of Theudius was the book that Aristotle had in mind (Karasmanis 2009, 160).

Bedu-Addo, J. T. 1983. "Sense-Experience and Recollection in Plato's Meno." In: *The American Journal of Philology* 104, 228–248.

Bluck, R. S. 1961. *Plato's Meno*. Cambridge.

Cherniss, H. 1944. *Aristotle's Criticism of Plato and the Academy*. Baltimore.

Cherniss, H. 1951. "Plato as Mathematician." In: *The Review of Metaphysics* 4, 395–425.

Cornford, F. M. 1971. "Anamnesis." In: M. Brown (ed.), *Plato's Meno*. Indianapolis, 108–127.

Crombie, I. M. 1963. *An Examination of Plato's Doctrines. Vol. II: Plato on Knowledge and Reality*. London.

Ebert, T. 1973. "Plato's Theory of Recollection Reconsidered: An Interpretation of Meno 80a – 86c." In: *Man and World* 6, 163–181.

Fine, G. J. 1979. "Knowledge and Logos in the Theaetetus." In: *The Philosophical Review* 88, 366–397.

Fine, G. J. 1992. "Inquiry in the Meno." In: R. Kraut (ed.), *The Cambridge Companion to Plato*. Cambridge, 200–226.

Fine, G. J. 2007. "Enquiry and Discovery: A Discussion of Dominic Scott's Plato's *Meno*." In: *Oxford Studies in Ancient Philosophy* 32, 331–367.

Franklin, L. 2001. "The Structure of Dialectic in the 'Meno'." In: *Phronesis* 46, 413–439.

Franklin, L. 2009. "Meno's Paradox, the Slave-Boy Interrogation and the Unity of Platonic Recollection." In: *The Southern Journal of Philosophy* 47, 349–377.

Gulley, N. 1962. *Plato's Theory of Knowledge*. London.

Irwin, T. 1977. *Plato's Moral Theory: The Early and Middle Dialogues*. Oxford.

Hare, R. M. 1960. "Philosophical Discoveries." In: *Mind* 69, 145–162.

Hoerber, R. G. 1960. "Plato's Meno." In: *Phronesis* 5, 78–102.

Karasmanis, V. 1987. *The Hypothetical Method in Plato's Middle Dialogues*. PhD Thesis. Oxford University, Oxford. Unpublished manuscript.

Karasmanis, V. 1990. "The Hypotheses of Mathematics in Plato's Republic and his Contribution to the Axiomatization of Geometry." In: P. Nicolacopoulos (ed.), *Greek studies in the Philosophy and History of Science*. Dordrecht, 121–136.

Karasmanis, V. 2009. "Περί των Προευκλειδείων Στοιχείων Γεωμετρίας." In: D. Anapolitanos (ed.), *Στιγμές και Διάρκειες*. Athens, 145–163.

Klein, J. 1965. *A Commentary on Plato's Meno*. Chapel Hill.

McCabe, M. M. 1994. *Plato's Individuals*. Princeton.

McCabe, M. M. 2009. "Escaping One's Own Notice Knowing: Meno's Paradox Again." In: *Proceedings of the Aristotelian Society* 109, 233–256.

Moravcsic, J. 1971. "Learning as Recollection." In: G. Vlastos (ed.), *Plato I, Metaphysics and Epistemology*. New York, 53–69.

Nehamas, A. 1999. "Meno's Paradox and Socrates as a Teacher." In: A. Nehamas (ed.), *Virtues of Authenticity*. Princeton, 3-26 (orig. 1985. In: J. Annas (ed.), *Oxford Studies in Ancient Philosophy 3*. Oxford, 1–30).

Nakhnikian, G. 1971. "Elenctic Definitions." In: G. Vlastos (ed.), *The Philosophy of Socrates: A Collection of Critical Essays*. New York, 125–157.

Quine, W. V. O. 1961. *From a Logical Point of View*. New York.

Scott, D. 2006. *Plato's Meno*. Cambridge.

Sharples, R. W. 1985. *Plato: Meno*. Warminster.

Tarrant, H. 2005. *Recollecting Plato's Meno*. London.

Thomas, J. E. 1980. *Musings on the Meno*. The Hague.

Tinger, S. S. 1970. "On the 'kinship' of 'all nature' in Plato's Meno." In: *Phronesis* 15, 1–4.

Vlastos, G. 1965. "Anamnesis in the Meno." In: *Dialogue* 4, 143–167.

Vlastos, G. 1994. "The Socratic Elenchus: Method is All." In: M. Burnyeat (ed.), *Gregory Vlastos: Socratic Studies*. Cambridge, 1–38.

Weiss, R. 2001. *Virtue in the Cave: Moral Inquiry in Plato's Meno*. Oxford.

Stelios Negrepontis

The Anthyphairetic Revolutions of the Platonic Ideas

(Θ)εὰ σκέδασ' ἠέρα, εἴσατο δὲ χθών.[1]

Abstract: My aim in this paper is to present some new arguments to support the view (a) that Theodorus' method of proof in the famous *Theaetetus* 147d3-8 passage is anthyphairetic; (b) that Theaetetus' brilliant idea, described in *Theaetetus* 147d7-148b4, resulted in the proof of the proposition that 'two lines commensurable only in square, have a palindromically periodic anthyphairesis' by employing the tools developed in the 'Theaetetean' Book X of Euclid's *Elements* (the conjugacy of the 'apotome' lines and the lines 'of two names'), and (c) that the method of 'division and collection', that occupies the whole of the *Sophist* and the *Statesman*, is a philosophical analogue to Theaetetus' periodic anthyphairesis. Through this prism, the present work, combined with our interpretation of the second hypothesis of the *Parmenides* 142b-155e, suggests a radical re-interpretation of Plato's philosophy.

Keywords: Theodorus; Theaetetus; Plato; Division and Collection; Anthyphairesis; Incommensurability.

I Introduction

This interdisciplinary study illuminates both the history of Greek Mathematics and Plato's philosophy: it provides a radical new interpretation of Plato's philosophy and of an important chapter in the history of ancient Greek mathematics. More particularly, (a) Plato's method of 'division and collection', for obtaining knowledge of Platonic Ideas, is revealed to be a close philosophical analogue of the (palindromically) periodic anthyphairesis of lines incommensurable only in length; and, (b) Theaetetus' mathematical contribution, as reported in *Theaet.* 147d7-148b4, is shown—surprisingly—to be the proposition that 'every pair of lines, commensurable in power only, possesses a palindromically periodic anthyphairesis', the proof of which is reconstructed solely with tools found in Book X of Euclid's *Elements*.

In section 1, we present some novel arguments, centered to the use of distributive plural, in favor of an anthyphairetic interpretation of Theodorus' method of proof of the incommensurability of powers reported in the famous *Theaetetus* 147d3-148d7

1 '(G)oddess [Athena] dispersed the mist, and the land was recognized'; Homer, *Od.* XIII. 352.

https://doi.org/10.1515/9783110565959-013

passage, while, in section 2, we obtain a provisional reading of Theaetetus' enigmatic method of Collection of the infinity of powers divided by Theodorus', as reported in *Theaet.* 147d7-148b4. Of crucial importance in reaching an understanding of both Theaetetus' mathematical method of collecting infinity and Plato's philosophical method of division and collection is Socrates' insistence, examined in Section 3, to pursue knowledge of the Platonic Ideas via imitating (πειρῶ μιμούμενος, 148d4-5) Theaetetus' method. Plato's method of 'division and collection', presented in the *Sophist* (and exemplified by the definition of the Angler and the Sophist), and the *Statesman* (and exemplified by the definition of the Statesman), are studied in sections 5–7, and shown to be philosophical analogues of the (palindromically) periodic anthyphairesis. In addition, the cause of this striking structure of the knowledge of Platonic Ideas is, according to *Stat.* 283b6-287a6, the double measurement, a condition shown (in section 8) to be analogous to the Theaetetus' double mathematical condition of incommensurability in length and commensurability in square. The results stated in the first paragraph of this Introduction now follow in sections 9 and 10.

II Part A: *Theaetetus* 147d3-148b4 (Sections 1–3)

Our starting point is the famous *Theaetetus* 147d3-148b4 passage, where Plato relates three, equally compelling, events: (a) first, Theodorus' lesson to Theaetetus (and his friend) on the proof of some quadratic incommensurabilities ('powers'); (b) second, Theaetetus' mathematical contribution to the theory of quadratic incommensurabilities, described by the enigmatic method of 'collection' into one of the 'powers', shown in Theodorus' lesson to be infinite in multitude; and, (c) third, Socrates' efforts to convince Theaetetus to attempt to imitate his method of 'collection', in order to obtain the philosophical method of 'division and collection', by which method knowledge of Platonic Ideas may be obtained. The method that Theodorus used in his lesson (a) for the proof of these incommensurabilities; the nature, statement and the method of proof of Theaetetus' mathematical contribution (b), described by the mathematically unclear term of 'collection'; and the philosophical method of 'division and collection' (c), which was developed in imitation, are, by no means, evident.

1 Theodorus' Method of Proof of Incommensurabilities

Theodorus' lesson on quadratic incommensurabilities is contained in the first part 147d3-6 of the passage, which could be paraphrased, as follows:

> Theodorus, in his lesson about 'powers' (δυνάμεών), namely about lines a such that $a^2 = Nb^2$ (where N is a non-square number and b is the line of length one foot), by examining each power separately, showed (ἀποφαίνων) that, if N is a number from N = 3 to N = 17, then a is incommensurable in length (μήκει οὐ σύμμετροι) to the one foot line.

The basic question here is: What was Theodorus' method of proof? Hardy and Wright (1938, 42) stated: "The question how Theodorus proved his theorems has exercised the ingenuity of every historian." There have been several scholarly suggestions, which I wish to review.

1.1 The Mathematical Possibilities for the Proofs of Quadratic Incommensurabilities

Before outlining the historical development of these suggestions, we will outline their mathematical possibilities. We begin by the definition of anthyphairesis. For two lines a > b, the anthyphairesis of a to b is the following process, analogous to the so-called Euclidean algorithm for numbers:[2] There is a unique number I_0 such that $I_0 b \leq a < (I_0 + 1)b$. If $a = I_0 b$, then the process ends; if $I_0 b < a$, then there is c_1, such that $a = I_0 b + c_1$, and $c_1 < b$. We repeat the process with the two lines $b > c_1$. There is a unique number I_1, $I_1 c_1 \leq b < (I_1 + 1)c1$. By continuing in this way, we obtain a sequence $a = I_0 b + c_1$, $b = I_1 c_1 + c_2$, ..., $c_n = I_{n+1} c_{n+1} + c_{n+2}$, ... If this process ends after a finite number of steps, then the anthyphairesis is finite; otherwise, it is infinite. *Elem*. X.2 states that "if the anthyphairesis of a to b is infinite, then a and b are incommensurable". This can be stated, in modern terminology, as: 'If a, b line segments, N not-square number, and $a^2 = Nb^2$, then a, b are incommensurable'. We now examine the possible proofs of this quadratic incommensurability.

Anthyphairetic proof. It consists in proving that the anthyphairesis of a to b is infinite, and then applying X. 2. But to apply X.2, one must first prove the infinity of anthyphairesis. It is impossible to verify this infinity by direct computation, as this would involve an infinite number of computations, one for each step. A realistic way to prove infinity for the ancients would be to prove the far stronger condition that the anthyphairesis is (eventually) periodic. We shall examine the ways in which the ancients could have established anthyphairetic periodicity.

2 This definition is implicitly described in the proofs of *Elem.*VII.1-2 and X.2. The term can be rendered as 'mutual subtraction' and appears in Aristotle's *Topics* as 'antanairesis'. Anthyphairesis of natural numbers, on the one hand, is always finite (cf. VII.1,2), while of magnitudes it may also be infinite (cf. X.2).

Arithmetical Proofs: Assume that a, b are commensurable, say a = mc, b = nc. By *Elem*.X.3, we may assume that m and n are relatively prime numbers. Then $m^2 = Nn^2$, hence certainly $m^2/n^2 = N/1$.

Arithmetical proof 1: By *Elem*.VII.27, m^2, n^2 are also relatively prime numbers, and, hence, both ratios are on lowest terms, therefore $N = m^2$, a square number; which is a contradiction.[3]

Arithmetical proof 2 (Hardy & Wright 1938, 42): Case by case proof, for every N separately, not assuming VII.27. E.g., for N = 5, m^2 is divisible by 5. We wish to verify that m is divisible by 5. We prove that m cannot be of the form $m = 5s + 1, = 5s + 2, = 5s + 3, = 5s + 4$ (e.g., if $m = 5s + 4$, then $m^2 = 25s^2 + 16 + 40s = 5t + 1$, not divisible by 5). Then $m = 5s$, hence $m^2 = 5^2s^2 = 5n^2$, hence $5s^2 = n^2$. Hence n^2 is divisible by 5. By previous argument, n is divisible by 5. Hence m, n not relatively prime; which is a contradiction.

Arithmetical (partial) proof 3 (Knorr 1975, 181-193; McCabe 1976; Bashmakova & Lapin 1986, only for numbers $N \neq 4^s(8k + 1)$):

Case 1: N odd, $N \neq 8k + 1$. Since N is odd, it is immediate that m, n are odd, say $m = 2r + 1$, $n = 2s + 1$. Thus, $(2r + 1)^2 = N(2s + 1)^2$. It follows that $4r(r + 1) - 4Ns(s + 1) = N - 1$. Since $r(r + 1)$ and $s(s + 1)$ are even numbers, it follows that the left hand side is a number divisible by 8, while right hand side by assumption is not divisible by 8; which leads to a contradiction.

Case 2:[4] N even, $N \neq 4^s(8k + 1)$. Since N is not square, there exists a number s (possibly 0), such that $N = 4^sM$, with M is a non-square number, not divisible by 4. Setting $a' = a/2^s$, $b' = b/2^s$ it is clear that $a'^2 = Mb'^2$, and that a, b are (in)commensurable iff a', b' are (in)commensurable. Thus, M is either an odd number, $M \neq 8k + 1$, that has been dealt with in Case 1, or an even number, $M = 4t + 2$. Clearly, it is enough to prove that the equation $m^2 = (4t + 2)n^2$, with m.n relatively prime, is impossible. This imitates the proof that $m^2 = 2n^2$ is impossible. In fact, m^2 is even, hence m is even, $m = 2p$. Then, $m^2 = 4p^2$, hence $4p^2 = (4t + 2)n^2$, hence n^2 is even, hence n is even; which is a contradiction.

3 In the same way, one can prove the more general proposition: if $Ma^2 = Nb^2$, and M/N is not as square number to square number, then a,b incommensurable, using *Elem*. X. 9 and VII.27. A special case is Proposition: if $Na^2 = (N + 1)b^2$, then a,b are incommensurable. A proof can be given employing the Archytas musical-arithmetical proposition , reported by Boethius and appearing as *Elem*.VIII.7, whose proof relies again on VII.27.

4 Case 2 mimics the traditional arithmetic proof for N = 2.

1.2 The History of the Suggested Reconstructions

Zeuthen (1910, 418,422-429) suggested that the fact that Theodorus proceeded with his proofs case by case appears to favor an anthyphairetic method—since an arithmetical proof proceeds for all N together, and not in cases. Evidently, Zeuthen had some form of arithmetical proof 1 in mind, using the key proposition *Elem.* VII.27. Hardy & Wright (1938, p. 42–45) observed that perhaps Theodorus did not know *Elem.* VII.27, or VII.30, and, for this reason, he would have to proceed case by case, with arithmetical proof 2. This argument demolished any advantage that the anthyphairetic method interpretation might have had over an arithmetical one. Van der Waerden (1954, 141–146) 1950's adoption of an anthyphairetic proof at a posterior date was, then, a matter of choice, not based on arguments originating with the *Theaetetus* passage. Knorr (1975, 62–83) argued that Theodorus stopped at the case of $N = 17$, because he found some difficulty,[5] thus excluding the anthyphairetic proof, which is particularly simple for $N = 17$, and suggested an ingenious reconstruction, which has since been streamlined to the arithmetical proof 3, above.[6]

The current dominant thesis is that the *Theaetetus* 147d3-6 passage forces us to a position of *methodologic neutrality*, in that it contains no information on the method employed by Theodorus. Cherniss (1951, 410–414) states that

> The sole evidence for Theodorus' demonstration is the passage of the Theaetetus, and that reveals only that he began with square root of 3 and selecting one surd after another up to square root of seventeen there somehow stopped. This tells us nothing of the method that he used,

while Burnyeat (1978, 494–505) writes:

> we are not told how [Theodorus] proved this result…there is no good reason to expect that the answer [about Theodorus' method] is to be squeezed out of one ambiguous sentence in Plato's dialogue.

1.3 The Distributive Plural Interpretation in *Theaet.* 147d7-8: *Every* Dynamis Possesses Division and Collection, *not* All Powers Collectively

We will present novel arguments showing that Plato and his anonymous commentators do, in fact, provide information that decides Theodorus' method of proof.

---

5 Not in accord with the existing ancient sources; cf. *Anonymi Commentarius in Platonis Theaetetum* (*ACIPT*) 34, 15-32 against omitting the case $N = 17$; *ACIPT* 34,32-35 and 35,13-21, Iamblichus, *Theology of Arithmetic*, 11,13-16 against not simply stopping but finding a difficulty at $N = 17$; and, the fact that 'mechri x' has normally the meaning 'till x, including x'.

6 The streamlining has revealed that Knorr's suggestion is not really a proof of incommensurability case by case.

The thesis of neutrality would be justified, if it were true that all the information that Plato is providing on Theodorus' method is contained in 147d3-6, but, as we shall see, this is not true.

In the very next sentence, 147d7-e3, Plato goes on to relate the mathematical idea that occurred to Theaetetus and his friend in the course of Theodorus lesson. The crucial initial sentence is (147d7-8): "ἡμῖν οὖν εἰσῆλθέ τι τοιοῦτον, *ἐπειδὴ* ἄπειροι τὸ πλῆθος αἱ δυνάμεις *ἐφαίνοντο*." In paraphrase: 'we had an idea, since it was shown that *such and such*'—'it was shown', certainly by Theodorus. Here is how Burnyeat (1978, 501) describes this:

> The key sentence is 147d 7-e 1, which I render as follows: 'Since the *δυνάμεις* were turning out to be unlimited in number, it occurred to us to attempt to collect them up into a single way of speaking [i.e., a formula or definition] of all these *δυνάμεις* together;' Theaetetus is recounting the thoughts suggested to himself and his companion by and during Theodorus' lesson....

The point is that Theaetetus had the mathematical idea *in the aftermath* of Theaetetus' lesson, so 147d7-8 is a transitive phrase that has to do not only with Theaetetus' new idea, but also with Theodorus' lesson that caused it. As such, it is our last and only hope to extract Theodorus' method from Plato's passage.

We will now attempt to clarify the meaning of the statement (147d7-8) 'ἄπειροι τὸ πλῆθος αἱ δυνάμεις ἐφαίνοντο'. We ask, what does Plato mean? (a) That 'αἱ δυνάμεις', as a totality, as a set, appeared to be 'ἄπειροι τὸ πλῆθος' (in which case the plural 'αἱ δυνάμεις' is 'collective'), *or* (b) That every power separately appeared to be infinite in multitude (in which case the plural 'αἱ δυνάμεις' is 'distributive')?[7]

The passage has been interpreted in the collective sense (a), essentially by all scholars. Most casual readers would agree—Why, on earth, would anyone think of a distributive reading? Thus, Cornford (1935, 23) translates "seeing that *these square roots were evidently infinite in number*"; van der Waerden (1950, 166) "Now it occurred to us (Theaetetus and another young man), *since the number of roots appeared to be infinite*, to try to collect them under one name, by which we could henceforth call all the roots"; Knorr (1975, 63) "Now this is what occurred to us: that, since we recognized *the powers to be unlimited in number*, we might try to collect them under a single name"; and Burnyeat (1978, 500) "Well, since *the powers seemed to be unlimited in number*, it occurred to us to do something on these lines: to try to collect the powers under one term by which we could refer to them all" (my emphasis). I will argue that, contrary to present interpretation and appearances, Plato means the plural in a distributive manner (b); I will present four arguments in that direction:

7 To give an example, the phrase 'the policemen wear uniform' is a distributive plural, whereas 'the policemen have surrounded the block' is a collective plural.

[A] The proponents of a collective rendering for the plural 'αἱ δυνάμεις' in 147d7-8 have a serious problem of making sense of the crucial statement. One of the principal proponents of the collective rendering writes (Burnyeat 1978, 501): "the idea that there is an endless series of whole number squares (or sides of such squares) *would hardly need to be prompted by a process as protracted as Theodorus' lesson*" (my emphasis). Thus, Burnyeat rejects the statement 'there is an infinite multitude of lines a_N corresponding to each non-square number N' as a possible meaning of the statement that he has rendered as 'since all the powers appeared infinite in multitude', because this is something obvious, and would in no way need Theodorus' lesson to be understood. Burnyeat (1978, 501–502) then suggests the rendering that he adopts:

> That there are *an indefinite, perhaps infinite, number of squares with incommensurable sides*, on the other hand, is precisely the hypothesis that would suggest itself as Theodorus proceeded from case to case proving more and yet more examples of incommensurability, perhaps by a method which could be endlessly reapplied. Therefore, it is likely that, in context, *'all these δυνάμεις' refers to squares with incommensurable sides rather than to squares generally* (my emphasis).

However, Burnyeat's interpretation of 147d7-8, i.e. that there are infinitely many incommensurable powers, is subject to his own objection noted above; namely, that it "would hardly need to be prompted by a process as protracted as Theodorus' lesson." This happens because the incommensurability of the power for N = 2, known to the Pythagoreans[8] and considered known by Theodorus, implies the incommensurability of infinitely many powers; namely, of all powers corresponding to N = 8 (=2.2^2), =18 (=2.3^2), =32 (=2.4^2), in general N = $2.n^2$ for n = 1,2,.... Beyond these two, there does not seem to be any other suggestion for the crucial sentence with rendering (a).

[B] Distributive plural for all occurrences of 'αἱ δυνάμεις' before, at 147d4-5, and after, at 148b1-2, the crucial sentence. The plural 'powers' is obviously distributive in the first occurrence (147d4-5) '[δυνάμεις] μήκει οὐ σύμμετροι τῇ ποδιαίᾳ'—since each power, separately, is incommensurable to the one foot line—and in the third double occurrence (148b1-2) '[δυνάμεις] μήκει μὲν οὐ συμμέτρους ἐκείναις, τοῖς δ' ἐπιπέδοις ἃ δύνανται [συμμέτρους]'—since each power, separately, is incommensurable to the one foot line (or to a line commensurable to it), and the square of each power, separately, is commensurable to the one square foot. Sandwiched between the first (147d4-5) and the third (double) (148b1-2) distributive occurrence of the plural 'αἱ δυνάμεις' is the crucial second occurrence (147d7-8) of the plural 'αἱ δυνάμεις ... ἄπειροι τὸ πλῆθος'. It is thus natural to conclude that this particular (second) plural is distributive as well. With the currently dominant interpretation of 147d7-8, we would have to assume that

---

8 As attested by the accounts by Plutarch, Iamblichus, Pappus, and Proclus of the discovery of incommensurability by the Pythagoreans, and by the accounts by Theon of Smyrna, Iamblichus, and Proclus of the Pythagorean origin of the side and diameter numbers.

Plato first uses 'δυνάμεις' in the distributive sense, then in the next line switches to a collective plural for 'δυνάμεις', and, then, after another few lines, switches back to a double distributive plural for 'δυνάμεις'. It appears more natural to accept that the second plural is distributive as well.

[C] Support for a distributive reading from *Scholia in Platonem*. Plato comments briefly on Theodorus' demonstrations in 162e6-7: "Imagine how utterly worthless Theodorus or any geometer would be if he were prepared to rely on probability to do geometry." In this passage, Plato clearly refers to the Theodorus' incommensurability demonstrations mentioned earlier (in 147d3-e1), the only Theodorus' demonstrations mentioned in the *Theaetetus*. An interesting commentary on the 162e6-7, found in the anonymous work *Scholia in Platonem* (abbreviated as SIP), runs as follows:

> If we accept the reasoning of the many as the dominant one in geometry, we would be ridiculous to claim that [SIP1] magnitudes are incommensurable to each other (ἀσύμμετρα ...ἀλλήλοις μεγέθη), and that [SIP2] the finite line is divisible ad infinitum (τὴν πεπερασμένην εὐθεῖαν διαιρετὴν εἶναι εἰς ἄπειρον), and the like (τὰ τοιαῦτα).

The following paraphrasis of the SIP commentary is suggested: Theodorus, when proving that each of the powers (for N = 3,5,...,17) and the one foot line are incommensurable to each other ([SIP1], as in statement 147d1-3), employed, not any considerations of probability (which would be worthless and make him ridiculous), but logically impeccable proofs (ἀπόδειξιν δὲ καὶ ἀνάγκην (162e4-5)) involving the division *ad infinitum* of each of these finite lines ([SIP2], so that each of these lines, each power, is infinite in multitude (as in statement 147d7-8), in fact by a division *ad infinitum* connected to incommensurability)). This reading leads to a distributive interpretation of 'αἱ δυνάμεις' in 147d7-8: each power, being incommensurable to the one foot line, is divided into an infinite multitude.

[D] The author of the *Anonymi Commentarius in Platonis Theaetetum* (ACIPT), possibly Eudorus[9], a writer of the first century BC, lends strong support to the distributive interpretation. Indeed, commenting on the infinity arising from the 'powers infinite in multitude' and their 'collection' achieved by Theaetetus and his companion in the *Theaetetus* 147d-148b passage, the anonymous author makes the following statements:

> [ACIPT1; 37, 4–12]:[10] An infinity in general must be collected, if possible.
> [ACIPT2; 36, 45–48]:[11] Lines admit of the infinite either by (infinite) division, or by (infinite) increase.

9 Cf. Tarrant 1985.

10 ἐπεὶ τὸ ἄπειρον ἀπερίληπτόν ἐστιν, καὶ ἀ[όρι-]στ[ος] ἐν τῷι τοιούτωι ἡ διάνο[ι]α, δεῖ καθ' ὅσον ἐνδ[έ]χεται καθολικῶ[ι] τινι περιλαμβάνειν.

11 ἐπεὶ αἱ γραμμαὶ ἐπιδέχονται τὸ ἀόριστον, εἴ τις αὐτὰς ἢ αὔξοι ἢ διαιροῖ.

From [ACIPT1] it is clear that what is being collected into one is an infinity; from [2], a statement commenting on 147d7-8: 'the powers are infinite in multitude', it is explained that every infinity about line segments—like the infinity in multitude of the powers—can arise in just one of two ways: either by an increase to infinity or by an infinite division. But each of these ways is obviously a way by which a single line-segment becomes infinite. Thus, from [ACIPT1] and [ACIPT2], it follows that: the plural in the expression 'the powers infinite in multitude' is distributive, and 'every power ... infinite in multitude' is meant.

1.4 Distributive Plural for αἱ δυνάμεις in 147d7-8 Suggests an Anthyphairetic Method for Theodorus' Proofs

From the *Theaetetus* text itself, we obtain a distributive reading for the 147d7-8 crucial statement (cf. [A] and [B] above). The distributive reading is fully confirmed by the two existing ancient anonymous commentators (cf. [C] and [D] above). Taking into account the arguments presented, 147d3-148b3 must be rendered as follows:

> Since, in the aftermath of Theodorus' lesson, every power (αἱ δυν άμεις) was shown (ἐφαίνοντο) to be [divided into] an infinite multitude of parts (ἄπειροι τὸ πλῆθος), Theaetetus and his companion had the idea and in fact succeeded in collecting into One (συλλαβεῖν εἰς ἕν) the infinite multitude of every power.

These findings strongly suggest that Theodorus' method of proof of the incommensurability of each of the powers he examined was anthyphairetic; namely, a proof employing proposition *Elem.* X.2. In fact, no other method of proof of incommensurability would conceivably have resulted in the division of each power in an infinite multitude of parts. Thus, Zeuthen and van der Waerden were correct in proposing an anthyphairetic method for Theodorus' proofs, but did not have the necessary arguments, provided in the present section, to support this proposal.

1.5 Two Possible Objections to an Anthyphairetic Reconstruction

This is the point to deal with two objections against an anthyphairetic reconstruction of Theodorus' proofs. The first is that the proof of *Elem.* X.2 makes use of Eudoxus' condition (*Elem.* V.df4). For this reason, Knorr (1975, 122–123) believed that it could not have been used by Theodorus. Nevertheless, an elementary proof of X.2 without involving Eudoxus' condition can be given with a simple argument, similar to the proof of *Elem.* VI.1, as suggested in Aristotle's *Topics* 158 b 29–35. This argument was known at the time of Theaetetus—since, according to all accounts[12], Theaetetus is the originator of the theory of ratios to which Aristotle refers—, so quite likely to Theodorus too.

The second problem with the anthyphairetic reconstruction of Theodorus' proofs is that the reconstructions up to now depend on the so-called 'logos' criterion to prove periodicity of the anthyphairesis; and, the 'logos' criterion itself depends on a theory of ratios of (incommensurable) magnitudes. Such a theory was developed by Theaetetus, at the earliest, but was certainly not available to Theodorus. To avoid anachronism, there must be a proof of periodicity by using more archaic tools than ratios of magnitudes. Van der Waerden (1950, 145) suggested the use of gnomons, and Fowler (1999, 74–83; 374–378) introduced gnomons together with ratios of magnitudes. A reconstruction with gnomons, instead of ratios, covering all aspects of the proof, has been carried by Negrepontis & Tassopoulos (2015). Instead of the logos criterion, we revert to the Pythagorean principle of the preservation of the gnomons.

2 Theaetetus' Contribution in Terms of the Philosophical Term 'Collection', and its Relation to the Platonic Division and Collection

We now turn to the more complex question of Theaetetus' contribution to the theory of incommensurable magnitudes. Our sources are: (1) *Theaet.* 147d7-148b3; (2) the ACIPT (cf. section 1.3 [D]); and (3) an anonymous Scholium (62) to *Elem.* X.9.

2.1 The Traditional Reading of *Theaet.* 147d7-148b3

Scholia in Euclidis Elementa X.62 connects Theaetetus' discovery, reported in *Theaet.* 147d7-148b3, with *Elem.* X.9.[13] This has led some scholars to believe, with some variations, that Theaetetus' contribution consists of an arithmetical proof of quadratic incommensurability using *Elem.* X.9 and propositions from Books VII and VIII of the *Elements*, similar to the arithmetical Proof 1 in section 1.1. Thus, Zeuthen (1910) believed that Theaetetus had discovered and proved several propositions from the arithmetical books VII and VIII which were needed for this proof . Van der Waerden (1950,166) wrote:

12 Van der Waerden 1954, 178; Knorr 1975, 261, 273.

13 Τὸ θεώρημα τοῦτο [i.e. *Elem.* X.9] Θεαιτήτειόν ἐστιν εὕρημα, καὶ μέμνηται αὐτοῦ ὁ Πλάτων ἐν Θεαιτήτῳ; *Scholia in Euclidis elementa* X.62.

> What is for us more important than these trivial definitions, is the proposition given at the end, very briefly but nevertheless clearly: Line segments, which produce a square whose area is an integer, but not a square number, are incommensurable with the unit of length

and (1950,168):

> Book VII is of older date and forms the foundation of the Pythagorean theory of numbers. But, once one has at his disposal the propositions of Book VII, the number theoretical part of Theaetetus' proof has no further difficulties. One can, for instance, reason as follows: q and p in (1)[14] can of course be taken to be relatively prime. Then it follows from VII 27 that p^2 and q^2... are relatively prime, so that, by VII 21, they are the least of the numbers in that ratio. But in the left member of (1) occur n and 1 which are also relatively prime and hence also the least in this ratio. It follows that $n = p^2$ and $1 = q^2$, so that n is a square. ..In my opinion the merit of Theaetetus lies therefore not in his contribution to the theory of numbers, but in his study of incommensurable line segments which produce commensurable squares.

Thus according to van der Waerden Theaetetus' proof of the incommensurability of a,b in case $a^2 = Nb^2$, N non-square, is Arithmetical proof 1. In view of van der Waerden's support of an anthyphairetic method for Theodorus' incommensurabilities, we are to believe that Theodorus' anthyphairetic case by case proofs inspired Theaetetus' general arithmetical proof, a decidedly non-convincing situation. Knorr (1975, 227–228) also supports, with some secondary modifications, an arithmetical proof by Theaetetus:

> There is general agreement on the type of principles Theaetetus used in demonstrating the arithmetic part of these theorems. H. Zeuthen, E.J. Dijksterhuis, K. Reidemeister and B. L. van der Waerden have each given reconstructions, drawing from various portions of the arithmetic books of the *Elements*

2.2 A Provisional Statement of Theaetetus' Mathematical Contribution According to a New Reading of *Theaet.* 147d7-148b4

At this point, I will embark toward a new reading of 147d7-148b4 and a new proposal for Theaetetus' method of proof of the quadratic incommensurabilities, not in agreement with the traditional interpretations, described in Section 2.1. The distributive interpretation of 'αἱ δυνάμεις' in 147d7-8 has repercussions not only for Theodorus' method, but also for the brilliant idea Theaetetus had, in the aftermath of Theodorus' lesson. Our conclusion about the distributive meaning of αἱ δυνάμεις in 147d7-8 sweeps along to a distributive interpretation Theaetetus' collection of the infinite multitude of the powers, as well. Thus, the still enigmatic 'sullabein' (147d8) and 'perilabein'(148d6) that Theaetetus thought—and successfully carried out—is also to

---

14 $n : 1 = p^2 : q^2$.

be applied on the infinity of every power. The comments of ACIPT on Theaetetus' 'collection', on which we now turn, fully confirm the distributive reading:

> [ACIPT3; 37,1-3]: [15] The collection[16] of an infinity in a line is achieved by passing over to numbers.
>
> [ACIPT4; 37, 39-44]:[17] But, numbers cannot achieve collection of the infinity in a line by being increased, since numbers increase to infinity.
>
> [ACIPT5; 26,13-18]:[18] In fact, numbers achieve the collection of the infinity in a line, because numbers are commensurable to each other.
>
> [ACIPT6; 41,40-42,18]:[19] The meaning of passing over to numbers is made clear here. Passing over to numbers simply means passing over to the square of the power, since the square of the power is commensurable to the one foot square, and thus it is as number to number.
>
> [ACIPT7; 44,50-45,3]:[20] But a power passing over to someplace (to its square) must come from some other place (from its length). It is clarified here that we pass over to the more clear numbers, coming from the less clear magnitudes, and to the commensurables and as number to number from the incommensurables and not as number to number. This clearly refers to the incommensurability in length of the power to the one foot line.

Although 'collecting' infinity is at this point still mathematically unclear,[21] it still must be an effect of the finitizing, limiting cause of commensurability in square. But commensurability is a finiteness condition only in the sense of anthyphairesis. We

---

15 ὁρίζονται δ' ὑπὸ τῶν ἀριθμῶν, μετέβ[η-]σαν ἐπ' αὐτούς ([and because the lines] are limited by numbers, Theaetetus and his companion passed over to numbers). Also, see ACIPT 42, 30–33.

16 The term 'collection' renders περιḷạμβάνειν (as in *Theaet.* 148d6, where it is clearly synonymous to συλλαβεῖν, 147d8, occurring also in ACIPT 26,11; 37,5; 37,10; 37,29; 37,43; 37,46; 45,48; 46,39), where the terms 'περιορίζειν'' (42,32) and 'ὁρίζειν' (37,1; 37,11) are also used.

17 Ὁ ἀριθμὸς ἄπειρός ἐστιν κατὰ τὸ αὔξεσθαι· οὐ δ̣ύναται οὖν τις τὸ ἐπ' ἄ̣πειρον προϊὸν περιλαβεῖν (number is infinite by increase; therefore one cannot collect that which increases *ad infinitum*); ACIPT 37, 39–44.

18 ἦλ]θ[ον] ο̣ὖν ἐπ̣ὶ τ̣ὸ[ν ἀριθμ]ὸν διὰ τὸ [.]κ[......]ο̣ν τῶι π̣ά̣ν̣τ̣α̣[ς] τοὺς ἀρι[θμ]ο[ὺ]ς σ̣υ̣μμ[έτ]ρους [εἶ] ν̣α[ι] πρ[ὸς] ἀ̣λλή[λους] (thus they came to number as a consequence of the fact that all the numbers are commensurable to each other).

19 My rendering. [ὡς τ]οίνυν τῶν τετ̣ρ̣[αγώ]νων σχημάτ̣ω̣ν ἃ μὲν ἦν σύμμε̣τ̣[ρ]α τῆι ποδιείαι δυνάμει καὶ μήκει καὶ [πλ]άτει, καὶ ταῦτα μή[κη] ὠ̣νόμασαν, ἃ δὲ πλάτει μὲν οὐκέτι δὲ κ̣α̣ὶ̣ [τ]ῆ̣ι πλευρᾶι, κα̣ὶ̣ ταῦ̣[τα] δυνάμεις [ἐ]κάλεσαν τῶι κοινῷ̣ι προσχρησάμενοι ὀνόματι, οὕτως καὶ ἐπὶ τῶν στερεῶν ἦλθον ἐπὶ τὰ κυβ̣ικ̣ὰ σχήματα καὶ ἐτίθεσαν κύβον, οὗ αἱ τρεῖς πλευραὶ ἑκάστη ποδός, καὶ γενόμεναι ἐπ' ἀλλήλας ποιοῦσι ἕνα στερεὸν πόδα. καὶ πρ̣ο̣αιροῦντες κύβον δύο ποδῶν καὶ ἄλλον τριῶν εἶτα τεσσάρων εὑρισκον αὐτὸ μὲν τὸ στερεὸν πρὸς τὸ στερεὸν σύμμετρον· ἔχει γὰρ λόγον, ὃν ἀριθμὸς πρὸς ἀριθμόν· τὰς δὲ πλευρὰς ἀσυμμέτρο[υς,].

20 My rendering. ἀπὸ τῶν ἀσαφεσ̣τ̣έρω̣ν ἐ̣π̣ὶ τὰ σαφέστερα δε̣ῖ μεταβαίνειν ὡς ἀπὸ τῶν μεγεθῶν μετέβησαν ἐπὶ τοὺς ἀριθμούς (from the less clear one must pass to the more clear as when passing from magnitudes to numbers). Also, cf. ACIPT 32,1-4. The same point is made in the *Scholion* X.39 to the *Elements*.

21 It will be clarified in section 9.1.4 below.

conclude that the infinity of the power that is being collected by this finitising cause is precisely its incommensurability, its infinity in anthyphairesis.

It follows that 147d7-148b3 should be paraphrased as follows: 'Theaetetus succeeded in the Collection into One of every (divided) power, by realizing that every power is not only a line segment incommensurable in length to the one foot line, but also commensurable in square'. The new information that we get from the ACIPT comments, in relation to the original Theaetetus' text, is, besides the distributive plural for the powers, is that (a) the cause of the collection of the anthyphairetic infinity and incommensurability of a power is the commensurability of the power in square, of a^2 to b^2; and (b) the effect is the collection.

The collection of the anthyphairetic infinity of a power (the effect, the conclusion of Theaetetus' proposition) is possible only because the square of the power is commensurable to the one foot square (the cause, the assumption of Theaetetus' proposition). Accordingly, we are guided to the following provisional statement of

'Theaetetus Proposition': 'If a is a power, namely a is a line incommensurable with respect to the one foot line b, and a^2 is commensurable to b^2, then the infinity of the anthyphairesis of a to b, which arises because of the incommensurability of a to b, can be "collected" into one.'

We will then make sense of this proposition, only if we can make mathematical sense of the still enigmatic meaning of 'collection'. But how can we expect to do this? Our plan will be explained in the next section. A full understanding of the meaning of 'collection' will be achieved only in Section 9.1.4, below.

3 Plato's Plan to Imitate (πειρῶ μιμούμενος, *Theaet.* 148d4-5) the Theaetetean Collection with his Method of Division and Collection in the *Sophist* and the *Statesman*

3.1 The Philosophical Problem of knowledge, Theaetetus' Mathematical Collection and the Plan to Imitate it, towards Dealing with the Philosophical Problem

Socrates, with a characteristic ironic understatement (μικρὸν δέ τι ἀπορῶ, 145d6), introduces the fundamental question of the trilogy *Theaetetus*, *Sophist*, and *Statesman*; namely, how is it possible to obtain full knowledge (λαβεῖν ἱκανῶς παρ' ἐμαυτῷ, ἐπιστήμη ὅτι ποτὲ τυγχάνει ὄν, 145e8-9) of a Platonic Idea. A careful reading of the first pages of *Theaetetus* reveals that Plato has built a close relation between Theaetetus' mathematical discovery on quadratic incommensurabilities, that is described by the term 'collection', and Plato's quest for a method to obtain full knowledge (ἐπιστήμη).

After some failed attempts, Theaetetus (147c7-d1) realizes that Socrates' question on the acquisition of knowledge is similar (οἷον) to his treatment of the mathematical problem, conceived during Theodorus' lesson on quadratic incommensurabilities. The account of Theodorus' lesson and Theaetetus' discovery follows (described in Sections 1 and 2). According to this account Theaetetus succeeded in showing that each power, divided into an infinite multitude of parts by Theodorus' proof of incommensurability, can be collected into One. Socrates then exhorts Theaetetus to try to imitate (πειρῶ μιμούμενος) this process of Collection towards the knowledge of Platonic Ideas.

> *Soc.* Then you must have confidence in yourself (θάρρει...περὶ σαυτῷ), and try earnestly (προθυμήθητι) in every way to receive 'logos' of knowledge (πέρι καὶ ἐπιστήμης λαβεῖν λόγον) as well as of other things. *Theaet.* If it is a question of earnestness, Socrates, the truth will come to light (*Theaet.* 148c9-d2).

> *Soc.* Well then—for you pointed out the way admirably just now—try to imitate (πειρῶ μιμούμενος)[22] your answer about the powers (περὶ τῶν δυνάμεων), and just as you embraced (περιέλαβες)[23] them in one kind (ἑνὶ εἴδει), though they were [infinitely] many (πολλὰς), try to designate the [infinitely] many pieces of knowledge (τὰς πολλὰς ἐπιστήμας) by one 'logos' (ἑνὶ λόγῳ) (*Theaet.* 148d4-7).[24]

Thus, at this point both Theaetetus[25] and Socrates agree that Theaetetus' mathematical method has a bearing on Socrates' philosophical question. It strikes us as strange that a proof of quadratic incommensurability can be considered as the mathematical result that should be imitated for the acquisition of the complete knowledge of the intelligible Beings. It is an indication that Theaetetus' method of 'collection' was not just a proof of incommensurability but a method that was leading to some sort of complete knowledge, akin to the complete knowledge sought. This will be explained in section 9.1.3, below.

22 *Theaet.*148d4-7 contradicts Burnyeat's (1978, 505) claim that "Plato has no motive to indicate to the reader whether he has in mind any definite method of proof." It would be most unlikely to be indifferent to a method he is about to imitate for his own method of Division and Collection, by which the full knowledge of Platonic Ideas may be obtained.

23 Thus, περιέλαβες, circularly embracing, has clearly the same force as συλλαβεῖν collecting.

24 Translation by Fowler, modified by the author.

25 Even though he cannot give the precise same answer for the 'ἐπιστήμη' of Platonic Beings that he gave 'collecting' a power; cf. Plato, *Theaet.* 148b5-8.

3.2 The Imitation is Realized with the Method of Division and Collection in the Dialogues *Sophist* and *Statesman*

In the two dialogues of the trilogy succeeding the *Theaetetus*, the *Sophist* and *Statesman*, the method of 'division and collection' (διαίρεσις καὶ συναγωγὴ), by which the human soul is able to obtain full knowledge of the Platonic Ideas is presented by the Eleatic Stranger in discussion with Theaetetus and his companion. Since,

(a) the description of Theaetetus' mathematical achievement of dealing with quadratic incommensurabilities is, according to our distributive reading (in Section 2), the collection of each power divided into an infinite multitude (collection being a rendering of 'συλλαβεῖν εἰς ἕν' (*Theaetetus* 147d8) and περιέλαβες' (*Theaetetus* 148d6); cf. also fn. 17 above).
(b) the description of Plato's philosophical method of acquiring knowledge (ἐπιστήμη) of Platonic Ideas, presented in the dialogues *Sophist* and *Statesman*, is 'division and collection' (collection being a rendering of συναγωγή, but also of συλλαβεῖν and περιλαμβάνειν[26]), and
(c) Plato's philosophical method was explicitly in *Theaet.* 148b to be an imitation of Theaetetus' mathematical method,

there can be little doubt that the philosophical Collection in the *Sophist* and the *Statesman* is an imitation of the geometrical Theaetetus' Collection.

So even though Plato does not provide any more information on the nature of Theaetetus' Collection, and it might thus appear difficult to break into its meaning (as indeed it has been the case thus far), there still is a window of hope, our one and only hope really, to reach an understanding of the double, mathematical and philosophical, meaning of Collection, and this is by examining the philosophical method in these two dialogues. Plato generously supplies us with the means to reach such an understanding. We will exploit these means supplied by Plato in Sections 5–8. After studying Plato's three principal examples of his method in sections 5, 6, and 7, the close philosophical imitation of Theaetetus' mathematical contribution (presented in *Theaet.* 147d7-148b4 and provisionally described in section 2) is presented in *Stat.* 283b6-287a6 and described in section 8.

---

26 The reader may find in 9.1.1 below a list of the terms used by Plato to describe his method of division and collection.

III Part B. Plato's Method of Division and Collection in the *Sophist* and the *Statesman* (Sections 4–8)

In the second part of the paper, we will examine Plato's method of 'division and collection', as presented in the *Sophist* and the *Statesman*. In particular, we will examine in detail the Division of the Angler (in Section 5) and of the Sophist (in Section 6) in the *Sophist*, the Division of the Statesman (in Section 7) and the double measurement (in Section 8) in the *Statesman*. We discover that (a) the two Divisions in the *Sophist* are analogues of periodic anthyphairesis; (b) the Division in the *Statesman* is an analogue of palindromically periodic anthyphairesis; while (c) the double measurement in the *Statesman* is a philosophic analogue of the mathematical Proposition: if a, b are two lines commensurable in square only, then the anthyphairesis of a to b is palindromically periodic. The periodicity of the anthyphairesis in all cases is established by using the Logos Criterion, established in Section 4, on the basis of Theaetetus' theory of ratios of magnitudes. The far-ranging consequences of these discoveries in the present Part B, combined with our findings in Part A, will be discussed in Section 9.

4 Theaetetus' Theory of Ratios of Magnitudes, the Logos Criterion for Periodic Anthyphairesis, the Abbreviated Tree-Like Form of Anthyphairesis

Before embarking on an examination of the dialogues *Sophist* and *Statesman*, attempting to discover the nature of the Platonic method of 'division and collection', we will need to establish the Logos Criterion for periodic anthyphairesis, a criterion that presupposes the theory of ratios of (incommensurable) magnitudes, discovered by Theaetetus.

4.1 The Theaetetean Theory of Ratios of Magnitudes

Aristotle, in *Top.* 158b24-159a2, informs us that before the Eudoxian theory of ratios of magnitudes presented in Book V of the *Elements*, there was a theory of ratios of magnitudes based on equality of anthyphairesis:

Definition. If a,b are homogeneous magnitudes and c, d are homogeneous magnitudes then a:b = c:d iff Anth (a,b) = Anth (c,d) (where Anth (a,b) denotes the sequence of successive quotients of the anthyphairesis of a to b).[27]

27 With the notation of Section 1.1, Anth (a,b) = $[I_0, I_1, I_2, \ldots, I_n, \ldots]$.

As already noted, it is generally accepted that the theory of ratios based on anthyphairesis is due to Theaetetus. From all we know this is the first theory of ratios of magnitudes, which went beyond commensurable ratios. Knorr (1975, Appendix B) believes that the theory was using in some informal way Eudoxus' definition 4 of *Elem.* V, otherwise there would be a problem with the proof of the fundamental Proposition analogous to *Elem.* V.8; and, Acerbi, 2003, essentially because of the same difficulty, believes that in fact there was never any theory of ratios with this definition. But in our opinion such a theory definitely existed, as Aristotle clearly states, but it was a theory, not for all ratios on which Eudoxus theory applies, but for a limited class of ratios on which Eudoxus' principle is not needed. A work on this subject, by Negrepontis and Protopappas, is under preparation. It is easily seen that the definition of proportion is a generalization of the definition of ratios of numbers, based on VII.1, 2 and 4, and of commensurable magnitudes, based on X.3.

4.2 The Logos Criterion for Periodic Anthyphairesis

The theory is particularly well suited, probably not by chance, to give a proof of a criterion for the periodicity of anthyphairesis, the Logos Criterion.

Proposition (Logos Criterion for the periodicity of an anthyphairesis): Let a,b be two line segments, with a>b, and let the anthyphairesis of a to b be described with the notation of Section 1.1. Assume that there are indices n<m such that $c_n:c_{n+1} = c_m:c_{m+1}$. Then the anthyphairesis of a to b is eventually periodic, in fact: $\mathrm{Anth}(a, b) = [\mu_1, I_1, \dots, I_n, \mathrm{period}(I_{n+1}, I_{n+2}, \dots, I_m)]$.

Proof: By the Theaetetean definition of analogy, $\mathrm{Anth}\,(c_n, c_{n+1}) = \mathrm{Anth}\,(c_m, c_{m+1})$. But $\mathrm{Anth}\,(a,b) = [\mu_1, I_1, \dots, I_n, \mathrm{Anth}\,(c_n, c_{n+1})] = [\mu_1, I_1, \dots, I_n, I_{n+1}, I_{n+2}, \dots, I_m, \mathrm{Anth}\,(c_m, c_{m+1})] = [\mu_1, I_1, \dots, I_n, I_{n+1}, I_{n+2}, \dots, I_m, \mathrm{Anth}\,(c_n, c_{n+1})] = [\mu_1, I_1, \dots, I_n, I_{n+1}, I_{n+2}, \dots, I_m, I_{n+1}, I_{n+2}, \dots, I_m, \mathrm{Anth}\,(c_m, c_{m+1})] = \dots = [\mu_1, I_1, \dots, I_n, \mathrm{period}(I_{n+1}, I_{n+2}, \dots, I_m)]$.

Even though the Criterion is natural and almost obvious, it is not stated explicitly in Greek mathematical sources.

4.3 An Abbreviated Tree-Like Representation of the Anthyphairesis and the Logos Criterion

We note that the Logos criterion is stated solely in terms of the remainders, and no use is made in its statement of the quotients. The Logos criterion may be adequately represented in the following abbreviated tree-like form:

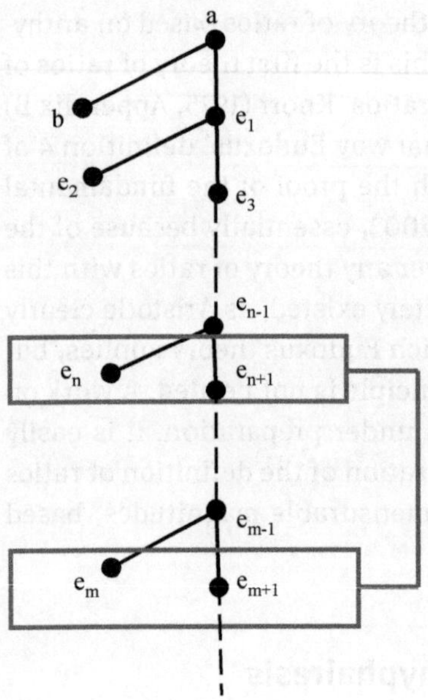

4.4 Example: The Anthyphairesis of a to b with $a^2 = 19b^2$

It was shown in Section 1 that the method of Theodorus was anthyphairetic, but Theodorus did not have at his disposal Theaetetus' theory of ratios and the Logos Criterion, thus probably found recourse to the Pythagorean theory of gnomons. Had he recourced to Theaetetus' theory, he might use the Logos Criterion. We exhibit the Logos Criterion for the first power that Theodorus did not prove in his lesson to Theaetetus and his companion.

(i) The Anthyphairesis of a to b
$a = 4b + e_1$, με $e_1<b$ (whence $e_1 = a - 4b$),
(and $b = 2e_1 + e_2$, $e_2<e_1$ (whence $e_2 = 9b - 2a$));
$e_1 = e_2 + e_3$, $e_3<e_2$ (whence $e_3 = 3a - 13b$),
(and $e_2 = 3e_3 + e_4$, $e_4<e_3$ (whence $e_4 = 48b - 11a$)),
$e_3 = e_4 + e_5$, $e_5<e_4$ (whence $e_5 = 14a - 61b$),
(and $e_4 = 2e_5 + e_6$, $e_6<e_5$ (whence $e_6 = 170b - 39a$)),
$e_5 = 8e_6 + e_7$, $e_7<e_6$ (whence $e_7 = 326a - 1421b$).

(ii) The Verification of the Logos Criterion $b:e_1 = e_6:e_7$
We simply verify that $b \cdot e_7 = b \cdot (326a - 1421b) = (a - 4b) \cdot (170b - 39a) = e_1 \cdot e_6$. The Logos Criterion results in the full knowledge of the ratio a:b, since Anth (a,b) = [4, period (2,1, 3,1,2,8)]. By X.2, of course, it is concluded that a,b are incommensurable.

(iii) Abbreviated Tree-Like Representation of the Logos Criterion

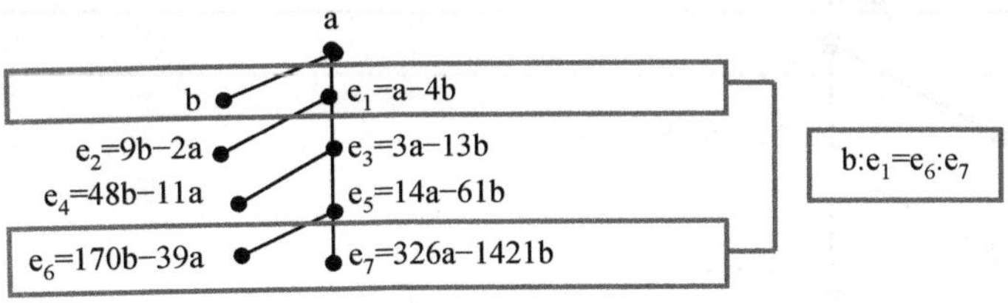

Again it should be pointed out that these are our reconstructions, not to be found explicitly in extant Greek mathematical sources.

5 The Division and Collection of the Platonic Idea 'the Angler' in *Soph.* 218b-221c Is the Philosophic Analogue of Periodic Anthyphairesis

The method of 'division and collection', also called 'Name and Logos' (cf. *Theaet.* 201e2-202b5, *Soph.* 218c1-5, 221a7-b2, 268c5-d5), is exemplified at the start of *Sophist* 218b-221c by the definition of the Being Angler.

5.1 The Division of the Platonic Idea 'the Angler'

In the scheme below, we reproduce the binary division process which leads to the Angler. The whole division scheme is as follows:

[G] all arts (τῶν τεχνῶν πασῶν) divided in [B_1] productive arts (ποιητικὴ) and [A_1] acquisitive arts (κτητικὴ) (219a8-d3)
[A_1] acquisitive arts divided in [B_2] voluntary arts (ἑκόντων πρὸς ἑκόντας μεταβλητικὸν) and [A_2] coercive arts (χειρωτικὸν) (219d4-8)
[A_2] coercive arts divided in [B_3] fighting (ἀναφανδὸν, ἀγωνιστικὸν) and [A_3] hunting (κρυφαῖον, θηρευτικόν) (219d9-e3)
[A_3] hunting divided in [B_4] hunting of lifeless things (ἀψύχου γένους) and [A_4] hunting animals (ἐμψύχου, ζῳοθηρικήν) (219e4-220a6)
[A_4] hunting animals divided in [B_5] hunting land animals (πεζοθηρικόν) and [A_5] hunting water animals (ἐνυγροθηρικόν) (220a7-11)
[A_5] hunting water animals divided in [B_6] fowling (ὀρνιθευτική) and [A_6] fishing (ἁλιευτική) (220b1-8)
[A_6] fishing divided in [B_7] fishing by enclosures (ἑρκοθηρικὸν) and [A_7] fishing by striking (πληκτικὴν) (220b9-d4)
[A_7] fishing by striking divided in [B_8] fire-fishing (πυρευτικὴ) and [A_8] barb-fishing (ἀγκιστρευτικόν) (220d5-e1)
[A_8] barb-fishing divided in [B_9] fishing by tridentry (τριοδοντία) and [A_9] angling (ἀσπαλιευτική) (220e2-221a7).

The Division of the Angler

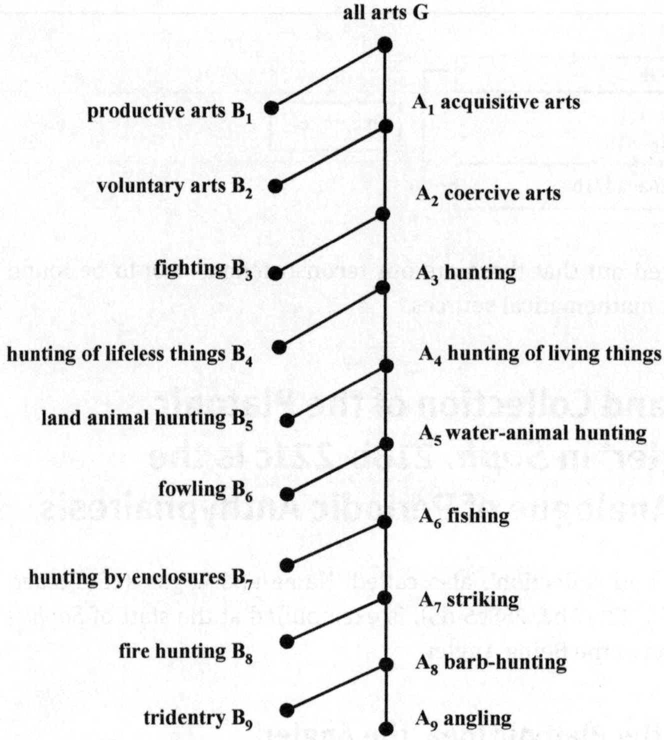

5.2 Collection-Logos of the Angler

The summary description of the Logos-Collection of the Angler, contained in *Soph.* 220e2-221c3, is naturally divided into two parts: [A] corresponding to 'name', and [B] corresponding to 'logos'.

> [A] 'name' (220e2-221b1): Stranger: Then of striking which belongs to barb-hunting [A8], that part which proceeds downward from above (ἄνωθεν εἰς τὸ κάτω), is called, because tridents are chiefly used in it, tridentry [B9],...Stranger: The kind that is characterized by the opposite sort of blow, which is practised with a hook and strikes,...and proceeds from below to the opposite upwards (κάτωθεν εἰς τοὐναντίον ἄνω)...By what name, Theaetetus, shall we say this ought to be called? Theaetetus: I think our search is now ended and we have found the very thing we set before us a while ago as necessary to find. Stranger: Now, then, you and I are only agreed about the name of angling, [A9]

> [B] 'logos' (221b1-c3): but we have acquired also a satisfactory 'Logos' of the thing itself. For (γὰρ) of art as a whole, half was acquisitive, and of the acquisitive, half was coercive, and of the coercive, half was hunting, and of hunting, half was animal hunting, and of animal hunting, half was water hunting, and of water hunting [A5] the whole part from below (τὸ κάτωθεν τμῆμα ὅλον) was fishing,[A6] and of fishing, half was striking, and of striking, half was barb-hunting,[A8] and

of this [A8] the part in which the blow is pulled from below upwards (τὸ περὶ τὴν κάτωθεν ἄνω) was angling.[A9]'.[28]

In [A], the opposing relation of Tridentry to Angling is carefully explained: all Fishing with a hook is divided into Tridentry (=Fishing with a trident), which is described as Fishing with a hook with an art that proceeds *from above downwards*, and Angling (=Fishing with a rod), which is described as Fishing with a hook with an art that proceeds *from below upwards*. We have pushed Division all the way to the Angling; thus, we have certainly found 'the name' of Angling.

But now, in [B], it is claimed that 'the Logos' of the Angling has been found, too. The justification—the proof—that we have indeed found the Logos, too, is contained in the remainder of [B], since this remaining part of [B] starts with a 'for' (γὰρ), and this justification can be seen to consist of: (i) an accurate recounting of all the division-steps, abbreviated in the sense that of the two species into which each genus is divided, only the one that contains the Angler is mentioned, while the species opposite to it is omitted; (ii) a reminder that the last species, the angling, is characterized as the part of its genus that proceeds *'from below upwards'*; and, (iii) the ONLY new information (since (i) and (ii) are repetitions of things already contained in the Division and in [A]), which concerns the species of fishing, three steps before angling, and which informs us for the first time that this species is *'the whole part from below'* of its genus.

This explanation clearly corresponds to common sense since fish is in the sea and can be obtained by the action of fishing only from below upwards. Since this is an abbreviated account, there is no explicit information on the species opposite to 'fishing'—namely 'fowling', but since 'fishing' was described not simply as 'the part from below' of its genus, but emphatically as 'the whole part from below', it follows, by exactly the same common sense, that the opposite species—'fowling'—must be characterized as 'the (whole) part from above' of the same genus. In fact, there can be no other justification for the presence of the term 'whole' in the description of 'fishing' with a view to arguing that we have obtained 'Logos', except to indicate and imply this description for its opposite species, 'fowling'.

We recall that the part of [B] from the word 'for' (gar) is explicitly a justification of the claim that we have succeeded in finding the 'Logos' of the Angling. We may then ask: what is the 'Logos' of the Angler that reasonably results from such a justification? There can really be only one answer: the 'Logos' we are looking for is the equality of the 'philosophic ratio' of Tridentry to Angling, namely the equality of the ratio 'of from above downwards to from below upwards', to the ratio 'of Fowling to Fishing'. Thus, fowling B6:fishing A6 = tridentry B9:Angling A9 = from above downwards: from below upwards. Since, the species Tridentry and Angling forms a pair of opposite species, and the species Fowling and Fishing form another pair of opposite species

28 Translated by Fowler, with modifications by the author.

in the Division Scheme for the Angler, the resulting 'Logos' bears a most uncanny similarity to the Logos Criterion for the periodicity of the anthyphairesis of geometric magnitudes, and of geometric powers in particular

5.3 The Division and Collection of the Angler

The 'division and collection' of the Angler thus takes the following form and it consists of the Division, which is in the abbreviated tree-like form, and of the Logos Criterion, for the periodicity of the anthyphairesis (given in section 4).

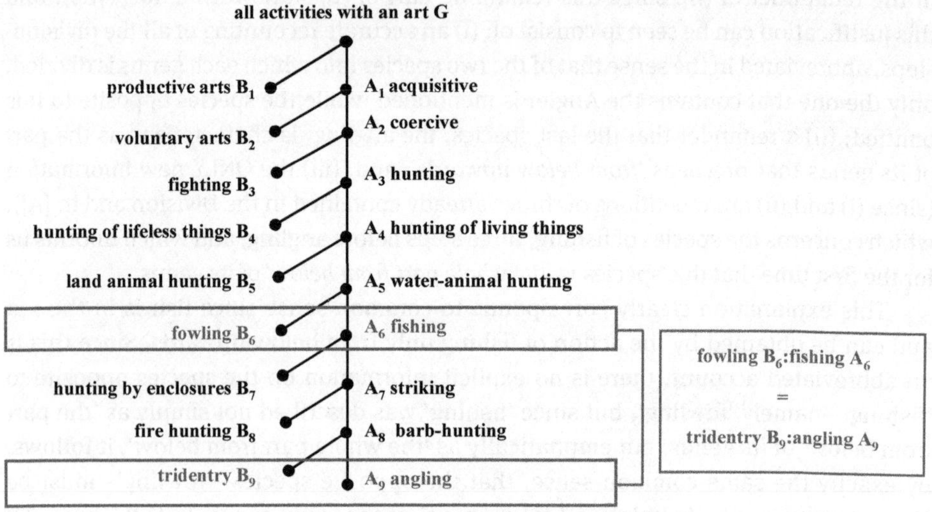

6 The Division and Collection of the Platonic Idea 'the Sophist' in *Soph*. 234e-236d and 264b-268d is the Philosophic Analogue of Periodic Anthyphairesis

It now appears that the 'division and collection' of a Platonic Idea—and the Angler is certainly a lowly paradigm of a Platonic Idea—is very much like the anthyphairetic Division and the Logos Criterion of a geometric 'power'. When Socrates expressed, in *Theaet*. 145c7-148e5, his exhortation to imitate the geometric situation, it seems that he meant a much closer imitation that anybody has thus far suspected! But before proceeding on to wide-ranging conclusions, it would be prudent to examine whether the 'division and collection' of the Sophist, in *Soph*. 264b-268d, has the same kind of structure and, in particular, whether there is a similar type of 'Logos'. So we shall take a look at the 'division and collection' of the Sophist.

6.1 The Division of the Sophist

The Division for the Sophist follows the same pattern as the Division of the Angler, starting with a Genus, in this case 'all the productive arts', proceeding by binary division of each Genus to two species, where the next Genus is that species of the previous step in the Division that contains the entity to be defined, in this case the Sophist, and ending with the division-step that produces the Sophist as a species. The whole division scheme is as follows:

[G] the productive arts (ποιητική) divided into [D_1] the divine productive arts (θεῖον) and [C_1] the human productive arts (ἀνθρώπινον) (265b4-e7).
[C_1] the human productive arts divided into [D_2] the arts of makng real things (αὐτοποιητικόν, αὐτουργική) and [C_2] the arts of makng images (εἰδωλοποιική) (266c7-d4).
[C_2] the arts of making images divided into [D_3] the likeness-making arts (εἰκαστική) and [C_3] the fantastic arts (φανταστική) (266d8-e5).
[C_3] the fantastic art divided into [D_4] art produced by instruments (δι' ὀργάνων) and [C_4] mimetic art αὐτοῦ ἑαυτὸν ὄργανον τοῦ ποιοῦντος (μιμητικὸς) (267a1-b3).
[C_4] the mimetic art divided into [D_5] mimetic art with knowledge (μετ' ἐπιστήμης μίμησις) and [C_5] mimetic art with opinion (μετὰ δόξης μίμησις, δοξομιμητικὴ) (267b4-e6).
[C_5] the mimetic art with opinion divided into [D_6] the simple imitator (ἁπλοῦς μιμητής) and [C_6] the dissembling imitator (εἰρωνικὸς μιμητὴς) (267e7-268a8).
[C_6] the dissembling imitator divided into [D_7] the popular orator (δημολογικός) and [C_7] the sophist (σοφιστής) (268a9-c4).

The Division of the Sophist

G **productive arts**

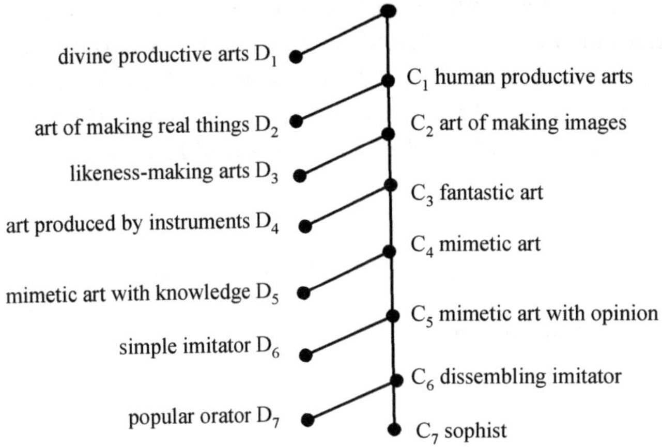

6.2 The Fundamental Analogy in the Divided Line of the *Resp.* 509d-510b and the Logos criterion-Collection of the Sophist

We will now prepare the ground for the Logos-Collection of the Sophist. A careful perusal of the Division of the Sophist may not reveal an obvious equality of proportion necessary for the Logos Criterion, as was the case with the Division of the Angler. It turns out that the fundamental analogy in the Divided Line of *Resp.* 509d-510b plays a central role in the Logos Criterion of the Sophist. This analogy may be rendered as follows:

> First, divide a line segment, say L, into two unequal sections, say A representing the intelligible Beings, identified with the Beings of which we may obtain knowledge (γνῶσις, ἐπιστήμη), the knowables (τὸ γνωστόν, 510a9) and B the sensible Beings, identified with the Beings of which we may only obtain opinion (δόξα), the opinables (τὸ δοξαστὸν, 510a9); next, divide (section A into two subsections, say C and D, and) section B into two subsections, say E, representing the real sensible Beings, and F, representing the images (εἰκόνες, 509a1), or likenesses ('to homoiothen') of the real entities. (The real sensible entities are described in relation to their images as 'that of which this is an image' (ᾧ τοῦτο ἔοικεν, 510a5), as 'that of which it is a likeness' (τὸ ᾧ ὡμοιώθη, 510a10)).

Plato's fundamental philosophical statement in the Divided Line is, then, the analogy B:A (=D:C) = F:E, i.e. the statement 'the opinable: the knowable = the likeness: that of which it is a likeness'.[29] The Logos Criterion for the Sophist is now immediately recognised as 'the art of making real things D2: the art of making images C2 = mimetic art with knowledge D5: mimetic art with opinion C5'.

A strong indication that Plato's intention was to use the *Republic*'s ratio of the Divided line is the extra care he takes to justify the equality of the *Republic* ratio, namely the ratio of all likenesesses-images to all real things, all meaning both divine and human, to the ratio in the *Sophist*, namely the ratio of human likenesses-images (C2) to human real things (D2). He does this meticulously in *Sophist* 265e8-266d7, in which there is a division not only of C1 into D2 and C2, but also of D1 into the arts producing real things (say D2') and into the arts producing images (C2'), a division of a 'rejected' species, something unexpected and unique in Plato's divisions. The proof of equality, involving 'breadth' and 'width', employs an analogue of *Elem.* VI.1.[30]

---

29 ὡς τὸ δοξαστὸν πρὸς τὸ γνωστόν, οὕτω τὸ ὁμοιωθὲν πρὸς τὸ ᾧ ὡμοιώθη; *Rep.* 509a9-10.

30 The correlation of the *Republic* ratio in the divided line with the Division and Collection of the Sophist in the *Sophist* is hinted at by Proclus, *In Resp. Com.* I, 290, 7–10.

6.3 Additional Evidence in Favor of an Anthyphairetic Logos Criterion

The definition of the Sophist presents an additional feature that ties it even closer to the mathematical anthyphairetic model. There are two ratios after the Logos Criterion, and if the mathematical anthyphairetic model is indeed followed in the 'division and collection', we would expect to have two further equalities of Logoi: namely the ratio of the third step should be equal to the ratio of the sixth division-step, and the ratio of the fourth division-step should be equal to the ratio of the seventh and final division-step. Indeed, the simple imitators (D6) do not distort their opinion, but rather express a likeness of their opinion, while the dissemblers (C6) distort and disguise their opinion behind a false appearance. Therefore we have:

Proposition: The ratio of the third step in the Division of the Sophist, namely the ratio of the likeness-making arts (D3) to the phantastic arts (C3), is equal to the ratio of the sixth step in the Division of the Sophist, namely the ratio of the simple imitator (D6) to the dissembling imitator (C6).

Furthermore, he who listens to a dissembler is deceived, but, if that dissembler is a demagogue (D7), he is contradicted not by himself but by another instrument of deceit (namely the demagogue himself), while if the dissembler is a sophist (C7), the listener is forced by the sophist to himself become the instrument of deceit

Proposition: The ratio of the fourth step in the Division of the Sophist, namely the ratio of the imitator who uses other instruments of imitation (D4) to the imitator who is himself the instrument of imitation (C4) is equal to the ratio of the seventh step in the Division of the Sophist, namely the ratio of the demagogue (D7), whose listener is contradicted and deceived not by himself but by another, to the sophist (**C7**), whose listener is forced to be contradicted and deceived by himself.

These observations provide additional evidence in favor of the anthyphairetic interpretation of 'division and collection'.

6.4 The Division and Collection of the Sophist

The complete Division and Collection of the Sophist can thus be summarized in the following Table (*Soph.* 264b-268d):

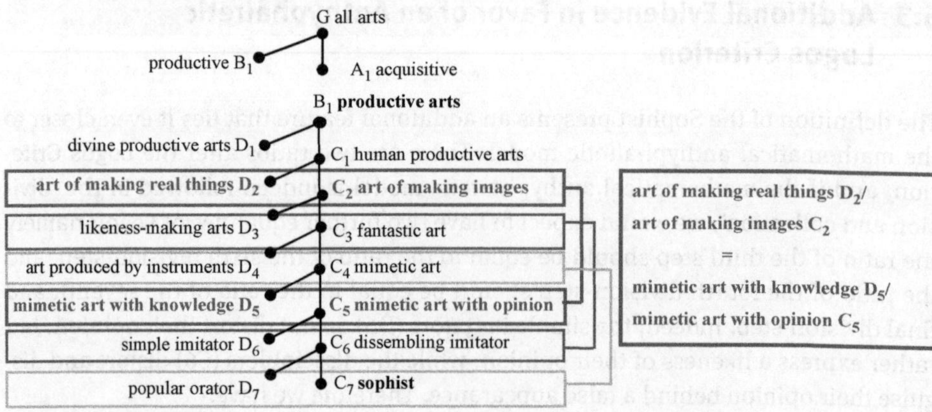

We thus have full confirmation for the anthyphairetic meaning of Logos that we proposed in the 'division and collection' of the Angler. This time, the Logos is not of the simple-minded, obvious type that appeared in the definition of the Angler (from above downwards/from below upwards), but of a more sophisticated and philosophical kind, playing a central role in the dialectics of the *Republic*.

6.5 Note

Most scholars describe the divisions in the *Sophist* as finite in length, while in our presentation we have shown them to be modelled after periodic anthyphairesis, and thus to be of infinite length. The division of the Statesman in the *Statesman*, examined in the next section, will quite clearly be of infinite length. See also note 9.1.2 below.

7 The Division and Collection of the Statesman in the *Statesman* is the Philosophic Analogue of Palindromically Periodic Anthyphairesis

In our account of the 'division and collection' of the Statesman in the *Statesman*, we will outline the new most striking feature of anthyphairetic palindromic periodicity, in relation to the mere anthyphairetic periodicity of the Division-and-Collections in the *Sophist*.

7.1 The First Part of the Division 258b-266d

The dialogue starts with the first ten Division steps towards the Division of the Statesman, in *Sophist* style. Here are the steps:

> [G] all arts divided into [K_1] practical arts (πρακτικὴν) and [L_1] scientific (γνωστικήν) arts (258b-259d);

[L_1] scientific arts divided into [K_2] appraising arts (κριτικὸν) and [L_2] commanding arts (ἐπιτακτικὸν) (259d-260c);
[L_2] commanding arts divided into [K_3] someone else's commands and [L_3] one's own commands (αὐτεπιτακτικὴν) (260c-261b);
[L_3] one's own commands divided into [K_4] lifeless herd (τῶν ἀψύχων ἐπιστατοῦν) and [L_4] herd of living beings (=animals) (τὸ περὶ τὰ ζῷα ἐπιτακτικόν) (261b-d);
[L_4] herd of animals divided into [K_5] herd of a single animal (μονοτροφίαν) and [L_5] herd of many animals (ἀγελαιοτροφίαν) (261d-e);
[L_5] herd of many animals divided into [K_6] herd of aquatic animals (ὑγροτροφικὸν) and [L_6] herd of land animals (ξηροτροφικόν) (264b-d);
[L_6] herd of many land animals divided into [K_7] herd of flying animals (πτηνῷ) and [L_7] herd of walking animals (πεζῷ, πεζονομικήν) (264e-265b);
[L_7] herd of many walking land animals divided into [K_8] herd of horned animals (ἐπὶ τῷ κερασφόρῳ) and [L_8] herd of hornless animals (ἐπὶ τῷ τῆς ἀκεράτου) (265b-d);
[L_8] herd of many hornless walking land animals divided into [K_9] herd of mixing animals and [L_9] herd of non-mixing animals (τῆς ἀμείκτου) (265d-e);
[L_9] herd of many non-mixing hornless walking land animals divided into [K_{10}] herd of four-footed (τῇ τῆς διαμέτρου διαμέτρῳ) (pigs) and [L_{10}] herd of two-footed (τῇ διαμέτρῳ) (humans, τἀνθρώπινον) (265e-266d).

Note that after [L_4] we are dealing with the arts of feeding an animal herd (ἀγελαιοτροφικὴ, 261e); what is really being divided is the nature of the commanded herd, narrowing it down, by successive binary division and exclusion, till we get to [L_{10}] the arts of commanding a human herd (265e-266d).

7.2 The Myth of Palindromic Periodicity with Kronos Era and Zeus Era 268d-274e

At this point, there is an *impasse* and it is not clear how to continue with the Division, (or even to decide that perhaps the Division has come to an end). In order to break the impasse and help complete the Division, the Eleatic Stranger takes a most striking and unexpected turn, he relates the myth of palindromic periodicity. According to this myth (268d-274e), there are two eras in the world,

the Kronos era	
	and the Zeus era.
The world, to be thought of as cyclical entity,	
rotates in a certain direction during the Kronos era. After a long time, when all the possible stages of the Kronos era are exhausted (τὸ γήινον ἤδη πᾶν ἀνήλωτο γένος, πάσας ἑκάστης τῆς ψυχῆς τὰς γενέσεις ἀποδεδωκυίας,[31] 272e1-2),	

31 The reason given by Plato for the change from the Kronos era to the Zeus era is the exhaustion of all possibilities in the Kronos era, essentially a pigeonhole argument. This is in interesting accord with the proof in Section 10.

	it is necessary to reverse the direction (πάλιν ἀνέστρεφεν, 272e5-6) and to go
from the Kronos era	
	to the Zeus era (272d6-e6). But again after a long time, when the Zeus era'might founder in the tempest of confusion and sink in the infinite sea of dissimilarity' (κηδόμενος ἵνα μὴ χειμασθεὶς ὑπὸ ταραχῆς διαλυθεὶς εἰς τὸν τῆς ἀνομοιότητος ἄπειρον ὄντα πόντον[32] δύῃ, 273d5-e1), God perceives that it is in trouble,
and he … reverses (στρέψας, 273e3),	
	whatever had become unsound and unsettled in the previous period when the world was left to itself,
restores it back again to the Kronos era thus completing a full period.	
In this way the world becomes immortal (ἀθάνατον) and ageless (ἀγήρων)'(273c4-e4). The world moves forever repeating this period, alternating	
the Kronos era	
	with the Zeus era.

The relation of the two eras is especially interesting. It is schematically described as follows (270d-271a):

	In the Zeus era (in which era we are presently) men come from non-existence **a** into existence **b**, then they grow **c**, then they get a beard **d**, then their beard becomes white **e**, then they grow old and are about to die **f**. If at this time
the world happens to switch to the Kronos era, then the human will go from stage **f** to **e** then to **d** then to **c** then to **b** and then they pass into nonexistence **a**	

32 The reason given by Plato for the change from the Zeus era to the Kronos era is the need to secure periodicity and avoid bad infinity. This is in interesting accord with the proof in Section 10.

Summarizing, according to the myth, the world is periodic, and the period consists of two half periods,	
the Kronos era followed	
	by the Zeus era,
in other words, it is palindromically periodic.	

7.3 The Instructions in *Stat.* 276a1-7: Follow the Kronos/Zeus Myth for a Palindromically Periodic Division and Collection

Next, in 276a1-7, crucial for our interpretation of the *Statesman*, the Eleatic Stranger makes clear that this myth must be employed for the continuation and completion of the 'division and collection' of the Statesman:

		Younger Socrates: ...but how would the next division be made? Stranger: In the same kinds (κατὰ ταὐτὰ)
as we divided before (καθ' ἅπερ ἔμπροσθεν διῃρούμεθα) the art of feeding herds (τὴν ἀγελαιοτροφικὴν) in		
	those that go on foot [L7]	
and the winged, [K7]		
	and the unmixed breeds [L9]	
	and the hornless, [L8]	
		we must divide in arts similar to (τοῖς αὐτοῖς... διαιρούμενοι)
those (τούτοις),		
		the art of tending herds too (καὶ τὴν ἀγελαιοκομικὴν),
embracing (περιειληφότες) in the logos (ἐν τῷ λόγῳ) both ὁμοίως		
		the present [Zeus] era
and the Kronos era.		

7.4 The Division of the Statesman in the *Statesman* is Guided by the Myth Towards an Anthyphairetic Palindromic Periodicity

From 7.1 and 7.3, the first ten division steps (ἀγελαιοτροφικὴ, where *the nature of the commanded herd* is the variable) is only the first half of the Division and corresponds to the Kronos era of the myth:

'ἀγελαιοτροφική', Kronos era			
Genus G all arts divided into↓			
K_1, practical arts,	L_1, scientific arts		
	L_1 divided into↓		
K_2, appraising arts	L_2, commanding arts		
	L_2 divided into↓		
K_3, someone else's commands	L_3, one's own commands		
	L_3 divided into↓		
K_4, lifeless herd	L_4, herd of living beings, animals		
	L_4 divided into↓		
K_5, herd of a single animal	L_5, herd of many animals		
	L_5 divided into↓		
K_6, herd of aquatic animals	L_6, herd of land animals		
	L_6 divided into↓		
K_7, herd of flying animals	L_7, herd of walking animals		
	L_7 divided into↓		
K_8, herd of horned animals	L_8, herd of hornless animals		
	L_8 divided into↓		
K_9, herd of mixing animals	L_9, herd of non-mixing animals		
	L_9 divided into↓		
K_{10}, herd of four-footed (pigs)	L_{10}, herd of twofooted (humans)		

From 7.1, 7.2, 7.3 The Division must be completed with the second half of the Division (ἀγελαιοκομική, where the commanded herd is the human herd, held fixed, and *the nature of the commanding herdsman* is the variable) that corresponds to the Zeus era of the myth. The definition of the Statesman must be completed, as in the myth, and the instructions with its second half, with last division steps, say W_{11} divided into Z_{10}, W_{10}, W_{10} divided into Z_9, W_9, ...W_2 divided into Z_1, W_1, corresponding to the Zeus era, palindromic to the Kronos era. The whole Division will comprise and be embraced by a cycle-full period, in precisely the same way that the Kronos era followed by the Zeus era comprise a cycle-full period. Since we have already shown, by studying the Sophist, two Platonic divisions are modelled after periodic anthyphairesis, it is now evident that the division of the Statesman in the Statesman is modelled after palindromically periodic anthyphairesis.

7.5 The Palindromicity of the Periodic-Anthyphairetic Division and Collection of the Statesman Realized in *Stat.* 289e4-305e11

The instructions given in 276a1-7 are fully realized in the final long stage of *Stat.* 289e4-305e11, and the palindromiciy of the 'division and collection' of the Statesman is shown schematically in the table below.

The Palindromically Periodic Division & Collection of the Statesman

'ἀγελαιοτροφική', Kronos era [Kronos column to be read from Genus G downwards to L_{10}]	'ἀγελαιοκομική', Zeus era [Zeus column to be read from W_{10} upwards to W_1]

	← ←		
Genus G all arts is divided into ↓ **[[W_1 divided into]]** ↓↓		**W_1 divided into**↑↑	
K_1, practical arts **[[K'_1, strategic art]]**	L_1 **[[L'_1, true statesman]**	$W_1 = W_2\backslash Z_1$[33]	Z_1, rhetoric[34]
	L_1 divided into↓	W_2 divided into↑	
K_2, appraising arts	L_2	$W_2 = W_3\backslash Z_2$	Z_2, judging[35]
	L_2 divided into↓	W_3 divided into↑	
K_3, someone else's commands	L_3	$W_3 = W_4\backslash Z_3$	Z_3, diviners, priests[36]
	L_3 divided into↓	W_4 divided into↑	
K_4, lifeless herds	L_4	$W_4 = W_5\backslash Z_4$	Z_4, government with laws[37]
	L_4 divided into↓	W_5 divided into↑	
K_5, herd of a single animal	L_5	$W_5 = W_6\backslash Z_5$	Z_5, democracy without laws[38]
	L_5 divided into↓	W_6 divided into↑	
K_6, herd of aquatic animals	L_6	$W_6 = W_7\backslash Z_6$	Z_6, merchants[39]
	L_6 divided into↓	W_7 divided into↑	

33 W_1, the last species of Zeus era [recall that the Zeus column is read from down upwards], is divided into K_1',L_1', forming the 'logos' criterion for periodicity, 305e2-e7, K_1'(strategic arts, 304e3-305a10): L_1' (the real statesman) = K_1 (practical arts):L_1(theoretic arts). We indicate the last division step, with which the full circle of an anthyphairetic period is achieved, with Arial Black.

34 Z_1 art of rhetoric, a practical art, 304a6-e2, is palindromic to K_1 practical arts, 258b-259d.

35 Z_2 art of judging, 305b1-c8, is palindromic to K_2 art of appraising, 259d-260c.

36 Z_3 art of priests and diviners, 290c3-291a2 is palindromic to K_3 commanding else's commands, 260c-261b.

37 Z_4 art of government with laws, 291c8-303b7, is palindromic to K_4 lifeless 'herds', 261b-d.

38 Z_5 democracy without laws, 291c8-303b7, is palindromic to K_5 herds of one animal, 261d-e.

39 Z_6 art of merchants, at sea, 289e4-290a3, is palindromic to K_6 herds of aquatic animals, 264b-d.

K_7, herd of flying animals	L_7	$W_7 = W_8 \setminus Z_7$	Z_7, heralds[40]
	L_7 divided into↓	W_8 divided into↑	
K_8, herd of horned animals	L_8	$W_8 = W_9 \setminus Z_8$	Z_8, satyrs of politics oligarchy[41]
	L_8 divided into↓	W_9 divided into↑	
K_9, herd of mixing animals	L_9	$W_9 = W_{10} \setminus Z_9$	Z_9, centaurs of politics, tyranny[42]
	L_9 divided into↓	W_{10} divided into↑	
K_{10} diameter's diameter, herd of fourfooted (pigs)	L_{10}	W_{10}	Z_{10}, sophists' sophists[43]
	…↓		…↑
	→ from L_{10} (eventually) to W_{10}		

Notes: Some of the palindromic relations are obvious. The Z_9 centaurs and Z_8 satyrs of politics are the two worst type, tyranny and oligarchy, of the worst government, namely that without laws, one step above the very worst pretenders to politics, the Z_{10} sophists, in playful palindomicity, the sophists' sophists, with the herds of diameter's diameter, the only four-footed tame animals left, the pigs. The relatively best goverment without laws, Z_5 democracy without laws, essentially anarchy, where every one is for himself, is palindromic to one animal herds (cf. *Laws* 942a-d).

The most interesting relation is the palindromicity between Z_4 all three kinds of government with fixed laws but without philosophic and dialectic knowledge, that a true statesman must possess (democracy, aristocracy, monarchy, in ascending order), with the lifeless 'herds; this correspondence is closely related to Plato's rejection of the written 'logos' in the *Phaedr.* 274c5-277a5 as lifeless. Finally, Z_1 rhetoric is the practical art palindromic to the practical arts, and K'_1 strategic art is the practical art which is last subtracted leaving alone the true statesman L'_1, to form a logos K'_1 to L'_1 equal to the initial logos K_1 to L_1, and thus satisfy the Logos Criterion for anthyphairetic periodicity.

Thus, the method of 'division and collection', of which the 'division and collection' of the Statesman is a paradigm, is here in the *Statesman* (276a) explicitly connected not simply to periodic anthyphairesis, as was the case with the Divisions in

40 Z_7 art of heralds, 290a8-c2, protected by flying Hermes, is palindromic to K_7 herds of flying animals, 264e-265b.

41 Z_8 satyrs of politics-oligarchy, 291c8-303b7, is palindromic to K_8 herds of horned animals, 265b-d.

42 Z_9 centaurs of politics-tyranny, 291a2-b6, 303c8- d3, 291c8-303b7, is palindromic to K_9 mixed breeding herds, 265d-e.

43 Z_{10} sophists' sophists 291b7-c7, 303b8-c8 is palindromic to K_{10} diameter's diameter, pigs, 265e-266d.

the *Sophist*, but with palindromically periodic anthyphairesis. Thus, the new striking element in the *Statesman* 'division and collection', in relation to the *Sophist*, is not only periodic anthyphairesis, as in the *Sophist*, but in addition palindromicity within the period; in other words, we expect that the 'division and collection' will be an imitation of an anthyphairesis not only with a period, as in the *Sophist*, but with a period possesing symmetry with respect to the middle of the period.

8 The Double Measurement in *Stat.* 283c8-284e8 and 286c8-d6 Completes the Imitation of Theaetetus' Proposition

8.1 The Double Measurement is the Philosophic Analogue, the Imitation of the Theaetetean Two Conditions for 'Dynamis'

We consider opposite entities, of the type numbers, magnitudes (lines, areas, volumes), qualities (velocities), say x, y. There is a double 'measurement': (a) x against y, and (b) x against 'the mean(x, y)' (Greek term: 'mesos') (284e7). Condition (a) on measurement: not all opposite pairs x, y are acceptable; in fact if the measurement of x against y produces a 'pleasure' (Greek term: 'hedone'), then this pair must be excluded (286d4-6). Thus we require that x,y is a dyad, whose measurement against each other does not produce pleasure. Condition (b) on measurement: we further assume that the second measurement, x against the mean(x, y), is a 'generation' (Greek term: 'genesis') (283d, 284b-c, d-e). We will now decipher these terms and conditions:

[measurement of opposites] Measurement is about length and shortness, or excess and deficiency, or greatness and smallness, or the greater and the less, or the great and the small (283c-e), collectively of a magnitude or number or quality against its opposite (284e).

[mean] We interpret 'the mean(x, y)' as the philosophic analogue of the geometric mean of two numbers or magnitudes. It is remarkable that in the *Elements* there are 738 occurrences of the term 'meson' (and its variants), and these exclusively and without exception refer to the 'geometric mean' (of two numbers or of two magnitudes), while NONE of them refers to the arithmetic or any other mean, or any other sense; of them, the great majority 624, occur in the Theaetetean Book X.[44]. The alternative term used 'μέτριος' (283e, 284a, c, d, e) has normally a meaning equivalent to 'μέσος'.

---

44 See Bassiakou 2004.

[pleasure] That 'the dyad x, y produces pleasure' is equivalent, as explicitly stated in the Platonic dialogue on pleasures *Phil.* 27e, 31a, 41d, with 'the dyad x, y is an Infinite'; thus condition (a) is equivalent with the condition 'the dyad x, y does not produces an infinite', namely equivalent with the condition 'the dyad x, y does not produces a (Philebean) finite' Thus the condition (a) is that the measurement of x against y produces a (Philebean) finite. According to the interpretation given in Negrepontis 2005, the Philebean finite is the anthyphairetic finite, namely the commensurable.[45]

[generation] In Plato there is 'generation only', as in *Phaed.* 70b-72e, *Theaet.* 156a-157b, *Soph.* 247d-248a, or *Tim.* 27d5-29d3, 57d7-58c4, and 'generation in substance', as in *Phil.* 27c, or *Parm.* 142c-143a[46], or *Stat.* 283d, 284c, d. Thus 'γένεσις' and 'generation' does not refer only to the sensibles, but also to the intelligibles.[47] That 'the dyad x, z is a generation' implies that 'the dyad x, z generates an infinite multitude of parts', either as an indefinite dyad only or within a Being, namely that 'the dyad x, z is either an Infinite or a mixture of Infinite and Finite', hence, according to our interpretation given in Section 3, that 'the dyad x, z produces an infinite anthyphairesis'. Thus the condition on the measurement (b) is that the pair x, mean(x, y) produces a (Philebean) infinite.[48]

The double measurement in *Stat.* 283a-287b is the philosophic analogue of a dyad of lines incommensurable in length only. Finally, in order to state the two measurement conditions (a) and (b) in the more familiar terms employed by Book X. of the *Elements*, we choose an assumed 'unit' line r (as with the 'protetheisa line' in *Elem.* X.df.3, and the one foot line in *Theaet.* 147d-148b), and, employing *Elem.* II.14, we construct lines a, b, such that $x\,r = a^2$, $y\,r = b^2$, so that mean(x, y) = a b. Then, the ratio x:y = the ratio xr:yr = $a^2:b^2$; hence (using the *Topics* definition, mentioned in Section 4, above), the ratio x: mean(x, y) = the ratio xr:mean(x,y) r = the ratio a^2: a b = the ratio a:b. In conclusion, the double measurement in the *Statesman* is revealed to be the exact philosophic analogue of the Theaetetus condition 'commensurability in power only'.

45 Anthyphairetic finite; cf. Negrepontis 2005.

46 Cf. Τὸ τοίνυν συχνῷ τῷ τῆς γενέσεως ὀνόματι χρῆσθαι περὶ τῆς προόδου τοῦ νοητοῦ πλήθους ἀναδιδάσκοντα the frequent use of words related to generation, such as 'γίγνεσθαι' (142d5), 'γίγνεται' (142e4), 'γένηται' (142e6), 'γιγνόμενον' (143a1), refers to 'the progress of the intelligible multitude' Proclus, *Platonic Theology* 3,91,9-24.

47 It is precisely the lack of realisation, by Cherniss and others, that there is intelligible 'genesis' as well, that has led them to mistakenly conclude that the mixtures of Infinite and Finite in the *Philebus* are not true Beings and Ideas.

48 Anthyphairetic infinite, according to Negrepontis 2005.

8.2 The Double Measurement as the Cause of all Division and Collection, *Stat.* 283a-287b

Plato next makes a most interesting claim about the fundamental significance of the condition of the double measurement we have just deciphered:

> The Beings as in the case of the examples examined in the *Sophist* and the *Statesman*, the Sophist, the Statesman, and the art of Weaving become known, with the method of Division and Collection-Logos (τῆς τῶν ὄντων λόγῳ δηλώσεως, 287a3-4) and the palindromic periodicity of every Being (καὶ τὴν περὶ τὴν τοῦ παντὸς ἀνείλιξιν,[49] 286b8-10).

An absolutely necessary condition that the Beings and the arts must satisfy in order that they have 'division and collection', by which they become known, is the double measurement:

- The statesman, the art of Weaving, and all the arts will be destroyed without the double measurement;[50]
- without the double measurement it is impossible to obtain the 'episteme' of the Statesman, or of any other Being;[51]
- the double measurement is a necessary (and sufficient) condition that all the arts are true Beings;[52]
- the arts must necessarily satisfy the condition of the double measurement;[53]
- when the double measurement is satisfied (δέον, 285a7), then one can apply Division and Collection (described in *Stat.* 285a7-b6—see 9.1.1, below).

This sometimes results in very long strings of logoi in order to attain the period (λόγων μήκη καὶ τὰς ἐν κύκλῳ περιόδους, 286e4-5) but it is necessary, and there is no shorter or easier method[54]. In short, Plato is essentially formulating the following:

---

49 The term 'ἀνείλιξις' occurs in the *Statesman* twice. The first occurrence (270d3) explains what is meant by 'ἀνείλιξις'; the *LSJ* defines it as 'reversal of motion', in fact palindromic periodicity. The second occurrence (286b9) tells us that the phenomenon of 'ἀνείλιξις', namely of palindromic periodicity, is in all arts caused by the fundamental condition of the double measurement.

50 'ΞΕ. Οὐκοῦν τὰς τέχνας τε αὐτὰς καὶ τἆργα αὐτῶν σύμπαντα διολοῦμεν τούτῳ τῷ λόγῳ, καὶ δὴ καὶ τὴν ζητουμένην νῦν πολιτικὴν καὶ τὴν ῥηθεῖσαν ὑφαντικὴν ἀφανιοῦμεν' (284a4-7)

51 'οὐ γὰρ δὴ δυνατόν γε οὔτε πολιτικὸν οὔτ' ἄλλον τινὰ τῶν περὶ τὰς πράξεις ἐπιστήμονα ἀναμφισβητήτως γεγονέναι τούτου [the double measurement] μὴ συνομολογηθέντος' (284c1-3)

52 'ἡγητέον ὁμοίως τὰς τέχνας πάσας εἶναι, μεῖζόν τε ἅμα καὶ ἔλαττον μετρεῖσθαι μὴ πρὸς ἄλληλα μόνον ἀλλὰ καὶ πρὸς τὴν τοῦ μετρίου γένεσιν. τούτου τε γὰρ ὄντος ἐκεῖνα ἔστι, κἀκείνων οὐσῶν ἔστι καὶ τοῦτο, μὴ δὲ ὄντος ποτέρου τούτων οὐδέτερον αὐτῶν ἔσται ποτέ' (284d4-8)

53 'μὴ πρὸς ἄλληλα τὰ μήκη κρίνοντες ἀλλὰ κατὰ τὸ τῆς μετρητικῆς μέρος ὃ τότε ἔφαμεν δεῖν μεμνῆσθαι, πρὸς τὸ πρέπον' (286c8-d2).

54 cf. *Philebus* 16a7-b7

'Double measurement Proposition': if an art, like that of the Sophist or the Statesman, satisfies the condition of double measurement, then this art becomes knowable by division and collection and palindromically periodic ἀνείλιξις.

In view of our interpretation of the double measurement, in 8.1, as the philosophic analogue of Theaetetus' mathematical condition of 'dynamis', namely of a line incommensurable in length with respect to an assumed line, and commensurable in square to its square, the double measurement Proposition is clearly the philosophic imitation of the Theaetetus Proposition, enunciated in Section 2.2:

'Theaetetus' Proposition': If a is a power, namely a is a line incommensurable with respect to the one foot line b, and a^2 is commensurable to b^2, then the infinite anthyphairesis of a to b can be 'collected' into one.

Philosophical cause	**Mathematical cause**
Measurement of x against y is not pleasure	commensurability of a^2, b^2
Measurement of x against the mean of x and y is generation	incommensurability of a, b

Philosophical effect	**Mathematical effect**
Division and Collection, aneilixis of the Platonic Ideas in the *Sophist* and the *Statesman*	Palindromically periodic anthyphairesis

9 The Double Illumination

We are now in position to put together Part A and Part B and obtain a double illumination, one on Plato's philosophy, the other on Theaetetus' true mathematical contribution. It is clear that (a) the 'division and collection' of the *Sophist* (Sections 5, 6) is a philosophical analogue of periodic anthyphairesis in abbreviated tree-like form; (b) the 'division and collection' of the *Statesman* (Section 7) is a philosophical analogue of palindromically periodic anthyphairesis in abbreviated tree-like form; (c) the double measument in the *Statesman* (Sections 8) is the philosophic analogue of the mathematial Proposition: if a, b are two lines commensurable in squre only, then the anthyphairesis of a to b is palindromically periodic.

9.1 Philosophical Illumination

It follows from our analysis of the *Sophist* and *Statesman* (Sections 5, 6, and 7) that Plato's philosophical method of 'division and collection' is an imitation of periodic anthyphairesis, and in fact the whole content of the *Statesman* is revealed to be nothing else but an exact philosophical imitation of the highly nontrivial Palindromically Periodic Theorem. The fundamental role of periodic anthyphairetic in Plato's 'division and collection' is not limited in the dialogues *Sophist* and *Statesman*, examined in the present paper, but is shown to be present in our parallel and independent analysis of the second hypothesis of Plato's *Parmenides* 142b-155e in Negrepontis (forthcoming a), where it is shown that the One of the second hypothesis consists of a dyad in periodic anthyphairesis. Let me note that the anthyphairesis appearing in the One of the second hypothesis in the *Parmenides* is in full, not in abbreviated tree-like form. In the same paper, Zeno's paradoxes and arguments, are shown to be in close relation to the second hypothesis of the *Parmenides*, and to be equally based on periodic anthyphairesis. In particular Zeno's Dichotomy argument, is based on an abbreviated, tree-like form of periodic anthyphairesis. It thus seems clear that Plato's presentation of Collection and Division in abbreviated, tree-like form in the *Sophist* and the *Statesman* has its origin in Zeno's Dichotomy paradox of motion.

9.1.1 Plato's Collection is Always Described as Cyclical-Periodical

Confirming the periodic nature of Plato's Collection we note that in every one of the descriptions, in the *Sophist*, *Statesman*, *Phaedrus*, and *Philebus*, Collection is invariably described as cyclical, periodical. Here is a list of such descriptions (translations by Fowler, modified by the author):

> we have almost got him [the sophist] into a kind of encircling net (περιειλήφαμεν[55] ἐν ἀμφιβληστρικῷ τινι), we will collect (συλλαβεῖν)[56] him [the sophist] by the orders of royal logos; *Soph.* 235a10-c6.

> many species differing from one another circularly comprehended (περιεχομένας) by an exterior one; *Sophist* 253d1-e3.

> let us divide in half the assumed genus (σχίζοντες διχῇ τὸ προτεθὲν γένος), and proceed always by way of the right-hand part of the thing divided, keeping communion to the species the sophist belongs, until having circularly stripped (περιελόντες) him of all common properties, we left him only his own peculiar nature; *Soph.* 264d10-265a1.

> Then why in the world did we not say in straight manner (εὐθὺς) that weaving is the intertwining of woof and warp, but we went round in circle (περιήλθομεν ἐν κύκλῳ) making very many divisions (πάμπολλα διοριζόμενοι), in vain? *Stat.* 283b1-c2.

55 Cf. ἑνὶ εἴδει περιέλαβες; *Theaet.* 148d6.
56 Cf. συλλαβεῖν εἰς ἕν; *Theaet.* 147d8.

when all sorts of dissimilarities (τὰς δὲ αὖ παντοδαπὰς ἀνομοιότητας) are seen in multitudes (ἐν πλήθεσιν), he must find it impossible to be discouraged or to stop until has gathered all the things into one circle of similarity which are related to each other (ἐντὸς μιᾶς ὁμοιότητος ἕρξας) and has included them in some sort of class on the basis of their essential nature (γένους τινὸς οὐσίᾳ περιβάληται); *Stat*. 285a4-b6.

anyone who finds fault with the length of 'logoi' (λόγων μήκη), and objects to the circular periods (τὰς ἐν κύκλῳ περιόδους) in these processes [of 'division and collection'], must not merely find fault with the method; *Stat*. 286d6-287a6.

one and many (ἓν καὶ πολλὰ) become the same (ταὐτὸν... γιγνόμενα) by logoi circulating (ὑπὸ λόγων περιτρέχειν) everywhere in each true Being. *Philebus* 15d4-8

cutting off the left-hand part [of madness], continued to divide this not returning to its previous position (οὐκ ἐπανῆκεν πρὶν ἐν αὐτοῖς)[57] found among its parts a sort of left-handed love which it very justly reviled, but the other leading us to the right-hand part of madness, found a love having the same name as the first, but divine, which it held up to view and praised as the author of our greatest blessings....Now I myself...am a lover of these processes of 'division and collection' (τῶν διαιρέσεων καὶ συναγωγῶν)...; and if I think any other man is able to see things that can naturally be collected into one and divided into many (εἰς ἓν καὶ ἐπὶ πολλὰ), him I follow after and "walk in his footsteps as if he were a god; *Phaedrus* 264e4-266c1.

unless some one of the hearers is able to count (διαριθμήσηται) the natures and to divide the genera in species and to be able to circularly comprehend (περιλαμβάνειν)[58] by means of one idea in every intelligible being (καθ' ἓν ἕκαστον) he will never be an expert of 'logoi'; *Phaedrus* 273d7-e4.

9.1.2 Note on the Seeming Finiteness of the Platonic Divisions

In Sections 5 and 6, we have argued that each of the two examples of 'division and collection' contained in the *Sophist*, that of the Angler and of the Sophist, has the structure of a philosophic analogue of periodic, hence infinite, anthyphairesis. But in these two Divisions in the *Sophist* there appears to be no indication that the Divisions are anything but finite, the language employed suggesting that the number of steps of these divisions is finite. However: (a) in the *Statesman* Plato definitely describes the 'division and collection' as periodically infinite (Section 7.2. & 3)[59]; and, (b) as noted in Section 9.1.1, all descriptions of the method of (Division and) Collection in the Platonic dialogues exhibit the cyclical, periodic nature of the method.

57 At this point our translation diverges from that by Fowler that runs as follows 'cutting off the left-hand part, continued to divide this until it found among its parts a sort of left-handed love', not realizing a recurring and periodical meaning in the phrase 'οὐκ ἐπανῆκεν πρὶν ἐν αὐτοῖς'.
58 Cf. ἑνὶ εἴδει περιέλαβες; *Theaet*. 148d6.
59 In particular, the Platonic Division and Collection, having a periodic form, is an infinite division, and is thus unlike the Aristotelian Division into a genus and differentia, clearly a strictly finite division; cf. Lamprinidis 2015.

9.1.3 Division and Collection Results in the Full Knowledge ('Episteme') of a True Being Exactly Because it is a Periodic Anthyphairesis

We recall that the import of Socrates' interest in Theaetetus' mathematical achievement and his urging to imitate it was precisely the quest for the full knowledge ('episteme') of Beings. The fact that 'division and collection' has been found to be an analogue of periodic anthyphairesis clearly explains how it is possible to obtain the full knowledge of a true Being, although it is infinitely divisible. The Logos Criterion, achieved after a finite number of stages and Divisions, provides the required full knowledge.

It is now understood how is it possible a method for proving quadratic incommensurabilities to qualify as a method for obtaining the full knowledge (of Plato's intelligible Beings). Certainly the arithmetical methods of proving incommensurability in no way provide full knowledge, only the method of proof by periodic anthyphairesis of a to b provides full knowledge of the ' logos' a:b.

By contrast the method of proof suggested by van der Waerden, Knorr, and others, is just a proof of incommensurability, not leading to the knowledge of the 'logos' a:b. To realize the difference consider that all the arithmetical proofs of the incommensurability of a to b, for $a^2 = Nb^2$, N non-square, simply show that the anthyphairesis of a to b is infinite, but otherwise unknown, while Theaetetus' method provides full knowledge of the infinite anthyphairesis of a to b. Thus, Theaetetus's anthyphairetic method, and his alone, can serve as a model for acquiring full philosophical knowledge.

9.1.4 How Periodic Anthyphairesis Achieves Collection into One

The reason that the method is called Collection in One is explained in the *Parmenides*: the periodic anthyphairesis of a,b results in the equalization of a and b, in the sense either that the (finite) number of the parts of a is equal to the number of the parts of b (*Parm.* 144d4-e3, with a = One, b = Being)[60]; or, equivalently, that the (finite) number of times that Being is not-Being is equal to the number of the not-Beings (*Soph.* 257a4-6, with a = Being, and b = not-Being), both following closely Zeno's Fragment B3. Thus, Collection, meaning equalisation of all parts produced by the infinite anthyphairetic Division, is a consequence of anthyyphairetic periodicity. The Oneness of a Platonic Idea is thus not the Oneness of a partless entity, such as the One of the first hypothesis in the Parmenides or the geometric point, but of a self-similar One with parts, in fact an infinite multitude of parts, but such that its parts are equalized to each other.

60 The same equalization works for all successive parts, generated in the anthyphairesis of a to b.

9.1.5 Why Plato Based his Theory of Ideas on Periodic Anthyphairesis?

We have already seen that periodic anthyphairesis explains two basic tenets of Plato's philosophy: the Oneness of a platonic Idea (9.1.4) and the possibility of full knowledge of a Platonic Idea by 'division and collection' (9.1.3). It is beyond the scope of the present work to explain the mechanism of participation of the sensibles in the Platonic Ideas, presented by Plato mainly in the *Timaeus*. The sensibles are just the anthyphairetic approximations-convergents, in the case of the diameter to the side of a square the side and diameter numbers, indeed incomplete images and likenesses, of the Idea in which they participate. The basic ideas of the mechanism of participation are outlined in Negrepontis 2005. A joint work on the *Timaeus* is in preparation by Negrepontis and Kalisperi.

9.2 Mathematical Illumination: The Mathematical Contribution of Theaetetus in the Theory of Incommensurability

On the one hand, as we saw, the philosophic method of 'division and collection' was expressly declared in *Theaet.* 148d (Section 3) to be an imitation of the mathematical method of Collection of the infinity in each power, by Theaetetus, reported in *Theaet.* 147d7-148b4. On the other hand, the philosophic method of 'division and collection' turned out to be, as a result of our study of the *Sophist* and *Statesman* Divisions and Collections (Sections 5, 6, and 7) and the double measurement (Section 8), an imitation of what we called, in provisional form, Theaetetus' Proposition Now we can make sense of the enigmatic statement of Theaetetus' Proposition: it states that if a, b are commensurable in square only, then the anthyphairesis of a to b is periodic, hence infinite, hence the incommensurability statement of X.9. But, in fact, he proved a far stronger and deeper result. Theaetetus' mathematical method of Collection, reported in *Theaet.* 147d-148b, was in fact a method for the proof of the following Palindromic Periodicity Theorem.

Proposition (Theaetetus' Palindromically Periodic Theorem PPT): If two lines, a and b, are incommensurable in length only, as in the case that b is the one foot line and a is a Theaetetean power, then the anthyphairesis of a to b is periodic, and in fact palindromically periodic. In fact, there are numbers n and $I_0, I_1, I_2, \ldots, I_{n-1}, (I_n)$, such that the sequence Anth(a,b) of successive quotients of the continued fraction of a to b has the form Anth(a,b) = $[\mu_1, \text{period } (I_1, I_2, \ldots, I_{n-2}, I_{n-1}, (I_n), I_{n-1}, I_{n-2}, \ldots, I_2, I_1, 2I_0)]$.

The PPT Theorem is the ancient analogue of a remarkable theorem[61] on the continued fraction of the square root of every non-square natural number, developed gradually in

61 Cf. Hardy & Wright 1938; Weil 1984; Kahane 1985; Fowler 1999.

the modern era (seventeenth and eighteenth century), by Fermat, Brouncker, Euler[62], Lagrange[63], Legendre, Galois[64]. Thus, Theaetetus' method of proof, reported in *Theaet.* 147d7-148b4, was then purely anthyphairetic, and we have to declare as mistaken the associations of Theaetetus' contribution to the theory of quadratic incommensurability with any form of arithmetic reconstructions (including those by van der Waerden and Knorr, mentioned in Section 2). The scholars that adopted arithmetical proofs for Theaetetus' method clearly did not take into account not only the distributive reading of the plural αἱ δυνάμεις in 147d3-148d4, but also the association of Theaetetean collection of each divided power with Plato's method of division and collection in the *Sophist* and *Statesman*. In particular, there is absolutely no evidence that Theaetetus has anything to do with the arithmetical proofs of quadratic incommensurability that make use of VII.27 or arithmetical propositions of Book VIII (especially VIII.7,8,14, which in fact follow from VII.27) in conjunction with *Elem.* X.9. These arithmetical propositions originated by Archytas,[65] were clearly inspired by music for epimoric ratios, and were adopted in the theory of music of *Sectio Canonis* (κατατομὴ κανόνος). These arithmetical proofs of incommensurability were produced shortly after Theodorus' and Theaetetus'[66] anthyphairetic proofs. Finally, the fact that the statement of X.9 has been associated with Theaetetus in Euclid's *Scholia* X.62 seems to be due to a misunderstanding, as the anonymous scholiast seems to regard X.9 as a proposition proving incommensurability, while in reality is only a useful but trivial proposition.

A lingering doubt might be the question how, with what method and what mathematical tools, an important theorem that exercised the ingenuity of some of the greatest seventeenth-eighteenth -century mathematicians could be proved in classical antiquity. We will show, in the next section 10, that the tools needed are indeed found in the Theaetetean Book X of the *Elements*.

10 A Reconstruction of the Palindromic Periodicity Theorem Using only Mathematical Tools from Book X of the *Elements*

Our next question is necessarily mathematical: Can we conceive the proof that Theaetetus discovered of the PPT? We will now show that indeed Theaetetus had developed all the means to achieve the proof of the PPT theorem. In fact, our

62 Euler 1765.
63 Lagrange 1769, 1771.
64 Galois 1829.
65 Reported by Boethius, *De Institutione Musica* III.11 and IV, 2.
66 Archytas date of birth is set at about 427 BC and Theaetetus' at about 417 BC, but Theaetetus was uncommonly precocious.

reconstruction will be based solely on tools known to be due to Theaetetus or in fact found in Book X of the *Elements*, a Theaetetean work according to ancient sources, notably Pappus (Thomson, 1930).

10.1 The tools

We first make a list of these tools:

(a) Commensurability-incommensurability. The definitions of commensurability and incommensurability appear in definition 1, while the fundamental definition of commensurability in power only appears in *Elem.* X.df.2. Of course commensurability in power only appears explicitly in the beginning of the Platonic trilogy, in *Theaet.* 147d-148d, considered in Sections 1, 2, and 3), and credited to Theaetetus (and younger Socrates).

(b) Anthyphairesis. The ancient concept equivalent to the modern concept of continued fraction expansion of (positive) reals is anthyphairesis of two homogeneous magnitudes; this concept is implicitly defined in the *Elements*, in the course of *Elem.* X.2. In order to fix the notation we will consider the anthyphairesis of two lines $\alpha>\beta$:

$\alpha = \mu_1\beta + \gamma_1$, with $\beta>\gamma_1$, $\beta = I_1\gamma_1 + \gamma_2$, with $\gamma_1>\gamma_2$, $\gamma_1 = I_2\gamma_2 + \gamma_3$, with $\gamma_2>\gamma_3$, … , $\gamma_n = I_{n+1}\gamma_{n+1} + \gamma_{n+2}$, with $\gamma_{n+1}>\gamma_{n+2}$, …

The natural number I_n is the $(n + 1)^{th}$ quotient, and the line γ_n is the nth anthyphairetic remainder of the anthyphairesis of α to β.

(c) Apotome line. One of the basic concepts in *Elem.* X, is the concept of an apotome line defined in the statement of *Elem.* X.73 (and studied in *Elem.* X.73-109):

Definition: Let β be an assumed line, ζ, η lines with $\zeta>\eta$, incommensurable, and such that ζ^2, η^2 are commensurable to β^2. Then the line ζ-η is called an apotome (ἀποτομή) line (with respect to the assumed line β). The lines ζ, η are called the names of the line ζ-η.

The basic reason for the introduction of the apotome line is the fact that all the remainders of the anthyphairesis of a quadratic irrational is an apotome line:

Proposition: Let N be a non-square natural number, and let α, β line segments such that $\alpha^2 = N\beta^2$. Then every anthyphairetic remainder γ_n is an apotome line wrt β for every n = 1,2,.., and in fact there is a double sequence of numbers (x_n) and (y_n), such that $x_1 = \mu_1$, $y_1 = 1$, and $\gamma_n = (-1)^{n+1}(y_n\alpha - x_n\beta)$ for n = 1,2,…

(d) Line of two names. Next, we note that in Book X. the other basic concept studied in a similar way is the concept of a line of two names, defined in the statement of the X.36 (and studied in X.36-72).

Definition. Let β be an assumed line, ζ, η lines with $\zeta>\eta$, incommensurable, and such that ζ^2, η^2 are commensurable to β^2. The line $\zeta+\eta$ is called line of two names (ἐκ δύο ὀνομάτων).

The basic reason for introducing the line of two names is the conjugacy of the binomial line to the apotome line.

(e) Conjugacy of apotomae and of lines of two names. The conjugacy between the lines of two names and the apotomae, of crucial importance to the proof of the Palindromic Periodicity Theorem, is described in the fundamental Propositions X.112-114:

> X.112: If β is a rational line, ζ is a line of two names, with names $\zeta1>\zeta2$, and η is a line such that $\zeta \cdot \eta = \beta^2$, then η is an apotome with names $\eta_1>\eta_2$, and $\zeta_1:\eta_1 = \zeta_2:\eta_2$ is a commensurable ratio.
>
> X.113: If β is a rational line, ζ is an apotome, with names $\zeta1>\zeta2$, and η is a line such that $\zeta \cdot \eta = \beta^2$, then η is a line of two names with names $\eta_1>\eta_2$, and $\zeta_1:\eta_1 = \zeta_2:\eta_2$ is a commensurable ratio.
>
> X.114: If ζ is a line of two names, with names $\zeta_1>\zeta_2$, η is an apotome with names $\eta_1>\eta_2$, $\zeta_1:\eta_1 = \zeta_2:\eta_2$ is a commensurable ratio, and $\eta \cdot \zeta = \beta^2$, then β is a rational line. The lines $\zeta + \eta$ and $\zeta - \eta$ are conjugate to each other and write $\zeta + \eta = (\zeta - \eta)^*$ and $\zeta - \eta = (\zeta + \eta).^*$

The presence in Book X of the lines of two names and of the three conjugacy Propositions X.112-114 between apotomae and lines of two names strongly suggest that Theaetetus had achieved a proof of the PPT, analogous to Euler's approach to the modern continued fractions of square roots of non-square natural numbers; and indeed the reconstruction of the proof of the PPT employs solely as tools the concepts of the lines 'apotome' and 'of two names' and their conjugacy. Theaetetean conjugacy, closely related to Eulerian conjugacy, plays in effect the same role that conjugation plays in complex numbers[67]: it is useful in computing inverses, a computation vital for ancient anthyphairetic and modern continued fraction expansions.[68]

---

67 A. Weil (1984, 202) beautifully describes these connections: "The first concept, motivated by the work of the Italian algebraists of the sixteenth century on equations of degree 3 and 4, had been introduced by Bombelli in Book I of his Algebra of 1572, in close imitation of Euclid's theory of irrationals of the form $a + \sqrt{b}$ in his Book X ('binomial' in the terminology of Campanus' Latin translation of Euclid; accordingly, binomio is Bombelli's word for a complex number $a + \sqrt{-b}$). Further developments belong to the history of algebra and analysis rather than to number theory; suffice it to say that Euler played a decisive role in extending to complex numbers the main operations of analysis."

68 But Weil (1984, 15), evidently not suspecting Theaetetus' deep knowledge of palindromically periodic anthyphairesis, as revealed by Plato's method of Division and Collection in the *Sophist* and *Statesman*, dismisses the importance of X.112-114: "Not only is Euclid himself well aware of the relation $(\sqrt{r} + \sqrt{s})(\sqrt{r} - \sqrt{s}) = r - s$ but even the identity $1/(\sqrt{s} + \sqrt{r}) = \sqrt{r}/(r - s) - \sqrt{s}/(r - s)$ may be regarded as the essential content of prop. 112 of that book. Unfortunately, Euclid's motivation in Book X seems to have been the wish to construct a general framework for the theory of regular polygons and polyhedra, and not, as modern mathematicians would have it, an algebraic theory of quadratic fields. So we are left to speculate idly whether, in antiquity or later, identities involving square roots may not have been used, at least heuristically, in arithmetical work." Knorr (1983, 60) provides further information on Weil's view. Knorr, after himself dismissing the role of Propositions X.112-114, "Certain potentials of the subject matter are entirely missed, for instance, insight into how products of the form $(a \pm b\sqrt{N})(c \pm d\sqrt{N})$ bear on the finding of solutions to integral relations of the form $x^2 - Ny^2 = \pm m$"; he then confirms Weil's view at footnote (68, fn. 53): 'At the time of his course of lectures on the history of the Pell equation (Institute for Advanced Study, 1978-79), Professor A. Weil expressed to me his disappointment over Euclid's utter failure to perceive this potentially fruitful development of the theory of Book X.'

10.2 The Increment Factor

But the conjugacy applied to the remainders γ_n, apotomae by Proposition 1, is unwieldy, so it is entirely possible that Theaetetus applied the conjugacy theory on the more manageable lines φ_n, which will turn out to be apotomae as well

Definition (Increment factors of the anthyphairetic remainders): Let $\alpha > \beta$ be two lines, with $\alpha > \beta > \gamma_1 > \gamma_2 > \gamma_3 > ... > \gamma_n > \gamma_{n+1} > ...$ the sequence of successive remainders. We define the sequence (φ_n) of increment factors by the equalities: $\varphi_1 = \gamma_1$, and $\gamma_{n-1} \cdot \varphi_n = \beta \cdot \gamma_n$ for $n > 1$.

Thus φ_n is the fourth proportional (*Elem.* I.44) of the three lines γ_n, β, and γ_{n-1}.

Proposition (elementary properties of the increment factors).
Let $\alpha > \beta$ be two lines, with $\alpha > \beta > \gamma_1 > \gamma_2 > \gamma_3 > ... > \gamma_n > \gamma_{n+1} > ...$ the sequence of successive remainders, $\mu_1, I_1, I_2, I_3, ..., I_n, I_{n+1}, ...$ the sequence of successive quotients, and $\varphi_1, \varphi_2, ..., \varphi_n, ...$ the sequence of increment factors of the anthyphairesis of α, β. Then

(a) $\varphi_n < \beta$ for every n,
(b) $\varphi_n (I_n \beta + \varphi_{n+1}) = \beta^2$ for every n,
(c) the anthyphairesis of β, φ_n is equal to $\text{Anth}(b, \varphi_n) = [I_n, I_{n+1}, ...]$ for every n,
(d) [Incremental Logos Criterion] if there exist n, m with n<m, such that $\varphi_n = \varphi_m$, then the anthyphairesis of α, β is eventually periodic, and in fact equal to $[\mu_1, I_1, ..., I_{n-1}, \text{period}(I_n, I_{n+1}, ..., I_{m-1})]$.

Proof. (a), (b) routine translation of the definition of anthyphairesis of magnitudes, implicit in *Elem.*X.2; (c) by the *Topics* Definition of proportion; (d) routine translation of the Logos Criterion (established in Section 4).

10.3 The Periodicity Theorem (PT) with the Tools of *Elem.* X

Due to space limitation, we shall not present the proof of the full palindromic periodicity theorem (PPT) but only the weaker periodicity theorem (PT). We emphasize however that the tools needed for the proof of the stronger PPT theorem are the same tools of Book X of the *Elements*.

The periodicity theorem (PT): Let N be a non-square natural number and α and β line segments such that $\alpha^2 = N\beta^2$, denoting by $\mu_1, I_1, I_2, ..., I_n, ...$ the sequence of successive quotients, and by $\varphi_1, \varphi_2, ..., \varphi_n, ...$ the sequence of successive increment factors of the anthyphairesis of α to β, then there are two sequences of natural numbers $\lambda_1(=1), \lambda_2, ..., \lambda_k, ...$ and $\mu_1, \mu_2, ..., \mu_k, ...$, such that for every natural number $k > 1$ (a_k) $\lambda_k \lambda_{k-1} = N - \mu_{k-1}^2$ (hence λ_k is a natural number smaller than N, and μ_k is a natural number whose square is smaller than N), (b_k) $\lambda_k (I_{k-1} \beta + \varphi_k) = \lambda_{k-1} (\varphi_{k-1})^*$, ($c_k$) $\mu_k + \mu_{k-1} = I_{k-1} \lambda_k$, ($d_k$) the line φ_k is an apotome and in fact $\lambda_k \varphi_k = \alpha\_\mu_k \beta$, and (e) the anthyphairesis of α and β is eventually periodic.

Proof of the periodicity theorem: We proceed by mathematical induction. We prove (a_2), (b_2), (c_2), (d_2), and assuming (a_k), (b_k), (c_k), (d_k) we prove (a_{k+1}), (b_{k+1}), (c_{k+1}), (d_{k+1}). The proof of (a_{k+1}) consists in the elementary verification, using ((d_k), (a_k), (c_k)), that the number λ_{k+1} defined by $\lambda_{k-1} + I_{k-1}(\mu_{k-1} - \mu_k)$ is indeed a natural number. Proof of (b_{k+1}): the equality (d_k) $\lambda_k \varphi_k = \alpha - \mu_k \beta$ implies that φ_k is an apotome line. We consider the line of two names $\lambda_k (\varphi_k)^\star = \alpha + \mu_k \beta$ conjugate to it; by (a_{k+1}) and Proposition X.114, we obtain that $\lambda_k{}^2 \varphi_k (\varphi_k)^\star = (\alpha - \mu_k \beta)(\alpha + \mu_k \beta) = \alpha^2 - \mu_k{}^2 \beta^2 = \lambda_{k+1}\lambda_k \beta^2$, whence, dividing both sides by λ_k, that (A) $\lambda_k \varphi_k (\varphi_k)^\star = \lambda_{k+1} \beta^2$. On the other hand, from the anthyphairetic relation (Proposition 2b) we obtain $\varphi_k (I_k \beta + \varphi_{k+1}) = \beta^2$, hence (B) $\lambda_{k+1} \varphi_k (I_k \beta + \varphi_{k+1}) = \lambda_{k+1}\beta^2$. Comparing (A) and (B), we get that $\lambda_{k+1}\varphi_k (I_k \beta + \varphi_{k+1}) = \lambda_k \varphi_k (\varphi_k)^\star$, hence $\lambda_{k+1}(I_k \beta + \varphi_{k+1}) = \lambda_k (\varphi_k)^\star$. The proof of ($c_{k+1}$) consists in the elementary verification, using ((b_{k+1}) and Proposition 2a, that the number μ_{k+1} defined by $I_k \lambda_{k+1} - \mu_k$ is indeed a natural number; and, the proof of (d_{k+1}) consists in an elementary verification, using (b_{k+1}) and (c_{k+1}). The proof of (b_{k+1}), relying on the conjugacy of Book X. (X.114), similar and simpler to that of (b_2), is left for the reader. The proof of (a_k), (b_k), (c_k), (d_k) is now complete. From (a_k), (b_k), (c_k), (d_k) just proved, and from the fact that the set of pairs (λ, μ) of natural numbers, with λ bounded above by N and μ by μ_1 is finite, it follows, by an application of the pigeonhole principle, that there are indices k and s, with $k<s$, such that $\varphi_k = \varphi_s$. The eventual periodicity now follows from the incremental Logos criterion (Proposition, d).[69]

Conclusion

The discovery of the fundamental role of periodic anthyphairesis in Plato's philosophy, something not understood by modern era scholars of Plato, constitutes a breakthrough in our understanding of Platonic philosophy; and, the discovery that the important Palindromic Periodic Theorem, normally attributed to seventeenth-eighteenth century Mathematics (Euler, Lagrange) was in fact already proved by Theaetetus, not understood by modern era historians of Greek Mathematics, constitutes a breakthrough in our assessment of Theaetetus' contribution to the theory of quadratic incommensurability and periodic anthyphairesis.

Our discovery that the anthyphairetic revolutions[70] govern both Theodorus' and Theaetetus' mathematics on quadratic incommensurabilities in *Theaet.* 147d3-148d7 (Sections 1-4, 10) and Plato's philosophy on the method of 'division and collection' in the *Sophist* and *Statesman* (Sections 5-8), coupled with our (independent) study of Plato's *Parmenides*[71], confirming that these revolutions form the basis for the paradigmatical true Platonic Being, the One of the second hypothesis 142b–155e, but also

69 Recall that in Section 7.2 we pointed out that Plato, in his description of the myth of palindromic periodicity, is mindful that the passage from the Kronos era to the Zeus era relies on a pigeon-hole argument, while the passage from the Zeus era to the Kronos era on a Logos Criterion of periodicity.

70 Revolution, a circular movement (Cambridge dictionary).

71 Negrepontis, forthcoming a; cf, also, Negrepontis, forthcoming b.

for the true Beings in Zeno's paradoxes of motion and arguments on multiplicity, the precursors of the second hypothesis in the *Parmenides* and really of all of Plato's philosophy, reveals the remarkable continuity that flows through the mainstream of Greek thought, from the Pythagoreans to the Eleatic philosophers, especially Zeno, to the geometers Theodorus and Theaetetus, culminating in Plato.

The anthyphairetic revolutions, acting resolutely according to their other meaning as well,[72] are producing illumination, hitherto unsuspected by Platonists and historians of Greek mathematics alike (outlined in Section 9), and, like the Homeric goddess, are dispersing the mist of modern era misconceptions both on Plato's philosophy and on Theaetetus' great mathematical contributions to the theory of periodic anthyphairesis.

Bibliography

Texts quoted

Aristotle, *Topics.*: Ross, W. D. 1958. *Aristotelis Topica et Sophistici Elenchi.* Oxford (repr. 1970).
Boethius: Friedlein, G. 1867. *De Institutione Musica*. Leipzig.
Euclid, *EOO.*: Heiberg, J. L. & H. Menge 1883–1916. *Euclidis Opera Omnia.* 8 vols. Leipzig.
Euler, L. 1765. "De usu novi algorithmiin problemate Pelliano solvendo." In: *Novi Commentarii Academiaescientarum Petropolitanae* 11, 28–66.
Galois, E. 1829 "Demonstration d'un Theoreme sur les Functions Continues Periodiques." In: *Annales de Mathematiques* 19, 294–301.
Iamblichus: Waterfield, R. 1988. *The Theology of Arithmetic.* Grand Rapids.
Lagrange, J. L. 1877. *Œuvres de Lagrange*. Vol. 7. Paris.
Plato: Fowler, H. N. 1921–1925. *Theaetetus, Sophist, Statesman, Phaedrus, Philebus.* London.
Plato, *Resp.*: Shorey, P. 1969. *Republic.* Cambridge.
Proclus, *Commentary on Plato's Republic.*: Kroll, W. 1899–1901. *Procli Diadochi in Platonis Rem Publicam Commentarii.* 2 vols. Leipzig (repr. Amsterdam 1965).
Proclus, *In Eucl.*: Friedlein, G. 1873. *Commentarii in Primum Euclidis Elementorum Librum*. Leipzig.
Proclus, *Platonic Theology*: Saffrey, D. & L. G. Westerink.1968–1987. *Proclus. Théologie Platonicienne.* 5 vols. Paris.

Works quoted

Acerbi, F. 2003. "Drowning by Multiples. Remarks on the Fifth Book of Euclid's Elements, with Special Emphasis on Prop. 8." In: *Archive for History of Exact Sciences* 57, 175–242.
Bashmakova, I. G. & A. I. Lapin 1986. "Pifagor." In: *Kvant* 1, 7–12.
Bassiakou, A. 2004. *Plato's Statesman and the Palindromic Periodicity of the Anthyphairesis of Quadatic Irrationals*, MSc Thesis, University of Athens.

72 Revolution, a very important change in the way that people do things (Cambridge dictionary).

Burnyeat, M. 1978. “The Philosophical Sense of Theaetetus’ Mathematics.” In: *Isis* 69, 489–513.
Burnyeat, M. *et al.* 1990. *The Theaetetus of Plato*. Indianapolis.
Cherniss, H. 1951. “Plato as Mathematician.” In: *The Review of Metaphysics* 4, 395–425.
Cornford, F. M. 1935. *Plato’s Theory of Knowledge: the Theaetetus and Sophist of Plato*. London.
Fowler, D. H. 1999. *The Mathematics of Plato’s Academy: A new Reconstruction*. Oxford.
Hardy, G. H. & E. M. Wright. 1938. *Introduction to the Theory of Numbers*. Oxford.
Kahane, J.-P. 1985. “La Theorie de Theodore des Corps Quadratiques Reels.” In: *L’ Enseignement Mathematique* 31, 85–92.
Knorr, W. R. 1975. *The Evolution of the Euclidean Elements: A Study of the Theory of Incommensurable Magnitudes and Its Significance for Early Greek Geometry*. Dordrecht.
Knorr, W. R. 1983. “‘La Croix des Mathématiciens’: The Euclidean Theory of Irrational Lines.” In: *Bulletin of the American Mathematical Society* 9, 41–69.
Lamprinidis, D. 2015. *The Anthyphairetic Nature of ‘Mathemata’ in Plato*, Ph.D. thesis, University of Athens.
McCabe, R. L. 1976. “Theodorus’ Irrationality Proofs” In: *Math. Magazine* 49, 201–203.
Negrepontis, S. 2005. “The Anthyphairetic Nature of the Platonic Principles of Infinite and Finite.” In: C. M. Society (ed.), *Proceedings of the Fourth Mediterranean Conference of Mathematics Education*. Nicosia, 3–27.
Negrepontis, S. forthcoming (a). “The Periodic Anthyphairetic Nature of the One of the Second Hypothesis in Plato’s *Parmenides*”. In: *Proceedings of the Conference Mathematiques et musique : des Grecs a Euler*, 10–11 September, 2015, Strasbourg, ed. X. Hascher and A. Papadopoulos, Hermann, Paris.
Negrepontis, S. forthcoming (b). “Plato on Geometry and the geometers”. In: *Geometry in History*, ed. S. G. Dani and A. Papadopoulos, Springer.
Negrepontis, S. & G. Tassopoulos 2015. “Theodorus’ Proofs of Incommensurabilities with Gnomons.” In: *BSHM Bulletin* 31, 15–30.
Tarrant, H. 1985. *Scepticism or Platonism? The Philosophy of the Fourth Academy*. Cambridge.
Thomson, W. (ed.) 1930. *The Commentary of Pappus on Book X of Euclid’s Elements*. Cambridge.
Van der Waerden, B. L. 1954. *Science Awakening* (1st Engl. ed). New York.
Weil, A. 1984. *Number Theory, An Approach Through History from Hammurapi to Legendre*. Boston.
Zeuthen, H. G. 1910. “Sur la constitution des livres arithmétiques des éléments d’Euclide et leur rapport à la question de l’irrationalité.” In: *Oversigt over det Kgl. Danske Videnskabernes Selskabs Forhandlinger* 5, 395–435.

Name index

https://doi.org/10.1515/9783110565959-014

General index

https://doi.org/10.1515/9783110565959-015